# Discrete Groups and Automorphic Functions

# Discrete Groups and Automorphic Functions

PROCEEDINGS OF AN INSTRUCTIONAL CONFERENCE
ORGANIZED BY THE LONDON MATHEMATICAL SOCIETY
AND THE UNIVERSITY OF CAMBRIDGE
(A NATO ADVANCED STUDY INSTITUTE)

*Edited by*
W. J. HARVEY
*Department of Mathematics, King's College*
*London, England*

1977

ACADEMIC PRESS
London New York San Francisco
*A Subsidiary of Harcourt Brace Jovanovich, Publishers*

ACADEMIC PRESS INC. (LONDON) LTD.
24/28 Oval Road,
London NW1

*United States Edition published by*
ACADEMIC PRESS INC.
111 Fifth Avenue,
New York, New York 10003

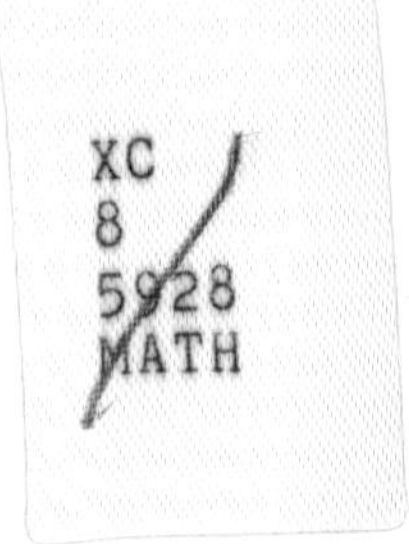




Library of Congress Catalog Card Number: 77 71823
ISBN: 0 12 329950 0

Printed in Great Britain by
PAGE BROS (NORWICH) LTD
Mile Cross Lane
Norwich

# Contributors

A. F. BEARDON *St Catharines College, Cambridge, CB2 1RL, England*

J. S. BIRMAN *Department of Mathematics, Columbia University, New York, NY 10025, USA*

C. J. EARLE *Department of Mathematics, Cornell University, Ithaca, NY 14853, USA*

L. GREENBERG *Department of Mathematics, University of Maryland, College Park, Maryland 20742, USA*

W. J. HARVEY *Department of Mathematics, King's College, Strand, London WC2, England*

J. LEHNER *Department of Mathematics, University of Pittsburgh, Pittsburgh, Pennsylvania 15260, USA*

O. LEHTO *Department of Mathematics, University of Helsinki, Hallituskatu 15, 00100 Helsinki 10, Finland*

A. M. MACBEATH *Department of Pure Mathematics, University of Birmingham, P.O. Box 363, Birmingham B15 2TT, England*

A. MARDEN *Department of Mathematics, University of Minnesota, Minneapolis Minn 55455, USA*

B. MASKIT *Department of Mathematics, State University of New York, Stony Brook, NY 11794, USA*

R. A. RANKIN *Department of Mathematics, University Gardens, University of Glasgow, Glasgow G12 8QW, Scotland*

H. P. F. SWINNERTON-DYER *St Catharines College, Cambridge CB2 1RL, England*

# Preface

This book contains the edited proceedings of an instructional conference on discrete groups and automorphic functions, held in Cambridge from 28th July to 15th August 1975 under the aegis of the London Mathematical Society. Invaluable financial support was provided by the Advanced Study Programme of N.A.T.O.

The extensive preparatory clerical work and organization of local arrangements were carried out by Dr A. F. Beardon with admirable efficiency.

The conference also owed much to the hospitality of our hosts in the University of Cambridge, and in particular to the Department of Pure Mathematics and Mathematical Statistics. A debt of thanks is of course due to all the participants, who made the conference worthwhile.

All of the formal lecture courses are reproduced here using drafts supplied by the speakers. The organizing committee wish to express here their gratitude to the lecturers, whose contributions ensured the progress of the meeting and the production of this volume.

Finally, the Editor is grateful to Professor A. M. Macbeath for editorial help, to the secretarial staff of King's College, London for clerical assistance and to our publishers for their advice and cooperation in turning the raw material into its present polished form.

*October 1977*

Organizing Committee:
A. F. Beardon
J. W. S. Cassels
W. J. Harvey
A. M. Macbeath
R. A. Rankin

# Contents

# Introduction

The Cambridge instructional conference was devoted to the study of discrete groups of Möbius transformations and automorphic functions of one variable. The formal programme of lectures is reproduced here with each chapter corresponding to a distinct course, and the structure of the book reflects the intention of the organizing committee to provide a comprehensive introduction to the subject suitable for second year graduate students or for non-specialist mathematicians, together with an account of recent progress. Particular attention was given to the geometry of Kleinian groups and to the moduli problem for Riemann surfaces and discontinuous groups. Both topics are currently rendered inaccessible by the lack of an adequate text covering the substantial advances made in the past two decades, beginning with Ahlfors' crystallization of Teichmüller's fundamental ideas and the resulting introduction of quasi-conformal mappings as a valuable tool in function theory.

We outline here the material covered in the book; for more details the reader should consult the table of contents and the introductions to individual chapters.

The first two chapters present material from topology and geometry respectively and are mainly foundational. Chapter 3 introduces Banach spaces of automorphic forms and surveys a selection of analytic techniques. Chapters 4 and 5 describe quasiconformal mappings in the plane and the closely-linked Teichmüller theory of deformations of Riemann surfaces. The rôle of the mapping class group (or Teichmüller modular group) in low-dimensional topology is surveyed in chapter 6, while chapters 7 and 8 describe the interrelation between the geometry of Fuchsian and Kleinian groups and the topology of the associated surfaces and 3-manifolds. The latter chapter concentrates on applications to deformations of discrete groups; this is also the subject of chapter 9, where a compactification of the space of moduli of Fuchsian groups is constructed. Chapter 10 considers the classification of Kleinian groups. In a brief final counterpoint to the main theme, arithmetic enters the picture with chapter 11, which treats modular forms, and chapter 12, which studies arithmetic Fuchsian groups in a classical fashion before surveying the contemporary viewpoint stemming from the theory of moduli of abelian varieties and representations of SL(2, $\mathbb{R}$).

As will become apparent to the careful reader, the pioneering work of Lars Ahlfors and Lipman Bers exerts a pervading influence on our field. It is hoped that these proceedings will provide a fitting tribute to their inspiring example by encouraging others to learn this beautiful chapter of mathematics.

# 1. Topological Background

A. M. Macbeath

*University of Birmingham, England*

## INTRODUCTION

The subject of our conference is a meeting point of topology, group theory and analysis. The relationship between analysis and topology is particularly close. Riemann surfaces were the first examples of manifolds, and homology theory grew out of the study of domains of integration on such surfaces. The universal covering surface and the fundamental group were invented in the context of analysis of one complex variable.

In the past twenty-five years, dramatic results and extensive technical progress in algebraic topology have eclipsed the parts of topology that are important for us. We need to concentrate on the study of the fundamental group more deeply than is usual in topology texts. The mapping class group is outside the field of interest of most algebraic topologists, but for us, its rôle as a group of transformations of the Teichmüller space gives it a central position. Its study has been given extra impetus by its connections with analysis; chapter 6 will bring us up to date in this rapidly developing area.

In the following lectures, only the basic groundwork immediately relevant for our purpose is covered; the fundamental group, covering spaces, van Kampen's theorem and Nielsen's theorem on the mapping class group. I hope that a self-contained account of this limited but essential selection of topics will be found useful.

Anyone with a background in Algebraic Topology which includes some elementary homotopy theory might omit Sections 1 and 2 and half of Section 3. As background references for surface topology we cite the textbooks [**1**, **5**, **2**, **7**].

## 1. PATHS, FUNDAMENTAL GROUP

### 1.1 The compact-open topology

Let $X$, $Y$ be topological spaces. In considering continuous mappings between $X$ and $Y$ it is natural, and turns out to be technically very fruitful, to consider continuous deformations of mappings. For this, and for other reasons, it is desirable to impose on the set of continuous mappings from $X$ into $Y$ a topology in which the neighbourhood concept is somehow compatible with our intuitive concept of "nearness" of mappings. The topology which has been found to be best suited for this purpose for maps between spaces in a fairly wide family is the *compact-open topology*, which can be loosely described as the least topology with the property that, given any compact subset $C$ of $X$, neighbouring mappings $f$, $g$ have neighbouring images $f(C)$, $g(C)$. The formal definition is as follows.

1.1.1 *Definition.* For each pair of subsets $A \subset X$, $B \subset Y$, let $T(A, B)$ denote the set of continuous mappings $f: X \to Y$ such that $f(A) \subset B$. Let $\mathcal{T}$ denote the topology generated by the sets of mappings $T(C, U)$, where $C$ ranges over all compact subsets of $X$ and $U$ ranges over all open subsets of $Y$. That is to say, a set of mappings is $\mathcal{T}$-open if and only if it is a union of finite intersections of sets $T(C, U)$. Then $\mathcal{T}$ is called the *compact-open topology* on the set of continuous mappings from $X$ into $Y$. We shall denote the space of continuous mappings from $X$ into $Y$, so topologized, by $Y^X$. Spaces of mappings in these lectures will always be endowed with the compact-open topology unless otherwise specified.

Let $x$ be any element of $X$. The map $\mathfrak{v}_X: Y^X \to Y$ defined by $\mathfrak{v}_x(f) = f(x)$ is called the *value map.* If $V \subset Y$ is open, then $\mathfrak{v}_x^{-1}(U) = T(\{x\}, U)$. Since $\{x\}$ is compact, $T(\{x\}, U)$ is open in $Y^X$, and it follows that, for each $x \in X$, the value map $\mathfrak{v}_x$ is continuous.

## 1.2 Paths

Let $X$ be a topological space, $I$ the unit interval $[0, 1]$ of the reals.

1.2.1 *Definition.* By a *path* in $X$ we mean a continuous map $f: I \to X$. $f(0)$ is called the *initial point* of the path and $f(1)$ the *end-point.* If $a$ is the initial point and $b$ the end point of $f$, we say that $f$ is a path *from a to b.* We shall use $P(a, b)$ to denote the set of all paths from $a$ to $b$, topologized as a subspace of $X^I$.

1.2.2 *Definition.* Let $p, q > 0$ be two real numbers such that $p + q = 1$. If $f$ and $g$ are two paths with $f(1) = g(0)$, the *$p - q$-product* of $f$ and $g$ is the path $h = fg[p, q]$ defined as follows:

$$h(px) = f(x), 0 \leqslant x \leqslant 1; h(p + qy) = g(y), 0 \leqslant y \leqslant 1.$$

This defines at least one image under $h$ for every $t \in I$, since each $t$ is either greater than $p$ and expressible in the form $p + qy$ or less than or equal to $p$ and expressible in the form $px$. There are two definitions for $h(p)$, namely $f(1)$ and $g(0)$, so that the path $h$ is defined only if $f(1) = g(0)$.

Triple and $n$-tuple products can be defined in the same way. In fact, let $p_1, \ldots, p_n$ be $n$ real numbers such that $p_1 + \ldots + p_n = 1, p_i > 0$ $(i = 1, \ldots n)$. Let $f_1, \ldots, f_n$ be $n$ paths satisfying the relations $f_r(1) = f_{r+1}(0)$ $(r = 1, \ldots, n-1)$. Then the $(p_1, \ldots, p_n)$-product of the $f$'s is the path $h$ defined as follows:

$$h(p_1 + p_2 + \ldots + p_{r-1} + p_r x) = f_r(x), 0 \leqslant x \leqslant 1, r = 1, \ldots, n.$$

We shall then denote $h$ by $f_1 f_2 \ldots f_n[p_1, p_2, \ldots, p_n]$.

From the relations

$$\text{(1.2.3)} \quad (fg[p, q])h[r, s] = fgh[pr, qr, s] \text{ and } f(gh[r, s])[p, q] = fgh[p, qr, qs]$$

it follows that every triple product can be expressed in terms of double products, and limited forms of associative law are valid. For instance:

$$\text{(1.2.4)} \qquad (fg[\tfrac{1}{2}, \tfrac{1}{2}])h[\tfrac{2}{3}, \tfrac{1}{3}] = fgh[\tfrac{1}{3}, \tfrac{1}{3}, \tfrac{1}{3}] = f(gh[\tfrac{1}{2}, \tfrac{1}{2}])[\tfrac{1}{3}, \tfrac{2}{3}],$$

if, of course, the products arising are defined, that is, if

$$f(1) = g(0) \quad \text{and} \quad g(1) = h(0).$$

Consider the relation $R$ between the points of $X$ defined as follows:

$$aRb \text{ if and only if } P(a, b) \neq \varnothing$$

We prove that $R$ is an equivalence relation.

(i) It is reflexive, since the *constant path* $c_a$ defined by $c_a(x) = a$ for all $x \in I$ is an element of $P(a, a)$.

(ii) It is symmetric, since, if $f \in P(a, b)$, the *reverse* path $f^{-1}$ defined by $f^{-1}(x) = f(1 - x)$ belongs to $P(b, a)$.

(iii) It is transitive, because if $f \in P(a, b)$ and $g \in P(b, c)$, then for any suitable $p, q, fg[p, q]$ belongs to $P(a, c)$.

The $R$-equivalence-classes are called *path-components* of $X$. A space consisting of a single path-component is said to be *path-connected*. A path-connected space is connected, but the converse does not necessarily hold.

## 1.3 Continuity of path multiplication

We now prove the following theorem, which may be summed up as an assertion that the path-product $fg[p, 1-p]$ is a continuous function of the triple $(f, g, p)$.

1.3.1 THEOREM. *Let $E$ denote the subset of $X^I \times X^I$ consisting of all pairs of paths $f, g$ with $f(1) = g(0)$. Let $I^\circ$ denote the open unit interval $\{p : 0 < p < 1\}$. Then the mapping $m : E \times I^\circ \to X^I$ defined by $m(f, g, p) = fg[p, 1-p]$ is continuous.*

*Proof.* Suppose that $K \subset I$ is compact, $U \subset X$ is open and that $h = m(f, g, p) = fg[p, 1-p] \in T(K, U)$. We shall prove that there exist compact sets $K_1$, $K_2$ and a positive number $\varepsilon$ such that $f \in T(K_1, U)$, $g \in T(K_2, U)$ and such that

$$f' \in T(K_1, U), \qquad g' \in T(K_2, U), \qquad p - \varepsilon < p' < p + \varepsilon \tag{1.3.2}$$

implies that

$$h' = f'g'[p', 1-p'] \in T(K, U). \tag{1.3.3}$$

From this it follows that $m^{-1}(T(K, U))$ is open and since the sets $T(K, U)$ generate $(\mathcal{T})$ the desired continuity follows.

Since $h(K) \subset U$, $K$ is contained in the open set $h^{-1}(U)$. As $K$ is compact, there is a positive number $\varepsilon$ such that, if the distance of $x$ from $K$ is less than $2\varepsilon$ then $x \in h^{-1}(U)$. Let $L$ denote the compact set consisting of all points in $I$ whose distance from $K$ does not exceed $\varepsilon$. Define

$$K_1 = \{t : pt \in L, t \in I\}, \qquad K_2 = \{t : p + (1-p)t \in L, t \in I\}.$$

Suppose now that $f', g', p'$ satisfy 1.3.2 We have to show that $h'(t) \in U$ if $t \in K$. There are two possibilities which must be considered separately.

(i) If $t \leqslant p'$, then $t = p'x (x \in I)$. Since $|(p - p')x| < \varepsilon$ and $p'x \in K$ we have $px \in L$ so $x \in K_1$. Thus $h'(t) = f'(x) \in U$.

(ii) If $p' \leqslant t \leqslant 1$, then $t = p' + (1-p')x$, $x \in I$. Let $u = p + (1-p)x$. Then $|t - u| = |p' - p|\,|1 - x| < |p' - p| < \varepsilon$. Since $t \in K$, $u \in L$ so $x \in K_2$. Thus $h'(t) = g'(x) \in U$. ■

## 1.4 Equivalence of paths–routes

1.4.1 *Definition.* Two paths in $X^I$ are called *equivalent* (or *homotopic*) if they have the same initial point $a$, the same endpoint $b$ and also if they belong to the same path-component of $P(a, b)$.

By 1.2.1 this is an equivalence relation. An equivalence class, that is, a path component of one of the subspaces $P(a, b)$ is called a *route* (from $a$ to $b$). If $f \in X^I$, the route containing $f$ will be written $[f]$.

$R(X)$ will denote the set of all routes in $X$, topologized with the quotient topology obtained from the compact-open topology on $X^I$. Since all the paths in a route $r$ have the same initial point and all have the same endpoint, we may refer to the initial point of $r$ and the end point of $r$. We shall extend our notation to denote the initial point of $r$ by $r(0)$, its endpoint by $r(1)$.

Let $f: I \to X^I$ be a path in the space $X^I$. Ignoring continuity for the moment, we have, for each $t \in I$, a path $f(t): I \to X$. Putting $F(t, u) = F(t)(u)$, we obtain a function from the unit square $I \times I$ into $X$. The question of its continuity is dealt with in the following theorem.

1.4.2 THEOREM. *There is a one-to-one correspondence between continuous maps $F: I \times I \to X$ and paths $f: I \to X^I$ defined by the relation*

$$F(t, u) = f(t)(u).$$

*Proof.* We have to show that the two statements below are equivalent.

(i) Each $f(t)$ is a continuous map and $f$ is a continuous map from $I$ into $X^I$.
(ii) $F$ is continuous as a map from the product space $I \times I$ into $X^I$.

(i) implies (ii). Let $t, u \in I$ and let $W$ be a neighbourhood of $F(t, u)$. Since $f(t): I \to X$ is a continuous function, there is a compact interval $K$ containing $u$ in its interior such that $f(t)(K) \subset W$. Since $f$ is continuous, the set $f^{-1}(T(K, W)) = \{v: v \in I, f(v)(K) \subset W\}$ is an open neighbourhood $U$ of $t$. Then $F$ maps the neighbourhood $U \times K$ of $(t, u)$ into $W$, so $F$ is continuous at $(t, u)$.

(ii) implies (i). Since $F(t, u)$ is continuous, it defines a continuous function of $u$ for each constant $t$, so each $f(t)$ is a path. To prove the continuity of $f$ it will be enough to show that, if $K \subset I$ is compact and $W \subset X$ is open, then the set $E = f^{-1}(T(K, W)) = \{t: t \in I, F(t, K) \subset W\}$ is open. Let $t_0 \in E$, the compact set $\{t_0\} \times K$ is contained in the open set $F^{-1}(W)$, so there exists $\varepsilon > 0$ such that $F^{-1}(W)$ contains all the points of $I \times I$ whose distance from

$\{t_0\} \times K$ does not exceed $\varepsilon$. In particular,

$$(t_0 - \varepsilon, t_0 + \varepsilon) \times K \subset F^{-1}(W),$$

so $t_0$ is an interior point of $E$ and $E$ is open.

1.4.3 Corollary. *Two paths $f_0, f_1$ belong to the same route if and only if there is a continuous map $F: I \times I \to X$ such that, for all $t \in I$,*

$$F(0, t) = f_0(t), \qquad F(1, t) = f_1(t), \qquad F(t, 0) = f_0(0) = f_1(0),$$

$$F(t, 1) = f_0(1) = f_1(1).$$

A map $F$ of this kind, describing a path in a subspace $P(a, b)$, is called a *homotopy* between the paths $f_0, f_1$. In specifying a homotopy it is often convenient to use the notation $f_t$ to denote the path defined by $f_t(u) = F(t, u)$.

## 1.5 Homotopy and isotopy

The concept of homotopy can be defined for mappings of any space into another.

1.5.1 *Definition.* Two mappings $f_0, f_1: X \to Y$ are called *homotopic* if there is a continuous mapping $F: I \times X \to Y$ such that, for all $x \in X$, $f_0(X) = F(0, x)$, $f_1(x) = F(1, x)$. Then the mapping $F$ is called a *homotopy*, and we write $f_0 \simeq f_1$.

A homotopy between mappings always defines a path in the space $Y^X$. Conversely, if the space $X$ is locally compact, a path in $Y^X$ defines a homotopy. The proof of 1.4.2 can be extended to prove these assertions.

Though the above concept of homotopy is fairly general, it does not quite cover the equivalence relation just introduced between paths, where the initial and end-points must remain fixed throughout the homotopy. The concept of relative homotopy is relevant here and elsewhere.

1.5.2 *Definition.* Let $X$, $Y$ be topological spaces, $A$ a subspace of $X$. Two mappings $f_0, f_1$ are said to be *homotopic relative to A* is there is a homotopy $F: I \times X \to Y$ such that $F$ is *stationary on A*, that is to say, for every point $a \in A$ and every pair $t, u \in I$, we have $F(t, a) = F(u, a)$. When this is so, we write $f_0 \simeq f_1$ rel $A$.

As a special case, two paths $f_0$, $f_1$ define the same route if $f_0 \simeq f_1 \text{ rel } \dot{I}$, where $\dot{I}$ is the subspace $\{0, 1\}$.

It is often necessary to take the subspace $A$ to consist of a single point. A *pointed space* $(X, x)$ is a pair consisting of a space $X$ and a distinguished point (or *base-point*) $x \in X$. A mapping $f: (X, x) \to (Y, y)$ between one pointed space and another is a continuous mapping $f: X \to Y$ such that $f(x) = y$. Two mappings $f_0, f_1: (X, x) \to (Y, y)$ are called *homotopic* if they are homotopic relative to $\{x\}$.

Sometimes it is desirable to restrict attention to other subspaces of $Y^X$, such as the homeomorphisms and the embeddings. A continuous mapping $f: X \to Y$ is called an *embedding* if it defines a homeomorphism between $X$ and $f(X)$. In the case of homeomorphisms and embeddings it is usual to replace the concept of homotopy by the stricter relationships of *isotopy* and *ambient isotopy*, which we now define. These concepts will be of importance in Chapter 6.

Let $p: I \times X \to I$ denote, for any $X$, the projection from the product to the first factor, so that $p(t, x) = t$. A *level-preserving mapping* $F: I \times X \to I \times Y$ is a mapping $F$ such that $p \circ F = p$.

1.5.3 *Definition.* Two embeddings $f_0, f_1: X \to Y$ are called *isotopic* if there is a level-preserving embedding $F: I \times X \to I \times Y$ such that, for all $x \in X$, $F(0, x) = (0, f_0(x))$, $F(1, x) = (1, f_1(x))$. Such a mapping $F$ is called an *isotopy*.

1.5.4 *Definition.* Two embeddings $f_0, f_1$ are said to be *ambient-isotopic* if there is a level-preserving homeomorphism $F: I \times Y \to I \times Y$ such that $F(0, y) = (0, y)$ for all $y \in Y$ and $F(1, f_0(x)) = (1, f_1(x))$ for all $x \in X$. Then the mapping $F$ is called an *ambient isotopy*. Ambient isotopy is the stronger concept, for if $F$ is an ambient isotopy between $f_0$ and $f_1$ ,then $F \circ (\text{id} \times f_0)$ is an isotopy. The last two definitions will not be used again until Chapter 6, where they will be applied to homeomorphisms of surfaces. In that special case, the following theorem holds.

1.5.5 THEOREM. *Two homeomorphisms of a compact surface are isotopic if and only if they are homotopic.*

For a proof see [**5**]. This result implies that, in the case of compact surfaces, the homotopy classification of homeomorphisms yields as much information as the apparently stricter isotopy classification.

## 1.6 Operations on the routes

Suppose now that $f_t$ is a homotopy from $f_0$ to $f_1$, and that $g_t$ is a homotopy from $g_0$ to $g_1$, where $f_0(1) = g_0(0)$, so that the products $h_t = f_t g_t[p, q]$ are defined. By continuity of multiplication $h_t$ defines a homotopy from $h_0$ to $h_1$. Thus the route of the product path depends only on the routes of the paths multiplied, and it is legitimate to define the product of routes $r, s$ with $r(1) = s(0)$ by $rs[p, q] = [fg[p, q]]$ for any $f \in r$, $g \in s$.

Moreover, the path $fg[p, q]$ is equivalent to the path $fg[p - a, q + a]$, the homotopy being defined by $h_t = fg[p - ta, q + ta]$. Thus if $r, s$ are routes with $r(1) = s(0)$, we have $rs[p, q] = rs[p', q']$. We can therefore drop the numbers $p, q$, which indicate the point of subdivision, and talk simply of the route product $rs$.

Analogously, if $f_t$ is a homotopy from $f_0$ to $f_1$, then $f_t^{-1}$ is a homotopy from $f_0^{-1}$ to $f_1^{-1}$. Thus it is legitimate to talk of the *inverse route* $r^{-1}$—the route consisting of the reverses of the paths in $r$ (see Section 2). Finally, the route containing a constant path $c_a$ is denoted by $1_a$. These notations are justified by the following theorem.

1.6.1 THEOREM.

(i) *If $r, s, t$ are routes with $r(1) = s(0)$, $s(1) = t(0)$, then $r(st) = (rs)t$.*
(ii) *If $r \in R(a, b)$, then $1_a r = r = r1_b$.*
(iii) *If $r \in R(a, b)$, then $rr^{-1} = 1_a$, $r^{-1}r = 1_b$.*

*Proof.* (i) Let $f \in r$, $g \in s$, $h \in t$. Then (i) follows from 1.2.4

$$(fg[\tfrac{1}{2}, \tfrac{1}{2}])\, h[\tfrac{2}{3}, \tfrac{1}{3}] = fgh[\tfrac{1}{3}, \tfrac{1}{3}, \tfrac{1}{3}] = f(gh[\tfrac{1}{2}, \tfrac{1}{2}])\,[\tfrac{1}{3}, \tfrac{2}{3}].$$

(ii) Let $f \in r$. A homotopy $h$ between $c_a f[\frac{1}{2}, \frac{1}{2}]$ and $f$ is given by

$$h_t = c_a f[\tfrac{1}{2} - \tfrac{1}{2}t, \tfrac{1}{2} + \tfrac{1}{2}t] \qquad (0 \leqslant t < 1), \qquad h_1 = f.$$

Similarly $fc_b[\frac{1}{2}, \frac{1}{2}]$ is homotopic to $f$.

(iii) We have to show that $ff^{-1}$ is homotopic to $c_a$. Put $f_t(u) = f(tu)$. Then $f_t f_t^{-1}[\frac{1}{2}, \frac{1}{2}]$ is a homotopy from $c_a$ to $ff^{-1}[\frac{1}{2}, \frac{1}{2}]$. This proves the first part of (iii). The second part follows from the first because $r = (r^{-1})^{-1}$. ■

## 1.7 Continuity of operations on $R(X) \times R(X)$

We now consider the continuity of multiplication of routes. Let $F \subset$

$R(X) \times R(X)$ denote the subspace consisting of the pairs of routes for which the product is defined, that is $F = \{(r, s): r(1) = s(0)\}$. There are two ways of topologizing $F$ which seem almost equally natural.

Topology $A$. $F$ has the quotient topology from the corresponding subspace $E \subset X^I \times X^I$.

Topology $B$. $F$ has the relative topology as a subspace of $R(X) \times R(X)$.

For many purposes Topology $B$ is more natural, but Topology $A$ is the one for which the continuity theorem seems easy to prove. I have not been able to prove the continuity theorem directly for Topology $B$, nor to determine what restrictions, on $X$ (if any) are necessary and sufficient for the topologies to be the same. However, the following results suffice for all our needs.

1.7.1 THEOREM. *If $X$ is a Hausdorff space, the $A$-topology and the $B$-topology on $F$ are the same.*

*Proof.* Let $p: X^I \times X^I \to R(X) \times R(X)$ denote the map $(f, g) \mapsto ([f], [g])$. If $(f, g) \notin E$, then $f(1)$, $g(0)$ have disjoint neighbourhoods $U$, $W$, so if $\mathfrak{v}_t$ denotes the value map (Section 1). we have

$$(\mathfrak{v}_1^{-1}(U) \times \mathfrak{v}_0^{-1}(W)) \cap E = \varnothing.$$

Thus the complement of $E$ is open so $E$ is closed. Since $p$ is a quotient map, $F$ is also closed in $R(X) \times R(X)$. Now suppose that a subset $D$ of $F$ is given. It is $A$-closed if and only if $p^{-1}(D)$ is closed in $E$, or, since $E$ is closed, if and only if $p^{-1}(D)$ is closed in $X^I \times X^I$. By a similar argument $D$ is $B$-closed if and only if it is closed in $R(X) \times R(X)$. Since $p$ is a quotient map, this is true if and only if $p^{-1}(D)$ is closed in $X^I \times X^I$. ■

Now let $\mu: F \to R(X)$ denote the map given by $\mu(r, s) = rs$. Also let $m$: $E \to X^I$ denote the map given by $m(f, g) = fg[\frac{1}{2}, \frac{1}{2}]$. The definition of route product is the assertion of commutativity in the following diagram:

$$\begin{array}{ccc} E & \xrightarrow{m} & X^I \\ {\scriptstyle p}\downarrow & & \downarrow{\scriptstyle p} \\ F & \xrightarrow{\mu} & R(X) \end{array} \qquad (1.7.2)$$

where the $p$ on the right is, of course, defined by $p(f) = [f]$.

1.7.3 THEOREM. *The map $\mu$ is continuous as a map from F with the A-topology into $R(X)$.*

*Proof.* Consider diagram 1.7.2. Suppose that $U \subset R(X)$ is open. By the continuity of $p$, $m$, the set $m^{-1}(p^{-1}(U))$ is open. By commutativity in 1.7.2, $p^{-1}(\mu^{-1}(U))$ is open. Since $p$ is a quotient map, $\mu^{-1}(U)$ is open and the continuity of $\mu$ is proved. ∎

Exactly the same argument enables one to deduce:

1.7.4 THEOREM. *The map $r \mapsto r^{-1}$ of $R(X)$ onto itself is continuous.*

## 1.8 The fundamental group

Though multiplication is only defined for certain pairs of routes, it is always defined on the subspaces $R(a, a)$. Moreover, $R(a, a)$ is a closed subspace of $R(X)$ if $X$ is a Hausdorff space. It then follows from the theorems of 1.6 and 1.7 that

1.8.1 THEOREM. *If $X$ is a Hausdorff space, $R(a, a)$ is a topological group under the operations of route multiplication and route inverse, the constant route $1_a$ being the unit element.*

This group is called the *fundamental group* of $X$ with base-point $a$, and it is usually denoted by $\pi_1(X, a)$.

*Proof.* Suppose now that $a, b$ belong to the same path-component of $X$, that is, $P(a, b) \neq \varnothing$, and hence $R(a, b) \neq \varnothing$. Let $t \in R(a, b)$. Define maps $\phi: \pi_1(X, a) \to \pi_1(X, b)$ and $\psi: \pi_1(X, b) \to \pi_1(X, a)$ by the relations $\phi(r) = t^{-1}rt$, $\psi(s) = tst^{-1}$. Now $\phi(r)\,\phi(s) = t^{-1}rtt^{-1}st = t^{-1}r1_ast = t^{-1}rst = \phi(rs)$. Thus $\phi$, and by the same argument $\psi$, is a homomorphism. Again $\psi(\phi(r)) = t\phi(r)\,t^{-1} = tt^{-1}rtt^{-1} = 1_ar1_a = r$, so $\psi \circ \phi = \text{id}$. Similarly $\phi \circ \psi = \text{id}$. Thus $\phi$ is an isomorphism, $\psi$ is its inverse and the groups $\pi_1(X, a)$, $\pi_1(X, b)$ are isomorphic. ∎

In studying the fundamental groups, and indeed in studying the algebra of the path and route spaces, there will be no interaction between different path components of the space. Moreover, most of the spaces to which these concepts find application are path-connected, so it is convenient to consider path-connected spaces only. Then we have

1.8.2 THEOREM. *If $X$ is path-connected, the fundamental groups $\pi_1(X, a)$, $\pi_1(X, b)$ are isomorphic.*

Though the groups are logically distinct and there is no canonical, or naturally preferred, isomorphism between them, one often refers (because of 1.8.2) simply to *the* fundamental group of the space $X$.

### 1.9 Simply connected spaces

Throughout this section $X$ is a path-connectd space.

1.9.1 *Definition.* A path-connected space is said to be *simply connected*, if, for one point $x \in X$, and therefore, by Theorem 1.8.2, for every point $x \in X$, the fundamental group $\pi_1(X, x)$ consists of the unit element only.

The property of being simply connected can be characterized in other ways. Let $S^1 = \{z: z \in \mathbb{C}, |z| = 1\}$ denote the unit circle in the complex plane and let $B^2 = \{z: z \in \mathbb{C}, |z| \leqslant 1\}$ denote the closed unit ball.

1.9.2 THEOREM. *$X$ is simply connected if and only if, for each continuous map $\phi: S^1 \to X$, there is a continuous extension $\Phi: B^2 \to X$, that is, a continuous map $\Phi$ such that, for all $z \in S^1$, $\Phi(z) = \phi(z)$.*

*Proof.* Suppose first of all that all maps of $S^1$ into $X$ can be extended to $B^2$. Note that there is a homeomorphism of the pair $(B^2, S^1)$ with the pair $(I^2, \dot{I}^2)$ where $I^2$ is the unit square $I \times I$ and $\dot{I}^2$ is its boundary as a subset of $\mathbb{R}^2$. Specifically, $\dot{I}^2$ is the set of pairs $(t, u)$ such that either $\max(t, u) = 1$ or $\min(t, u) = 0$. Suppose that $f \in P(a, a)$, $a \in X$. Define a map $\phi: \dot{I}^2 \to X$ by putting $\phi(0, t) = f(t)$, and, for all $t$, $\phi(t, 0) = \phi(t, 1) = \phi(1, t) = a$. By hypothesis, $\phi$ can be extended to a map $\Phi: I^2 \to X$, and by Corollary 1.4.3, $f$ is homotopic to the constant map $c_a$. Thus $R(a, a)$ is trivial and $X$ is simply connected.

Suppose conversely that $X$ is simply connected and let $\phi: S^1 \to X$. Let $f: I \to X$ be the map defined by $f(t) = \phi(\exp(2\pi i t))$. Then $f \in P(a, a)$, where $a = \phi(1)$. If $F(u, t)$ is a homotopy from $f$ to $c_a$, then an extension $\Phi$ to $B^2$ of the map $\phi$ is defined by the equation $\Phi(u \exp(2\pi i t)) = F(u, t)$ $(0 \leqslant t, u \leqslant 1)$. ■

1.9.3 THEOREM. *If $X$ is simply connected, then, for any pair of points $a, b \in X$,* *there is only one route from $a$ to $b$. Conversely, if, for a certain single pair of points $a, b \in X$, there is only one route from $a$ to $b$, then $X$ is simply connected.*

*Proof.* (i) Suppose $c, d \in R(a, b)$ where $a, b \in X$ and $X$ is simply connected. Then $cd^{-1} \in R(a, a)$, so $cd^{-1} = 1_a$ and $c = cd^{-1}d = 1_a d = d$.

(ii) If $X$ is not simply connected, let $e \neq 1_b$ in $R(b, b)$. If $c \in R(a, b)$, then $c, ce$ are distinct elements of $R(a, b)$ since $c = ce$ implies $1_b = c^{-1}c = c^{-1}ce = e$. ■

1.9.4 *Definition.* A subset $K$ of $\mathbb{R}^n$ is *convex*, if, for any two points $a, b \in K$ and any $t \in I$, we have $ta + (1 - t)b \in K$.

1.9.5 THEOREM. *A convex subset of $\mathbb{R}^n$ is simply connected.*

*Proof.* Let $f: I \to K$ with $f(0) = f(1) = a$. The map $F: I \times I \to K$ defined by $F(t, u) = ta + (1 - t) f(u)$ defines a homotopy between $f$ and $c_a$. ■

## 1.10 Local properties of spaces

Whenever $P$ is a property of topological spaces, the term *locally P* usually describes a space in which every point has arbitrarily small neighbourhoods with the property. In this spirit we make the following definitions.

1.10.1 *Definition.* A space $X$ is called *locally path-connected* if, for each $x \in X$ and each neighbourhood $V$ of $x$, there is a path-connected neighbourhood $W$ of $x$ such that $W \subset V$.

1.10.2 *Definition.* A space $X$ is called *locally simply connected* if, for each $x \in X$ and each neighbourhood $V$ of $x$, there is a simply connected neighbourhood $W$ of $x$ such that $W \subset V$.

The term "locally simply connected" is sometimes used to describe the following weaker property. Every point $x \in X$ has a neighbourhood $V$ such that every path in $V$ is homotopic *in* $X$ to a constant path. This concept is certain enough to give a satisfactory theory of fundamental groups and covering spaces. We believe however that the term "locally simply connected" ought to be reserved for the concept 1.10.2, and if one wishes to deal with more general spaces one should use a different term. Steenrod's term is "semi-locally 1-connected", while André Weil calls such spaces "homotopically flat".

We shall use the obvious abbreviations:

p.c. path-connected, l.p.c. locally path-connected,
s.c. simply connected, l.s.c. locally simply connected.

### 1.11 Route spaces for locally simply connected $X$

Suppose that $X$ is Hausdorff, p.c., l.p.c. and l.s.c. In this case we can give an alternative description of the topology on $R(X)$. Let $a \in X$ and let $V$ be a neighbourhood of $a$. Let $P(a, V)$ denote the set of paths $f: I \to V$ with $f(0) = a$. Let $R(a, V)$ denote the set of all routes in $X$ corresponding to such paths, that is, $R(a, V) = \{[f]: f \in P(a, V)\}$. In precisely the same way we denote by $R(V, a)$ the set of all routes defined by paths in $V$ whose *end* point is $a$.

1.11.1 THEOREM. *Let $r \in R(a, b)$ and let $V_1$, $V_2$ be simply connected neighbourhoods of $a$, $b$ respectively. Then the set of routes*

$$R(V_1, a)\, rR(b, V_2) = \{qrs: q \in R(V_1, a),\ s \in R(b, V_2)\}$$

*is a neighbourhood of $r$ in $R(X)$.*

*Proof.* Let $f \in r$. We have to show that there is a neighbourhood $W$ of $f$ in the compact-open topology such that $g \in W$ implies $[g] \in R(V_1, a)\, rR(b, V_2)$. Since $X$ is locally simply connected, each point of $f(I)$ has a simply connected neighbourhood $U$. The interiors of all such $U$ cover the compact set $f(I)$ so it is possible to subdivide $I$ into a finite set of sub-intervals $J_i = [t_{i-1}, t_i]$ $(i = 1, \ldots, k)$ where

$$0 = t_0 < t_1 < \ldots < t_k = 1,$$

such that $f(J_i) \subset \operatorname{int} U_i$, $i = 1, \ldots, k$, with $U_1, \ldots, U_k$ simply connected. We may of course assume that $U_1 = V_1$, $U_k = V_2$. For $i = 1, \ldots, k-1$, let $U_i^*$ be a path-connected set such that $f(t_1) \in \operatorname{int} U_i^* \subset U_i \cap U_{i+1}$. Set $U_0^* = V_1$, $U_k^* = V_2$.

Let $W_1 \subset X^I$ denote the set

$$W_1 = T(J_1, U_1) \cap \ldots \cap T(J_k, U_k),$$

and let

$$W_2 = T(\{t_0\}, U_0^*) \cap \ldots \cap T(\{t_k\}, U_k^*).$$

Then $W = W_1 \cap W_2$ is a neighbourhood of $f$ in $X^I$ which we shall prove has the desired property.

If we put $p_1 = t_1, p_2 = t_2 - t_1, \ldots, p_k = t_k - t_{k-1}$, so that $p_r$ is the length of the sub-interval $J_r$, then we have

$$f = f_1 f_2 \ldots f_k[p_1, \ldots, p_k]$$

where, for $i = 1, \ldots, k$, $f_i$ is defined by $f_i(t) = f(t_{i-1} + p_i t)$, so that $f_i(I) = f(J_i) \subset U_i$.

Now suppose that $g \in W$. Again defining $g_i$ by the relation $g_i(t) = g(t_{i-1} + p_i t)$ we have $g = g_1 g_2 \ldots g_k[p_1, \ldots, p_k]$, where $g_r(I) \subset U_r$, $g_r(0) = g(t_{r-1}) \in U^*_{r-1}$, $g_r(1) = g(t_r) \in U^*_r$.

For $r = 0, \ldots, k$ let $h_r$ be a path from $f(t_r)$ to $g(t_r)$ lying completely in $U^*_r$. Then $h^{-1}_{r-1} f_r h_r[\frac{1}{3}, \frac{1}{3}, \frac{1}{3}]$ lies in the simply connected set $U_r$ and has the same initial and end points as $g_r$. Thus by 1.9.2 $[h^{-1}_{r-1}][f_r][h_r] = [g_r]$,

$$\begin{aligned}[g] &= [g_1][g_2]\ldots[g_k] \\ &= [h_0^{-1}][f_1][f_2]\ldots[f_k][h_k] = [h_0^{-1}][f][h_k].\end{aligned}$$

Since $[h_0^{-1}] \in R(V_1, a)$, $[h_k] \in R(b, V_2)$ and $f \in r$, the result follows. ■

1.11.2 COROLLARY. *For each pair $a, b \in X$, the space $R(a, b)$ is discrete. In particular, the fundamental group $R(a, a)$ is discrete.*

The topology defined by the system of neighbourhoods 1.11.1 is sometimes referred to as the "dumb-bell topology". The next theorem, valid even without the restriction that $X$ should be locally simply connected, combines with 1.11.1 to show that, when $X$ is Hausdorff, p.c. and l.s.c. the topology induced on $R(X)$ by the compact-open topology on the set of paths coincides with the dumb-bell topology.

1.11.3 THEOREM. *Let $W \subset R(X)$ be an open subset of $R(X)$ and let $r \in W$, $r(0) = a$, $r(1) = b$. Then there exist neighbourhoods $V_1$ of $a$, $V_2$ of $b$ such that*

$$R(V_1, a)\, r R(b, V_2) \subset W.$$

*Proof.* Let $\Phi$ denote the set of all $g \in X^I$ with $[g] \in W$. Then $\Phi$, as the inverse image of $W$ under the quotient map, is open. Let $f \in r$. Since $f \in \Phi$, there exist finitely many compact sets $K_1, \ldots, K_r \subset I$ and open sets $U_1, \ldots, U_r \subset X$ with $f(K_i) \subset U_i$ and such that $g(K_i) \subset U_i$ $(i = 1, \ldots, r)$ implies $g \in \Phi$, Define two subsets of the integers between 1 and $r$ as follows. Let

$$E = \{i: 1 \leqslant i \leqslant r, 0 \in K_i\}; \quad F = \{j: 1 \leqslant j \leqslant r, 1 \in K_j\}.$$

Let $V_1 = \bigcap\{U_i: i \in E\}$, and let $V_2 = \bigcap\{U_j: j \in F\}$. Then $a \in V_1$ and $b \in V_2$. Let $\varepsilon$ be a positive number so small that the following are all true.

(i) $t \in K_i$, $i \notin E$, implies $t \geqslant \varepsilon$;
(ii) $t \in K_i$, $i \notin F$, implies $t \leqslant 1 - \varepsilon$;
(iii) $f([0, \varepsilon]) \subset V_1$;
(iv) $f([1 - \varepsilon, 1]) \subset V_2$.

Define the paths $f_1, f_2, f_3$ by the relation $f = f_1 f_2 f_3 [\varepsilon, 1 - 2\varepsilon, \varepsilon]$.

Suppose now that $s \in R(V_1, a)\, r R(b, V_2)$, so that $s = [h_1]\, r [h_2]$ where $h_1(I) \subset V_1, h_2(I) \subset V_2$. Define paths $k_1, k_2$ by $k_1 = h_1 f_1 [\frac{1}{2}, \frac{1}{2}]$, $k_2 = f_3 h_2 [\frac{1}{2}, \frac{1}{2}]$. Set $g = k_1 f_2 k_2 [\varepsilon, 1 - 2\varepsilon, \varepsilon]$. Then $g \in \bigcap_{1 \leqslant i \leqslant r} T(K_i, U_i) \subset \Phi$ for if $t \in K_i$ we have one of three possibilities:

(A) $0 \leqslant t < \varepsilon$, so $i \in E$ by (i) and $g(t) \in V_1 \subset \bigcap_{\nu \in E} U_\nu \subset U_i$;
(B) $1 - \varepsilon < t \leqslant 1$, so $i \in F$ by (ii) and $g(t) \in V_2 = \bigcap_{\mu \in F} U_\mu \subset U_i$;
(C) $\varepsilon \leqslant t \leqslant 1 - \varepsilon$, so $g(t) = f(t) \in U_i$ since $f(K_i) \subset U_i$.

Moreover $[g] = [k_1][f_2][k_2] = [h_1][f_1][f_2][f_3][h_2] = [h_1]\, r [h_2] = s$. Since $g \in \Phi$, $s \in W$, and the result is proved. ∎

## 1.12 The homomorphism induced by a continuous mapping

If $f: X \to Y$ is a continuous mapping, then there is an induced mapping $f_\#: X^I \to Y^I$ defined by $f_\#(\alpha) = f \circ \alpha$. If $K$ is a compact subset of $I$ and $U$ is open in $Y$, then $f_\#^{-1}(T(K, U)) = T(K, f^{-1}(U))$, so $f_\#$ is continuous.

Now if $\alpha_0, \alpha_1 \in X^I$ and $H$ is a homotopy from $\alpha_0$ to $\alpha_1$, then $f \circ H$ is a homotopy from $f \circ \alpha_0$ to $f \circ \alpha_1$, so that $[f_\#(\alpha_0)] = [f_\#(\alpha_1)]$. Hence we may define, for $r \in R(X)$, $f_*(r) = [f \circ \alpha]$, where $\alpha$ is any path in $r$, the result being independent of the choice of path $\alpha \in r$. The map $f_*: R(X) \to R(Y)$, being derived from the continuous map $f_\#: X^I \to Y^I$ by compatible equivalence relations, is itself continuous.

1.12.1 THEOREM. *The map $f_*$; $R(X) \to R(X)$ has the following properties.*

(i) *$f_*$ maps $R(a, b)$ into $R(f(a), f(b))$.*

(ii) *If $rs$ is defined, so is $f_*(r)\, f_*(s)$ and we have $f_*(rs) = f_*(r)\, f_*(s)$.*
(iii) $f_*: \pi_1(X, x) \to \pi_1(Y, f(x))$ *is a group homomorphism.*
(iv) $\mathrm{id}_* = \mathrm{id}$
(v) *If $f: X \to Y$ and $g: Y \to Z$ are continuous mappings then*

$$(g \circ f)_* = g_* \circ f_*.$$

(vi) *If $f_0, f_1: (X, x) \to (Y, y)$ are homotopic as maps of pointed spaces, then*

$$(f_0)_* = (f_1)_*: \pi_1(X, x) \to \pi_1(Y, y).$$

*Proof.* Assertions (i), (iv) are immediate, while (iii) follows from (i) and (ii). (ii) follows from the easily verified formula

$$f \circ (\alpha\beta[p, q]( = (f \circ \alpha)(f \circ \beta)[p, q].$$

while (v) follows from $g \circ (f \circ \alpha) = (g \circ f) \circ \alpha$ for $\alpha \in X^I$. Finally we prove (vi). If $f_t$, $t \in I$, is a homotopy from $f_0$ to $f_1$, where, for each $t$, $f_t(x) = y$, then $f_t \circ \alpha$ is a homotopy from $f_0 \circ \alpha$ to $f_1 \circ \alpha$, for any $\alpha \in R(x, x)$. Thus $f_{0*}([\alpha]) = [f_0 \circ \alpha] = [f_1 \circ \alpha] = f_{1*}([\alpha])$. ■

1.12.2 *Definition.* A mapping $f: X \to Y$ is called a *homotopy inverse* of a mapping $g: Y \to X$ if $f \circ g \simeq \mathrm{id}_Y$ and $g \circ f \simeq \mathrm{id}_X$. If $f$ has a homotopy inverse, it is called a *homotopy equivalence.* If the set of homotopy equivalences from $X$ to $Y$ is non-empty, then $X$, $Y$ are said to have the same *homotopy type.*

1.12.3 *Definition.* Let $X$ be a path-connected space, and let $x_0$, $x_1$ be two points of $X$. A *path-isomorphism* between $\pi_1(X, x_0)$ and $\pi_1(X, x_1)$ is an isomorphism of the form $r \mapsto s^{-1} rs$, where $s$ is some fixed route from $x_0$ to $x_1$.

1.12.4 THEOREM. *If $f: X \to X$ is homotopic to the identity, then $f_*: \pi_1(X, x_0) \to \pi_1(X, x_1)$ is a path-isomorphism.*

*Proof.* Let $f_t$ be a homotopy from $\mathrm{id} = f_0$ to $f = f_1$. Let $\alpha_i (i = 0, 1)$, $\beta_j (j = 0, 1)$ denote the paths in the unit square $I \times I$ defined by $\alpha_i(t) = (t, i)$, $\beta_j(t) = (j, t)$. Since $I \times I$ is simply connected we have

$$[\beta_0] = [\alpha_0][\beta_1][\alpha_1]^{-1}. \tag{1.12.5}$$

Now let $\gamma \in P(x_0, x_1)$. Define $F: I \times I \to X$ by $F(t, u) = f_t(\gamma(u))$. Then, for $i = 0, 1$, $F \circ \alpha_i(t) = F(t, i) = f_t(\gamma(i)) = f_t(x_0)$. Thus $F \circ \alpha_0 = F \circ \alpha_1 = q$, say.

We also have $F \circ \beta_0 = \gamma$, $F \circ \beta_1 = f_{\#}(\gamma)$. Operating with $F_*$ on 1.12.5 we deduce

$$[\gamma] = [F \circ \beta_0] = [F \circ \alpha_0][F \circ \beta_1][F \circ \alpha_1]^{-1} = [q] f_*(\gamma)[q]^{-1}. \quad \blacksquare$$

1.12.6 COROLLARY. *If $f: X \to X$ is homotopic to the identity and $f(x_0) = x_0$, then $f_*$ is an inner automorphism of $\pi_1(X, x_0)$.*

1.12.7 THEOREM. *If $f: X \to Y$ is a homotopy equivalence between the path-connected spaces $X$, $Y$, and if $x \in X$, then $f_*: \pi_1(X, x) \to \pi_1(Y, f(x))$ is an isomorphism.*

*Proof.* We note first that it is enough to prove 1.12.7 for one particular point $x = x_0 \in X$. For if $x_1$ is another point, if $r \in R(x_0, x_1)$, $s \in R(x_0, x_0)$, then $f_*(r^{-1}sr) = f_*(r)^{-1} f_*(s) f_*(r)$. Thus $f_* \circ \phi = \psi \circ f_*$, where $\phi$, $\psi$ are the path isomorphisms associated with $r \in R(X)$, $f_*(r) \in R(Y)$. Let us choose $x_0 = g(y_0)$ for some $y_0 \in Y$. Consider the homomorphisms

$$g_* \circ f_*: \pi_1(X, x_0) \to \pi_1(X, g(f(x_0))),$$
$$f_* \circ g_*: \pi_1(Y, y_0) \to \pi_1(Y, f(x_0)).$$

By 1.12.4, both are isomorphisms. From the first we see that Ker $f_* = 1$. From the second we see that $f_*$ is onto. Thus 1.12.7 is proved. $\blacksquare$

## 2. THEORY OF COVERING SPACES

### 2.1 Basic concepts

Let $X$, $\tilde{X}$ be two spaces and $p: \tilde{X} \to X$ a continuous mapping.

2.1.1 *Definition.* A subset $V$ of $X$ is called *p-canonical* if there is a family $\mathscr{F}$ of subsets of $\tilde{X}$ such that

(i) $p^{-1}(V) = \bigcup \mathscr{F}$.
(ii) Each element of $\mathscr{F}$ is relatively open in $p^{-1}(V)$.
(iii) If $W_1, W_2 \in \mathscr{F}$, $W_1 \neq W_2$, then $W_1 \cap W_2 = \varnothing$.
(iv) For each $W \in \mathscr{F}$, $p$ maps $W$ homeomorphically onto $V$.

It will sometimes be convenient to say that $W$ is *p-homeomorphic to $V$*, in situations like this where $W$ is a subset of the domain of $p$, $V$ is a subset of the range of $p$ and the restriction of $p$ to $W$ is a homeomorphism of $W$ onto $V$.

2.1.2 *Definition.* If $X$ is a union of $p$-canonical open sets, then $p$ is called a *covering map* and the pair $(X, p)$ is called a *covering space* of $X$.

## 2.2 The lifting problem, maps with small image

Let $f: A \to X$ be a continuous map and let $p: \tilde{X} \to X$ be a covering map. It is often important to know whether a map $\tilde{f}: A \to \tilde{X}$ exists such that $p \circ \tilde{f} = f$. The process of finding such a map $\tilde{f}$ is known as "lifting" the map $f$. To lift the map $f$ one must find a map $\tilde{f}$ to belong to the dotted arrow making the diagram below commutative.

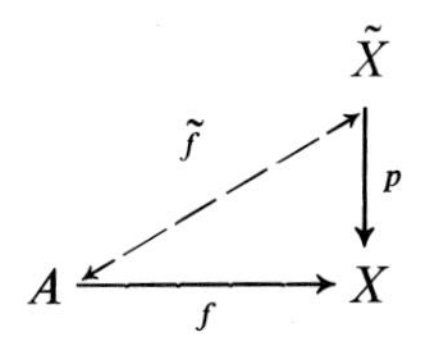


If $p$ is a homeomorphism, the unique solution to the lifting problem is given by $\tilde{f} = p^{-1} \circ f$. Since a covering map is, at least locally, a homeomorphism, lifting is always possible for maps with small enough image. This is formalized in our first result.

2.2.1 THEOREM

Hypothesis. (i) *$p: \tilde{X} \to X$ is a covering map.*

(ii) *$f: A \to X$ is continuous and $A$ is connected.*

(iii) *There is a p-canonical neighbourhood $V$ such that $f(A) \subset V$.*

(iv) *$a$, $x$, $\tilde{x}$ are points in $A$, $X$, $\tilde{X}$ such that $x = p(\tilde{x}) = f(a)$.*

Conclusion. *There is a unique map $\tilde{f}: A \to \tilde{X}$ such that $p \circ \tilde{f} = f, \tilde{x} = f(a)$.*

*Proof.* Let $\mathscr{F}$ be a partition of $p^{-1}(V)$ into open disjoint sets, each $p$-homeomorphic to $V$. Let $W$ be the element of $\mathscr{F}$ to which $\tilde{x}$ belongs. Since $A$ is connected and $W$ is relatively open and closed in $p^{-1}(V)$, any continuous map $\tilde{f}: A \to p^{-1}(V)$ with $\tilde{f}(a) = \tilde{x}$ must satisfy $\tilde{f}(A) \subset W$. Thus if $p_1$ denotes the restriction to $W$ of the map $p$, the unique solution of our lifting problem is $\tilde{f} = p_1^{-1} \circ f$. ■

## 2.3 Lifting of paths and maps of the square

We now extend 2.1 by allowing the set $A$ to be a finite union of subsets with small image. From this point onwards we formulate results in terms of the category of pointed spaces (see Section 1.5) since this is the best vehicle for a general theory of covering spaces.

2.3.1 THEOREM

Hypothesis. *$p: (\tilde{X}, \tilde{x}) \to (X, x)$ is a covering map and $f: (A, a) \to (X, x)$ is a continuous map. The set $A$ is a finite union $A = A_1 \cup \ldots \cup A_N$ of connected closed subsets, with $a \in A_1$, such that*

(i) *For $r = 1, \ldots, N$, there is a $p$-canonical open subset $V_r \subset X$ such that $f(A_r) \subset V_r$.*

(ii) *For $r = 1, \ldots, N - 1$, if we set $B_r = A_1 \cup \ldots \cup A_r$, then $B_r \cap A_{r+1}$ is connected and non-empty.*

Conclusion. *There is a unique map $\tilde{f}: (A, a) \to (\tilde{X}, \tilde{x})$ such that $p \circ \tilde{f} = f$.*

*Proof.* We use induction on $N$. The result for $N = 1$ is 2.2.1, so we may assume $N \geqslant 2$ and the induction hypothesis is that there is a unique map $\tilde{f}_1: (B_{N-1}, a) \to (\tilde{X}, \tilde{x})$ such that

$$\text{for } t \in B_{N-1}, f(t) = p(\tilde{f}_1(t)). \tag{2.3.2}$$

Let $c \in B_{N-1} \cap A_N$ and let $\tilde{f}_2$ be the map of $A_N$ into $\tilde{X}$ (which exists and is unique by 2.1) such that

$$\text{for } u \in A_N, f(u) = p(\tilde{f}_2(u)), \tilde{f}_2(c) = \tilde{f}_1(c). \tag{2.3.3}$$

Then $\tilde{f}_1, \tilde{f}_2$ agree at the point $c$ of the connected set $B_{N-1} \cap A_N$ whose $f$-image is $p$-canonical. By 2.2.1 (uniqueness) $\tilde{f}_1$ and $\tilde{f}_2$ agree on the whole of $B_{N-1} \cap A_N$. A map $\tilde{f}$ is thus defined by $\tilde{f}(t) = \tilde{f}_1(t)$ $(t \in B_{N-1})$, $\tilde{f}(t) = \tilde{f}_2(t)$ $(t \in A_N)$ and this satisfies our requirements.

Suppose $\tilde{g}$ were another such map. By the uniqueness part of the induction hypothesis we should have $\tilde{g}(t) = \tilde{f}(t)$ for $t \in B_{N-1}$. In particular, $\tilde{g}(c) = \tilde{f}(c)$. Then by the uniqueness part of 2.1 again, $\tilde{g}(t) = \tilde{f}_2(t)$ for all $t \in A_N$. Thus $\tilde{g}$ and $\tilde{f}$ agree on the whole of $A$ and uniqueness is proved. ■

2.3.4 COROLLARY. (Path Lifting Property). *Let $f: (I, 0) \to (X, x)$ and let*

*$(\tilde{X}, \tilde{x}) \to (X, x)$ be a covering map. Then there is a unique path $\tilde{f}: (I, 0) \to (\tilde{X}, \tilde{x})$ such that $p \circ \tilde{f} = f$.*

*Proof.* The family of open sets $\{f^{-1}(V): V\ p\text{-canonical}\}$ covers the compact set $I$. Let $N$ be a positive integer so large that any subinterval of length $1/N$ is contained in some $f^{-1}(V)$. Let $A_r$ denote the closed interval $[r - 1/N, r/N]$. 2.3.4 now follows from 2.3.1. ∎

2.3.5 Corollary. *Let $F: (I \times I, (0, 0)) \to (X, x)$ and let $p: (\tilde{X}, \tilde{x}) \to (X, x)$ be a covering map. Then there is a unique map $\tilde{F}: (I \times I, (0, 0)) \to (\tilde{X}, \tilde{x})$ such that $p \circ \tilde{F} = F$.*

*Proof.* Let $\varepsilon$ be a Lebesgue number for the covering of the compact set $I \times I$ by the family of open sets $\{F^{-1}(V): V \text{ open, } p\text{-canonical}\}$. Subdivide the square $I \times I$ into $n^2$ small squares by lines parallel to the $x$- and $y$-axes. Call these squares $J(1, 1), \ldots, J(n, n)$, where $J(r, s)$ is the set of points $(t, u) \in I \times I$ such that

$$r - 1 \leqslant nt \leqslant r, s - 1 \leqslant nu \leqslant s.$$

Each square has diameter $n^{-1}\sqrt{2}$. Choose $n$ so large that this diameter does not exceed the Lebesgue number $\varepsilon$, so that for each pair $(r, s)$ there is a $p$-canonical neighbourhood $V(r, s)$ such that $F(J(r, s)) \subset V(r, s)$.

Rename and renumber the sets $J(r, s)$ in lexical order as follows with $N = n^2$: $A_1 = J(1, 1), \ldots, A_n = J(1, n), \ldots, A_{n(r-1)+s} = J(r, s), \ldots, A_N = J(n, n)$. Similarly renumber the sets $V(r, s)$ as $V_{n(r-1)+s}$. The hypotheses of 2.3.1 now hold. 2.3.1(ii) consists of the connectedness of $B_k \cap A_{k+1}$. This is always either a single side of $A_{k+1}$ or a union of two adjacent sides. The three cases are illustrated below. The other hypotheses are straightforward, and 2.3.5 is proved.

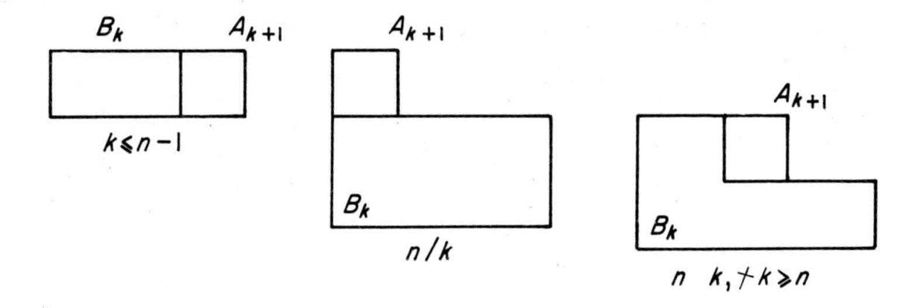

2.3.6 COROLLARY. *Let $p\colon(\tilde{X},\tilde{x})\to(X,x)$ be a covering map and let $f_0, f_1\colon (I,0)\to(X,x)$ be paths such that $[f_0]=[f_1]$. Let $\tilde{f}_0, \tilde{f}_1\colon (I,0)\to(\tilde{X},\tilde{x})$ be the unique paths such that $p\circ\tilde{f}_i=f_i (i=1,2)$. Then $\tilde{f}_0(1)=\tilde{f}_1(1)$ and $[\tilde{f}_0]=[\tilde{f}_1]$.*

*Proof.* Let $y=f_0(1)$. Since $[f_0]=[f_1]$. there is a map $F\colon(I\times I,(0,0))\to(X,x)$ such that, for all $t\in I$, $F(t,0)=x$, $F(0,t)=f_0(t)$, $F(1,t)=f_1(t)$, $F(t,1)=y$. Let $\tilde{F}\colon(I\times I,(0,0))\to(\tilde{X},\tilde{x})$ be the unique map (see 3.5) such that $p\circ\tilde{F}=F$. Define $\tilde{\phi}_0,\tilde{\phi}_1\colon I\to\tilde{X}$ by $\tilde{\phi}_0(t)=\tilde{F}(t,0)$ and $\tilde{\phi}_1(t)=F(t,1)$. Then $p\circ\tilde{\phi}_0(t)=x$ for all $t\in I$ and $p\circ\tilde{\phi}_1(t)=y$ for all $t\in I$. By the uniqueness of path lifting, $\tilde{\phi}_0,\tilde{\phi}_1$ are constant maps. Thus $\tilde{f}_0(1)=\tilde{\phi}_1(0)=\tilde{\phi}_1(1)=\tilde{f}_1(1)$ and $\tilde{F}$ defines a homotopy from $\tilde{f}_0$ to $\tilde{f}_1$. ■

## 2.4 The general lifting theorem

2.4.1 THEOREM. *If $p\colon(\tilde{X},\tilde{x})\to(X,x)$ is a covering map, then $p_*\colon\pi_1(\tilde{X},\tilde{x})\to\pi_1(X,x)$ is a monomorphism.*

*Proof.* We have to show that $\mathrm{Ker}(p_*)=\{1\}$. Suppose then that $[\alpha]\in\pi_1(\tilde{X},\tilde{x})$, with $p_*([\alpha])=1$, so that $p\circ\alpha$ is homotopic to the constant map $c_x$. By 2.3.6, $[\alpha]=1$. ■

2.4.2 THEOREM. *Suppose that $A$ is a path-connected locally path-connected space, $f\colon(A,a)\to(X,x)$ a continuous map, and $p\colon(\tilde{X},\tilde{x})\to(X,x)$ a covering map. Then there is a continuous mapping $\tilde{f}\colon(A,a)\to(\tilde{X},\tilde{x})$ if and only if $f_*(\pi_1(A,a))\subset p_*(\pi_1(\tilde{X},\tilde{x}))$. If such a map $\tilde{f}$ exists, it is unique.*

*Proof.* The uniqueness of $\tilde{f}$, if it exists, follows from the uniqueness of 2.3.4 and the fact that $X$ is path-connected. Also the condition given for the existence of $\tilde{f}$ is obviously necessary. We now assume that it is satisfied and show how to construct $\tilde{f}$.

Let $c\in A$. For any path $\phi\in P(a,c)$, let $\psi=f\circ\phi$ and let $\tilde{\psi}$ be the unique path (see 2.3.4) such that $p\circ\tilde{\psi}=\psi$, $\tilde{\psi}(0)=\tilde{x}$. Set $F_\phi(c)=\tilde{\psi}(1)$. Theorem 2.4.2 will now follow from Lemmas 2.4.3–2.4.6 below.

2.4.3 LEMMA. *$F_\phi(c)$ depends only on $c$ and not on the particular path $\phi$ chosen in $P(a,c)$.*

*Proof.* Suppose that $\phi_1, \phi_2 \in P(a, c)$. Then $[\phi_1][\phi_2]^{-1} \in \pi_1(A, a)$, so

$$[f \circ \phi_1][f \circ \phi_2]^{-1} \in f_*(\pi_1(A, a)) \subset p_*(\pi_1(\tilde{X}, \tilde{x})).$$

Thus the unique lift $(f \circ \phi_1)(f \circ \phi_2)^{-1}[\frac{1}{2}, \frac{1}{2}]$ has endpoint at $\tilde{x}$, showing that the unique lifts with initial point $\tilde{x}$ of each of the paths $f \circ \phi_1, f \circ \phi_2$ have the same end-point, i.e. $F_{\phi_1}(c) = F_{\phi_2}(c)$.

In view of Lemma 2.4.3, $\tilde{f}(c) = F_\phi(c)$ is a well-defined function of $c$ with domain $A$ and range $X$, such that $p \circ \tilde{f} = f$. To complete the proof of 2.4.2 we must show that $f$ is continuous. Two more lemmas are needed first.

2.4.4 LEMMA. *If $\phi \in P(a, c)$, where $c \in A$, then $\tilde{f} \circ \phi$ is the unique lift of $f \circ \phi$ with $\tilde{x}$ as initial point.*

*Proof.* If $\psi, \tilde{\psi}$ are as above and we define $\phi_t, \psi_t, \tilde{\psi}_t$, by $\phi_t(u) = \phi(tu)$, etc., then $\tilde{\psi}_t$ is the unique lift of $\psi_t$ with $\tilde{x}$ as initial point, so

$$\tilde{f}(\phi(t)) = F_{\phi_t}(\phi(t)) = \tilde{\psi}_t(1) = \tilde{\psi}(t).$$

2.4.5 LEMMA. *If $\phi_1 \in P(c, d)$, $c, d \in A$, then $\tilde{f} \circ \phi_1$ is a path.*

*Proof.* Let $\phi \in P(a, c)$. Apply the previous lemma to the path $\phi^* = \phi\phi_1[\frac{1}{2}, \frac{1}{2}]$. Since $\tilde{f} \circ \phi^*$ is a path by 2.4.4, the expression $\tilde{f}(\phi_1(t)) = \tilde{f}(\phi^*(\frac{1}{2} + \frac{1}{2}t))$ defines a continuous function of $t$.

2.4.6 LEMMA. *$\tilde{f}$ is continuous.*

Here is where we need the local path-connectedness of $A$. In the proof we use the letters $U, V, W$, with or without suffixes to denote open subsets of $A, X, \tilde{X}$. Suppose $\tilde{f}(c) \in W$. We have to show that there is an open set $U$, with $c \in U$ such that $\tilde{f}(U) \subset W$. Let $V_1$ be a $p$-canonical open neighbourhood of $f(c)$, $\mathscr{F}$ a partition of $p^{-1}(V_1)$ into open sets each $p$-homeomorphic to $V_1$. Let $W_1$ be the element of $\mathscr{F}$ which contains $\tilde{f}(c)$. Let $p_1$ be the restriction of $p$ to $W_1$. Since $p$ is a homeomorphism and $W \cap W_1$ is relatively open in $W_1$, $p(W \cap W_1) = p_1(W \cap W_1)$ is relatively open in $V_1$. Since $V_1$ is open in $X$, $p_1(W \cap W_1)$ is open in $X$. Since $f$ is continuous and $A$ is locally path-connected, there is a path-connected neighbourhood $U$ of $c$ such that

$$f(U) \subset p_1(W \cap W_1) \subset V_1. \tag{2.4.7}$$

Let $d \in U$, $\phi \in P(c, d)$, $\phi(I) \subset U$. By Lemma 2.4.4, $\tilde{f} \circ \phi$ is the unique lift of $f \circ \phi$ with initial point $f(c)$. Since $W_1$ is open and closed in $p^{-1}(V_1)$, $\tilde{f}(\phi(I)) \subset W_1$, so $\tilde{f}(d) = \tilde{f}(\phi(1)) \in W_1$. Using 2.4.7, $\tilde{f}(d) \in p_1^{-1}(f(U)) \subset W \cap W_1 \subset W$. Thus $\tilde{f}(U) \subset W$ and 2.4.6 is established. This completes the proof of 2.4.2. ■

2.4.8 Corollary. *If $A$ is path-connected and simply connected, then* any *map of $A$ into $X$ can be lifted to a map of $A$ into $\tilde{X}$.*

2.4.9 Theorem. *Suppose that $p: \tilde{X} \to X$ is a covering map. Suppose that $X$ is locally path-connected, path-connected and simply connected. Then $\tilde{X}$ is a disjoint union of open sets each $p$-homeomorphic to $X$.*

*Proof.* Let $a \in X$. For each $\tilde{a} \in p^{-1}(a)$ let $S_{\tilde{a}}: X \to \tilde{X}$ denote the unique lift of the identity map of $X$ on itself such that $S_{\tilde{a}}(a) = \tilde{a}$. Let $\tilde{X}_{\tilde{a}} = S_{\tilde{a}}(X)$. We claim

(2.4.10) *If* $\tilde{a}_1, \tilde{a}_2 \in p^{-1}(a)$, $\tilde{a}_1 \neq \tilde{a}_2$, *then* $\tilde{X}_{\tilde{a}_1} \cap \tilde{X}_{\tilde{a}_2} = \phi$. For suppose $\tilde{c} \in \tilde{X}_{\tilde{a}_1} \cap \tilde{X}_{\tilde{a}_2}$, $c = p(\tilde{c})$. Since there is only one lift $t$ of the identity map with $t(c) = \tilde{c}$, we must have $t = S_{\tilde{a}_1} = S_{\tilde{a}_2}$, so $\tilde{a}_1 = \tilde{a}_2$. Moreover

(2.4.11) $X_{\tilde{a}}$ *is open in* $\tilde{X}$.

For let $\tilde{c} \in \tilde{X}_{\tilde{a}}$, $= p(\tilde{c})$, and let $V$ be a $p$-canonical open set containing $c$. Let $\mathscr{F}$ be a partition of $p^{-1}(V)$ into open sets each $p$-homeomorphic to $V$, let $W$ be the element of $\mathscr{F}$ which contains $\tilde{c}$, $p_1$ the restriction of $p$ to $W$. Since $X$ is locally path-connected, there is a path-connected neighbourhood $V_1$ of $c$ with $V_1 \subset V$. Then $S_{\tilde{a}}$ must map $V_1$ into $W$, so $S_{\tilde{a}}(x) = p_1^{-1}(x)$ for all $x \in V_1$. Thus $p^{-1}(V_1)$ is a neighbourhood of $\tilde{c}$ in $X_a$, so the latter is open. 2.4.10 and 2.4.11 combine to prove 2.4.9. ■

2.4.12 Corollary. *If $p: \tilde{X} \to X$ is a covering map, and $A \subset X$ is p.c., l.p.c. and simply connected, then $A$ is $p$-canonical.*

*Proof.* Apply Theorem 2.4.9 to the mapping $p_1: p^{-1}(A) \to A$, the restriction of $p$ to the subspace $p^{-1}(A)$ of $X$. ■

2.4.13 Corollary. *If $p: \tilde{X} \to X$ is a covering map, where $X$ is p.c., l.p.c. and simply connected, and if $\tilde{X}$ is connected, then $p$ is a homeomorphism.*

2.4.14 Corollary. *Suppose that $X$ is l.p.c. and l.s.c. If $p: Y \to X$ and $q: Z \to Y$ are covering maps, so is $p \circ q: Z \to X$.*

*Proof.* Any point $x \in X$ has a simply connected neighbourhood, which must, by 2.4.12, be $p \circ q$-canonical. ■

## 2.5 Classification theory

In the rest of this chapter all spaces will be assumed to be p.c., l.p.c. and l.s.c. For such spaces a complete classification of covering spaces is possible.

2.5.1 *Definition.* Let $p_i\colon (\tilde{X}_i, \tilde{x}_i) \to (X, x)$, be covering maps $(i = 1, 2)$. Then $p_1$ is said to be *equivalent* to $p_2$ if there is a homeomorphism $h\colon (\tilde{X}_1, \tilde{x}_1) \to (\tilde{X}_2, \tilde{x}_2)$ such that $p_2 \circ h = p_1$.

2.5.2 THEOREM. *The covering map $p_1$ is equivalent to $p_2$ if and only if*

$$p_{1*}(\pi_1(\tilde{X}_1, \tilde{x}_1)) = p_{2*}(\pi_1(\tilde{X}_2, \tilde{x}_2)).$$

*Proof.* Let $\Gamma_i = \pi_1(\tilde{X}_i, \tilde{x}_i)$ $(i = 1, 2)$. If $p_1 = p_2 \circ h$, then by 13.1, $p_{1*} = p_{2*} \circ h_*$ and $p_{1*}(\Gamma_1) = p_{2*}(h_*(\Gamma_1)) = p_{2*}(\Gamma_2)$ since $h$ is bijective. Conversely, if $p_{1*}(\Gamma_1) = p_{2*}(\Gamma_2)$, then by 2.4.2 there are mappings $h_1\colon(\tilde{X}_1,\tilde{x}_1)\to(\tilde{X}_2,\tilde{x}_2)$ and $h_2\colon (\tilde{X}_2, \tilde{x}_2) \to (\tilde{X}_1, \tilde{x}_1)$ such that $p_2 \circ h_1 = p_1, p_1 \circ h_2 = p_2$. Then $p_1 \circ h_2 \circ h_1 = p_1$, i.e. $h_2 \circ h_1$ lifts $id_x$ and by uniqueness $h_2 \circ h_1 = \mathrm{id}_{\tilde{X}_1}$. Similarly $h_1 \circ h_2 = \mathrm{id}_{\tilde{X}_2}$, and 2.5.2 is proved. ■

To complete the picture, it is necessary to show that, for every subgroup $\Gamma$ of $\pi_1(X, x)$ there exists a covering space $((\tilde{X}, \tilde{x}), p)$ such that $p_*(\pi_1(\tilde{X},\tilde{x})) = \Gamma$. We deal first with the case $\Gamma = \{1\}$ by using the route spaces of Section 1. Let $R(x, X)$ denote the space of routes $r$ in $X$ such that $r(0) = x$. For each $r$ define $p(r) = r(1)$. It is an easy consequence of Section 1.1 that $p$ is a covering map.

2.5.3 *Definition.* The pair $(R(x, X), p)$ is called the *universal covering space* of $X$.

2.5.4 THEOREM. *$R(x, X)$ is simply connected.*

*Proof.* It is enough to show that, if $f\colon I \to R(x, X)$ is a path with $f(0) = f(1) = 1_x$, then $f$ is homotopic to the constant path at $1_x$. For $t \in I$, let $g(t) = f(t)(1)$, and, for each $t \in I$, define the path $g_t\colon I \to X$ by $g_t(u) = g(tu)$. Moreover, if $0 \leqslant p < p + q \leqslant 1$, define $g_{p,p+q}$ to be the path described by the relation

$$g_{p,p+q}(t) = g(p + qt).$$

It is easily verified that $[g_{p+q}] = [g_p][g_{p,p+q}]$.

2.5.5 LEMMA. *For* $t \in I$, $[g_t] = f(t)$.

*Proof.* Let $E$ be the set of $t \in I$ such that $[g_t] = f(t)$. Then $E$ is a closed set and since $0 \in E$, $E \neq \varnothing$. Suppose if possible, that $E \neq I$, and let $m = \inf(I \setminus E)$. Since $E$ is closed, $m \in E$. Set $g(m) = g_m(1) = b$. Since $X$ is locally simply connected, $b$ has a simply connected neighbourhood $V$. By the continuity of $f, g$, there exists $\varepsilon > 0$ such that, if $0 < u < \varepsilon$ then $f(m + u) \in f(m)\, R(b, V)$ and $g(m + u) \in V$. Then $[g_{m,m+u}] \in R(b, V)$, so, since there is only one route in $R(b, V)$ with given initial and end point, we must have

$$[g_{m,m+u}] = f(m)^{-1} f(m + u).$$

Hence

$$[g_{m+u}] = [g_m][g_{m,m+u}] = f(m)\, f(m)^{-1} f(m + u) = f(m + u).$$

Thus $m + u \in E$ for $0 < u < \varepsilon$, contrary to the definition of $m$. This proves 2.5.5.

From 2.5.5 with $t = 1$, the path $g = g_1$ belongs to the route $1_x = f(1)$. Let $G(t, u)$ be a homotopy of the path $g$ to the constant path $c_x$. We have $G(0, t) = x$, $G(t, 0) = G(t, 1) = x$, $G(1, t) = g(t)$. Let $\gamma_t$ be the path defined by $\gamma_t(u) = G(t, u)$. Then $\gamma_t \in P(x, x)$ for each $t$, so $t \mapsto \gamma_t$ is a path in $P(x, x)$ and all $\gamma_t$ belong to the same path-component of $P(x, x)$, so $[\gamma_t] = [\gamma_0] = 1_x$. Now let $F(u, t)$ be the path in $X^I$ defined by the relation $F(u, t)(v) = G(u, tv)$. Then $F(0, t) = c_a$, $F(1, t) = g_t$, $F(t, 0) = c_a$, $F(t, 1) = \gamma_t$. Since $\gamma_t = 1_x$, the function $\Phi: I^2 \to R(x, X)$ defined by $\Phi(t, u) = [F(t, u)]$ yields the desired homotopy of $f$ to the constant route $1_x$. ■

## 2.6 Action of the fundamental group; completion of the classification

Let $\Gamma$ denote the fundamental group $R(x, x) = \pi_1(X, x)$. If $t \in \Gamma$, $r \in R(x, X)$, then the route $tr$ is defined and belongs to $R(x, X)$, so we have a group action of $\Gamma$ on $R(x, X)$. If $r_1 = tr_2 (t \in \Gamma)$, then $r_1(1) = r_2(1)$ and conversely if $r_1(1) = r_2(1)$ then $t = r_1 r_2^{-1}$ is defined and $r_1 \in \Gamma r_2$. Thus the $\Gamma$-orbits are in one-to-one correspondence with the points of $X$, and it can be checked that this correspondence is consistent with the topologies of $X$ and $R(x, X)$. This proves the following theorem.

2.6.1 THEOREM. *The map* $p: R(x, X) \to X$ *induces a homeomorphism of* $R(x, X)/\Gamma$ *onto* $X$.

We are now in a position to complete the classification theory. Let $\Gamma_1 \subset \Gamma$. Then the map $p: R(x, X) \to X$ is constant along $\Gamma_1$-orbits, so $p$ induces a mapping $p_1: R(x, X)/\Gamma_1 \to X$. It is easily verified that $p_1$ is a covering map and that $p_{1*}(\pi_1(R(x, X)/\Gamma_1), p_1(1_x)) = \Gamma_1$. Thus there is a one-to-one correspondence between pointed equivalence classes of covering spaces of $(X, x)$ and subgroups $\Gamma_1$ of $\pi_1(X, x)$.

## 2.7 Group actions and coverings

Let $\Gamma$ be a group of homeomorphisms of $X$ such that every point $x \in X$ has a neighbourhood $V$ which is a $\Gamma$-*packing*, that is to say, the sets $\{\gamma V: \gamma \in \Gamma\}$ are disjoint. Then the quotient map $q: X \to X/\Gamma$ is a covering map. If $X$ is simply connected then by 2.5.2 $X$ must be equivalent to the universal covering space of $X/\Gamma$ and so $\Gamma$ must be isomorphic to the fundamental group of the quotient space. The following generalization is due to Armstrong.

2.7.1 THEOREM. *If* $\Gamma$ *is a properly discontinuous group of homeomorphisms of a simply connected space* $X$, *then* $\pi_1(X/\Gamma) \cong \Gamma/\Gamma_0$ *where* $\Gamma_0$ *is the subgroup of* $\Gamma$ *generated by those transformations in* $\Gamma$ *which have fixed points.*

*For the proof see* [**1**].

If $p: \tilde{X} \to X$ is a covering map, the homeomorphisms $h: \tilde{X} \to \tilde{X}$ such that $p \circ h = p$ are called covering transformations with respect to $p$. Thus if $X = R(x, X)$ is the universal covering, $\Gamma$ itself is the group of covering transformations. This is not the only example where the covering map can be expressed as the quotient map of a group action. In fact, if $\Gamma_1$ is any normal subgroup of $\pi_1(X, x) = \Gamma$, then $\Gamma/\Gamma_1$ acts on $R(x, X)/\Gamma$ and the quotient map is equivalent to the covering $R(x, X)/\Gamma_1 \to X$ described in Section 6. The converse is also true: a covering map $p: \tilde{X} \to X$ can be obtained as the quotient of a group action only if $p_*(\pi_1(\tilde{X}, \tilde{x}))$ is normal in $\pi_1(X, x)$.

# 3. THE FUNDAMENTAL GROUP OF A SIMPLICIAL COMPLEX

## 3.1 Introduction

The spaces we shall deal with can be represented as polyhedra of simplicial complexes. For such spaces there is an algorithm for computing a presentation of the fundamental group. The algorithm is cumbersome except for the very simplest spaces, but closer study of it yields the remarkable theorem of van Kampen describing the fundamental group of a complex obtained by pasting together complexes with known fundamental groups.

## 3.2 Simplicial complexes and simplicial maps

3.2.1 *Definition.* A finite set of points $x_0, \ldots, x_m$ in $\mathbb{R}^n$ is called *affinely independent* if the set of points $(x_0, 1), \ldots, (x_m, 1)$ in $\mathbb{R}^n \times \mathbb{R}$ is linearly independent.

3.2.2 *Definition.* If $\{x_0, \ldots, x_m\}$ is an affinely independent set, then the set

$$\{\alpha_0 x_0 + \alpha_1 x_1 + \ldots + \alpha_m x_m : \alpha_r > 0\,(r = 0, \ldots, m), \alpha_0 + \alpha_1 + \ldots + \alpha_m = 1\}$$

is called the *m-simplex spanned* by $x_0, \ldots, x_m$. The points $x_0, \ldots, x_m$ are called the *vertices* of the simplex.

3.2.3 *Definition.* The simplex spanned by a subset of the vertices of a given simplex is called a *face* of that simplex. Note that a face of a simplex is not a subset of it.

3.2.4 *Definition.* A (finite) *simplicial complex* is a finite set $K$ of simplexes in some Euclidean space with the properties:

(i) If $\sigma \in K$ and $\tau$ is a face of $\sigma$, then $\tau \in K$;
(ii) If $\sigma, \tau \in K$ and $\sigma \neq \tau$, then $\sigma \cap \tau = \varnothing$.

3.2.5 *Definition.* If $K$ is a complex, the polyhedron $|K|$ of $K$ is defined by

$$|K| = \bigcup\{\sigma : \sigma \in K\}$$

Since the vertices of a simplex determine it, a complex is determined by the set $v(K)$ of its vertices together with a list indicating which finite subsets of $v(K)$ span a simplex. Denote the family of subsets of $v(K)$ which span a simplex by $\sigma(K)$.

3.2.6 *Definition.* If $K$, $L$ are simplicial complexes, a *simplicial map* between $K$ and $L$ is a map $f: v(K) \to v(L)$ such that, if $A \in \sigma(K)$, then $f(A) \in \sigma(L)$.

Let $f: v(K) \to v(L)$ be a simplicial map. We can define a continuous map $|f|: |K| \to |L|$ by the rule

$$|f|(\alpha_0 x_0 + \ldots + \alpha_m x_m) = \alpha_0 f(x_0) + \ldots + \alpha_m f(x_m)$$

for $\{x_0, \ldots, x_m\} \in \sigma(K)$, $\alpha_r \geqslant 0\ (r = 0, \ldots, m)$, $\alpha_0 + \alpha_1 + \ldots + \alpha_m = 1$. It is easily verified that $|f|$ is continuous. For $v \in v(K)$, $f(v) = |f|(v)$.

The assignment $f \mapsto |f|$ is functorial; that is to say, if the maps $f\colon v(K) \to v(L)$ and $g\colon v(L) \to v(M)$ are simplicial, then so is $g \circ f$, and we have $|g \circ f| = |g| \circ |f|$. Moreover, if $f$ is the identity on $v(K)$, then $|f|$ is the identity on $|K|$.

## 3.3 Examples

3.3.1 *Example.* Let $I_n$ denote the complex in $\mathbb{R}$ with vertices

$$0, \frac{1}{n}, \frac{2}{n}, \ldots, 1$$

and 1-simplexes

$$\left\{\frac{r}{n}, \frac{r+1}{n}\right\} : 0 \leqslant r \leqslant n-1.$$

Then $|I_n| = I$.

3.3.2 *Example.* Let $S_n$ denote the complex in $\mathbb{R}^2$ with vertices

$$\left\{\left(\frac{r}{n}, \frac{s}{n}\right) : 0 \leqslant r \leqslant n, 0 \leqslant s \leqslant n\right\}$$

and with 2-simplexes of the two types:

$$\left\{\left(\frac{r}{n}, \frac{s}{n}\right), \left(\frac{r+1}{n}, \frac{s}{n}\right), \left(\frac{r+1}{n}, \frac{s+1}{n}\right)\right\} 0 \leqslant r \leqslant n-1, 0 \leqslant s \leqslant n-1.$$

$$\left\{\left(\frac{r}{n}, \frac{s}{n}\right), \left(\frac{r}{n}, \frac{s+1}{n}\right), \left(\frac{r+1}{n}, \frac{s+1}{n}\right)\right\}$$

The 1-simplexes of $S_n$ consist of the faces of its 2-simplexes and those only. Clearly $|S_n| = I \times I$.

## 3.4 Simplicial paths

3.4.1 *Definition.* Let $K$ be a simplicial complex. A *simplicial path* $f$ of length $n$ is a simplicial map $f\colon v(I_n) \to v(K)$.

An $(n+1)$-tuple $v_0v_1 \ldots v_n$ of vertices of $K$ such that, for $r = 0, \ldots, n$, $\{v_r, v_{r+1}\} \in \sigma(K)$, determines a simplicial path if we put $v_r = f(r/n)$; and conversely, a simplicial path of length $n$ defines such an $(n+1)$-tuple. The simplicial path so determined is denoted simply by $v_0v_1 \ldots v_n$. For each $v \in v(K)$, it is convenient to regard the 1-tuple $v$ as a *constant* simplicial path of length zero.

If $f: v(I_n) \to v(K)$ is a simplicial path, then $|f|: I \to |K|$ is a path. Thus the simplicial paths yield a subset of the family of all paths. A finite complex is a combinatorial object for which, by purely algebraic means, we can define a *simplicial* fundamental group. This will be done in the next paragraph. Having done this, we shall then show that the correspondence $f \mapsto |f|$ just defined induces an isomorphism between the two fundamental groups.

### 3.5 The algebra of simplicial paths

Let $P_n(K)$ denote the family of simplicial paths of length $n$ in $K$, and let $P(K) = \bigcup P_n(K)$. For $p \in P(K)$, $p(0)$ is the *initial point* and $p(1)$ the *end-point.*

3.5.1 *Definition.* If $p = v_0v_1 \ldots v_m \in P_m(K)$ and $q = w_0w_1 \ldots w_n \in P_n(K)$, then if $v_m = w_0$ we define the *simplicial product* (s-product) $pq$ of the s-paths by $pq = v_0v_1 \ldots v_m w_1 \ldots w_n \in P_{m+n}(K)$.
The s-product is associative whenever the relevant triple products have a meaning, and the constant paths of length zero are unit elements. The s-inverse of a simplicial path $p$ can be defined by $p^{-1} = v_m v_{m-1} \ldots v_0$. The s-operations are related to the corresponding relations on continuous paths as follows:

(3.5.2)
$$|pq| = |p|\,|q|\left[\frac{m}{m+n}, \frac{n}{m+n}\right],$$
$$|p^{-1}| = |p|^{-1}.$$

3.5.3 *Definition.* Two s-paths $p, q$ are called *neighbours* if there are paths $r, s$ (possibly of zero length) and a triple $u, v, w$ of vertices (not necessarily distinct) of a single simplex in $K$ such that either $p = r(uv)s$ and $q = r(uwv)s$, or vice versa.

3.5.4 *Definition.* Two s-paths $p$, $q$ are called *simplicially homotopic* (s-homotopic) if there is a sequence of s-paths

$$p = p_0, p, \ldots, p_m = q$$

such that, for $r = 0, \ldots, m-1$, $p_r$, $p_{r+1}$ are neighbours.

3.5.5 *Example.* Consider the complex $S_n$ of 3.3.2. Use $v_{ij}$ to denote the vertex $(i/n, j/n)$. Define s-paths as follows.

$$p_1 = v_{00}v_{01} \cdots v_{0n},$$

$$p_2 = v_{0n}v_{1n} \cdots v_{nn},$$

$$p_3 = v_{nn}v_{nn-1} \cdots v_{n0},$$

$$p_4 = v_{n0}v_{n-10} \cdots v_{00},$$

$$p_0 = p_1p_2p_3p_4.$$

The path $p_0$ is s-homotopic to the constant path $v_{00}$. This can be seen by the succession of moves indicated in the diagram.

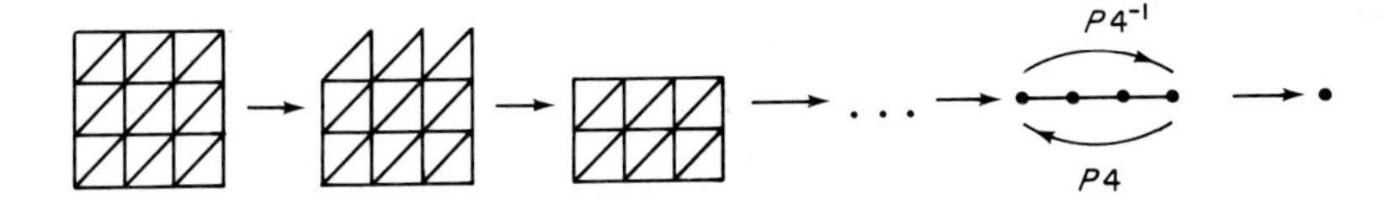


The relation of s-homotopy is the equivalence relation generated by the relation between s-paths of being neighbours. Equivalence classes are called s-*routes*, and the s-route containing the s-path $p$ is written $[p]$. It is easily checked that the relation of s-homotopy is compatible with the operations s-product and s-inverse. Also, if $p \in P(K)$, then $pp^{-1}$ is s-homotopic to the constant s-path of length zero $p(0)$. We deduce, as in Theorem 1.8.1, the result below.

3.5.6 THEOREM. *Let $v_0 \in v(K)$. The s-product induces a group structure on the set $R_s(v_0, v_0)$ of simplicial routes with initial and endpoint both located at $v_0$.*

3.5.7 *Definition.* The group $R_s(v_0, v_0)$ is called the *simplicial fundamental group* of the complex $K$ with base-point $v_0$. It is also denoted by $\pi_s(K, v_0)$.

Since the closure of the simplex with vertices $u, v, w$ is convex and therefore simply connected, it follows from 1.9.3, that $[|uv|] = [|uwv|]$. Thus, if $p, q$ are neighbours, then $[|p|] = [|q|]$. Since (topological) homotopy is an equivalence relation, we deduce that, if $[p] = [q]$, then $[|p|] = [|q|]$. Combined with 3.5.2 this yields

3.5.8 THEOREM. *The assignment $p \mapsto |p|$ yields a homomorphism*

$$\theta: \pi_s(K, v_0) \to \pi_1(|K|, v_0).$$

## 3.6 Simplicial approximation

3.6.1 *Definition.* Let $v \in v(K)$. The S*tar* of $v$ is the subset of $|K|$ defined by

$$\text{st}\, v = \bigcup \{\sigma : \sigma \in K, v \in v(\sigma)\}$$

It is immediate from the definition that $\{v_1, \ldots, v_r\} \in \sigma(K)$ if and only if $\text{st}\, v_1 \cap \ldots \cap \text{st}\, v_r \neq \emptyset$. The star is an open subset of $|K|$ and

$$|K| = \bigcup \{\text{st}\, v : v \in v(K)\}.$$

3.6.2 *Definition.* Let $K$, $L$ be complexes, $f: |K| \to |L|$ a continuous map. A simplicial map $\phi: v(K) \to v(L)$ is called a *simplicial approximation* to $f$ if, for every $v \in v(K)$, $f(\text{st}\, v) \subset \text{st}\, \phi(v)$.

3.6.3 THEOREM. *If $\phi: v(K) \to v(L)$ is a simplicial approximation to $f: |K| \to |L|$ then, for $0 \leqslant t \leqslant 1$ and for $x \in |K|$, the point $f_t(x) = t|\phi|(x) + (1 - t)f(x)$ belongs to $|L|$.*

*Proof.* Let $x = \alpha_0 v_0 + \ldots + \alpha_r v_r, \alpha_0, \ldots, \alpha_r > 0, \alpha_0 + \ldots + \alpha_r = 1, \{v_0, \ldots, v_r\} \in \sigma(K)$. Then $x \in \text{st}\, v_0 \cap \ldots \cap \text{st}\, v_r$, so by the simplicial approximation condition, $f(x) \in \operatorname{st} \phi(v_0) \cap \ldots \cap \operatorname{st} \phi(v_r)$. It follows that $\phi(v_0), \ldots, \phi(v_r)$ are vertices of the simplex $\sigma \in K$ to which $f(x)$ belongs. Now $\alpha_0 \phi(v_0) + \ldots + \alpha_r \phi(v_r) = |\phi|(x) \in \bar{\sigma}$. The result follows since $\bar{\sigma}$ is convex. ■

3.6.4 *Definition.* A complex $K_1$ is called a *subdivision* of the complex $K$ if $|K_1| = |K|$ and every simplex of $K_1$ is contained in some simplex of $K$.

3.6.5 *Example.* $S_{mn}$ is a subdivision of $S_n$.

The polyhedron $|K|$ of a complex $K$, being a subset of $\mathbb{R}^n$, is metrized by the Euclidean metric in $\mathbb{R}^n$. With this metric we have

3.6.6 *Definition.* The *mesh* $\mu(K)$ of the complex $K$ is defined by

$$\mu(K) = \max\{\operatorname{diam} |\sigma| : \sigma \in K\}$$

For instance, $\mu(S_n) = \sqrt{2}/n$.

3.6.7 Simplicial Approximation Theorem. *Suppose that $f: |K| \to |L|$ is a continuous map. Then there exists a number $\varepsilon > 0$ such that, for any subdivision $K_1$ of $K$ with $\mu(K_1) < \varepsilon$, there exists a simplicial approximation $\phi: v(K_1) \to v(L)$ of the mapping $f$.*

*Proof.* For each $v \in v(L)$ define the real-valued function $d_v$ on $|K|$ by

$$d_v(x) = \min\{|x - y| : y \in |K| \quad \operatorname{st} v\}.$$

Since $d_v$ is continuous for every $v$, the function $d(x) = \max\{d_v(x) : v \in v(L)\}$ is also continuous. Since $|L| = \bigcup\{\operatorname{st} v : v \in v(L)\}$, the attained minimum of $d(x)$ on the compact set $|K|$ is a positive number $\varepsilon$.

Suppose now that $K_1$ is a subdivision of $K$ with $\mu(K_1) < \varepsilon$. For each vertex $u \in v(K)$ we have $d_v(u) \geqslant \varepsilon$ for some $v \in v(L)$. Choose one such $v$ arbitrarily and set $v = \phi(u)$. If $y \in \operatorname{st} u$, then $|y - u| < \varepsilon$, and, by the definition of $d_v$, $y \in f^{-1}(\operatorname{st} v) = f^{-1}(\operatorname{st} \phi(u))$. Thus $\phi$ is a simplicial approximation to $f$. ■

## 3.7 The isomorphism theorem

In this section it is proved that the homomorphism $\theta$, whose existence was established in 3.5.8, is an isomorphism. This is a consequence of 3.7.1, 3.7.2 below.

3.7.1 ($\theta$ *is onto*). *If* $f: I \to |K|$ *is a continuous path with* $f(0), f(1) \in v(K)$, *then there is a number n and a simplicial path* $\phi: v(I_n) \to v(K)$ *such that* $[|\phi|] = [f]$.

*Proof.* By 3.6.7 there exists $n$ such that $f$ admits a simplicial approximation $\phi: v(I_n) \to v(K)$. Since $f(0) = f(1) = v$, and since $v$ does not belong to the star of any vertex other than itself, we must have $\phi(0) = \phi(1) = v$. Using 3.6.3, the family of paths $f_t$ yields a homotopy from $f_0 = f$ to $|\phi|$. ■

3.7.2 ($\operatorname{Ker}\theta = 1$). *Let K be a complex,* $v \in v(K)$. *If* $\phi$ *is a simplicial path such that* $|\phi|$ *is homotopic to the constant path at v in* $|K|$, *then* $\phi$ *is s-homotopic to the constant simplicial path of length zero.*

*Proof.* Let $\phi$ have length $n$, so we have $\phi(r/n) = v_r \in v(K)$, where $v_0 = v_n = v$. Since $|\phi|$ is homotopic to the constant path at $v$, there is a continuous map $F: I \times I \to |K|$ such that, for $t \in I$,

$$F(0, t) = |\phi|(t);\ F(1, t) = F(t, 0) = F(t, 1) = v.$$

By 3.6.7, there is a natural number $n$ and a simplicial approximation

$$\Phi: v(S_{mn}) \to v(K)$$

where $S_n$ denotes the complex of 3.3.2. It follows from 3.5.5 that $\Phi \circ p_0$ is s-homotopic to the constant path at $v$ of length zero. 3.7.2 will now follow at once from

3.7.3 Lemma. $[\Phi \circ p_0] = [\phi]$.

*Proof.* In the notation of 3.5.5 let $v_{ij}$ be any vertex in any of the s-paths $p_2, p_3$ or $p_4$. Then $F(v_{ij}) = v$. Since the only star to which $v$ belongs is st $v$, and since $\Phi$ is a simplicial approximation to $F$, we have $\Phi(v_{ij}) = v$. Thus $\Phi \circ (p_2 p_3 p_4)$ is $vv \ldots v$ which is s-homotopic to $v$.Now consider $p_1$ in the complex $S_{mn}$. We have $p_1 = q_0 q_1 \ldots 1_{n-1}$, $q_r$ being the path $w_{rm} w_{rm+1} \ldots w_{rm+r}$, where $w_i$ denotes the vertex $(0, i/mn)$. If $w$ is a vertex in the path $q_r$, then $F(w)$ belongs to the simplex $v_r v_{r+1}$, which does not meet the star of any vertex $v$ of $K$ other than st $v_r$, st $v_{r+1}$. Since $\Phi$ is a simplicial approximation to $F$, $\Phi \circ q_r$ is a sequence of vertices of which the first is $v_r$, the last is $v_{r+1}$ and every intermediate one is either $v_r$ or $v_{r+1}$. Thus $\Phi \circ q_r$ is s-homotopic to $v_r v_{r+1}$, from which it follows that $\Phi \circ p_1 = \Phi \circ (q_0 q_1 \ldots q_{n-1})$ is s-homotopic to $v_0 v_1 \ldots v_n = \phi$. Thus $[\Phi \circ p_0] = [\Phi \circ p_1][\Phi \circ (p_2 p_3 p_4)] = [\phi v_0] = [\phi]$. This completes the proof of 3.7.3 and hence of 3.7.2.

## 3.8 A presentation for the fundamental group

3.8.1 *Definition.* A complex $K$ is *connected* if each pair of vertices can be joined by a simplicial path.

It is easy to verify that $K$ is connected if and only if $|K|$ is a connected space.

3.8.2 *Definition.* A *circuit* in a complex is a simplicial path $v_0 v_1 \ldots v_n v_0$ with $n \geqslant 2$ and all the vertices $v_0, v_1, \ldots, v_n$ distinct.

3.8.3 *Definition.* A complex $T$ is called a *tree* if

(i) $T$ contains only 0- and 1-simplexes,

(ii) $T$ is connected,

(iii) $T$ contains no circuit.

If $T$ is a tree and $v_0$, $w$ are vertices of $T$ then there is a unique *irreducible* s-path from $v_0$ to $w$, that is, a simplicial path $v_0 v_1 \ldots v_r$ with $v_r = w$ and with all the $v_i$ distinct. There must be one such path by connectedness, and there cannot be two since $T$ contains no circuit. Let $p(w)$ denote this unique irreducible path, and let $\phi(w) = |p(w)|$. The map $\phi\colon v(T) \to |T|^I$ can be extended to a map $\phi\colon |T| \to |T|^I$ as follows. Suppose that $x$ belongs to the simplex $\{w_1, w_2\}$, where $w_1$ is nearer to $v_0$ than $w_2$ is, that is, $w_2$ is not a vertex in the path $p(w_1)$. Suppose that $n$ is the length of the path $p(w_1)$ and that $x = (1 - \alpha) w_1 + \alpha w_2$. Let $q$ be the path defined by $q(t) = (1 - t\alpha) w_1 + t\alpha w_2$, and let

$$\phi(x) = \phi(w_1)\, q\left[\frac{n}{n+\alpha}, \frac{\alpha}{n+\alpha}\right].$$

It is easily checked that $\phi\colon |T| \to |T|^I$ is continuous. The homotopy $F(t, w) = \phi(w)(t)$ shows that the identity map of $(|T|, v_0)$ is homotopic to the constant map of $|T|$ to $v_0$. Since the constant map induces the trivial homomorphism of $\pi_1$ to the identity element, it follows, by Theorem 1.12.1(vi) that $\pi_1(|T|, v_0) = \{1\}$, that is.

3.8.4 THEOREM. *A tree is simply connected.*

If $K$ is a finite connected complex and $T$ is a tree in $K$ which is maximal with respect to set inclusion, then $T$ must contain all the vertices of $K$. For if $v \in v(K) \setminus T$, one could join $v$ to some vertex of $T$ by a simplicial path without repeated vertices, and the union of $T$ with the initial segment of this path up to its first intersection with $T$ would be a larger tree, contradicting the maximality Thus we have the following theorem.

3.8.5 THEOREM. *If $K$ is a connected complex and $T$ is a subcomplex which is a tree, then there is a tree $T^* \supset T$ such that $v(T^*) = v(K)$.*

3.8.6 THEOREM. *Let $K$ be a connected complex, $T$ a maximal tree in $K$, $v_0 \in v(T)$. Then the following is a presentation of $\pi_s(K, v_0)$.*

Generators: *one generator $E_{uv}$ for each pair of vertices $u$, $v$ with $\{u, v\} \in \sigma(K)$.*

Relations: (i) $E_{uv} = 1$ *for* $\{u, v\} \in \sigma(T)$;
(ii) *For all ordered triples* $\{u, v, w\} \in \sigma(K)$, $E_{uv}E_{vw} = E_{uw}$.

*Proof.* Let $v \in v(K)$. Let $p(v)$ denote, as above, the unique irreducible simplicial path in $T$ from $v_0$ to $v$. For each pair $\{u, v\} \in \sigma(K)$, let $E_{uv}$ denote the simplicial route

$$E_{uv} = p(u)\,(uv)\,p(v)^{-1}$$

If $p = v_0 v_1 \ldots v_n v_0$ is a simplicial path from $v_0$ to $v_0$, then the s-route $E_{v_0 v_1} E_{v_1 v_2} \ldots E_{v_n v_0}$ is the s-homotopy class of the s-path $p$. Thus the $E_{uv}$ do indeed generate $\pi_s(K, v_0)$. Clearly, too, they satisfy the relations (i), (ii).

We must prove conversely that every true relation between words in the $E_{uv}$ is a consequence of (i), (ii). To do this we introduce a mapping $\gamma$ from the set of s-paths beginning and ending at $v_0$ into the set of words in the $E_{uv}$ as follows:

$$\gamma(v_0 v_1 \ldots v_n v_0) = E_{v_0 v_1} E_{v_1 v_2} \ldots E_{v_n v_0}.$$

Suppose now that the simplicial route

$$E_{u_1 v_1} E_{u_2 v_2} \ldots E_{u_n v_n} \tag{3.8.7}$$

is equal to 1. We have to show that the corresponding word in the $E$'s, which we also denote by

$$w = E_{u_1v_1}E_{u_2v_2}\dots E_{u_nv_n}$$

can be transformed as a consequence of relations (i), (ii) into the empty word. The route defined by the word $w$ contains the s-path

$$q = p(u_1)(u_1v_1)\,p(v_1)^{-1}\,p(u_2)(u_2v_2)\,p(v_2)^{-1}\dots p(u_n)(u_nv_n)\,p(v_n)^{-1}.$$

By 3.8.7, $q$ is simplicially homotopic to the constant path. Now the replacement of a path $q_1$ by a neighbour $q_2$ corresponds to a relation between $\gamma(q_1)$ and $\gamma(q_2)$ which is a consequence of one of the relations (ii). Since simplicial homotopy is the result of a succession of such transformations, it follows that the word $\gamma(q)$ can be reduced to 1 as a consequence of the relations (ii).

On the other hand $\gamma(q)$ is obtained from $w$ by inserting symbols $E_\sigma$ corresponding to simplexes $\sigma \in T$. Thus $\gamma(q)$ is equivalent to $w$ as a consequence of relations (i). It follows that, as a consequence of both sets of relations, (i) and (ii), $w$ can be reduced to 1. ■

*Note.* In actual computations, not all the generators and relations need be written out. All the generators corresponding to simplexes in $T$ can be forgotten, as can the generators $E_{uu}$, which must, by (ii) be equal to 1. Also, since $E_{uv} = E_{vu}^{-1}$, by (ii) only one generator from each pair $E_{uv}$, $E_{vu}$ is needed.

3.8.8 *Example.* Let $\dot{\sigma}_2$ denote the complex consisting of all proper faces of a 2-simplex $\sigma_2$. Clearly the space $|\dot{\sigma}_2|$ is homeomorphic to $S^1$. Let $u$, $v$, $w$ be the vertices of $\dot{\sigma}_2$. Then $uv$, $uw$ define a maximal tree. This leaves only one generator $E_{vw}$. There is no non-trivial relation of type (ii), since there is no 2-simplex. Thus $\pi_1(S^1)$ is infinite cyclic.

## 3.9 van Kampen's Theorem

Let $G_1$, $G_2$ and $G_{12}$ be three groups, $G_1$ and $G_2$ being considered disjoint. Suppose also that we are given homomorphisms

$$\theta_1\colon G_{12} \to G_1, \qquad \theta_2\colon G_{12} \to G_2.$$

We can construct a new group

$$H = (G_1 * G_2/G_{12}, \theta_1, \theta_2)$$

given by the following presentation:

(i) multiplication table of $G_1$,
(ii) multiplication table of $G_2$,
(iii) the relations $R_g$: $\theta_1(g) = \theta_2(g)$, for all $g \in G_{12}$.

The construction of $H$ is a sort of pasting operation—identifying images of the same element of $G_{12}$. $H$ is sometimes described as a free product of $G_1, G_2$ with amalgamation, but in group theory that term is usually reserved for the situation where $\theta_1$ and $\theta_2$ are injective. Clearly the set of relations we have given to define $H$ could be replaced by others. (i) could be any presentation of $G_1$: (ii) could be any presentation of $G_2$; and in (iii) the relations $R_g$ for a set of generators $g$ of $G_{12}$ would suffice.

3.9.1 VAN KAMPEN'S THEOREM. *Suppose that a connected complex $K$ is a union of two connected complexes $K_1$, $K_2$ with connected non-empty intersection $K_{12}$. Let $v_0 \in K_{12}$ and let $G_1 = \pi_s(K_1, v_0)$, $G_2 = \pi_s(K_2, v_0)$, $G_{12} = \pi_s(K_{12}, v_0)$. Let $i_1$, $i_2$ denote the inclusion maps $i_1 : K_{12} \subset K_1$, $i_2 : K_{12} \subset K_2$. Then*

$$\pi_s(K, v_0) \cong (G_1 * G_2/G_{12}, i_{1*}, i_{2*}).$$

*Proof.* Choose a maximal tree $T_{12}$ in $K_{12}$ and extend it to a maximal tree $T_1$ in $K_1$. Also extend it to a maximal tree $T_2$ in $K_2$. Then $T = T_1 \cup T_2$ is a maximal tree in $K$.

Now the fundamental group $\pi_s(K_1, v_0)$ is given by a presentation

$$[E_{vw}; R(E_{vw})], \{v, w\} \in K_1 \tag{3.9.2}$$

and the group $\pi_s(K_2, v_0)$ is given by

$$[F_{vw}; R(F_{vw})] \{v, w\} \in K_2, \tag{3.9.3}$$

where we use different symbols $E$, $F$ for the generators in $K_1$, $K_2$ because the groups are formally distinct. The fundamental group of $K_{12}$ is generated by elements $[G_{vw}]$, $\{v, w\} \in K_{12}$, and the relations need not concern us. Because of the relationship between the trees $T_1, T_2, T_{12}$, $G_{vw}$ is the simplicial route in $K_{12}$ which contains the same path $p(v)(vw)\,p(w)^{-1}$ as represents $E_{vw}$ in $K_1$. Thus for $(v, w) \in K_{12}$ we have $i_{1*}(G_{vw}) = E_{vw}$. Similarly $i_{2*}(G_{vw}) = F_{vw}$. If we now consider the relations defining $(G_1 * G_2/G_{12}, i_{1*}\, i_{2*})$, these are 3.9.2, 3.9.3, together with $E_{vw} = F_{vw}((vw) \in K_{12})$. These are the same as the presentation 3.8.6 of $\pi_s(K, v_0)$. ■

3.9.4 COROLLARY. *If $K_1$, $K_2$ are connected complexes and $K_1 \cap K_2$ is simply connected (in particular, if $K_1 \cap K_2$ is a single vertex) then $\pi_s(K_1 \cup K_2)$ is the free product $\pi_1(K_1) * \pi_1(K_2)$. Moreover the maps $i^1_*$, $i_{2*}$ are the natural isomorphisms of the groups into their free product.*

3.9.5 *Example.* We saw in 3.8.8 that $\pi_1(S^1)$ is an infinite cyclic group, the generator being the route of the map $t \mapsto \exp(2\pi i t)$.

The *n-leafed rose* $h_1(S^1) \cup \ldots \cup h_n(S^1)$ is the space formed by $n$ homeomorphic images of $S^1$, with a single point in common which is the common image of the point 1 under all the homeomorphisms. Thus $h_r(1) = h_s(1)$, but if $r \neq s$ then $h_r(S^1 \setminus \{1\}) \cap h_s(S^1) = \varnothing$. By repeated application of 3.9.4 we deduce that the fundamental group of the rose is free on $n$ generators. Moreover the free generators are the routes given by $t \mapsto h_r(\exp(2\pi i t))$.

## 3.10 Compact surfaces

3.10.1 *Definition.* Let $P$ be a convex closed plane polygon with an even number of sides. A *pairing* of the sides of $P$ consists of a partition of the set of sides of $P$ into disjoint pairs together with, for each pair $s, s'$, a homeomorphism $T(s, s')$ called the *pairing map* of $s$ onto $s'$; where the pairing maps must satisfy the restriction $T(s, s') = T(s', s)^{-1}$. The configuration consisting of a polygon and a pairing of its sides is called a *marked polygon.*

Let $\mathfrak{q}$ denote the equivalence relation induced by identifying each point $p \in P$ with its images under pairing maps. Thus each inner point of $P$ is a whole $\mathfrak{q}$-class by itself. If $x$ belongs to a side of $P$, but is not a vertex, then its $\mathfrak{q}$-class consists of the two-elements $\{x, T(s, s')\, x\}$. The $\mathfrak{q}$-class of a vertex consists only of vertices.

The quotient space $P/\mathfrak{q}$ is called the *surface defined* by the marked polygon $P$. (It is easy to verify that it is a surface, i.e. locally homeomorphic to the plane.) If $K$ is a complex such that $|K|$ is a surface, it is possible to produce a marked polygon defining the surface $|K|$. This can be done by building up the polygon one 2-simplex at a time, always adding a new simplex which has an edge in common with the union of the simplexes that have so far been included. If the new simplex has only one edge in common with the preceding complex, and the preceding polygon is a disc, then the complex obtained by uniting the new simplex is still a disc. However, if the new simplex has more than one edge in common, we keep the second (and, if necessary the third) pair of edges formally distinct but paired. When we have added all the 2-simplexes, we shall have no free unpaired edges left, and thus we shall have a marked polygon.

The pairing of the edges in a marked polygon can be conveniently expressed by means of a *surface symbol*, obtained as follows. Pick one side from each pair of sides, and assign a letter to the selected side, choosing different letters for different pairs. If the selected edge $s$ in the pair $(s, s')$ is assigned the letter $a$, say, then assign a letter to $s'$ according to the following rule.

(i) If $T(s, s')$ maps the initial point of $s$, in the anticlockwise order round $P$, on the initial point of $s'$, assign the *same* letter $a$ to $s'$.

(ii) If $T(s, s')$ maps the initial point of $s$ on the final point of $s'$, assign to $s'$ the "letter" $a^{-1}$.

Finally, write down the letters assigned to the edges of $P$ in order anticlockwise. The sequence of letters so obtained is called a surface *symbol* for the marked polygon $P$.

The surface symbol is not unique, since the letters may be permuted and there is a choice of starting point. Moreover, the same surface may be described by many essentially distinct marked polygons, which can be obtained from one of them by cutting and pasting. By such processes one can show [2] that a compact surface can be defined by a surface symbol of one of the following two *canonical forms*.

(3.10.2)
(i) (orientable) $a_1 b_1 a_1^{-1} b_1^{-1} \ldots a_g b_g a_g^{-1} b_g^{-1}$,
(ii) (non-orientable) $a_1 a_1 a_2 a_2 \ldots a_q a_q$.

In each of these forms the whole set of vertices of $P$ forms a single equivalence-class. Thus the q-quotient of the frontier $\dot{P}$ of $P$ is an $n$-leafed rose ($n = 2g$ in case i, $n = q$ in case ii). The situation is described by saying that the surface is an "$n$-leafed rose with a 2-cell attached".

### 3.11 Fundamental group of a surface

Consider a surface $S$ represented as the quotient of a marked polygon 3.10.2. Let $v_0$ be the vertex of $P$ which is the initial point of the side labelled $a_1$. Assume that the convex polygon $P$ contains the origin $O$ of the vector space $\mathbb{R}^2$ and let $P_2 = \alpha P$, where $0 < \alpha < 1$. Let $L$ be the line-segment joining $v_0$ to $\alpha v_0$ and let $P_1$ be the closure of $P \setminus P_2$. (The case $n = 4$ is illustrated in Fig. 3.11.1.)

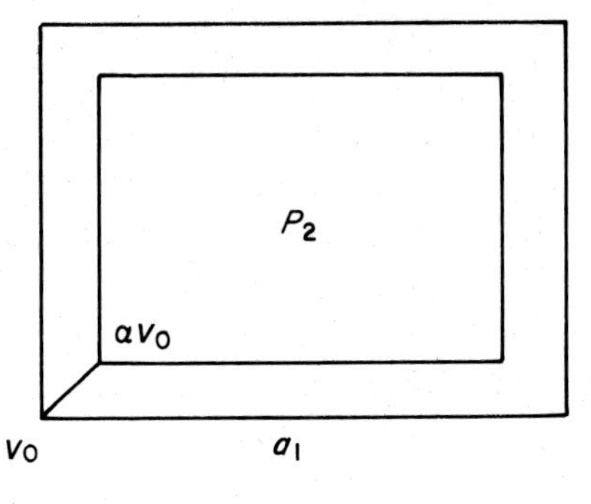


*Fig.* 3.11.1

We shall compute the fundamental group of $S$ by applying van Kampen's theorem to the spaces $K_1 = P_1/\mathfrak{q}$, $K_2 = P_2 \cup L/\mathfrak{q}$, $K_{12} = \dot{P}_2 \cup L/\mathfrak{q}$. For $x \in P_1$, let $f(x)$ denote the unique point where the line $Ox$ meets $\dot{P}_1$. The map $f: P_1 \to P_1$ is homotopic to the identity, since we can put $f_t(x) = (1-t)f(x) + tx$, with $f_0 = f$, $f_1 = id$. Since $f_t$ is $\mathfrak{q}$-compatible for every $t$, it follows that the inclusion map $i: \dot{P}/\mathfrak{q} \to K_1$ is a homotopy equivalence. Now $\dot{P}/\mathfrak{q}$ is an $n$-leafed rose, so its fundamental group is free on the generators $[a_1], \ldots, [b_g]$ (in case i), $[a_1], \ldots, [a_q]$ (in case ii). (To save complicated symbolism, we use $[a_r]$, etc., to denote the route of the path in $P/\mathfrak{q}$ determined by the oriented side $a$, after the identification $\mathfrak{q}$.)

Similarly $K_{12}$ has the homotopy type of $S^1$, the generating element of its infinite cyclic fundamental group being the route $z$ represented by the path obtained by travelling in succession from $v_0$ to $\alpha v_0$, then from $\alpha v_0$ round $P_2$ and finally from $\alpha v_0$ back to $v_0$ along $L$. Clearly, in $K_1$ we have

$$\text{(in case i) } i_{1*}(z) = [a_1] \ldots [b_g]^{-1},$$

$$\text{(in case ii) } i_{1*}(z) = [a_1] \ldots [a_q].$$

However $K_2 \simeq P_2$ is simply connected ($P_2$ is convex and the segment $L$ does not effect its homotopy type). Hence

$$i_{2*}(z) = 1.$$

It follows from van Kampen's theorem that $\pi_1(K, v_0)$ is obtained from the free group on $n$ generators by adding a single relator—$[a_1] \ldots [b_g]^{-1}$ in the orientable case, or $[a_1][a_1] \ldots [a_q][a_q]$ in the non-orientable case.

### 3.12 Nielsen's Theorem on surfaces

Suppose $(X, x)$ is a pointed space. Let $\mathrm{Homeo}(X, x)$ denote the group of homeomorphisms of $X$ itself which fix $x$. Let $H_0$ denote the subgroup consisting of those which are homotopic, relative to $x$, to the identity. Then the assignment $f \mapsto f_*$ yields a homomorphism (see 1.12)

$$\mathrm{Homeo}(X, x)/H_0 \to \mathrm{Aut}(\pi_1(X, x)).$$

In general, this homomorphism is neither surjective nor injective. However, if $X$ is a compact surface the situation is different. The sphere and the projective plane are exceptional because they have non-vanishing higher homotopy groups. Therefore, in the rest of this paragraph, $X$ will denote a closed surface other than the sphere or the projective plane.

3.12.1 Nielsen's Theorem. *The assignment* $f \mapsto f_*$ *induces an isomorphism between* $\mathrm{Homeo}(X, x)/H_0$ *and* $\mathrm{Aut}(\pi_1(X, x))$.

It is often inconvenient to consider only homeomorphisms and homotopies which fix the base-point. If we allow *free* homotopies (not necessarily fixing the base-point) between homeomorphisms, the associated automorphism of the fundamental group will be subject to variation by composing it with an inner automorphism. This leads us to the more usual statement.

3.12.2 Corollary. *If* $\mathrm{Homeo}(X)$ *is the group of all homeomorphisms of* $X$ *and* $K$ *is the subgroup consisting of all those which are freely homotopic to the identity; and if* $\mathrm{Aut}_0$ *denotes the subgroup of* Aut *consisting of the inner automorphisms, then* $f \mapsto f_*$ *induces an isomorphism between the groups*

3.12.3 $$\mathrm{Homeo}(X)/K, \qquad \mathrm{Aut}(\pi_1(X, x))/\mathrm{Aut}_0(\pi_1(X, x)).$$

3.12.4 *Definition.* Either of the groups 3.12.3 is called the *mapping class group* of the surface $X$.

There is insufficient space here for a complete proof of Nielsen's theorem. Nielsen's original proof [**8**] is very long, but has the merit from our point of view of using hyperbolic geometry. We give an account based on a paper by Mangler [**5**]. Nielsen's theorem can be thought of as a special case of a theorem about geometrical realizability of isomorphisms between Fuchsian groups with compact quotient space. An extension to non-compact quotient spaces, when one must restrict to the "type-preserving" isomorphisms, is due to Harvey and Maclachlan [**3**]. See also Tukia [**10**] and Chapter 9, Section 1.9.

We do not treat here the case where $X$ has genus 1 as this admits a simpler proof.

Nielsen's theorem amounts to two assertions, 3.12.5 and 3.12.6.

3.12.5 *If two homeomorphisms of* $(X, x)$ *induce the same automorphism of* $\pi_1(X, x)$, *then they are homotopic.*

*Proof*†. We assume (see for example Chapter 7, Section IH, or Siegel [**9**]) that $X$ is the quotient of the hyperbolic plane $\mathscr{H}$ by a group $\Gamma$ of isometries without fixed point. Let $z$ be a point, determined once for all, in the $\Gamma$-orbit which represents the base-point $x$, and let $p\colon \mathscr{H} \to X$ be the quotient map. Let $\tilde{f}, \tilde{g}$ be the lifts of the maps $f \circ p, g \circ p\colon \mathscr{H} \to X$, uniquely specified by the initial conditions $\tilde{f}(z) = z$, $\tilde{g}(z) = z$. These exist and are unique by Chapter 2, Section 4. In this situation, $\mathscr{H}$ is considered to be the universal covering space, $p$ the covering map and $\Gamma$ the group of covering transformations, identified with the fundamental group. If $w \in \mathscr{H}$, and $\gamma' = f_*(\gamma)$, $\gamma \in \Gamma$, then $\gamma'$ is characterized by the relation $f(\gamma w) = \gamma' f(w)$. Since $f_* = g_*$, we also have $g(\gamma w) = \gamma' g(w)$.

For each point $w \in \mathscr{H}$, $t \in I$, define $\tilde{F}_t(w)$ to be the point dividing the non-Euclidean line-segment joining $\tilde{f}(w)$, $\tilde{g}(w)$ in the ratio $t{:}1 - t$. Then $\tilde{F}_t(\gamma w) = \gamma' \tilde{F}_t(w)$, so $\tilde{F}_t$ induces a map $F_t\colon X \to X$. This is the desired homotopy between $f$ and $g$ and proves 3.12.5.

3.12.6. *If* $\alpha \in \mathrm{Aut}(\pi_1(X, x))$, *then there exists* $f \in \mathrm{Homeo}(X, x)$ *such that* $f_* = \alpha$.

The proof of 3.12.6 is in two steps

3.12.7. *If* $(Y, y)$ *is any pointed space and* $\alpha$ *is a homomorphism of* $\pi_1(X, x)$ *into* $\pi_1(Y, y)$, *then there is a mapping* $f\colon (X, x) \to (Y, y)$ *such that* $f_* = \alpha$.

3.12.8. *If* $f\colon (X, x) \to (X, x)$ *is a mapping and* $f_*$ *is bijective, then* $f$ *is homotopic to a homeomorphism.*

*Proof of 3.12.7.* Let $X$, as in Section 10, be regarded as a $q$-leafed rose $h_1(S^1) \cup$

† My attention was drawn to this method by Marden.

$\ldots \cup h_q(S^1)$ with a disc attached. Then $\pi_1(X, x)$ is generated by the routes $a_1 = [h_1 \circ \lambda], \ldots, a_q = [h_q \circ \lambda]$, where $\lambda: I \to S^1$ is given by $\lambda(t) = \exp(2\pi i t)$. There is a single relation $x_1 x_2 \ldots x_{2q} = 1$, where each $x_r$ is of the form $a_s^{\pm 1}$. This relation amounts geometrically to the assertion that the map $\phi: S^1 \to X$ given by

$$(\exp(2\pi i t)) = h_s(\exp(\pm 2\pi i(2qt - r + 1))), \text{ for } \frac{r-1}{2q} \leqslant t \leqslant \frac{r}{2q},$$

where the sign and the choice of $s$ is given by $x_r = a_s^{\pm 1}$, can be extended to a map $F: D^2 \to X$.

Let $g_1, \ldots, g_q: I \to Y$ be paths belonging to the routes $\alpha(a_1), \ldots, \alpha(a_q) \in \pi_1(Y, y)$. Define a map $f$ of the $q$-leafed rose into $Y$ by $f(h_r(\lambda(t))) = g_r(t)$ $(r = 1, \ldots, q)$. The route of the path $f \circ \phi$ is then $\alpha(x_1)\alpha(x_2)\ldots\alpha(x_{2q})$. Since $\alpha$ is a homomorphism, this is the trivial route, that is to say the map $f \circ \phi$ can be extended to a map of the disc $D^2$ into $Y$. The same extension yields an extension of $f$ from the $n$-leafed rose to the whole of $X$, with $f_* = \alpha$. ■

*Discussion of 3.12.8*. The most difficult step in this proof of Nielsen's theorem is 3.12.8, and for this Mangler refers to two papers of Kneser on the smoothing of simplicial maps between surfaces. By a complicated sequence of simplicial subdivisions and modifications of simplicial mappings, Kneser proves that every mapping $f: X \to Y$ between two closed surfaces contains in its homotopy class a simplicial map under which every 2-simplex in $Y$ is covered equally often by simplexes in $X$. He also shows in [**5b**] that if $f: X \to X$ is a self-mapping, then either the homotopy class of $f$ includes a homeomorphism or it includes a mapping $g$ such that $X \setminus g(X)$ has non-empty interior. It is easy to show that if $f_*$ is an automorphism, then $f$ has degree $\pm 1$ (if $X$ is orientable—if not one can pass to the orientable cover). This fact and Kneser's theorem give us 3.12.8. For further details the reader is referred to [**5a, 5b, 5c**].

## REFERENCES


1. M. A. Armstrong, On the fundamental group of an orbit space. *Proc. Camb. Philos. Soc.* **64** (1968), 299–301.
2. S. Lefschetz, "Introduction to Topology". Princeton University Press, Princeton, 1949.
3. A. M. Macbeath, Geometrical realisation of isomorphisms between plane groups. *Bull. Amer. Math. Soc.* **71** (1965), 629–630.
4. C. Maclachlan and W. J. Harvey, On mapping class groups and Teichmüller spaces. *Proc. London Math. Soc.* **3**(30) (1975), 495–512.

5. W. Mangler, Die Klassen von topologischen Abbildungen einer geschlossene Fläche auf sich. *Math. Z.* **44** (1939), 541–554.
5. (a) H. Hopf, Beiträge zur Klassifizierung der Flächenabbildungen. *J. Reine Angew. Math.* **185** (1931), 225–236.
5. (b) H. Kneser, Die kleinste Bedeckungszahl innerhalb einer Klasse von Flächenabbildungen. *Math. Annlen,* **103** (1930), 347–358.
5. (c) H. Kneser, Glättung von Flächenabbildungen. *Math. Ann.* **100** (1928),
6. A. Marden, Isomorphisms between Fuchsian groups, *in* "Advances in Complex Function Theory". Springer Lecture Notes No. 505, Springer-Verlag pp. 56–77.
7. W. Massey "Algebraic Topology: an Introduction". Harcourt, Brace and World, 1967.
8. J. Nielsen, Untersuchungen zur Topologie der geschlossenen zweiseitigen Flächen. *Acta Math.* **50** (1927), 189–358.
9. C. L. Siegel, "Topics in Complex Function Theory", Vol. II, pp. 83–87. Wiley–Interscience, 1971.
10. P. Tukia, On discrete subgroups of the unit disc and their isomorphisms. *Ann. Acad. Sci. Fenn. Ser A1*, **504** (1972), 1–42.

# 2. The Geometry of Discrete Groups

A. F. BEARDON

*University of Cambridge, England*



## INTRODUCTION AND NOTATION

This course is designed to provide the basic facts about the algebra and geometry of the $2 \times 2$ complex matrices viewed as a normed space and of the associated Möbius transformations.

We first list the notation used in this and subsequent courses.

Euclidean $n$-space is denoted by $\mathbb{R}^n$: we write $\mathbb{R}$ for $\mathbb{R}^1$ and denote the complex plane by $\mathbb{C}$. The one-point compactifications of these spaces are denoted by $\mathbb{R}^n_\infty$ and $\mathbb{C}_\infty$ respectively. We shall also discuss the upper half space $\mathscr{H}^n$ defined by

$$\mathscr{H}^n = \{(x_1, \ldots, x_n) \in \mathbb{R}^n : x_n > 0\}:$$

for brevity we use $\mathscr{H}$ for $\mathscr{H}^2$. We shall frequently (and sometimes without explicit mention) identify $x + iy$ in $\mathbb{C}$ with the point $(x, y, 0)$ in $\mathbb{R}^3$ and $\infty$ in $\mathbb{C}_\infty$ with $\infty$ in $\mathbb{R}^3_\infty$. With this identification $\mathbb{C}_\infty$ is the boundary of $\mathscr{H}^3$ in $\mathbb{R}^3_\infty$.

We denote by **Möb** the group of *Möbius transformations* $V$ of the form

$$V(z) = \frac{az + b}{cz + d}, \qquad ad - bc \neq 0:$$

such a $V$ is a one-to-one analytic map of $\mathbb{C}_\infty$ onto itself. For each subset $E$ of $\mathbb{C}_\infty$ we write

$$\mathbf{M\ddot{o}b}(E) = \{V \in \mathbf{M\ddot{o}b}: V(E) = E\}.$$

The algebra of all $2 \times 2$ matrices with complex entries is denoted by $M(2, \mathbb{C})$. The subset of non-singular matrices is the group $GL(2, \mathbb{C})$: the subset of matrices with determinant one is the group $SL(2, \mathbb{C})$. We use $\mathbb{Z}$ to denote the integers and $SL(2, \mathbb{Z})$ and $SL(2, \mathbb{R})$ are then defined in the obvious way.

## 1. COMPLEX MATRICES

Let $A = \begin{pmatrix} a & b \\ c & d \end{pmatrix}$ be any complex matrix, then

$$\bar{A} = \begin{pmatrix} \bar{a} & \bar{b} \\ \bar{c} & \bar{d} \end{pmatrix}, \qquad A^t = \begin{pmatrix} a & c \\ b & d \end{pmatrix}$$

and these operations commute.

Our discussion is based on the trace function:

$$\operatorname{tr}(A) = a + d.$$

Obviously

$$\operatorname{tr}(\bar{A}) = \overline{\operatorname{tr}(A)}, \qquad \operatorname{tr}(A^t) = \operatorname{tr}(A)$$

$$\operatorname{tr}(A + B) = \operatorname{tr}(B + A), \qquad \operatorname{tr}(AB) = \operatorname{tr}(BA);$$

if $B^{-1}$ exists then

$$\operatorname{tr}(BAB^{-1}) = \operatorname{tr}(AB^{-1}.B) = \operatorname{tr}(A).$$

For any $A$ and $B$ in $M(2, \mathbb{C})$ we now define

$$(A, B) = \operatorname{tr}(A\bar{B}^t).$$

This is a *complex scalar product*, explicitly

(i) $(A, B) = \overline{(B, A)}$;
(ii) $(A, A) \geqslant 0$ with equality if and only if $A = 0$;
(iii) $(A, B)$ is a linear function of $A$.

Any function $(A, B)$ satisfying (i), (ii) and (iii) defines a norm, $\| \ \|$, on $M(2, \mathbb{C})$ which in our case is given by

$$\|A\|^2 = (A, A) = |a|^2 + |b|^2 + |c|^2 + |d|^2.$$

In terms of the metric $\|A - B\|$,

$$\begin{pmatrix} a_n & b_n \\ c_n & d_n \end{pmatrix} \to \begin{pmatrix} a & b \\ c & d \end{pmatrix}$$

if and only if $a_n \to a$, $b_n \to b$, $c_n \to c$ and $d_n \to d$, [**10**], p. 52.

The following inequalities are valid:

(iv) $|(A, B)| \leqslant \|A\| \,.\, \|B\|$,
(v) $\|AB\| \leqslant \|A\| \,.\, \|B\|$,
(vi) $2|\det(A)| \leqslant \|A\|^2$.

Inequality (iv) is the Cauchy–Schwarz inequality and this holds for any scalar product.

To prove (v) we write

$$A = \begin{pmatrix} a & b \\ c & d \end{pmatrix}, \qquad B = \begin{pmatrix} \alpha & \beta \\ \gamma & \delta \end{pmatrix},$$

compute $\|AB\|^2$ and use the Cauchy–Schwarz inequality, for example

$$|a\alpha + b\gamma|^2 \leqslant (|a|^2 + |b|^2)(|\alpha|^2 + |\gamma|^2).$$

Finally, (vi) is valid, as

$$2|ad - bc| \leqslant 2|ad| + 2|bc| \leqslant (|a|^2 + |d^2) + (|b|^2 + |c|^2).$$

From now on we shall only be interested in SL(2, $\mathbb{C}$). Observe that if $A \in \mathrm{SL}(2, \mathbb{C})$, then $\|A\|^2 \geqslant 2$. This inequality is best possible (consider $a = b = d = -c = 1/\sqrt{2}$).

## 2. MÖBIUS TRANSFORMATIONS

We assume familiarity with the elementary properties of Möbius transformations. The proof of the following result is easy and is omitted.

2.1 THEOREM. *The map* $\phi\colon \mathrm{SL}(2, \mathbb{C}) \to$ **Möb** *given by*

$$\phi\begin{pmatrix} a & b \\ c & d \end{pmatrix} = \left[ z \mapsto \frac{az + b}{cz + d} \right]$$

*is a group homomorphism with kernel* $\{\pm I\}$.

Thus **Möb** is isomorphic to SL(2, $\mathbb{C}$)/$\{\pm I\}$; each $V$ in **Möb** determines a pair of matrices $\{\pm A\}$ in SL(2, $\mathbb{C}$) such that $\phi(B) = V$ if and only if $B = \pm A$.

The functions det and $\| \ \|$ on SL(2, $\mathbb{C}$) now convert naturally into functions on **Möb**. If $\phi(A) = V$ define

$$\det(V) = \det(A) = \det(-A),$$

$$\|V\| = \|A\| = \|-A\|.$$

The function tr satisfies $\mathrm{tr}(-A) = -\mathrm{tr}(A)$: thus we define

$$\mathrm{tr}^2(V) = [\mathrm{tr}(A)]^2 = [\mathrm{tr}(-A)]^2.$$

We wish to classify the elements in **Möb** geometrically and this is best understood by examining the finer classification into conjugacy classes. Two elements $V$ and $V^*$ in **Möb** are conjugate if and only if there is a $U$ in **Möb** with $V^* = UVU^{-1}$. The identity $I$ is conjugate only to itself.

2.2 THEOREM. *The elements* $V$ *and* $V^*$ *in* **Möb** $\setminus \{I\}$ *are conjugate if and only if* $\mathrm{tr}^2(V) = \mathrm{tr}^2(V^*)$.

*Proof.* One implication is easy. We assume that $V^* = UVU^{-1}$ and select $A$ and $B$ in SL(2, $\mathbb{C}$) with $\phi(A) = V$, $\phi(B) = U$. Then

$$\begin{aligned} \operatorname{tr}^2(V^*) &= [\operatorname{tr}(BAB^{-1})]^2 \\ &= [\operatorname{tr}(A)]^2 \\ &= \operatorname{tr}^2(V). \end{aligned}$$

To prove the converse we first consider the following elements in **Möb**:

$$M_k(z) = kz, \qquad k \neq 0, 1.$$

It is also convenient to define $M_1$ by $M_1(z) = z + 1$: then for all $k \neq 0$ (including $k = 1$),

$$\operatorname{tr}^2(M_k) = k + 2 + 1/k.$$

It is now easy to see that $\operatorname{tr}^2(M_k) = \operatorname{tr}^2(M_q)$ if and only if $k = q$ or $kq = 1$. It is also easy to see that $M_k$ and $M_q$ are conjugate if and only if $k = q$ or $kq = 1$. We conclude that $M_k$ and $M_q$ are conjugate if and only if $\operatorname{tr}^2(M_k) = \operatorname{tr}^2(M_q)$.

The proof is now completed by showing that any $V$ in **Möb** is conjugate to some $M_k$ for if

(i) $V$ and $M_k$ are conjugate,
(ii) $V^*$ and $M_q$ are conjugate and
(iii) $\operatorname{tr}^2(V) = \operatorname{tr}^2(V^*)$

then $\operatorname{tr}^2(M_k) = \operatorname{tr}^2(M_q)$, hence $M_k$ and $M_q$ are conjugate and so are $V$ and $V^*$.

If $V$ has only one fixed point, say $\alpha$, select $U$ so that $U(\alpha) = \infty$. Then $UVU^{-1}$ has $\infty$ as its only fixed point and so is of the form $z \mapsto z + b$. If $B(z) = z/b$, then $(BU)V(BU)^{-1} = M_1$.

If $V$ has two fixed points $\alpha$ and $\beta$ select $U$ so that $U(\alpha) = \infty$ and $U(\beta) = 0$, then $UVU^{-1} = M_k$ for some $k$. ∎

This proof shows how to compute the $n$th iterate $V^n$ of $V$. For example, if $V$ had fixed points $\alpha$ and $\beta$ write $U(z) = (z - \beta)/(z - \alpha)$; thus for some $k \neq 0, 1$, $UVU^{-1}(z) = kz$ and $UV^nU^{-1} = (UVU^{-1})^n$ so

$$V^n(z) = U^{-1}(k^n U(z)).$$

We now give the geometric classification referred to earlier. If $V \neq I$,

(i) $V$ is *parabolic* if $V$ is conjugate to $M_1$ so $\mathrm{tr}^2(V) = 4$;
(ii) $V$ is *hyperbolic* if $V$ is conjugate to some $M_k, k > 0, k \neq 1$, so $\mathrm{tr}^2(V) > 4$;
(iii) $V$ is *elliptic* if $V$ is conjugate to some $M_k$, $|k| = 1, k \neq 1$, so $0 \leqslant \mathrm{tr}^2(V) < 4$;
(iv) $V$ is *loxodromic* in all other cases.

It follows that $V$ can be classified simply by knowing $\mathrm{tr}^2(V)$.

## 3. HYPERBOLIC GEOMETRY

We begin by constructing the hyperbolic geometry of the upper half-plane $\mathscr{H}$. The subgroup **Möb**($\mathscr{H}$) consists of those $V$ which can be written in the form

$$(3.1) \qquad V(z) = \frac{az + b}{cz + d};\ a, b, c, d \in \mathbb{R}, \qquad ad - bc = 1.$$

We deduce that if $V(\mathscr{H}) = \mathscr{H}$ then $\mathrm{tr}^2(V) \geqslant 0$ and so $V$ cannot be loxodromic. If $V$ in **Möb**($\mathscr{H}$) is parabolic or hyperbolic the fixed points of $V$ lie in $\mathbb{R}_\infty$: if $V$ is elliptic the fixed points are of the form $\alpha$ and $\bar{\alpha}$, $\alpha \in \mathscr{H}$. These facts follow because the fixed points of $V$ (given above) are (using $ad - bc = 1$ and $c \neq 0$)

$$\alpha, \beta = \frac{(a - d) \pm \sqrt{[(a + d)^2 - 4]}}{2c}$$

(the case $c = 0$ is left to the reader).

We now introduce some further notation. Let $z$ and $w$ be distinct points in $\mathscr{H}$. There is a unique circle (or straight line) which contains $z$ and $w$ and which is orthogonal to $\mathbb{R}_\infty$: we denote the part of this circle in $\mathscr{H}$ by $[z, w]$. The end points of $[z, w]$ (which is a semicircle or half-line) are labelled $z^*$ and $w^*$ so that $z$ separates $z^*$ and $w$.

It is easy to see that there is a unique $V$ in **Möb**($\mathscr{H}$) which maps $z^*, z, w^*$ to $0, i, \infty$ respectively. It follows that $V(w) = ri$ for some positive $r$, $r > 1$. This simple result is exploited now in our discussion of certain invariant functions.

The *cross-ratio* of any four distinct points $z_1, z_2, z_3, z_4$ in the extended complex plane is defined as

$$[z_1, z_2, z_3, z_4] = \frac{(z_3 - z_1)(z_4 - z_2)}{(z_2 - z_1)(z_4 - z_3)}$$

and (as is easily checked) this is invariant under any $V$ in **Möb**, that is

$$[V(z), V(z_2), V(z_3). V(z_4)] = [z_1, z_2, z_3, z_4].$$

We deduce that $\eta$ defined by

$$\eta(z, w) = [z^*, z, w, w^*]$$

is also invariant, and so

$$\eta(z, w) = [0, i, ri, \infty] = r. \tag{3.2}$$

Another useful invariant is

$$\tau(z, w) = \left|\frac{z - w}{z - \overline{w}}\right|, \qquad z, w \in \mathscr{H}.$$

If $V \in \mathbf{Möb}(\mathscr{H})$ then $V(\overline{w}) = \overline{V(w)}$ and we find (by direct computation) that

$$\tau(z, w) = \tau(V(z), V(w)).$$

We conclude that

$$\tau(z, w) = \tau(i, ri) = \frac{r - 1}{r + 1}. \tag{3.3}$$

We next consider an invariant differential on $\mathscr{H}$. If $V \in \mathbf{Möb}(\mathscr{H})$ then

$$\mathrm{Im}[V(z)] = |V'(z)| \,.\, \mathrm{Im}[z]$$

and so we have the invariant differential

$$\mathrm{d}s = \frac{|\mathrm{d}z|}{\mathrm{Im}[z]} = \frac{|\mathrm{d}V(z)|}{\mathrm{Im}[V(z)]}.$$

We deduce that

$$\int_\sigma \mathrm{d}s = \int_{V(\sigma)} \mathrm{d}s$$

over all (smooth) curves $\sigma$ in $\mathscr{H}$.

We now define

$$\rho(z, w) = \inf_{\sigma} \int_{\sigma} ds$$

the infimum being over all (smooth) curves $\sigma$ in $\mathscr{H}$ which join $z$ to $w$. Thus $\rho$ is an invariant metric on $\mathscr{H}$:

$$\rho(z, w) = \rho(V(z), V(w)), \qquad V \in \mathbf{M\ddot{o}b}(\mathscr{H}).$$

We wish to obtain an explicit expression for $\rho$ and to do this we observe that

$$\rho(z, w) = \rho(i, ri) = \log r \tag{3.4}$$

(a simple proof is given in [**9**], p. 24).

We may now express the invariants $\eta$, $\tau$ and $\rho$ in terms of each other by eliminating $r$ from 3.2, 3.3 and 3.4. In particular

$$\tau(z, w) = \tanh[\tfrac{1}{2}\rho(z, w)] \tag{3.5}$$

(this shows that $\tau$ is also a metric on $\mathscr{H}$) and

$$\rho(z, w) = \log\left\{\frac{|z - \bar{w}| + |z - w|}{|z - \bar{w}| - |z - w|}\right\}$$

$$= \log[z^*, z, w, w^*].$$

The proof of 3.4 in [**9**] also solves the problem of finding the $\rho$-geodesics. The only curves $\sigma$ joining $i$ to $ri$ for which

$$ds = \rho(i, ri)$$

are those satisfying $\sigma(t) = x(t) + iy(t)$, $x(t) \equiv 0$ and $\dot{y}(t) \geqslant 0$. Thus there is a unique geodesic segment joining $i$ to $ri$ and this is the segment of the imaginary axis between these points. In general, the arc $[z, w]$ is the unique $\rho$-geodesic through $z$ and $w$.

The metric $\rho$ induces a topology on $\mathscr{H}$ and by virtue of 3.5 $\tau$ induces the same topology on $\mathscr{H}$: *in fact this is the Euclidean topology*. To see this we consider the hyperbolic circle ($h$-circle) with $h$-centre $i$ and $\rho$-radius $\delta$. The equation of this circle is

$$\tau(z, w) = d$$

where $d = \tanh(\frac{1}{2}\delta)$ is given by 3.5. This equation simplifies to

$$|z|^2 + 1 = 2y\cosh(\delta) \qquad (z = x + iy). \tag{3.6}$$

We conclude that the hyperbolic and Euclidean circles in $\mathscr{H}$ coincide (of course the hyperbolic and Euclidean centres and radii differ): the same is true of open discs and hence the two topologies are the same.

There is another application of 3.6. If $V$ is given by 3.1 observe that

$$|V(i)|^2 = \frac{a^2 + b^2}{c^2 + d^2}, \qquad \operatorname{Im}[V(i)] = \frac{1}{c^2 + d^2}$$

Equation 3.6 must be satisfied with $z = V(i)$ and $\delta = \rho(i, V(i))$. This proves Theorem 3.7.

3.7 THEOREM. *If* $V \in \mathbf{M\ddot{o}b}(\mathscr{H})$ *then* $\|V\|^2 = 2\cosh[\rho(i, V(i))]$.

The differential d$s$ gives rise to the *hyperbolic area* of a (measurable) set $E$, namely

$$h - \operatorname{area}(E) = \iint_E \frac{\mathrm{d}x\,\mathrm{d}y}{y^2}.$$

We then have (see [**9**], p. 48) the following theorem.

3.8 THE GAUSS–BONNET THEOREM. *The hyperbolic area of a hyperbolic triangle with angles* $\alpha$, $\beta$ *and* $\gamma$ *is* $\pi - (\alpha + \beta + \gamma)$.

This is true even if some or all of the vertices of the triangle lie on $\mathbb{R}_\infty$ (the extended real line). If a vertex $v$ lies on $\mathbb{R}_\infty$ the angle at $v$ is necessarily zero (both sides from the vertex are orthogonal to $\mathbb{R}_\infty$ at $v$): the triangle has area $\pi$ if all the vertices lie on $\mathbb{R}_\infty$, otherwise the area is less than $\pi$.

We now turn our attention to the hyperbolic geometry of the unit disc $\Delta$. The transformation

$$Q(z) = \frac{z - i}{z + i}$$

is a conformal homeomorphism of $\mathscr{H}$ onto $\Delta$ and using $Q$ the entire discussion of the hyperbolic geometry in $\mathscr{H}$ can be interpreted in terms of $\Delta$. The corresponding results can, of course, be derived directly (see [**10**], p. 78 and [**13**], p. 509). Here, we shall simply state the main results.

We remark first that $Q$ is particularly suited to our purpose as

$$\tau(z, i) = |Q(z)|$$

and so $h$-circles with centre $i$ in $\mathscr{H}$ map to $h$-circles in $\Delta$ with centre 0.

Next it is clear that $V \in \mathbf{M\ddot{o}b}(\mathscr{H})$ if and only if $QVQ^{-1} \in \mathbf{M\ddot{o}b}(\Delta)$ and, moreover, a direct computation shows that if $V \in \mathbf{M\ddot{o}b}(\mathscr{H})$ then

$$\|V\| = \|QVQ^{-1}\|.$$

We know ([**6**], p. 31) that $V \in \mathbf{M\ddot{o}b}(\Delta)$ if and only if we can write

$$V(z) = \frac{az + \bar{c}}{cz + \bar{a}}, \qquad |a|^2 - |c|^2 = 1, \tag{3.9}$$

and for these $V$ we have the invariant differential

$$ds_1 = \frac{|dz|}{1 - |z|^2} = \frac{|dV(z)|}{1 - |V(z)|^2}$$

(to see this, simply compute $|V^{(1)}(z)|$). If, however, we transform the differential $|dz|/y$ to $\Delta$ we find that it gives rise to the differential $ds^*$ where $ds^* = 2\,ds_1$ and for this reason we use $ds^*$ to define a metric $\rho^*$ on $\Delta$. Obviously

$$\rho^*(z, w) = \rho(Q^{-1}(z), Q^{-1}(w)), \qquad z, w \in \Delta.$$

If $V \in \mathbf{M\ddot{o}b}(\Delta)$ we have

$$\begin{aligned} \|V\|^2 &= \|Q \,.\, Q^{-1}VQ \,.\, Q^{-1}\|^2 \\ &= \|Q^{-1}VQ\|^2 \\ &= 2\cosh[\rho(i, Q^{-1}VQ(i))] \\ &= 2\cosh[\rho^*(Q(i), VQ(i))] \\ &= 2\cosh[\rho^*(0, V(0))]; \end{aligned} \tag{3.10}$$

this is the analogue of Theorem 3.7 in $\Delta$.

A direct proof is also available. We have

$$\rho^*(0, w) = \int_0^{|w|} \frac{2\,dt}{1 - t^2} = \log\left\{\frac{1 + |w|}{1 - |w|}\right\}.$$

If $V$ is given by 3.9, $|V(0)| = |c/a|$ and so

$$\rho^*(0, V(0)) = \log\left\{\frac{|a| + |c|}{|a| - |c|}\right\}$$

which yields 3.10 as $\|V\|^2 = 2(|a|^2 + |c|^2)$.

The invariant $\tau$ in $\mathscr{H}$ yields the invariant

$$\tau^*(z, w) = \tau(Q^{-1}(z), Q^{-1}(w)) = \left|\frac{z - w}{1 - z\bar{w}}\right|$$

in $\Delta$ and this too is an invariant metric in $\Delta$.

Our next task is to see how the action of a general Möbius transformation can be extended from $\mathbb{C}_\infty$ (viewed as a subset of $\mathbb{R}^3_\infty$) to $\mathbb{R}^3_\infty$.

We consider the *quaternion*

$$\alpha + \beta i + \gamma j + \delta k = (\alpha + i\beta) + (\gamma + i\delta)j = z + wj$$

say, where $\alpha, \beta, \gamma \in \mathbb{R}$ and $z, w \in \mathbb{C}$. Observe that $wj = j\,\bar{w}$ and

$$(z + wj)(\bar{z} - wj) = |z|^2 + |w|^2.$$

We now identify $(\alpha, \beta, t)$ in $\mathbb{R}^3$ with $z + tj$ and extend $V$ to $\mathbb{R}^3$ by

$$V(z + tj) = [a(z + tj) + b]\,.\,[c(z + tj) + d]^{-1}$$

$$= \frac{[(az + b) + taj]\,.\,[\overline{cz + d} - tcj]}{|cz + d|^2 + |c|^2t^2}$$

$$\text{(3.11)} \qquad = \frac{(az + b)(\overline{cz + d}) + t^2a\bar{c} + tj}{|cz + d|^2 + |c|^2t^2}.$$

Thus if

$$V(z, t) = (z^*, t^*)$$

then

$$z^* = \frac{a\bar{c}(|z|^2 + t^2) + a\bar{d}z + b\bar{c}\bar{z} + b\bar{d}}{|cz + d|^2 + |c|^2 t^2},$$

(3.12) $$t^* = \frac{t}{|cz + d|^2 + |c|^2 t^2}.$$

Observe from 3.11 that $V(z, 0) = (V(z), 0)$, thus this definition agrees with the original definition of $V$ on $\mathbb{C}_\infty$. Also 3.12 shows that each $V$ preserves $\mathscr{H}^3$.

As an illustration of the use of quaternions let us discuss the fixed points (if any) of $V$ in $\mathscr{H}^3$. First, $(0, 0, 1)$ in $\mathbb{R}^3$ is identified with $j$ and is a fixed point of $V$ if and only if

$$aj + b = j(cj + d),$$

or equivalently

(3.13) $$a = \bar{d},\ b = -\bar{c}.$$

The condition $ad - bc = 1$ becomes $|a|^2 + |c|^2 = 1$ and so we must have

$$0 < \operatorname{tr}^2(V) = (a + \bar{a})^2 < 4$$

unless $V = I$. We conclude (by conjugation) that apart from $I$ only the elliptic elements can fix points in $\mathscr{H}^3$.

If $V$ is any elliptic element we may consider a conjugate $V^*$ of $V$ with fixed points $0$ and $\infty$, so $V^*(z) = az/\bar{a}$, $|a| = 1$, and $V^*$ fixes all points $(0, t)$ and no others in $\mathscr{H}^3$. In general the set of fixed points in $\mathscr{H}^3$ of an elliptic $V$ is precisely the *axis* of $V$ (the semicircle orthogonal to $\mathbb{C}_\infty$ and joining the fixed points of $V$ in $\mathbb{C}_\infty$).

The discussion of the geometry in $\mathscr{H}$ is easily modified and remains valid in $\mathscr{H}^3$ (the exception is the Gauss–Bonnet Theorem). The differential

(3.14) $$\mathrm{d}s = \frac{|\mathrm{d}(z + tj)|}{t}$$

is an invariant differential (this can be seen from 3.12) which gives rise to an invariant metric $\rho$ in $\mathscr{H}^3$. The geodesics are the Euclidean semicircles which are orthogonal to $\mathbb{C}_\infty$: the hyperbolic planes are Euclidean hemispheres also orthogonal to $\mathbb{C}_\infty$. Finally,

$$\|V\|^2 = 2\cosh[\rho(j, V(j))]$$

where, of course, $j = (0, 0, 1)$ in $\mathscr{H}^3$, [**5**].

There is an alternative (but equivalent) method of extending the action of $V$ from $\mathbb{C}_\infty$ to $\mathbb{R}^3_\infty$. Each $V$ in **Möb** is the composition $I_1, \ldots, I_{2m}$ of an even number of inversions $I_n$ in circles $Q_n$ in $\mathbb{C}_\infty$. Each circle $Q_n$ is a subset of a sphere $\Sigma_n$ in $\mathbb{R}^3_\infty$ which is orthogonal to $\mathbb{C}_\infty$ and we may regard $I_n$ as an inversion in $\Sigma_n$ rather than in $Q_n$. In this way $V = I_1 \ldots I_{2m}$ can be extended to a map of $\mathbb{R}^3_\infty$ onto itself and it is easy to see this extension is independent of the representation $I_1 \ldots I_{2m}$.

At this point we observe that we can define the entire group of Möbius transformations acting in $\mathbb{R}^3_\infty$ as those transformations which are a composition of an even number of inversions in any spheres in $\mathbb{R}^3_\infty$. As before we can find a Möbius transformation mapping $\mathscr{H}^3$ to $\Delta_3$ (the unit ball in $\mathbb{R}^3$). The invariant differential in $\Delta_3$ becomes

$$\mathrm{d}s = \frac{2|\mathrm{d}x|}{1 - |x|^2}, \qquad x \in \Delta_3.$$

There is one representation of $V$ as a product of inversions which warrants particular mention. We assume that

$$V(z) = \frac{az + b}{cz + d}, \qquad ad - bc = 1,$$

and also that $c \neq 0$. Then the action of $V$ in $\mathbb{R}^3_\infty$ is an inversion in the *isometric sphere*

$$\{x \in \mathbb{R}^3 : |x - V^{-1}(\infty)| = 1/|c|\}$$

of $V$ followed by an Euclidean isometry $z \to Az + B$, $|A| = 1$. The isometry is known explicitly [**6**] p. 28, and this action is precisely that given by 3.11. The isometric sphere is so called as it is the sphere

$$\{\zeta \in \mathbb{R}^3 : |V'(\zeta)| = 1\}:$$

the intersection of this sphere with $\mathbb{C}$ is the *isometric circle* of $V$. This representation of $V$ gives one an explicit formula for the change of *Euclidean* distance under $V$, namely

$$|V'(\zeta)| = |cz + d|^2 + |c|^2 t^2, \qquad \zeta = z + tj$$

and this with 3.12 proves the invariance of the differential 3.14.

## 4. DISCRETE GROUPS

We again consider $SL(2, \mathbb{C})$ as a metric space with metric $\|A - B\|$. Let $G$ be a subgroup of $SL(2, \mathbb{C})$ and suppose for the moment that there are distinct elements $A$ and $A_1, A_2, \ldots$ in $G$ with $A_n \to A$ (equivalently, $\|A_n - A\| \to 0$). Then for each $B$ in $G$, $A_n A^{-1} B \to B$ as

$$\begin{aligned} \|A_n A^{-1} B - B\| &= \|(A_n - A) A^{-1} B\| \\ &\leqslant \|A_n - A\| \,.\, \|A^{-1} B\|. \end{aligned}$$

This shows that either all of the elements of $G$ are isolated points of $G$ (as a metric space) or none are.

4.1 *Definition.* A subgroup $G$ of $SL(2, \mathbb{C})$ is *discrete* if and only if each element of $G$ is an isolated point of $G$.

In fact, the elements of a discrete group $G$ cannot even converge to an element of $SL(2, \mathbb{C})$ which is not in $G$ (if $A_n \in G$ and $A_n \to A$, then $A_n A_{n+1}^{-1} \in G$ and $A_n A_{n+1}^{-1} \to I$): this simply says that $G$ is a closed subset of $SL(2, \mathbb{C})$.

The following facts are easily established.

(i) A subgroup of a discrete group is discrete.
(ii) If $G$ is discrete so is any conjugate group $UGU^{-1}$.
(iii) A subgroup $G$ of $SL(2, \mathbb{C})$ is discrete if and only if for each positive $k$, $\|A\| < k$ for only a finite number of $A$ in $G$.
(iv) A discrete group $G$ is countable and $\|A_n\| \to +\infty$ for any enumeration $\{I, A_1, A_2, \ldots\}$ of $G$.

A refinement of (iv) is that if $G$ is discrete, then

$$\Sigma \|A_n\|^{-t} < +\infty$$

for each $t > 4$ [**5**].

We now wish to relate discrete subgroups of SL(2, $\mathbb{C}$) to subgroups of **Möb**. Let $D$ be a subdomain of $\mathbb{C}_\infty$ or of $\mathbb{R}^3_\infty$ and let $G$ be a subgroup of **Möb**($D$). We say that $G$ is *(properly) discontinuous* in $D$ if and only if each compact subset of $D$ meets only a finite number of its $G$-images.

If $G$ is a discrete subgroup of SL(2, $\mathbb{C}$) we select any compact subset $K$ of $\mathscr{H}^3$, so $K$ lies in an $h$-ball with centre $j$ and radius $d$. If $A \in \phi(G)$ and $A(K) \cap K \neq \varnothing$, then $\rho(j, A(j)) \leqslant 2d$ and by Theorem 3.7, $\|A\|^2 \leqslant 2\cosh[2d]$. As $G$ is discrete this can only occur for a finite number of $A$ and so $\phi(G)$ is discontinuous in $\mathscr{H}^3$.

The argument is reversible and we have

4.2 THEOREM. (i) *If $G$ is a discrete subgroup of* SL(2, $\mathbb{C}$) (*or* SL(2, $\mathbb{R}$)) *then* $\phi(G)$ *is discontinuous in* $\mathscr{H}^3$ (*or* $\mathscr{H}$).

(ii) *If $G$ is a subgroup of* **Möb** *which is discontinuous in* $\mathscr{H}^3$ (*or* $\mathscr{H}$) *then* $\phi^{-1}(G)$ *is a discrete subgroup of* SL(2, $\mathbb{C}$) (*or* SL(2, $\mathbb{R}$)).

In (i) $\phi(G)$ may or may not be discontinuous in some domain in $\mathbb{C}_\infty$ (or $\mathbb{R}_\infty$). For example the Picard group, SL(2, $\mathbb{Z} + i\mathbb{Z}$), is discrete but not discontinuous in any subdomain in $\mathbb{C}_\infty$, [**10**], p. 96.

## 5. STABILIZERS

For brevity we shall say a subgroup $G$ of **Möb** is discrete if the group $\phi^{-1}(G)$ is discrete in SL(2, $\mathbb{C}$) and we shall frequently perform computations with matrices instead of Möbius transformations.

Let $G$ be any subgroup of **Möb** and denote by $G_\zeta$ the *stabilizer* in $G$ of $\zeta$: thus

$$G_\zeta = \{V \in G \colon V(\zeta) = \zeta\}.$$

If $G$ is discrete the nature of the stabilizers $G_\xi$ is severely restricted and we shall now examine the possible cases.

First, let us consider a point $\zeta$ in $\mathscr{H}^3$. We have already seen that the stabilizer $G_\zeta$ contains (apart from $I$) only elliptic elements. Further, by considering a group conjugate to $G$ we may assume $\zeta = j$ (for example, $(x, y, t)$ can be mapped to $(0, 0, t)$ by a translation and then to $j = (0, 0, 1)$ by a hyperbolic element which fixes 0 and $\infty$). By Theorem 3.7,

$$G_j = \{V \in G \colon \|V\|^2 = 2\}$$

and so $G_j$ (and hence $G_\zeta$) is finite (recall that $G$ is discrete). The following result gives a more complete description.

5.1 THEOREM. *Let $G$ be a discrete subgroup of* **Möb**. *The following conditions are equivalent:*

(i) *apart from $I$, $G$ contains only elliptic elements;*
(ii) *the elements of $G$ have a common fix-point $\zeta$ in $\mathscr{H}^3$;*
(iii) *$G$ is finite,*

*Proof.* The proof that (ii) implies (iii) is given above: the proof that (iii) implies (i) is simply that if $V$ is parabolic, hyperbolic or loxodromic then the elements $V, V^2, V^3, \ldots$ are distinct.

We now prove that (i) implies (ii). We assume that (i) holds and (by considering a conjugate group) we may assume that

$$A(z) = \alpha z/\bar{\alpha} \in G, \qquad |\alpha|^2 = 1, \qquad \alpha^2 \neq 1.$$

Now consider any element $V$ in $G$ with

$$V(z) = \frac{az + b}{cz + d}, \qquad \text{and} \quad ad - bc = 1.$$

The elements $V, AV, AVA^{-1}V^{-1}$ are all in $G$ and have real traces (to within a factor $-1$)

$$\lambda = a + d, \qquad \mu = \alpha a + \bar{\alpha} d, \qquad \gamma = 2 - bc(\alpha - \bar{\alpha})^2 = 2 + 2bc\,|\alpha - \bar{\alpha}|^2$$

respectively (use $ad - bc = 1$ to compute the last trace). We solve for $a$ and $d$ and (as $\lambda$ and $\mu$ are real) we find $a = \bar{d}$. Next, as $AVA^{-1}V^{-1}$ is either $I$ or is elliptic we have $\gamma \leqslant 2$ and so $bc \leqslant 0$. If $AVA^{-1}V^{-1} = I$ then $V$ leaves the pair $\{0, \infty\}$ invariant (for example $AV(0) = VA(0) = V(0)$) and so either

(i) $V$ fixes 0 and $\infty$ so $b = c = 0$, or
(ii) $V(0) = \infty, \quad V(\infty) = 0 \quad$ so $\quad a = d = 0, \quad bc \neq 0$.

If $AVA^{-1}V^{-1} \neq I$ then $bc \neq 0$ so again $bc < 0$. We conclude that in all cases either $b = c = 0$ or $bc < 0$.

We can now show that the elements of $G$ have a common fix-point of the form $sj$, $s > 0$. First if $V$ is such that $b = c = 0$, then $V(tj) = tj$ for every real $t$.

If $V$ is such that $bc < 0$ we may select a positive $t$ with $t^2 = |b/c|$ so $b = -t^2\bar{c}$ and a simple computation shows that $V(tj) = tj$.

If we consider another such element, $U$ say, which we now know to be of the form

$$U(z) = \frac{a_1 z - t_1^2 \bar{c}_1}{c_1 z + \bar{a}_1}, \qquad t_1 > 0,$$

then the restriction corresponding to "$a = \bar{d}$" is satisfied by $UV$ and we find $t = t_1$ so $U(tj) = tj$ also. Thus each element in $G$ fixes $tj$ and the proof is complete. ■

We now consider the nature of the stabilizer $G_\zeta$ of a point $\zeta$ in $\mathbb{C}_\infty$. By considering a conjugate group we may, of course, assume that $\zeta = \infty$.

First, if $G_\infty$ consists only of elliptic elements and $I$ then all elements of $G_\infty$ fix $\infty$ and have a common fix-point in $\mathscr{H}^3$ (Theorem 5.1). Thus all elliptic elements in $G_\infty$ have a common axis and hence the same *pair* of fixed points. It is now easy to see that $G_\infty$ is cyclic.

Let us now suppose that $G_\infty$ contains a hyperbolic or loxodromic element which we may assume also fixes 0: say

$$K(z) = kz/k^{-1}, \qquad \mathrm{tr}^2(K) = k^2 + 2 + k^{-2}, \qquad k^2 \neq 1.$$

Now let

$$V(z) = \frac{az + b}{d}, \qquad ad = 1$$

be any other element in $G_\infty$. A computation shows that

$$K^n V K^{-n} V^{-1}(z) = z + ab(k^{2n} - 1)$$

and so

$$\|K^n V K^{-n} V^{-1}\|^2 = 2 + |ab|^2 . |k^{2n} - 1|^2.$$

These members are distinct and bounded as $n \to +\infty$ or as $n \to -\infty$ (because $|k| \neq 1$) unless $ab = 0$. As $G_\infty$ is discrete and $ad = 1$ we have $b = 0$. Thus every element of $G_\infty$ fixes 0 and $\infty$: in particular $G_\infty$ cannot contain parabolic elements. Further as each element in $G_\infty$ fixes 0 and $\infty$, $G_\infty$ is abelian. The general element in $G_\infty$ is

$$A(z) = az/a^{-1}, \qquad |a| \neq 1:$$

by discreteness, there is a minimal $a$ satisfying $|a| > 1$ and $G_\infty$ is generated by a hyperbolic or loxodromic element and possibly an elliptic element (of finite order).

Finally, we consider the case when $G_\infty$ contains a parabolic element $P$ (which is necessarily a translation). The subgroup of translations in $G_\infty$ is the maximal parabolic subgroup of $G_\infty$: this is abelian and has one or two generators (see any introduction to elliptic functions). By virtue of the preceding paragraph, $G_\infty$ cannot contain any hyperbolic or loxodromic elements: $G_\infty$ may contain elliptic elements but these are of a restricted type [**6**], p. 140.

We now consider any discrete group $G$ containing a parabolic element $P$ which (by conjugation) we assume is the transformation $P(z) = z + 1$. If $V \in G$ and $V(\infty) \neq \infty$ we define $V_n$, $n = 1, 2, \ldots$, inductively by

$$V = V_1, \qquad V_{n+1} = V_n^{-1} P V_n.$$

In matrix form,

$$V = \begin{pmatrix} a & b \\ c & d \end{pmatrix}, \qquad P = \begin{pmatrix} 1 & 1 \\ 0 & 1 \end{pmatrix}, \qquad ad - bc = 1, \quad c \neq 0$$

and, say,

$$V_n = \begin{pmatrix} a_n & b_n \\ c_n & d_n \end{pmatrix},$$

so

$$\begin{pmatrix} a_{n+1} & b_{n+1} \\ c_{n+1} & d_{n+1} \end{pmatrix} = \begin{pmatrix} 1 + d_n c_n & d_n^2 \\ -c_n^2 & 1 - d_n c_n \end{pmatrix}.$$

We can also solve the recurrence relation for $c_n$ explicitly and if $|c| = |c_1| < 1$ then $c_n = o(1/n)$. Thus for sufficiently large $n$, $|d_{n+1}| \leqslant 1 + |d_n|$ which shows $d_n = O(n)$: these estimates now show that $d_{n+1} = O(1)$. We conclude that the distinct elements $V_n$ have bounded norm and this is a contradiction. Thus we have proved ([**11**] and [**12**]) Theorem 5.2.

5.2 THEOREM. *Let $G$ be discrete and let $P: z \to z + 1$ be in $G$. If $V = \begin{pmatrix} a & b \\ c & d \end{pmatrix} \in G$ then either $c = 0$ or $|c| \geqslant 1$.*

Now suppose that $V \in G$ and $V(\infty) \neq \infty$. By Theorem 5.2, $|c| \geqslant 1$ and so if $V(z, t) = (z^*, t^*)$ then

$$t^* = \frac{t}{|cz + d|^2 + |c|^2 t^2} \leqslant \frac{t}{|c^2 t^2} \leqslant \frac{1}{t}.$$

Thus if $t > 1$ then $t^* < 1$ and we have

5.3 COROLLARY. *Let $\Sigma = \{(z, t): t > 1\}$. If $V \in G$ then either $V(\infty) = \infty$ and $V(\Sigma) = \Sigma$ or $V(\infty) \neq \infty$ and $V(\Sigma) \cap \Sigma = \varnothing$.*

A *horosphere at* $w$, $w \in \mathbb{C}$, is an open ball in $\mathscr{H}^3$ which has $\mathbb{C}_\infty$ as a tangent plane at $w$: a *horosphere at* $\infty$ is a set $\{(z, t): t > t_0\}$. If $G$ is discrete and if $\zeta$ is a point in $\mathbb{C}_\infty$ which is fixed by some parabolic element in $G$ we can hence find a horosphere $\Sigma$ at $\zeta$ such that the distinct horospheres $V(\Sigma)$, $V \in G$, are mutually disjoint.

These results remain true in the case of Fuchsian groups (discrete subgroups of **Möb**($\mathscr{H}$) or **Möb**($\Delta$)) although one uses the word *horocycle* instead of horosphere. In fact the Fuchsian case may be considered as a special case of the general discussion by considering the action of a group $G$ on $\{x + yj: y > 0, x \in \mathbb{R}\}$ rather than on $\{x + iy: y > 0, x \in \mathbb{R}\}$.

An analogous situation holds for elliptic (rather than parabolic) fix-points, [**4**].

## 6. THE GEOMETRY OF DISCRETE GROUPS

We now consider a discrete subgroup $G$ of **Möb** and focus our attention on the action of $G$ in $C_\infty$. It is advantageous to consider first certain exceptional elementary groups and then to exclude these from our discussion.

First, if $G$ contains only $I$ and elliptic elements then it is finite. Now suppose that $G$ contains only $I$, elliptic and parabolic elements, and that there is a parabolic element in $G$: we may assume (by conjugation) that $P \in G$ where $P(z) = z + 1$. If $V \in G$ (with coefficients $a, b, c, d$) then $P^n V \in G$ for all integers $n$ and necessarily $\mathrm{tr}^2(P^n V) \in [0, 4]$ for all $n$. Thus $c = 0$, every element of $G$

fixes $\infty$ and we refer the reader to the earlier discussion of the stabilizers in discrete groups.

Finally, we suppose that $G$ contains hyperbolic or loxodromic elements but that all such elements $V$ have a common pair of fixed points, say 0 and $\infty$. If $U \in G$, then $UVU^{-1}$ is a hyperbolic or loxodromic element in $G$ which fixes $U(0)$: thus $U(0) = 0$ or $U(0) = \infty$. Similarly $U(\infty)$ is 0 or $\infty$ and we see that each element of $G$ leaves the pair $\{0, \infty\}$ invariant. Observe that if $U(0) = \infty$ and $U(\infty) = 0$ the $U^2$ fixes $0, \infty$ and also the fixed points of $U$: thus $U$ is elliptic and of order two.

The groups considered above are elementary in character and are well understood ([**6**], Chapter 6). We call these the *elementary groups* and, except where we explicitly state to the contrary, we shall assume that the discrete groups which we consider are not of this type. *If $G$ is not elementary each orbit*

$$G(z) = \{V(z): V \in G\}$$

*is infinite* (as $G$ must contain a hyperbolic or loxodromic element not fixing $z$).

The discussion of the action of a discrete group $G$ in $\mathbb{C}_\infty$ centres around a partitioning of $\mathbb{C}_\infty$ into two sets $L$ and $\Omega$ with quite different properties. We let $F$ be the set of points fixed by some hyperbolic or loxodromic element of $G$ and define the *limit set* $L$ of $G$ as the closure of $F$ in $\mathbb{C}_\infty$ (observe that as $G$ is not elementary $F \neq \varnothing$ and so $L \neq \varnothing$). We define the set $\Omega$ of *ordinary points* of $G$ by $\Omega = \mathbb{C}_\infty - L$.

We now list the essential properties of $L$ and $\Omega$.

(1) *First, $F$ (and hence $L$) is $G$-invariant: $V(L) = L$ and $V(\Omega) = \Omega$ for all $V$ in $G$.* It is only necessary to show that $F$ is $G$-invariant and this is trivial as $V$ fixes $\omega$ if and only if $UVU^{-1}$ fixes $U(w)$.

(2) By definition, *$L$ is closed and $\Omega$ is open.*

(3) *$L$ is the smallest non-empty closed $G$-invariant set.* To see this let $K$ be any non-empty closed $G$-invariant set. As each orbit is infinite, $K$ is an infinite set. Let $V$ be a hyperbolic or loxodromic element in $G$ and suppose $V$ has a fixed point $w$ not in $K$. There exists a neighbourhood $N$ of $w$ disjoint from $K$ and a point $z$ in $K$ not fixed by $V$ and $V^n(z) \to w$ as $n \to +\infty$ or $n \to -\infty$. Thus (for large $n$) $V^n(K)$ $(=K)$ meets $N$, a contradiction.

(4) A family of functions meromorphic in a domain $D$ is said to be *normal* in $D$ if and only if every infinite subfamily contains a sequence of distinct functions which converges uniformly (in the chordal metric) on each compact subset of $D$. A family is normal in $D$ if the family omits three values in $D$ (see Hille [**8**], p. 248 or Lehner [**10**], p. 97).

We now prove that $G$ is normal in $\Omega$ and not normal in any domain containing a point of $L$: *thus $\Omega$ is the largest domain of normality of $G$ and $L$ is the set of non-normality of $G$*. As $G$ is non-elementary, $F$ is infinite; if $D \subseteq \Omega$, then as $\Omega$ is $G$-invariant the family $G$ omits in $D$ all values in $F$. Thus $G$ is normal in $D$. If $D$ is a domain meeting $L$ then $D$ contains a fix-point $\alpha$ of a hyperbolic or loxodromic element $V$ in $G$. Let $\beta$ be the other fix-point of $B$ and suppose that for $z \neq \alpha$, $V^n(z) \to \beta$ as $n \to +\infty$. Then the infinite sequence $V, V^2, V^3, \ldots$ converges pointwise in $D$ to $f$ where $f(\alpha) = \alpha$ and $f(z) = \beta$ when $z \neq \alpha$. As $f$ is not meromorphic in $D$, $G$ is not normal in $D$.

(5) Next, we prove that $G$ is discontinuous in $\Omega$ and not in any domain which meets $L$: *thus $\Omega$ is the maximal domain of discontinuity of $G$*. First, if $D$ is a domain meeting $L$ then $D$ contains a fix-point $w$ of a hyperbolic or loxodromic $V$ in $G$. We take a compact disc $K$ in $D$ and containing $w$ and observe that $V^n(K)$ meets $K$ for all $n$. Thus $G$ is not discontinuous in any domain which meets $L$.

To show that $G$ is discontinuous in $\Omega$ we first assume not: thus there is a compact subset $K$ of $\Omega$ which meets infinitely many of its own $G$-images. This means there is a sequence $z_n$ in $K$ and a sequence $V_n$ in $G$ such that for all $n$, $V_n(z_n) \in K$. By considering subsequences we may assume that $V_n(z_n) \to w$ so $w \in K$ and as $G$ is normal in $\Omega$ we may assume that the sequence $V_n$ converges uniformly on compact subsets of $\Omega$, say to a function $f$. It is known, by the theorem of Hurwitz (see Hille [**8**], p. 205), that either $f$ is a Möbius transformation (which would contradict the discreteness of $G$) or a constant which must, of course, be $w$. Thus there is a sequence $V_n$ in $G$ converging uniformly to $w$ (in $\Omega$) on compact subsets of $\Omega$. We conclude that $w \in F$ and this is again a contradiction (see Lehner [**10**], pp. 73, 74, 97–99).

(6) *For each $z$ in $\mathbb{C}_\infty$ the orbit $G(z)$ accumulates at $L$ and only at $L$.* Obviously if $w$ is fixed by any hyperbolic or loxodromic $V$ we may find a point $z_1$ in $G(z)$ which is not fixed by $V$ and $V^n(z_1) \to w$ as $n \to +\infty$ or $n \to -\infty$. As $G(z_1) = G(z)$, $G(z)$ accumulates at the set $F$ and hence at $L$. Finally, if $G(z)$ accumulates at a point $w$ in $\Omega$, say $V_n(z) \to w$ where the $V_n$ are distinct, then $z \in \Omega$ (else $z \in L$ and so $w \in L$) and a compact set $K$ consisting of the union of $\{z\}$ and a compact neighbourhood of $w$ meets its image $V_n(K)$. This contradicts the discontinuity of $G$ in $\Omega$.

(6′) We conclude that *a necessary and sufficient condition for a discrete group $G$ to be discontinuous in some subdomain of $\mathbb{C}_\infty$ is that some orbit $G(z)$ is not dense in $\mathbb{C}_\infty$.*

(7) An immediate consequence of (6) is that $L$ *is a perfect set* ($L$ is closed and has no isolated points) and hence $L$ is uncountable. The elementary groups are precisely those groups for which $L$ has at most two points.
(8) *Either* $L = \mathbb{C}_\infty$ *or* $L$ *is nowhere dense in* $\mathbb{C}_\infty$ (if $L \neq \mathbb{C}_\infty$, then $\Omega \neq \varnothing$ and we use (6) with $z$ in $\Omega$).
(9) The pairs $(\alpha, \beta)$ of fix-points of hyperbolic or loxodromic elements in $G$ are dense in $L \times L$, [7].

This is an appropriate place to define Fuchsian and Kleinian groups. Let $G$ be any discrete subgroup of **Möb**, in particular $G$ may be elementary. We say that $G$ is a *Fuchsian* group if there is an open disc (or half plane) $\Delta$ in $\mathbb{C}_\infty$ with $G \subset$ **Möb**$(\Delta)$. If so, then for all $V$ in $G$ we have

$$V(\Delta) = \Delta \quad \text{and} \quad V(\partial\Delta) = \partial\Delta:$$

thus by (3), $L \subset \partial\Delta$. We say that $G$ is of the *first kind* if $L = \partial\Delta$, otherwise $G$ is of the *second kind.*

Next, we shall say that $G$ is a *Kleinian* group if $\Omega \neq \varnothing$. [The reader should be warned, however, that some authors call all discrete sub-groups of **Möb** Kleinian and classify these groups as of the first kind if $L = \mathbb{C}_\infty$ and $\Omega = \varnothing$ and of the second kind if $L \subset \mathbb{C}_\infty$ and $\Omega \neq \varnothing$.]

For the remainder of this section we shall assume that $G$ is a Kleinian group. With this assumption we can strengthen an earlier result by showing that

$$\sum_{V \in G} \|V\|^{-4} < +\infty. \tag{6.1}$$

This can be proved directly by selecting a disc $Q$ in $\Omega$ which does not meet any of its $G$-images (such a disc must exist) and then observing that with respect to the chordal metric in $\mathbb{C}_\infty$

$$\sum_{V \in G} \text{area } V(Q) \leqslant \text{area } \mathbb{C}_\infty < +\infty.$$

Explicitly

$$\sum_{V \in G} \iint_Q \frac{|V^{(1)}(z)|^2 \, dx \, dy}{(1 + |V(z)|^2)^2} < +\infty$$

and this yields the desired result (see also [5]). If $\infty \in \Omega$ and is fixed only by

$I$ in $G$ then (13) reduces to the classical result

$$\sum_{V \neq I} |c|^{-4} < +\infty: \tag{6.2}$$

however (6.1) is invariant under conjugation whereas (6.2) is not.

The quotient space $S = \Omega/G$ has a natural analytic structure and is a union of Riemann surfaces: the natural projection $\pi: \Omega \to \Omega/G$ is analytic and the components $S_i$ of $S$ are Riemann surfaces. We define $\Omega_i = \pi^{-1}(S_i)$ and note that $\Omega_i$ need not be connected. We let $\Omega_{i,j}$, $j \in J_i$, be the components of $\Omega_i$ and so

$$\Omega = \bigcup_i \Omega_i = \bigcup_i \left( \bigcup_j \Omega_{ij} \right).$$

The elements of $G$ leave each $\Omega_i$ invariant and permute the components $\Omega_{i,j}$, $j \in J_i$.

Obviously the $\Omega_{i,j}$ are the components of $\Omega$ and $\partial\Omega_{i,j} \subset L$. In fact,

$$\bigcup_{i,j} \partial\Omega_{i,j} \subset L$$

but we need not have equality (see [**1**]): however for each $i$, $\partial\Omega_i = L$ (this follows from (6)).

It is sometimes possible to enlarge $\Omega$ by the addition of certain parabolic orbits. We call a parabolic fix-point $p$ a *cusp* if there is an open disc $Q$ in $\Omega$ which contains $p$ on its boundary and for which $V(Q) = Q$ if $V$ is parabolic in $G$ and fixes $p$ while $V(Q) \cap Q = \varnothing$ for all other $V$ in $G$. In this case we may regard $\pi(p)$ as a puncture on $S$ and $\pi(Q)$ as a deleted neighbourhood of that puncture [2]. The discs $Q$ play the part of the horocycles introduced earlier for Fuchsian groups and these too are called horocycles.

## 7. FUNDAMENTAL DOMAINS

Let $\Sigma$ be a subdomain of $\mathbb{C}_\infty$ or of $\mathbb{R}^3_\infty$: if $E \subset \Sigma$ we denote the closure of $E$ in $\Sigma$ by $\tilde{E}$. We consider a discrete subgroup $G$ of **Möb** such that $\Sigma$ is $G$-invariant and $G$ is discontinuous in $\Sigma$.

7.1 *Definition.* A subset $D$ of $\Sigma$ is a fundamental domain for $G$ in $\Sigma$ if and only if

(a) $D$ is an open subset of $\Sigma$;
(b) each orbit $G(z)$, $z \in \Sigma$, meets $\tilde{D}$ at least once;
(c) each orbit $G(z)$, $z \in \Sigma$, meets $D$ at most once.

Observe that (b) and (c) imply that

$$\Sigma = \bigcup_{V \in G} V(\tilde{D})$$

and

$$V(D) \cap D = \varnothing \quad \text{if} \quad V \in G \quad \text{and } V \neq I.$$

In the case $\Sigma = \mathscr{H}$ there is a classical construction of a fundamental domain $D$ in $\Sigma$ (a similar construction is available when $\Sigma$ is $\mathscr{H}_3$, $\Delta$ or $\Delta_3$). We select any $z^*$ in $\mathscr{H}$ which is not fixed by any element of $G$ except $I$ and define the *Dirichlet region* $D$ with centre $z^*$ by

$$D = \{z \in \mathscr{H} : \rho(z, z^*) < \rho(z, V(z^*)), V \in G, V \neq I\}.$$

This construction is carried out in detail in [**9**], pp. 27–31.

Now let $z_1$ and $z_2$ be distinct points of $\mathscr{H}$: then

$$L(z_1, z_2) = \{z \in \mathscr{H} : \rho(z, z_1) = \rho(z, z_2)\}$$

is a hyperbolic line with $z_1$ and $z_2$ as inverse points with respect to the Euclidean circle determined by this line [**9**], p. 26. The complement of this line in $\mathscr{H}$ is the disjoint union of the half-planes $H(z_1, z_2)$ and $H(z_2, z_1)$, where

$$H(z_1, z_2) = \{z \in \mathscr{H} : \rho(z, z_1) < \rho(z, z_2)\}$$

and $H(z_2, z_1)$ is defined similarly, and each of these half-planes is convex.

Considering again the Dirichlet region $D$ we may now observe that

$$D = \bigcap_{V \in G, V \neq I} H(z^*, V(z^*))$$

and this shows that $D$ is hyperbolically convex.

The boundary of $D$ in $\Sigma$ consists of a countable number of maximal geodesic segments (called *sides*) and these can only accumulate at $L$.

A similar construction is valid when $\Sigma = \Delta$ and in this case a simple calculation shows that $H(O, V^{-1}(O))$ is the set of points in $\Delta$ which lie outside the isometric circle of $V$.

Quite generally, if $\Sigma$ is a $G$-invariant subset of $\Omega$ and if $\infty$ is in $\Omega$ and is fixed by no element of $G$ except $I$ then the *Ford region* (the interior of the set of points which lie outside every isometric circle) is indeed a fundamental domain for $G$, [**6**]. In this case, of course, hyperbolic geometry is not directly available.

The fundamental domains constructed by the method given above have certain desirable properties, for example their boundary consists of circular arcs which are pairwise identified by elements of $G$ and the set of identifying elements generates $G$.

There is another important property which a fundamental domain may or may not have.

7.2 *Definition.* A fundamental domain $D$ is *locally finite* if and only if each compact subset of $\Sigma$ intersects only a finite number of images of $\tilde{D}$.

The constructions given above always yield a locally finite fundamental domain.

It is quite natural to consider $\tilde{D}$ and identify the $G$-equivalent points on the boundary of $\tilde{D}$ thus forming the quotient space $\tilde{D}/G$: however $\tilde{D}/G$ and $\Omega/G$ need not be topologically equivalent (even if $D$ is hyperbolically convex). There is a natural map

$$\theta: \tilde{D}/G \to \Omega/G$$

induced by the inclusion $\tilde{D} \to \Omega$, which is always a 1–1 continuous map of $\tilde{D}/G$ onto $\Omega/G$ and we have the following result.

7.3 THEOREM. *$D$ is locally finite if and only if $\theta$ is a homeomorphism.*

In particular, if $D$ is the Dirichlet region with centre $z^*$ then $\tilde{D}/G$ is topologically the same as $\Omega/G$ and so is independent of $z^*$. The details of these results can be found in [**3**].

## REFERENCES

1. W. Abikoff, Some remarks on Kleinian groups, *in* Ann. Math. Stud. No. 66 (1971).
2. L. V. Ahlfors, Finitely generated Kleinian groups, *Amer. J. Math.* **86** (1964), 413–429.

3. A. F. Beardon, Fundamental domains for Kleinian groups, *in* Ann. Math. Stud. No. 79 (1974) pp. 31–44.
4. A. F. Beardon, On the isometric circles of elements in a Fuchsian group, *J. London Math. Soc.* **10** (1975), 329–337.
5. A. F. Beardon and P. J. Nicholls, On classical series associated with Kleinian groups, *J. London Math. Soc.* **5** (1972), 645–655.
6. L. R. Ford, "Automorphic Functions". Chelsea New York, 1951.
7. L. Greenberg, Discrete subgroups of the Lorentz groups, *Math. Scand.* **10**, (1962), 85–107.
8. E. Hille, "Analytic Function Theory", Vol. II. Ginn and Co., Boston, 1962.
9. J. Lehner, "A Short Course in Automorphic Functions". Holt, Rinehart and Winston, New York, 1966.
10. J. Lehner, "Discontinuous Groups and Automorphic Functions". American Mathematical Society, Rhode Island, 1964.
11. A. Leutbecher, Über Spitzen diskontinuierlicher Gruppen von lineargebrochenen Transformationen, *Math. Z.* **100** (1967), 183–200.
12. H. Shimizu, On discontinuous groups operating on the product of the upper half planes, *Ann. Math.* **77** (1963), 33–71.
13. M. Tsuji, "Potential Theory in Modern Function Theory". Maruzen, Tokyo, 1959.

# 3. Automorphic Forms

J. LEHNER

*University of Pittsburgh, U.S.A.*



## INTRODUCTION

The theory of automorphic forms is the study of functions and differentials that are invariant on discontinuous groups of Möbius transformations acting on the complex sphere. One naturally requires a good deal of information about the groups themselves, but this will be mostly covered in Chapter 2, from which I shall quote freely. This material can be found, for example, in the textbooks of Ford [**9**], Kra [**12**], Lehner [**14**] and Siegel [**30**].

The theory of discontinuous groups in the plane has now grown to vast proportions, and these lectures will consider only Fuchsian groups. Even so, only a small part of the known theory can be covered here, and the selection of topics necessarily reflects the writer's personal taste. The material that follows is approximately what was presented in lectures at the Cambridge Conference in July 1975. An expanded version of this material will be published in the University of Pittsburgh Lecture Notes Series.

The author acknowledges with thanks the critical remarks and suggestions of T. A. Metzger.

## 1. FUCHSIAN GROUPS

**1.1** We consider, almost always, Fuchsian groups on the unit disc $\mathscr{U} = \{z = x + iy : |z| < 1\}$. The basic definitions and results are contained in Chapter 2. Corresponding notations are:

| | Beardon | Lehner |
|---|---|---|
| unit disc | $\Delta$ | $\mathscr{U}$ |
| upper half-plane | $\mathscr{H}$ | $\mathscr{H}$ |
| hyperbolic- | $h$- | $H$- |
| $H$-metric in $\mathscr{U}$ | $2\lvert dz\rvert/(1 - \lvert z\rvert^2)$ | $\lvert dz\rvert/(1 - \lvert z\rvert^2)$ |
| Dirichlet region with centre $z_0$ | $D$ | $\Omega(z_0)$; $\Omega(0) \equiv \Omega$ |
| $H$-distance from $z$ to $w$ | $\rho(z, w)$ | $d(z, w)$ |
| Euclidean disc centre $a$, radius $r$ | — | $\Delta(a, r)$ |
| $H$-disc, centre $a$, radius $\rho$ | — | $D(a, \rho)$ |
| $H$- area of $E$ | — | $\sigma(E)$ |

We shall require the hyperbolic area ($H$-area) in $\mathscr{U}$ which is denoted by $\sigma$. Thus $\sigma(E) = \iint_E (1 - |z|^2)^{-2}\, dx\, dy$. We have

$$\sigma(\Delta(0, r)) = \frac{\pi r^2}{1 - r^2}, \qquad \sigma(D(0, \rho)) = \pi \sinh^2 \rho. \tag{1.1.1}$$

**1.2** The material of this Section is not contained in Chapter 2 and we give complete proofs.

We introduce the orbital counting function

$$n_G(S, z) \equiv n(S, z) := \{gz : gz \in S\}, \qquad S \subset \mathscr{U}, \qquad z \in \mathscr{U}. \tag{1.2.1}$$

In particular, for $S = \Delta(0, r) = \{|z| < r\}$ we write

$$n(\Delta(0, r), z) \equiv n(r, z) = \{gz : |gz| < r\}. \tag{1.2.2}$$

Since

$$n(r, z) = \operatorname{card}\{Gz \cap \Delta_r\},$$

we see that $n(r, z)$, regarded as a function of $z$, is $G$-invariant.

An estimate of $n(r, z)$ due to Tsuji ([**31**], p. 516) will prove quite useful. It holds for all Fuchsian groups.

Let $\rho = d(0, r)$ and let $D_\rho \equiv D(0, \rho)$ be the $H$-disc of radius $\rho$; then $n(r, z) = \{gz: gz \in D_\rho\}$. Let $\delta > 0$ be so small that $D_\delta$ is contained in $\Omega$, the fundamental region. $Az \in D_\rho$ if and only if $d(Az, 0) = d(z, A^{-1}0) < d(0, \rho)$, so $n(r, z) = \{A0: A0 \in D(z, \rho)\}$. If $A0 \in D(z, \rho)$, then $A(D_\delta) \subset D(0, \rho + \delta) = D_{\rho+\delta}$. Moreover, all $A(D_\delta)$ are disjoint. Hence $\sigma(D_\delta). n(r, z) \leqslant \sigma(D(0, \rho + \delta))$. or

$$(1.2.3) \qquad n(r, z) < \sinh^2(\rho + \delta)/\sinh^2\delta < e^{2\rho}/(1 - e^{-2\delta})^2;$$

hence we get:

1.2.4 THEOREM. *For $z \in \mathscr{U}$, $0 < r < 1$, we have*

$$n(r, z) < m/(1 - r), \qquad \textit{where} \quad m = 2(1 - e^{-2\delta})^{-2}.$$

It is noteworthy that $m$ does not depend on $z$. We distinguish between $n(r, z)$ and

$$(1.2.5) \qquad n_0(r, z) = \{A \in G: |Az| < r\}.$$

These two functions are the same if $z$ is not a fixed point. In any case we have

$$(1.2.6) \qquad n_0(r, z) = |(\operatorname{stab} z)| \,.\, n(r, z),$$

where $|(\operatorname{stab} z)|$ is the order of the stabilizer of $z = \{A \in G: Az = z\}$ and is finite, because of discontinuity (See Chapter 2, Section 5.1).

For groups with Dirichlet region of finite area we can get a lower bound.

1.2.7 THEOREM. ([**31**], p 518). *If $\sigma(\Omega) < \infty$,*

$$n(r, z) > C(z)/(1 - r), \quad \textit{with} \quad C(z) > 0.$$

*If $|z| \leqslant \rho$ we can replace $C(z)$ by $C_0(\rho)$.*

*Proof.* Let $\varepsilon > 0$, $0 < r < 1$. Define

$$\textstyle\sum_0 := \{z \in \Omega: |z| > r_0\},$$

where $r_0$ is chosen so that

$$\rho < r_0 < 1, \qquad \sigma(\Sigma_0) < \varepsilon.$$

Put

$$\Sigma = \bigcup_{A \in G} A(\Sigma_0), \Sigma_r = \Sigma \cap \Delta_r.$$

By Theorem 1.2.4

$$\sigma(\Sigma_r) = \iint_{\Sigma_0} n(r, z)\, d\sigma(z) \leqslant \frac{m}{1-r}\sigma(\Sigma_0) \leqslant m(\rho)\frac{\varepsilon}{1-r}. \tag{1.2.8}$$

Now let

$$T_0 = \Omega - \Sigma_0, \qquad T = \bigcup_{A \in G} A(T_0), \qquad T_r = T \cap \Delta_r.$$

Since $\Sigma_r \cup T_r = \Delta_r$ and $\Sigma_r \cap T_r = \varnothing$, we get

$$\sigma(\Sigma_r) + \sigma(T_r) = \sigma(\Delta_r) \geqslant \frac{m}{1-r},$$

so that by 1.2.8,

$$\sigma(T_r) \geqslant \frac{m(\rho)}{1-r}$$

if $\varepsilon$ is sufficiently small.

Let $\delta_0 = \sup_{z, w \in T_0} d(z, w)$ the *H-diameter* of $T_0$ and let $\rho$ be such that $D_\rho = \Delta_r$. If $A(T_0)$ intersects $\partial\Delta_r = \partial D_\rho$, then $A(T_0)$ lies within $D(0, \rho + \delta_0)$. Let $r'$ be such that $d(0, \rho + \delta_0) = \Delta(0, r')$; an easy calculation yields

$$1 - r' \geqslant m(1 - r).$$

Hence

$$m \,.\, n(r', z) \geqslant n(r', z)\, \sigma(T_0) \geqslant \sigma(T_r) \geqslant \frac{m(\delta_0)}{1-r} \geqslant \frac{m(\delta_0)}{1-r'}.$$

Replace $r'$ by $r$. ∎

## 2. POINCARÉ SERIES

**2.1** In his search for analytic invariants of Fuchsian groups Poincaré was led to introduce his famous series. Let $q > 0$ be an integer and

$$f(z) = \sum_{A \in G} (A'z)^q, \qquad A'z \equiv A'(z) \tag{2.1.1}$$

the order of the terms being specified. If the series converges absolutely, the order of the terms is irrelevant and we may perform the following calculation for $L \in G$,

$$\begin{aligned} f(Lz)\,(L'z)^q &= \sum_{A \in G} (A'L(z))^q\,(L'z)^q \\ &= \sum_{A} ((AL)'(z))^q = f(z). \end{aligned} \tag{2.1.2}$$

If the series converges uniformly on compact subsets of $\mathscr{U}$, $f$ is analytic in $\mathscr{U}$. Thus, in the presence of absolute and uniform convergence, $f(z)\,(dz)^q$ is a $G$-invariant holomorphic differential in $\mathscr{U}$. These are called automorphic forms of weight $q$ and will be studied systematically in a later section. At this point we consider the Poincaré series for its own sake.

**2.2** *Definition*

$$\Theta_q(z) = \sum_{A \in G} (A'z)^q, \qquad \Phi_s(z) = \sum_{A \in G} |A'z|^s, \tag{2.2.1}$$

$$\Psi_s(z) = \sum_{A \in G} (1 - |Az|^2)^s$$

which are called the Poincaré series, absolute Poincaré series, and invariant Poincaré series, respectively. Here $q$ is an integer but $s$ is merely real. Note the relations (see Chapter 2, p. 46.)

$$(1 - |z|^2)^q\,|\Theta_q(z)| \leqslant (1 - |z|^2)^q\,\Phi_q(z) = \Psi_q(z). \tag{2.2.2}$$

2.2.3 THEOREM. *If the series for $\Phi_s(z)$ converges for one $z$ in $\mathscr{U}$, it converges for all $z$ in $\mathscr{U}$ and uniformly on compact subsets. The same statement is true of $\Psi_s(z)$.*

*Proof.* We first prove.

2.2.4 LEMMA. *If* $Tz = (\alpha z + \bar{\gamma})/(\gamma z + \bar{\alpha}) \in \mathbf{Möb}\,\mathscr{U}$, *then*

$$\frac{1-|z|}{2}(1-|T0|^2) \leqslant 1-|Tz|^2 \leqslant \frac{2}{1-|z|}(1-|T0|^2),\ z \in \mathscr{U}.$$

In fact, since $1-|Tz|^2 = (1-|z|^2)|T'(z)|$,

$$\frac{1-|Tz|^2}{1-|T0|^2} = (1-|z|^2)\left|\frac{T'z}{T'0}\right| = (1-|z|^2)\left|\frac{\bar{\alpha}}{\gamma^z+\bar{\alpha}}\right|^2$$

$$= (1-|z|^2)/|\gamma z/\bar{\alpha}+1|^2,$$

and since $\alpha\bar{\alpha} - \gamma\bar{\gamma} = 1$,

$$1-|z| \leqslant |1+\gamma z/\bar{\alpha}| \leqslant 1+|z|.$$

If the series for $\Psi_s$ converges at $z_0$, it therefore converges at 0 and so at any $z$. Since $1-|z|$ is bounded away from zero on compact subsets, the convergence is uniform on compact subsets, and the theorem is proved for $\Psi_s$ and so, by 2.2.2, for $\Phi_s$.

*Definition.* We say $G$ is of *divergence type* (DT) or *convergence type* (CT) according as

$$\Psi_1(z) = \sum_{A \in G}(1-|Az|^2)$$

diverges or converges. By the preceding theorem the above series converges uniformly on compact subsets when $G$ is CT.

Tsuji has proved ([**31**], *p.* 522): *G is* CT *if and only if the Riemann surface* $\mathscr{U}/G$ *has a Green's function (i.e. is in the class* $P_G$).

*Definition.* We define the *exponent of convergence* $s_0$ of $\Phi_s$:

$$s_0 = \inf\{s > 0 : \Phi_s(z) \text{ converges for } z \in \mathscr{U}\}.$$

Of course $s_0$ is also the exponent of convergence of $\Psi_s$.

In order to investigate convergence we make use of a relation between $\Psi_s$ and $n(r, z)$. Instead of $\Psi_s$ we shall often treat $\Sigma(1 - |Az|)^s$. Since

$$1 - |Az| \leqslant 1 - |Az|^2 \leqslant 2(1 - |Az|),$$

the two series have the same convergence properties.

2.2.5 THEOREM. For $s > 0$, $|z| < 1$,

$$n(r, z)(1 - r)^s - n(0, z) + s \int_0^r n(t, z)(1 - t)^{s-1}\, dt = h(z) \sum_{|Az|<r} (1 - |Az|)^s,$$

*where* $h(z) = 1/|(\text{stab } z)|$.

*Proof.* The formula follows by integration by parts from

$$\int_0^r (1 - t)^s \, dn_0(t, z) = \sum_{\substack{A \\ |Az|<r}} (1 - |Az|)^s$$

and 1.2.6. ■

2.2.6 THEOREM. *If $s > 1$ the series for $\Psi_s$ converges uniformly on compact subsets of $\mathscr{U}$. Moreover,*

$$\Psi_s(z) \leqslant m \cdot 2^s s/(s - 1) \qquad \textit{for} \quad |z| < 1,$$

*where m is a positive constant depending only on G.*

*Proof.* By Tsuji's, estimate, 1.2.4, we have

$$\int_0^r n(t, z)\,(1 - t)^{s-1}\, dt < m \int_0^r (1 - t)^{s-2}\, dt \leqslant m/(s - 1);$$

hence by 2.2.4 we have for all $z$ in $\mathscr{U}$,

$$h(z) \sum_{\substack{A \\ |Az|<r}} (1 - |Az|)^s \leqslant m(1 - r)^{s-1} + ms/(s - 1) < m + ms/(s - 1). \tag{2.2.7}$$

Thus the series converges at each point of $\mathscr{U}$ and, by Theorem 2.2.3 the convergence is uniform on compact subsets. Hence $\Psi_s(z)$ is a continuous function of $z$ in $\mathscr{U}$.

Now let $F$ be the set of $z \in \mathscr{U}$ that are fixed by some element $\neq 1$ of $G$. $F$ is a countable set and $1/(h(z) \equiv |(\text{stab } z)| = 1$ for $z \in \mathscr{U} - F$. For these $z$ we let $r \to 1$ in 2.2.7 to find that

$$|\Psi_s(z)| \leqslant m \,.\, 2^s \, s/(s-1).$$

By continuity this applies to all $z$ in $\mathscr{U}$. ■

Thus $s_0 = 1$ for a group of DT. On the other hand we know only that $s_0 \leqslant 1$ for a CT group. In fact, Beardon [**1**] has shown that $s_0 < 1$ for every finitely generated group of the second kind. By definition a group of DT diverges at its exponent of convergence. Very recently Patterson [**19**] has shown that this is also true for certain groups of CT, namely, finitely generated groups of the second kind.

**2.3** Let

$$e_0 = \overline{\Omega} \cap \partial\mathscr{U}.$$

$e_0$ is measurable; let $\mu(e_0)$ be the linear measure of $e_0$.

2.3.1. THEOREM. *If* $\mu(e_0) > 0$, $G$ *is* CT *and*

$$\sum_{A \in G} (1 - |A0|) < 4\pi/\mu(e_0).$$

*In particular, a group of the second kind is* CT.

*Proof.* We have $Ae_0 \subset \partial\mathscr{U}$ and $\mu(e_0) = \mu(Ae_0)$ for all $A \in G$. Let

$$P(z, \zeta) = \frac{1 - |z|^2}{|z - \zeta|^2} \tag{2.3.2}$$

be the Poisson kernel. It is easy to check that $P|d\zeta|$ is **Möb** $\mathscr{U}$-invariant:

$$P(Tz, T\zeta)\,|dT\zeta| = P(z, \zeta)\,|d\zeta|. \tag{2.3.3}$$

Set

$$u(z) = \int_{e_0} P(z, \zeta)\,|d\zeta|, \qquad z \in \mathcal{U}.$$

Then

$$u(0) = \mu(e_0) = \int_{e_0} P(0, \zeta)\,|d\zeta| = \int_{Ae_0} P(0, A^{-1}\zeta)\,|dA^{-1}\zeta|$$

$$= \int_{Ae_0} P(A0, \zeta)\,|d\zeta| \leqslant \frac{1 - |A0|^2}{(1 - |A0|)^2}\,\mu(Ae_0)$$

$$= \frac{1 + |A0|}{1 - |A0|}\,\mu(Ae_0) < \frac{2\mu(Ae_0)}{1 - |A0|}.$$

Hence

$$\sum_A (1 - |A0|) < \frac{2}{\mu(e_0)} \sum_A \mu(Ae_0) \leqslant \frac{4\pi}{\mu(e_0)}. \qquad \blacksquare$$

*Exercise.* Prove that $\sum_A |A'z|$ converges for almost all $z$ in $e_0$. (Show that the integral of the series over $e_0$ is finite.)

2.3.4 THEOREM. *A group of finite invariant area is* DT.

For the proof we use Tsuji's lower bound, 1.2.7, in the formula 2.2.5. Clearly, when $s = 1$ the integral diverges for fixed $z$ as $r \to 1$. ■

2.3.5 THEOREM. *If G is* DT *there is no bounded nonconstant analytic automorphic function on G. There is no bounded nonconstant harmonic automorphic function on G.* (*In another notation* $\mathcal{U}/G \subset O_{AB}$, $\mathcal{U}/G \subset O_{HB}$.)

The first assertion is easy, for a bounded analytic function $f$ is in class $N$ and its zeros $z_n$ satisfy $\sum(1 - |z_n|) < \infty$ (*See* P. Duren, "Theory of $H^p$ Spaces)", Academic Press, New York, 1970, p. 18). Let $g(z) = f(z) - f(0)$; $g$ is bounded, automorphic, and $\{A0\}$ are zeros of $g$. Hence $\sum(1 - |A0|) < \infty$ and $G$ is CT.

The proof of the second statement can be found in [**31**], p. 515. ■

2.3.6 THEOREM. *If $G$ is* DT, *there is no $G$-invariant measurable set $S \subset \partial U$ with* $0 < \mu(S) < 2\pi$.

This is a weak form of "metric transitivity". Let $S$ be such a set and let

$$u(z) = \frac{1}{2\pi}\int_S P(z, \zeta)\,|d\zeta|.$$

Then $u$ is harmonic in $\mathcal{U}$ and bounded, $|u| \leqslant 1$, and $u$ is automorphic:

$$u(Az) = \frac{1}{2\pi}\int_S P(Az, \zeta)\,|d\zeta| = \frac{1}{2\pi}\int_{A^{-1}S} P(Az, A\zeta)\,|dA\zeta| = u(z),$$

by 2.3.3 and the invariance of $S$. By Theorem 2.3.5, $u$ = constant, and $u \equiv 1$ since $u(z) \to 1$ as $z$ tends radially to a point of $S$. Thus

$$1 = u(0) = \frac{1}{2\pi}\int_S |d\zeta| = \frac{\mu(S)}{2\pi}. \qquad \blacksquare$$

## 3. AUTOMORPHIC FORMS

**3.1** *Historical. An automorphic form* of integral weight $q$ on the group $G$ is a function $f$ that

(i) is meromorphic in $\mathcal{U}$,
(ii) satisfies the functional equation

$$f(Az)\,(A'z)^q = f(z), \qquad z \in \mathcal{U}, \qquad A \in G. \tag{3.1.1}$$

The second property may be expressed by saying that the differential $f\,dz^q$ is $G$-invariant. An *automorphic function* is a form of weight 0.

This definition is too broad. It would allow very strong singularities on approach to $\partial\mathcal{U}$. The classical restriction is to require that $f(z)$ tends to a definite limit, finite or infinite, as $z$ tends to a parabolic cusp within a parabolic sector.

This restriction makes more sense if we consider it on the Riemann surface $\mathcal{U}/G$. An automorphic form $f$ of weight $q$ on $\mathcal{U}$ projects to a meromorphic differential $\hat{f}$ of weight $q$ on $\mathscr{U}/G$ by the formula

$$\hat{f}(\zeta)\,(\mathrm{d}\zeta)^q = f(z)\,(\mathrm{d}z)^q, \qquad \text{with} \quad \zeta = \pi z \tag{3.1.2}$$

where $\pi: \mathscr{U} \to \mathscr{U}/G$ is the projection map. Conversely, a meromorphic differential of weight $q$ on $\mathscr{U}/G$ lifts to an automorphic form of weight $q$ on $\mathscr{U}$. Let $\zeta_0 = \pi z_0$ be the projection of a parabolic cusp $z_0 \in \partial\mathscr{U}$. A pole of $\hat{f}(\zeta)$ at $\zeta_0$ translates into a left-finite expansion for $f$ in the correct "local variable" at $z_0$. The latter condition, however, is equivalent to the restriction stated above. Thus, in the classical treatment, an automorphic form was defined as a $G$-invariant function meromorphic in $\mathscr{U}$ and at the parabolic cusps on $\partial\mathscr{U}$.

The existence problem for automorphic forms was solved classically in two different ways. Klein used the existence theorem for meromorphic differentials on a Riemann surface, a theorem that was the all-consuming interest of a series of mathematicians between 1870 and 1910. As we have already noted, a meromorphic differential on $\mathscr{U}/G$ lifts to a meromorphic automorphic form on $\mathscr{U}$.

Poincaré invented the series that now bears his name. If $F$ is holomorphic on $\mathscr{U}$, define

$$\mathbf{\Theta}_q F = \sum_{A \in G} F(Az)\,(A'z)^q, \qquad q \text{ an integer} \geqslant 1, \tag{3.1.3}$$

called the *Poincaré* or $\mathbf{\Theta}$ *series* of $F$. Assuming the series converges absolutely we calculate exactly as in 2.1.2:

$$\begin{aligned} \mathbf{\Theta}_q F(Lz)\,(L'z)^q &= \sum_{A \in G} F(ALz)\,(A'Lz)^q\,(L'z)^q \\ &= \sum_{A} F(ALz)\,((AL)'z)^q = \mathbf{\Theta}_q F(z), \qquad \text{for } L \in G. \end{aligned} \tag{3.1.4}$$

If the series converges uniformly on compact subsets of $\mathscr{U}$, $\mathbf{\Theta}_q F$ will be holomorphic on $\mathscr{U}$. This would seem to settle the question of existence.

However, $\mathbf{\Theta}_q F$ might be identically zero! This can actually happen and Poincaré gave examples [**14**], p. 184. He overcame this difficulty by allowing $F$ to have poles. Suppose $\alpha \in \mathscr{U}$ is not a fixed point of the group $G$ and let $F(z) = (z - \alpha)^{-k}$. The term in $\mathbf{\Theta}_q F$ corresponding to the identity transformation of $G$ is $(z - \alpha)^{-k}$ and no other term in the series has a singularity at $\alpha$. Thus $\mathbf{\Theta}_q F$ has a pole at $\alpha$ and cannot be identically zero. Two forms of the same weight with poles of different orders at non-equivalent points yield, on division, a non-constant automorphic function.

After the classical period the theory was greatly expanded by Hecke and Petersson. They confined themselves to finitely generated groups of the first kind. These are exactly the cases in which the group has finite invariant area. Moreover, $q$ was positive and mostly $q > 1$.

*Definition.* A function that is holomorphic in $\mathscr{U}$ and vanishes at all cusps is called a *cusp form.* The vector space of cusp forms of weight $q$, denoted by $C^0(G, q)$, is of finite dimension. In fact, the dimension is easily calculated by applying the Riemann–Roch theorem to a suitable divisor on the surface $\mathscr{U}/G$ (see Shimura [**29**], pp. 45–50):

$$\text{(3.1.5)}\quad \dim C^0(G,q) = (2q-1)(g-1) + (q-1)m + \sum_j \left[ q\left(1 - \frac{1}{\ell_j}\right) \right], \quad q > 1$$

where $g$ is the genus of $\mathscr{U}/G$, $m$ is the number of inequivalent cusps and $\{\ell_j\}$ the orders of the inequivalent elliptic vertices in a fundamental region for $G$.

In 1939 Petersson [**21**] made a decisive advance by introducing a scalar product in the space of cusp forms, thereby making it a *Hilbert* space. The new space is still finite-dimensional and was treated by Petersson by the methods of linear algebra. His first success was a notable strengthening and simplification of the theory of Hecke's operators on the modular group (see Chapter 11). Petersson showed that every cusp form is the Poincaré series of a polynomial, and that a linear relation among $\{\boldsymbol{\Theta}_q z^k\}$ is induced by a similar relation among its Taylor coefficients; in particular this applies to the identical vanishing of the series $\boldsymbol{\Theta}_q z^k$. Many other results were proved.

In 1960 [**3**] and more definitively in 1965 [**4**], Bers introduced what are now called the Bers spaces $A_q^p(G)$, $1 \leqslant p \leqslant \infty$. These are *Banach* spaces of holomorphic automorphic forms on *arbitrary* groups $G$. When $G$ has infinite invariant area (and $q > 1$), these spaces are infinite-dimensional. The classical restriction on boundary behaviour of a form $f$, mentioned above, is replaced by a growth condition on $f$, namely, the existence of a certain integral that acts as a norm. The previous theory is the case $p = 2$.

From now on we confine ourselves to the Bers spaces.

**3.2** *Definition.* Let $q \geqslant 1$ be an integer. We say $f \in A_q^p(G)$, $1 \leqslant p \leqslant \infty$, if $f$ is holomorphic in $\mathscr{U}$,

$$\text{(3.2.1)}\qquad f(Az)(A'z)^q = f(z) \qquad \text{for} \quad A \in G, \quad z \in \mathscr{U},$$

and

$$\|f\|_p^p := \iint_W \lambda^{-pq}(z)\,|f(z)|^p \, d\omega(z) < \infty, \qquad p < \infty \tag{3.2.2}$$

$$\|f\|_\infty := \sup_{z \in \mathscr{U}} \lambda^{-q}(z)\,|f(z)| < \infty,$$

where $W$ is an arbitrary fundamental region having the properties of $\Omega$ (see pp. 70–71). It is seen that the integral in 3.2.2 is independent of the choice of $W$. $A_q^p(G)$ is a Banach space with norm $\|f\|_p$. (To see this for $p < \infty$, note that convergence in $\mathscr{L}^p$ (with the measure $\lambda^{-pq}\,d\omega$) implies uniform convergence on compact subsets, and every compact subset can be covered by a finite number of fundamental regions. When $p = \infty$ we have uniform convergence directly.) $A_q^p(G)$, $p < \infty$, is called the space of *p-integrable* forms (of weight $q$); $A_q^\infty(G)$, the space of *bounded* forms. If necessary we write $\|f\|_{p,G}$, or even $\|f\|_{q,p,G}$. $A_q^1(G)$ is abbreviated to $A_q(G)$. Some authors use $A_q^p(\mathscr{U}, G)$ instead of $A_q^p(G)$.

The above definitions apply to any Fuchsian group $G$. When $G = \{\mathrm{id}\}$ we write $A_q^p$ or $A_q^p(\mathscr{U})$ for $A_q^p(\mathrm{id})$. Also we usually write $B_q(G)$, $B_q$ for $A_q^\infty(G)$, $A_q^\infty$. Clearly $A_q^p(G) \cap A_q^p = \{0\}$ if $G$ is an infinite group, whereas $B_q(G)$ is a closed subspace of $B_q$.

The restriction that $q$ be integral was introduced to avoid difficulty with the branches of the functions in equations like

$$(A \circ B)'(z))^q = (A' \circ B(z))^q \,(B'(z))^q.$$

In the case of $A_q^p$ this problem does not arise and we can define $A_t^p$ for all $t > 0$. Trivially we have

$$A_t^p \subset A_{t+h}^p \qquad \text{for} \quad h > 0.$$

The elements of $A_q^p(G)$ are *cusp forms* in the following sense.

3.2.3 THEOREM. *If $f \in A_q^p(G)$, $q \geqslant 1$, $1 \leqslant p \leqslant \infty$, then*

$$f(z) \to 0$$

*as $z \to z_0$, a parabolic cusp, within a cusp sector.*

*Proof.* Let stab $z_0 = \langle P \rangle$. Map $\mathscr{U}$ to $\mathscr{H}$, the upper half-plane, by a Möbius transformation $w = Tz = u + iv$ with $Tz_0 = \infty$. Then $G_1 = TGT^{-1}$ is a Fuchsian group acting on $\mathscr{H}$, $\infty$ is a cusp on $G_1$, and stab $\infty = \langle S \rangle$, where $S = TPT^{-1}$ is a translation, which we may assume is $Sz = z + 1$. Also $T\Omega \equiv \Omega_1$ is a fundamental region for $G_1$ and contains a strip $\{w: \xi < u < \xi + 1, v > v_0 \geqslant 1\}$. To make $T$ an isometric map, define the Poincaré metric on $\mathscr{H}$ by

$$\lambda_{\mathscr{H}}|\mathrm{d}w|: = \lambda|\mathrm{d}z| = (2v)^{-1}|\mathrm{d}w|; \; \mathrm{d}\omega_{\mathscr{H}}(w) = \lambda_{\mathscr{H}}^2\, \mathrm{d}u\, \mathrm{d}v = (4v^2)^{-1}\, \mathrm{d}u\, \mathrm{d}v. \tag{3.2.4}$$

$T$ carries $A_q^p(G)$ onto $A_q^p(G_1)$ by

$$\hat{f}(w): = f(z)(\mathrm{d}z/\mathrm{d}w)^q. \tag{3.2.5}$$

Finally $T$ is norm preserving

$$\|\hat{f}\|_{p,G_1} = \|f\|_{p,G}.$$

Let $\hat{f} \in A_q^p(G_1)$, $1 \leqslant p \leqslant \infty$. Since $S \in G_1$ we have $\hat{f}(w + 1) = f(w)$ and so

$$\hat{f}(w) = \sum_{-\infty}^{\infty} \hat{a}_n\, \mathrm{e}^{2\pi i n w}.$$

Here

$$\hat{a}_n = \int_{\xi}^{\xi+1} \hat{f}(w)\, \mathrm{e}^{-2\pi i n w}\, \mathrm{d}w,$$

and for $n \leqslant 0$, $h > h_0 > v_0$, $m = m(G, p)$,

$$|\hat{a}_n| \int_{h_0}^{h} v^{q-2}\, \mathrm{d}v \leqslant \int_{h_0}^{h} \int_{\xi}^{\xi+1} v^{q-2} |\hat{f}(w)|\, \mathrm{e}^{2\pi n v}\, \mathrm{d}u\, \mathrm{d}v \leqslant m\|\hat{f}\|_{p,G}, \tag{3.2.6}$$

by Hölder's inequality ($1 < p < \infty$) or directly ($p = 1, \infty$).

Also

$$C_q(h): = \int_{h_0}^{h} v^{q-2}\, \mathrm{d}v = (h^{q-1} - h_0^{q-1})/(q-1), \qquad q > 1$$

$$= \log h/h_0, \qquad q = 1$$

and in both cases $C_q(h) \to \infty$ for $h \to \infty$. Thus

$$\hat{a}_n = O(1/C_q(h)) \to 0, \qquad h \to \infty. \tag{3.2.7}$$

It follows that $\hat{a}_n = 0$ for $n \leqslant 0$, so $\hat{f} = O(e^{-2\pi v})$, $v \to \infty$. Hence

$$f(z) = \hat{f}(w)(dw/dz)^q \to 0 \quad \text{as} \quad z \to z_0. \qquad \blacksquare$$

The important *scalar product* will now be defined.

*Definition.* The scalar product of $f \in A_q^p(G)$ and $g \in A_g^{p'}(G)$, $1/p + 1/p' = 1$, $1 \leqslant p < \infty$, is defined by

$$(f, g; \Omega) \equiv (f, g) := \iint_{\Omega} f(z)\, \bar{g}(z)\, \lambda^{-2q}(z)\, d\omega(z). \tag{3.2.8}$$

It is seen that $(f, g)$ is independent of the choice of $\Omega$. The existence of 3.2.8 follows from Hölder's inequality applied to $f\lambda^{-q}$, $g\lambda^{-q}$ when $p > 1$, and directly when $p = 1$. If we write $(f, g) = \ell_g(f)$, then $\ell$ is a continuous linear functional on $A_q^p(G)$ of norm $\leqslant \|g\|$. Let $X^*$ denote the space dual to $X$.

The main theorem here is

3.2.8 THEOREM. *For $1 \leqslant p < \infty$ the map $\psi : A_q^{p'}(G) \to (A_q^p(G))^*$ defined by*

$$g \mapsto \ell_g(f) := (f, g), \tag{3.2.9}$$

*is an anti-linear (surjective) isomorphism (see Bers [4] and Earle [7]).*

**3.3** Before proving Theorem 3.2.8 we shall develop the main facts about Poincaré series.

*Definition.* Let $F$ be a complex-valued function on $\mathcal{U}$, $q$ an integer $\geqslant 1$. We say $\mathbf{\Theta}_q F$, the Poincaré series of $F$, exists if

$$\mathbf{\Theta}_q F(z) := \sum_{A \in G} F(Az)(A'z)^q \tag{3.3.1}$$

converges absolutely and uniformly on compact subsets of $\mathcal{U}$.

Until further notice we assume

$$q > 1. \tag{3.3.2}$$

3.3.3 THEOREM. *The map* $\Theta_q : F \mapsto \Theta_q F$ *is a surjective continuous linear operator from* $A_q$ *to* $A_q(G)$ *of norm* $\leqslant 1$.

3.3.4 THEOREM. *Let* $F \in A_q$, $g \in B_q(G)$. *Then*

$$(\Theta_q F, g; \Omega) = (F, g; \mathscr{U}) := \iint_{\mathscr{U}} F(z)\, \bar{g}(z)\, \lambda^{-2q}(z)\, d\omega(z).$$

3.3.5 THEOREM. (Completeness). *For* $1 \leqslant p \leqslant \infty$, *every* $f \in A_q^p(G)$ *is of the form*

$$f = \Theta_q F$$

*for some* $F \in A_q^p$.

3.3.6 THEOREM (Scalar Product Formula). *For* $g = \Sigma b_n z^n \in A_q^p(G)$, $1 \leqslant p \leqslant \infty$

$$(\Theta_q z^k, g) = \pi k!\, \frac{\Gamma(2q+1)}{\Gamma(2q+k)}\, \bar{b}_k.$$

3.3.7 THEOREM *Let* $\mathscr{P}$ *be the set of polynomials. Then* $\Theta_q \mathscr{P}$ *is dense in* $A_p^q(G)$, $1 \leqslant p < \infty$, $q > 1$.

**3.4** The proofs to be given in this and the next few sections make use of material in [**4**], [**5**], [**6**], [**7**], and [**12**].

We begin the proof of Theorem 3.3.3. The following method of integration is frequently applied.

Let $F \in A_q$. We have, by absolute convergence

$$\infty > \|F\|_1 = \iint_{\mathscr{U}} \lambda^{-q}(z)\, |F(z)|\, d\omega(z) = \sum_{A \in G} \iint_{A\Omega} \cdots \tag{3.4.1}$$

$$= \sum_{A} \iint_{\Omega} \lambda^{-q}(Az)\, |F(Az)|\, d\omega(Az)$$

$$\|F\|_1 = \iint_{\Omega} d\omega(z) \sum_A \ldots = \iint_{\Omega} d\omega(z) \lambda^{-q}(z) \sum_A |F(Az)|\, |A'z|^q$$
$$\geqslant \iint \lambda^{-q}(z) |\mathbf{\Theta}_q F(z)|\, d\omega(z).$$

Thus $\mathbf{\Theta}_q F$ converges in $\mathscr{L}^1(\Omega, d\omega)$—the Lebesgue space with domain $\Omega$ and measure $d\omega$. By the argument following 3.2.2, $\mathbf{\Theta}_q F$ converges absolutely and uniformly on compact subsets of $\mathscr{U}$. This implies that $A_q$ is in the domain of $\mathbf{\Theta}_q$. Moreover, by absolute convergence

$$\mathbf{\Theta}_q F(Lz)(L'z)^q = \sum_A F(ALz)(A'Lz)^q(L'z)^q = \sum_A F(ALz)((AL)'(z))^q$$
$$= \mathbf{\Theta}_q F(z),$$

so that $\mathbf{\Theta}_q F$ is automorphic on $G$ of weight $q$. By uniform convergence, $\mathbf{\Theta}_q F$ is analytic in $\mathscr{U}$. Finally, since the rightmost member of 3.4.1 is the series of $\mathbf{\Theta}_q F$, we have that $\mathbf{\Theta}_q F \in A_q(G)$. Equation 3.4.1 states that $\|\mathbf{\Theta}_q F\|_{1,G} \leqslant \|F\|_1$ and completes the proof of all statements of Theorem 3.3.3 except for the surjectivity.

We shall need the following inclusions.

3.4.2 LEMMA. (i) $A_q^p(G) \subset A_{q+h}$ *for* $h > 1$, $1 \leqslant p \leqslant \infty$,

(ii) $A_q(G) \subset A_{q+h}$ *for* $h \geqslant 1$ *if* $G$ *is* $CT$,

(iii) $A_q^p(G) \subset A_{2q}$, $q > 1$, $1 \leqslant p \leqslant \infty$.

*Proof.* Assume $h > 1$, $1 < p < \infty$. (i) is a direct verification using Hölder's inequality and the convergence properties of the absolute Poincaré series (2.2.6). When $p = 1$ or $\infty$, Hölder's inequality is not needed. (iii) follows from (i). (ii) depends on the convergence of $\Sigma(1 - |Az|^2)$ when $G$ is CT, a known result stated here without proof. (See Tsuji [**31**], p. 517). ■

(ii) is a slight improvement of (i), a result originally proved by Earle [**6**]. We now introduce a *kernel function*

$$K(z, \zeta) = (2q - 1)\pi^{-1}(1 - z\bar{\zeta})^{-2q}, \qquad z, \zeta \in \mathscr{U}. \tag{3.4.3}$$

This is essentially the $q$th power of the Bergman kernel function for $\mathscr{U}$.

3.4.4 LEMMA. *For* $z, \zeta \in \mathscr{U}$, *we have*

$$K(z, \zeta) = \bar{K}(\zeta, z) \tag{3.4.5}$$

$$K(Az, A\zeta)(A'\zeta)^q\,(\overline{A'\zeta})^q = K(z, \zeta). \tag{3.4.6}$$

(3.4.7) *As a function of* $z$, $K(z, \zeta) \in A_q^p$, $\quad 1 \leqslant p \leqslant \infty$.

*If* $\phi \in A_q^p(G)$, *we have the reproducing formula*

$$\phi(\zeta) = \iint_{\mathscr{U}} \phi(z)\,\bar{K}(z, \zeta)\,\lambda^{-2q}(z)\,d\omega(z). \tag{3.4.8}$$

*Proof.* 3.4.5 and 3.4.6 are immediate verifications, and 3.4.7 follows readily if we use the boundedness of $K$: namely $|K(z, \zeta)| \leqslant m(1 - |\zeta|)^{-q}$, $|z| < 1$. To prove the reproducing formula let $\phi \in A_{2q}$; then

$$\phi(0) = \iint_{\mathscr{U}} \phi(z)\,\bar{K}(z, 0)\,\lambda^{-2q}(z)\,d\omega(z). \tag{*}$$

This formula is easily proved, starting with the mean value property for holomorphic functions; the condition $\phi \in A_{2q}$ is needed for convergence of the integral.

Now observe that (*) is true for every Fuchsian group $G_1$ and every function in $A_{2q}$. We choose $G_1 = TGT^{-1}$ and $f = (\phi \circ T)T'^q$, where $T \in \mathbf{M\ddot{o}b}\,\mathscr{U}$ is so far unspecified. (It is clear that $f \in A_{2q}$ since our Banach spaces are invariant under conformal self-maps of $\mathscr{U}$.) Hence ($S = T^{-1}$)

$$\begin{aligned}
f(0) = \phi(T(0))(T'(0))^q &= \iint_{\mathscr{U}} \phi(Tz)(T'z)^q\,\bar{K}(z, 0)\,\lambda^{-2q}(z)\,d\omega(z) \\
&= \iint_{T\mathscr{U}} \phi(w)(T'Sw)^q\,\bar{K}(Sw, 0)\,\lambda^{-2q}(Sw)\,d\omega(w) \\
&= \iint_{\mathscr{U}} \phi(w)(S'w)^{-q}\bar{K}(Sw, 0)(\overline{S'w})^q\,(S'w)^q\,\lambda^{-2q}(w)\,d\omega(w).
\end{aligned}$$

For fixed $\zeta$ choose $Tw = (w - \zeta)/(\bar{\zeta}w - 1)$; then $T(0) = \zeta$, $0 = S\zeta$, $S'\zeta =$

$1/T'0$. Hence

$$\phi(T(0))(T(0))^q = \iint_{\mathscr{U}} \phi(w)\,\bar{K}(Sw, S\zeta)\,(\overline{S'w})^q\,(S'\zeta)^q\,(T'0)^q\,\lambda^{-2q}(w)\,d\omega(w),$$

and cancelling $(T'(0))^q = (1 - |\zeta|^2)^2 \neq 0$ from both sides and using 3.4.6 we get

$$\phi(\zeta) = \iint_{\mathscr{U}} \phi(w)\,\bar{K}(w, \zeta)\,\lambda^{-2q}(w)\,d\omega(w),$$

which is 3.4.8, but proved under the assumption $\phi \in A_{2q}$.

Our hypothesis was $\phi \in A_q^p(G)$. By the previous lemma, $A_q^p(G) \subset A_{2q}$, and this completes the proof of Lemma 3.4.4. ■

Let $L_q^p(G)$, $1 \leqslant p \leqslant \infty$, be the space of measurable functions $f$ satisfying 3.2.1 and 3.2.2. Suppose $f_n \to ff$ in $L_q^p(G)$. Clearly $f$ satisfies 3.2.4 and

$$(A'z)^q f(Az) = \lim (A'z)^q f_n(Az) = \lim f_n(z) = f(z).$$

Thus $f$ filfils 3.2.1 and so $f \in L_q^p(G)$. It follows that $L_q^p(G)$ is a Banach space with norm given by 3.2.2. When $G = \{\text{id}\}$ we write $L_q^p$.

*Remark.* In 3.2.2 it is necessary to allow $W$ to be an *arbitrary* fundamental region. The sequence $f_n = f$ for all $n$. with $f$ satisfying 3.2.1 and 3.2.2 for $W = \Omega$, converges to the function $g = f$ on $\Omega$, $g = 0$ otherwise; but $g$ is not automorphic.

For the present we need the following lemma only in the case $G = \{\text{id}\}$, but we prove it for all $G$ as we shall need the general result later.

3.4.9 LEMMA. *Let $P$ be the operator defined by*

$$Pf(\zeta) := \iint_{\mathscr{U}} f(z)\bar{K}(z, \zeta)\lambda^{-2q}(z)\,d\omega(z).$$

*For $1 \leqslant p \leqslant \infty$, $P$ is a projection from $L_q^p(G)$ onto $A_q^p(G)$. In particular, $P$ is a projection from $L_q^p$ onto $A_q^p$.*

*Proof.* The identity

$$\iint_{\mathscr{U}} \lambda^{-q}(\zeta)\,|K(z,\zeta)|\,d\omega(\zeta) = \frac{2q-1}{q-1}\lambda^{q}(z), \qquad q > 1 \tag{3.4.10}$$

can be proved by the substitution $\zeta = (w+z)/(\bar{z}w+1)$, which preserves $\mathscr{U}$ and under which $d\omega$ is invariant.

We consider $p = 1, \infty$. Let $g = Pf$, $B = A^{-1}$. Using 3.4.10

$$\|g\|_{1,G} \leqslant \iint_{\Omega} \lambda^{-q}(\zeta)\,d\omega(\zeta) \iint_{\mathscr{U}} f(z)\,|K(z,\zeta)|\,\lambda^{-2q}(z)\,d\omega(z).$$

Interchange the order of integration, replace $\mathscr{U}$ by $\bigcup A\Omega$, etc., and finally use 3.4.10 to get

$$\|g\|_{1,G} \leqslant c_q \|f\|_{1,G}.$$

The proof for $p = \infty$ is a direct consequence of 3.4.10.

Next, $g$ is automorphic, as is easily verified from 3.4.9.

Thus $P$ is from $L_q^p(G)$ to itself when $p = 1, \infty$ and so for $1 \leqslant p \leqslant \infty$ by the Riesz convexity theorem (see for example Dunford-Schwartz, "Theory of Linear Operators", Vol. I, Interscience, New York, 1958, p. 526, Cor.). Now it is easily proved that $Pf$ is analytic. In fact

$$\iint_{\mathscr{U}} f(z)\,\lambda^{-2q}(z) \frac{\partial}{\partial \zeta} \bar{K}(z,\zeta)\,d\omega$$

converges absolutely, since $L_q^p(G) \subset L_{2q}$ (the proof being exactly that of Lemma 3.4.2) and $\partial \bar{K}/\partial\zeta$ is bounded in $z$. This shows that $P$ is from $L_q^p(G)$ to $A_q^p(G)$. Finally, if $f \in A_q^p(G)$, $f = pf$ by 3.4.8: the map is onto. The second assertion of the lemma is the case $G = \{\mathrm{id}\}$. ∎

We return to the surjectivity of $\boldsymbol{\Theta}_q$. Let $f \in A_q^p(G)$. Define the operator $\boldsymbol{\alpha}$ by

$$\boldsymbol{\alpha} f(\zeta) = \iint_{\Omega} f(z)\,\bar{K}(z,\zeta)\,\lambda^{-2q}(z)\,d\omega(z). \tag{3.4.11}$$

If we set $\hat{f} = f$ on $\Omega$, $\hat{f} = 0$ otherwise, then $\hat{f} \in L^p_q$ and

$$\boldsymbol{\alpha} f(\zeta) = \iint_{\mathscr{U}} \hat{f}(z)\, \bar{K}(z, \zeta)\, \lambda^{-2q}(z)\, d\omega = P\hat{f}(\zeta).$$

By Lemma 3.4.9, $\boldsymbol{\alpha} f \in A^p_q$.

It remains to show that $f$ is the Poincaré series of $\boldsymbol{\alpha} f$, i.e. $\Theta_q(\boldsymbol{\alpha} f) = f$. Indeed, with 3.4.6, 3.4.8,

$$\Theta_q \boldsymbol{\alpha} f(\zeta) = \sum_{A \in G} (A'\zeta)^q \iint_{\Omega} f(z)\, \bar{K}(z, A\zeta)\, \lambda^{-2q}(z)\, d\omega$$

$$= \sum_{A} (A'\zeta)^q \iint_{A^{-1}\Omega} f(Az)\, \bar{K}(Az, A\zeta)\, \lambda^{-2q}(Az)\, d\omega$$

$$= \sum_{A} \iint_{A^{-1}\Omega} f(Az)\,(A'z)^q\, \bar{K}(Az, A\zeta)\,(\overline{A'z})^q\,(A'\zeta)^q\, \lambda^{-2q}(Az)\,(A'z)^{-2q}\, d\omega$$

$$= \sum_{A} \iint_{A^{-1}\Omega} f(z)\, \bar{K}(z,\zeta)\, \lambda^{-2q}(z)\, d\omega$$

$$= \iint_{\mathscr{U}} f(z)\, \bar{K}(z, \zeta)\, \lambda^{-2q}(z)\, d\omega = f(z).$$

We have proved Theorem 3.3.3 and the first statement of Theorem 3.3.5.

**3.5** *Proof of Theorem 3.3.4.* We have formally ($B = A^{-1}$),

$$(\Theta_q F, g; \Omega) = \iint_{\Omega} \sum_{A \in G} F(Az)\,(A'z)^q\, \bar{g}(z)\, \lambda^{-2q}(z)\, d\omega(z)$$

$$= \sum_{A} \iint_{A\Omega} F(z)\,(A'Bz)^q\, \bar{g}(Bz)\, \lambda^{-2q}(Bz)\, d\omega$$

$$(\Theta_q F, g; \Omega) = \sum_A \iint_{A\Omega} F(z)\,(B'z)^{-q}\bar{g}(Bz)\,(\overline{B'z})^q\,(B'z)^q\,\lambda^{-2q}(z)\,d\omega$$

$$= \sum_A \iint_{A\Omega} F(z)\,\bar{g}(z)\,\lambda^{-2q}(z)\,d\omega = \iint_{\mathcal{U}} \ldots = (F, g; \mathcal{U}).$$

The interchange of sum and integral is justified by absolute convergence. ■

**3.6** *Proof of Theorem 3.3.6.* This was first established by Petersson [**21**] for groups of finite invariant area and $p = 2$. Let now $G$ be an arbitrary Fuchsian group and let $1 < p < \infty$. Put $F = z^k \in A_q$ in the first equation of Section 3.5 to get

$$(\Theta_q z^k, g; \Omega) = \iint_{\mathcal{U}} z^k \bar{g}(z)\,\lambda^{-2q}\,d\omega = \lim_{r \to 1} \iint_{\Delta_r} \ldots,$$

which equals $\bar{b}_k \Gamma(k + 1)\Gamma(2q - 1)/\Gamma(2q + k)$, after a computation using the uniform convergence on $\Delta_r$. When $p = 1$ or $\infty$ the argument is easier. This completes the proof of Theorem 3.3.6. ■

Theorem 3.3.7 is a corollary. Let $g = \sum a_k z^k$ be in $A_q^{p'}$, $1 < p' \leqslant \infty$. By the theorem just proved, $(\Theta_q z^k, g)$ exists for every $g$. Since $A_q^{p'}(G)$ and $A_q^p(G)$ are conjugate by Theorem 2.2.5 we can apply the converse of Hölder's theorem and deduce that $\Theta_q z^k$, and hence $\Theta_a \mathscr{P}$, is in $A_q^p(G)$, $1 \leqslant p < \infty$. Now suppose $(\theta_q z^k, g) = 0$ for all $k \geqslant 0$; then by Theorem 3.3.6 $a_k = 0$ for all $k \geqslant 0$ so that $g = 0$. ■

**3.7** We shall now prove Theorem 2.2.5. Let $1 \leqslant p < \infty$ and let $\psi$ be the map

$$\psi : g \to \ell_g(f) := (f, g), \qquad g \in A_q^{p'}(G), \qquad f \in A_q^p(G). \tag{3.7.1}$$

It is clear that $\psi$ is a homomorphism from $A_q^{p'}(G)$ to $(A_q^p(G))^*$. What we have to show is that $\psi$ is surjective and its kernel is 0.

So let $\ell^* : A_q^p(G) \to \mathbb{C}$ be a linear continuous map. Consider the Lebesgue space $\mathscr{L}^p(\Omega, d\omega)$ of functions $f$ defined and measurable on $\Omega$ (fixed) with measure $d\omega$. Since $A_q^p(G) \subset \mathscr{L}^p(\Omega, d\omega)$, we can by the Hahn–Banach theorem extend $\ell^*$ to $\mathscr{L}^p(\Omega, d\omega)$; we still call the extended function $\ell^*$. By the Riesz representation theorem there is a unique function $h \in \mathscr{L}^{p'}(\Omega, d\omega)$ such that

$$\ell^* f = (f, h; \Omega), \qquad f \in \mathscr{L}^p(\Omega, d\omega).$$

Now extend $h$ to all of $\mathscr{U}$ by imposing the relation

$$h(Az)(A'z)^q = h(z), \qquad A \in G;$$

the extended function (still called $h$) is measurable and belongs to $L_q^{p'}(G)$, since the integral 3.2.2 defining the norm of $h$ is independent of the choice of fundamental region. Thus we can write

$$\ell^* f = (f, h; \Omega), \qquad f \in A_q^p(G).$$

If $h$ were in $A_q^{p'}(G)$, the surjectivity of $\psi$ would be proved. However, $Ph$ is in $A_q^{p'}(G)$ by Lemma 3.4.9. Also

$$(3.7.2) \qquad (f, h; \Omega) = (f, Ph; \Omega), \qquad f \in A_q^p(G), \qquad h \in L_q^{p'}(G).$$

In fact

$$(f, Ph) = \iint_{\Omega} f(\zeta)\, \lambda^{-2q}(\zeta)\, d\omega(\zeta) \iint_{\mathscr{U}} \bar{h}(z)\, K(z, \zeta)\, \lambda^{-2q}(z)\, d\omega(z),$$

and we proceed by manipulations very similar to those used in the proof of Lemma 3.4.9. This yields

$$\ell^* f = (f, Ph; \Omega) = \ell_{Ph}(f), \qquad f \in A_q^p(G)$$

proving the surjectivity of $\psi$.

Finally, suppose $\ell_g(f) = (f, g; \Omega) = 0$ for all $f \in A_q^p(G)$ and some $g$ in $A_q^{p'}(G)$. Let $k \in L_q^p(G)$, then $Pk \in A_q^p(G)$ and so $(Pk, g; \Omega) = 0$. An equation analogous to 3.7.2 and proved in the same way is

$$(3.7.3) \qquad (f, h; \Omega) = (Pf, h; \Omega), \qquad f \in L_q^p(G), \qquad h \in A_q^{p'}(G)$$

and this shows that

$$(k, g; \Omega) = (Pk, g; \Omega) = 0, \qquad k \in L_q^p(G).$$

That is, $\psi(g) = \ell_g$ is the zero functional on $L_q^p(G)$. It is well known that $\psi$ is an isomorphism from $L_q^p(G)$ onto $L_q^{p'}(G)$ and this implies $g = 0$. ■

## 4. INCLUSION RELATIONS BETWEEN BERS SPACES

We consider first the spaces $A_q^p$. Note that these are defined for real $q > 0$. Since $G = \{\mathrm{id}\}$, there is no equation of invariance (3.1.1) here involving the branches of $(A'z)^q$.

4.1.1 THEOREM. (i) $A_q^p \subset B_q$, $\quad p \geqslant 1$.
(ii) $A_q^{p_1} \subset A_q^{p_2}$, $\quad 1 \leqslant p_1 \leqslant p_2 \leqslant \infty$.

*Proof.* The second assertion is a consequence of the first. For, with $F \in A_q^{p_1}$,

$$\|F\|_{p_2}^{p_2} = \iint_{\mathscr{U}} |F|^{p_1} \lambda^{-p_1 q}\, d\omega \,.\, |F|^{p_2 - p_1} \lambda^{-(p_2 - p_1)q}$$

$$\leqslant \|F\|_\infty^{p_1 - p_2} \|F\|_{p_1}^{p_1}.$$

Now let $F \in A_q^p$. Use the area mean value formula for analytic functions and Hölder's inequality ($p > 1$):

$$F(z) = \frac{1}{\pi\rho^2} \iint_{\Delta(z,\rho)} F(\zeta)\, dx\, dy, \qquad \rho = \frac{1 - |z|}{2}$$

$$\lambda^{-2}(z)\, \lambda^{-q+2}(z)\, |F(z)| \leqslant m \iint_{\Delta} \lambda^{-q}(\zeta)\, |F(\zeta)|\, d\omega \,.\, \sup_{\zeta \in \Delta} \lambda^{-q}(z)\, \lambda^{q}(\zeta)$$

$$\leqslant mc(q) \iint_{\Delta} \lambda^{-q}(\zeta)\, |F(\zeta)|\, d\omega$$

$$\leqslant m(q) \left\{ \iint_{\Delta} \lambda^{-qp}\, |F(\zeta)|^p\, d\omega \right\}^{1/p} \left\{ \iint_{\Delta} d\omega \right\}^{1/p'}$$

$$\leqslant m(q)\, \|F\|_p\, \sigma(\Delta(z, \rho))^{1/p'}.$$

Since the $H$-radius of $\Delta(z, \rho)$ is $\sim \log 3$ as $|z| \to 1$, $\sigma(\Delta(z, \rho))$ is bounded for $z$ in $U$. When $p = 1$, Hölder's inequality is unnecessary. ■

4.1.2 THEOREM. (i) $L_q^p(G) \subset L_{q+h}$, *where* $h > 1$, $p \geqslant 1$; $h \geqslant 1$, $p = 1$ *for* $G$ *of* CT.
(ii) $L_q^p \subset L_{2q}$, $q > 1$.
(iii) $A_q^p(G) \subset A_{q+h}$.
(iv) $A_q^p(G) \subset A_{2q} \subset A_{2q}^p \subset B_{2q}$, $p \geqslant 1$.
(v) $A_q^p(G) \subset A_{q+k/p}$, *where* $k > 1$, $1 \leqslant p < \infty$; $k \geqslant 1$, $1 \leqslant p < \infty$ *for* $G$ *of* CT.

*Proof.* Lemma 3.4.2 proves (iii) and the proof serves also for (i). (Also (iii) is obtained from (i) by intersecting with the analytic functions on $\mathscr{U}$.) (ii) is an obvious consequences of (i), and (iv) follows from (iii) and Theorem 4.1.1. (v) is proved by the equation

$$\|F\|_{q+k/p,p}^p = \iint_{\mathscr{U}} \lambda^{-k}\lambda^{-pq}|F|^p\,d\omega = \iint_{\Omega} \sum_A (1 - |Az|^2)^k \lambda^{-pq}|F|^p\,d\omega$$
$$\leqslant m(k, G)\|F\|_{p,G}^p. \qquad \blacksquare$$

**4.2** When we consider the spaces $A_q^p(G)$, we again assume $q$ is an integer.

4.2.1 THEOREM. $A_q^p(G) \subset B_q(G)$ *for all* $p \geqslant 1$ *implies*

$$A^{p_1}(G) \subset A^{p_2}(G), \qquad 1 \leqslant p_1 \leqslant p_2 \leqslant \infty.$$

See the proof of Theorem 4.1.1.

Rajeswara Rao [25] has proved a sharper result:

THEOREM A. *Let* $q > 1$. $A_q^p(G) \subset B_q(G)$ *for some* $p$, *where* $1 \leqslant p < \infty$ *implies*

$$A_q^{p_1}(G) \subset A_q^{p_2}(G) \text{ for } 1 \leqslant p_1 \leqslant p_2 \leqslant \infty.$$

A good deal of work has been done on the conjecture

$$A_q^p(G) \subset B_q(G) \tag{4.2.2}$$

for every group $G$, but we now know this to be false.

4.2.3 THEOREM (Pommerenke [23]). *There exists a Fuchsian group G such that* $A_1^2(G) \not\subset B_1(G)$.

We sketch the construction, which is a generalization to infinitely generated groups of the classical treatment of uniformization (for example, in [9], Chapters 9, 10) combined with a theorem on "Bloch functions".

THEOREM B. (See for example [24].) *Let f be analytic in* $\mathscr{U}$. *Then* $f' \notin B_1$ *if and only if*

(4.2.4) *for each* $R < \infty$, *f is a one-to-one map from some region in* $\mathscr{U}$ *onto a disc of radius R.*

In view of this theorem it suffices to exhibit a function $f$ and a group $G$ for which $f'$ is automorphic on $G$ of weight 1 and

$$\text{(i)}\quad \|f'\|_{1,2,G}^2 = \iint_\Omega |f'(z)|^2\, dx\, dy < \infty, \text{ (ii) f satisfies 4.2.4.}$$

Define the domain $(w = u + iv)$

$$B = \bigcup_{n=1}^{\infty} \{0 \leqslant u < 3^{-n}, 2^n < v < 2^{n+1}\} \bigcup_{n=2}^{\infty} \{0 < u < 3^{-n}, v = 2^n\}. \tag{4.2.5}$$

$B$ has free sides, for example, $\{u = 3^{-n}, 2^n < v < 2^{n+1}\}$. Let $\{\sigma_1, \sigma_2, \ldots\}$ be elements of a free group $T = \langle\sigma_1, \sigma_2, \ldots\rangle$. An element of $T$ is uniquely of the form

$$\tau = \sigma_{n_1}^{k_1} \ldots \sigma_{n_t}^{k_t}; \qquad k_\nu = \pm 1, \qquad k_{\nu+1} \neq -k_\nu \qquad \text{if } n_\nu = n_{\nu+1}; \tag{4.2.6}$$

call $t$ the length of $\tau$. With each $\tau$ we associate a set $(B, \tau)$ called a "copy" of $B$ and regard $(B, \tau_1)$ as disjoint from $(B, \tau_2)$ if $\tau_1 \neq \tau_2$. Let

$$R = \bigcup_{\tau \in T} (B, \tau).$$

Every point of $R$ can be written uniquely as $\tilde{w} = (w, \tau)$ with $w \in B$ and $\tau \in T$.

Let $p: R \to \mathbb{C}$ be defined by

$$p(\tilde{w}) = w + \sum_{\nu=1}^{t} k_\nu 3^{-n_\nu}, \qquad \tilde{w} = (w, \tau) \in R$$

where $\tau$ is of the form 4.2.6. It can be verified that

$$p((B, \tau_1)) \cap p((B, \tau_2)) = \varnothing \quad \text{if} \quad \tau_1 \neq \tau_2. \tag{4.2.7}$$

This motivates the definition of $B$. We have $p(B) = B$ and $p((B, \tau))$ is a horizontal translate of $B$; these translates are mutually disjoint. Moreover, $p$ is one-to-one from $(B, \tau)$ to its image.

It is easy to define a neighbourhood of each $(w, \tau)$ in $R$ that is mapped one-to-one by $p$ onto a disc in $\mathbb{C}$, and $R$ becomes a Riemann surface. Furthermore, it is not hard to show $R$ is simply connected. Since $R$ has free boundaries (for example, $0 \leqslant u < 3^{-1}, v = 2$), there is a function $h(z)$ mapping $\mathscr{U}$ conformally and one-to-one onto $R$.

We make $T$ act on $R$ by defining, for $\lambda \in T$, the map $\lambda^*: R \rightarrow R$ by

$$\lambda^*(\tilde{w}) \equiv \lambda^*(w, \tau) := (w, \lambda\tau), \qquad \lambda = \sigma_{m_1}^{j_1} \ldots \sigma_{m_s}^{j_s} \in T.$$

Each element of $T^* = \{\lambda^*\}$ is a one-to-one conformal self-map of $R$ and $T^*$ is isomorphic to $T$. Clearly

$$p \circ \lambda^*(\tilde{w}) = p(\tilde{w}) + \sum_{\nu=1}^{s} j_\nu 3^{-m_\nu} \tag{4.2.8}$$

Now define

$$G = h^{-1} \circ T^* \circ h.$$

The elements of $G$ are conformal one-to-one maps of $\mathscr{U}$ on itself and so are Möbius transformations. An easy consideration shows that $G$ is discontinuous in $\mathscr{U}$. Hence $G$ is a Fuchsian group. A fundamental region for $G$ is $\Omega = h^{-1}(\operatorname{Int} B)$.

The complex-valued function $f = p \circ h$ is analytic in $\mathscr{U}$ and we have, for $\gamma = h^{-1} \circ \lambda^* \circ h \in G$, $\lambda = \sigma_{m_1}^{j_1} \ldots \sigma_{m_s}^{j_s}$,

$$f(\gamma z) = p \circ h \circ \gamma z = p \circ \lambda^*(hz) = p \circ h(z) + c(\gamma) = f(z) + c(\gamma), \tag{4.2.9}$$

where $c(\gamma) = \sum_{\nu=1}^{s} j_\nu 3^{-m_\nu}$ depends only on $\gamma$. ($f$ is an *automorphic integral.*) Hence the derivative $f'$ satisfies

$$f'(\gamma z)\gamma' z = f'(z), \qquad \gamma \in G,$$

and $f'$ is automorphic on $G$ of weight 1. Also

$$f(\Omega) = p \circ h(\Omega) = p(\operatorname{Int} B) = \operatorname{Int} B;$$

hence

$$\|f\|^2_{1,2,G} = \iint_{\Omega} \lambda^{-2}(z)|f'(z)|^2 \, d\omega(z) = \iint_{\Omega} |f'(z)|^2 \, dx \, dy$$

$$= \operatorname{area} B = \sum_1^{\infty} 3^{-n}(2^{n+1} - 2^n) = \sum_1^{\infty} (2/3)^n < \infty.$$

That is, $f' \in A_1^2(G)$.

But $f' \notin B_1(G)$. For let $V_n$ be the horizontal strip $\{w: 2^n < v < 2^{n+1}\}$. Then if $\zeta \in V_n$, we can write $\zeta = \zeta_0 + 3^{-n}k$ with an integer $k$ and $\zeta_0 \in B$. There is a $z_0 \in \Omega$ for which $f(z_0) = \zeta_0$. Set $V = h^{-1} \circ (\sigma_n^k)^* \circ h$. Then $f(\gamma z_0) = f(z_0) + 3^{-n}k = \zeta_0 + 3^{-n}k = \zeta$. Moreover, if $f(z) = \zeta = p \circ h(z) = p \circ h(\gamma z_0)$, it follows from 4.2.7 that $h(z)$ and $h(\gamma z_0)$ lie in the same $(B, \tau)$ and have the same projection; so $z = \gamma z_0$. Hence $f$ maps some domain in $\mathscr{U}$ one-to-one onto $V$. Since $n$ is arbitrary, we see that $f$ satisfies 4.2.4. By Theorem B, $f' \notin B_1(G)$. ■

Combining this result with Theorem A (above) we get

(4.2.10) $$A_2^p(G) \not\subset B_2(G), \qquad 1 \leqslant p < \infty.$$

To see this, let $g \in A_1^2(G) - B_1(G)$. Then $g^2 \in A_2(G) - B_2(G)$. That is $A_2(G) \not\subset B_2(G)$, and 4.2.10 follows.

**4.3** In many cases, however, the inclusion 4.2.2 is valid. For example, if $\sigma(G) < \infty$, the spaces $A_q^p(G)$, $1 \leqslant p \leqslant \infty$, are all equal.

4.3.1 THEOREM, *If* $\sigma(G) < \infty$, $A_q^p(G) = B_q(G)$ *for* $1 \leqslant p \leqslant \infty$.

*Proof.* The fundamental region $\Omega = K \cup \Pi$, where $K$ is compact and $\Pi$ consists of a finite number of parabolic cusp sectors. In $K$, $f \in A_q(G)$ is bounded, and in $\Pi$, $f \to 0$ (Theorem 3.2.3). Hence $\lambda^{-q}|f|$ is bounded in $\Omega$ and so in $U$. Therefore $A_q(G) \subset B_q(G)$ and, by Theorem 3.2.3, $A_q(G) \subset A_q^p(G) \subset B_q(G)$, $1 \leqslant p < \infty$. But $B_q(G) \subset A_q(G)$ since $\sigma(B) < \infty$.

Already when $G$ is of the second, even though finitely generated, the situation is non-trivial and many proofs of 3.2.3 have been given. In the general case we have

THEOREM C. [13]. *Let $G$ be a group satisfying the condition*

(4.3.2) $\quad |\text{trace } A| \geqslant 2 + m$ *for all hyperbolic* $A \in G$, $m = m(G)$.

*Then* $A_q(G) \subset B_q(G)$ *with a continuous inclusion map.*

This result includes the previous cases, since it is known (Chapter 7, Theorem 1.9) that 4.3.2 is satisfied for a finitely generated group.

The proof of Theorem C in [13] made use of Marden's fundamental results on the intersector subgroup of a Fuchsian group [15]. Following conversations with Beardon, I saw that Marden's theorem could be completely avoided and the proof greatly simplified. It is the simplified proof that we shall now sketch.

The proof proceeds in two stages. In the first stage we make no assumption whatever about $G$. Let $p$ be a cusp of $G$, $|p| = 1$. Define $w = u + iv = T_p z \equiv Tz = -i(z + p)/(z - p)$; $T$ carries $\mathscr{U}$ into $\mathscr{H} = \{w : v > 0\}$ and carries $p$ to $i\infty$. The map is isometric if we define the Poincaré metric in $H$ by $\lambda_1 |dw|$, $\lambda_1 = 1/2v$. The group $G_1 = TGT^{-1}$ acts on $H$ and if $(\text{stab } p)_G = \langle P \rangle$, $(\text{stab } (\infty)_{G_1} = \langle P_1 \rangle$, where $P_1 = TPT^{-1} = \{w \mapsto w + h\}$. Let $f \in A_q(G)$; then $f_1(w) := f(z))\,(dz/dw)^q \in A_q(G_1)$ and $\lambda^{-q}(z)|f(z)| = \lambda_1^{-q}(w)|f_1(w)|$, $\|f\|_1 = \|f_1\|_1$, the last symbols being the norms in $A_q(G)$, $A_q(G_1)$.

Now $f_1(P_1 w)(P_1' w)^q = f_1(w + h) = f_1(w)$, so $f_1$ has a Fourier series $f_1(w) = \sum_1^\infty a_k e^{2\pi i k w/h}$ having only terms with positive exponents (3.2.7). The $a_k$ can be estimated and this gives

$$\lambda^{-q}(z)|f(z)| = \lambda_1^{-q}(w)|f_1(w)| \leqslant m\|f_1\|_1 = m\|f\|_1, \qquad v \geqslant 3\lambda.$$

Let $\Pi_p = T^{-1}\{v \geqslant 3h\}$, a horocycle at $p$ in $\mathscr{U}$, and let $\Pi = \bigcup_p \Pi_p$. Thus

$$\lambda^{-q}(z)|f(z)| \leqslant m\|f\|_1, \qquad z \in \Pi. \tag{4.3.3}$$

Next, let $\omega$ be an elliptic vertex of $G$ of order $l_w = l$, $|\omega| < 1$, $l \geqslant 2$. Define $w = T_\omega(z) \equiv Tz = (z - \omega)/(\bar{\omega}z - 1)$, a map of $\mathscr{U}$ on itself. Let $G_1 = TGT^{-1}$, $f_1(w) = f(z)\,(dz/dw)^q$; $(\text{stab } \omega)_G = \langle E \rangle$, $(\text{stab } 0)_{G_1} = \langle E_1 \rangle$ with $E_1$: $w \mapsto \varepsilon^2 w$, $\varepsilon = e^{\pi i/l}$. Because $w^q f_1(w)$ is $E_1$-invariant we get an expansion

$w^q f_1(w) = \sum_1^\infty b_k w^{kl}$. Estimating the $b_k$ we find

$$\lambda^{-q}(z)|f(z)| = \lambda_1^{-q}(w)|f_1(w)| \leqslant m\|f_1\|_1 = m\|f\|_1, w \in D_\omega$$

where $D_\omega = \{w: |w| < 1 - \gamma/l\}$ when $l \geqslant l_0 = l_0(\gamma)$ and $D_\omega = \varnothing$ when $1 < l_0$; $\gamma =$ absolute const. Letting $\Lambda_\omega = T^{-1}D_\omega$, $\Lambda = \bigcup_\omega \Lambda_\omega$, we therefore have

$$\lambda^{-q}(z)|f(z)| \leqslant m\|f\|_1, \qquad z \in \Lambda. \tag{4.3.4}$$

For every group $G$, then, there is a noncompact region $\Pi \cup \Lambda$ within which $\lambda^{-q}|f|/\|f\|_1$ is bounded.

In the second stage of the proof we must of course make use of the trace condition 4.3.2 and we can confine ourselves to the complementary region $\Sigma = \mathscr{U} - (\Pi \cup \Lambda) = (\mathscr{U} - \Pi) \cap (\mathscr{U} - \Lambda)$. The basic formula used to estimate $f \in A_q(G)$ is obtained from the area mean-value formula for $f$:

$$\begin{aligned} \lambda^{-q}(z)|f(z)| &\leqslant m \iint_{\Delta(z)} \lambda^{-q}(w)\,|f(w)|\,d\omega\,.\ \sup_{w \in \Delta(z)} \lambda^{-q}(z)\,\lambda^q(w) \\ &\leqslant m \iint_{\Delta(z)} \lambda^{-q}(w)\,|f(w)|\,d\omega,\ m = m(G, c), \end{aligned} \tag{4.3.5}$$

where $\Delta(z) \equiv \Delta_c(z) = \Delta(z, c(1 - z))$ with $c, 0 < c < 1$, still to be chosen. The integral can be extended over $\Omega$ if we insert the factor $n_0(\Delta(z), w)$; $n_0$ is defined in 1.2.5. Since $n_0$ and $n$ differ only on a countable set (the elliptic vertices) we get

$$\lambda^{-q}(z)\,|f(z)| \leqslant m \iint_\Omega n(\Delta(z), w)\,\lambda^{-q}(w)\,|f(w)|\,d\omega(w). \tag{4.3.6}$$

This equation is valid for $z \in \mathscr{U}$ but we are now interested only in $z \in \Sigma$. Let $m_1 > 0$ satisfy the condition that $\Delta(0, m_1 + c(1 - m_1))$ lies in $\Omega$, where $c, 0 < c < 1$, will be defined presently. With 4.3.3 and 4.3.4, Theorem C is now implied by the two equations

$$\lambda^{-q}(z)\,|\mathrm{f}(z)| \leqslant m\|f\|_1, \qquad |z| < m_1 \tag{4.3.7}$$

(4.3.8) $\lambda^{-q}(z)\,|f(z)| \leqslant m\|f\|_1, \qquad z \in \Sigma_1 := \Sigma \cap \{|z| > m_1\}.$

Equation 4.3.8 follows from the lower bound

(4.3.9) $d(z, Az) \geqslant m_2 > 0, \qquad z \in \Sigma_1, \qquad 1 \neq A \in G.$

That is, each element $A \neq 1$ in $G$ moves $z$ a hyperbolic distance at least $m_2 > 0$, where $m_2$ does not depend on $z$ or $A$, but may depend on $G$. If we then choose $c$ so that $\Delta_c(z) \equiv \Delta(z)$, regarded as an $H$-disc, has $H$-radius smaller than $m_2/5$, we shall have

$$n(\Delta(z), w_1) = 1, \qquad w_1 \in \Delta(z), \qquad z \in \Sigma.$$

If $w \in \mathscr{U}$ has no $G$-equivalent in $\Delta(z)$, $n(\Delta(z), w) = 0$; otherwise $w$ is equivalent to $w_1 \in \Delta(z)$ and then $n(\Delta(z), w) = n(\Delta(z), w_1)$. Hence

$$n(\Delta(z), w) \leqslant 1, \qquad w \in \mathscr{U}, \qquad z \in \Sigma_1.$$

This relation substituted in 4.3.6 proves 4.3.8. We have therefore to prove 4.3.9.

Let $p$ be a cusp and $P_1 : w \mapsto w + \lambda$ a generator of the transformed group $(G_1)_\infty$, as previously defined. Since $z \in \Pi'_p$, $T_p z = w$ lies in the region $v < 3\lambda$, $w = u + iv$. Thus with $w = u + \lambda/2 + \rho e^{i\theta}$, $\phi = \arctan(\lambda/2v)$,

$$d(w, w + \lambda) = \int_{\pi/2-\phi}^{\pi/2} \frac{\rho\, d\theta}{\rho \sin\theta} > \phi = \arctan\frac{\lambda}{2v} > \arctan\frac{1}{6}.$$

Next, suppose $\omega$ is an elliptic vertex of order $l \geqslant l_0$ fixed by $E$ (primitive); $\Lambda_\omega$, $D_\omega = T_\omega \Lambda_\omega$ discs about $\omega$ and 0, respectively; and $E_1 : w \to w\,.\exp(2\pi i/l)$ a generator of the transformed group $(G_1)_0$; these have already been mentioned. Let $z \in \Sigma$, $z_1 = T_\omega z$, $z_2 = E_1 z_1$, and $\rho = |z_1| = |z_2|$. Then $1 - \gamma/l \equiv \rho_0 \leqslant \rho < 1$. Write $z_1 = z_1(\rho)$, etc. We also require $l \geqslant 2\gamma$.

Let $J$ be the circle orthogonal to $\partial\mathscr{U}$ and passing through $z_1, z_2$, and let the bisector of angle $z_1 0 z_2$ meet $J$ at $Q = Q(\rho)$. The triangle $T = OQz_2$ is a hyperbolic right triangle. By well known formulas of hyperbolic trigonometry, we find

$$\sinh 2d(Q, z_2) = \sinh 2d(0, z_2) \sin\frac{\pi}{l},$$

$Q$ and $z_2$ being functions of $\rho$. This shows that as $\rho \to 1$, $d(Q, z_2)$ increases. Hence

$$\text{(4.3.10)} \quad \begin{aligned} \sinh d(z, Ez) &= \sinh d(z_1, z_2) \geqslant \sinh d(z_1(\rho_0), z_2(\rho_0)) \\ &= \sinh 2d(Q(\rho_0), z_2(\rho_0)), \end{aligned}$$

$$\sinh d(z, Ez) \geqslant \sinh 2d(0, \rho_0) \sin\frac{\pi}{l} > \frac{2}{l}\sinh\left(\log\frac{2-\gamma/l}{\gamma/l}\right)$$

$$\geqslant \frac{2}{l}\cdot\frac{2l}{3\gamma} = \frac{4}{3\gamma},$$

as $l \geqslant 2\gamma$.

Finally, we must treat the case of hyperbolic $A$. Let $\xi_1, \xi_2$ be the fixed points of $A$, let $C$ be the circle through these points and $z \in \mathscr{U}$, and let $D$ be the geodesic connecting the fixed points. When the figure is mapped to the upper half-plane by a Möbius transformation $V$ with the fixed points going to $0, \infty$, then $D$ becomes the positive imaginary axis while $C$ becomes an oblique line through 0; also $A$ goes into $A_1 : w \to \kappa w$, $\kappa > 1$, $w = Vz$. It is now easy to check that

$$d(z, Az) = d(w, A_1 w) \geqslant d(i|w|, i|A_1 w|) = \tfrac{1}{2}\log\kappa.$$

By the trace condition 4.3.2 we have $\kappa \geqslant 1 + m$, or $d(z, Az) > m_3$. Combining this with the previous results yields 4.3.9 for all $A$ in $G$ except elliptic $A$ of orders $l < l_1 \equiv l_0 + 2\gamma$.

We now consider the cases $l < l_1$. Recall that $m_1$ satisfies the condition that $\Delta(0, m_1 + c(1 - m_1))$ lies in $\Omega$, with the $c$ defined in the lines following 4.3.9. Furthermore, let

$$m_4 = \inf_{z, E_1} d(z, E_1 z), \qquad |z| \geqslant m_1, \qquad m_4 = m_4(G),$$

where $E_1$ runs over the transforms $T_\omega E T_\omega^{-1}$ of elliptic elements fixing the $\omega$'s of orders $2 \leqslant l < l_1$. Using the same considerations that led to 4.3.10 we see that $m_4 > 0$. Thus we have proved 4.3.9 for

$$z \in \Sigma_1 := \Sigma \cap \{|z| \geqslant m_1\},$$

and we have 4.3.8 when $z \in \Sigma_1$.

Finally, let $z \in \Delta(0, m_1)$. Since $\Delta(0, m_1 + c(1 - m_1)) \subset \Omega$, we have that $\Delta_c(z) \subset \Omega$ and so $n(\Delta_c(z), w) \leqslant 1$, $w \in \Omega$. Hence 4.3.6 yields 4.3.7 for $z \in \Delta(0, m_1)$. This completes the proof of Theorem C.

COROLLARY. *If* 4.3.2 *is satisfied,* $A_q^p(G) \subset B_q(G)$, $1 \leqslant p < \infty$.

This follows immediately from Theorem A.

## 5. TAYLOR COEFFICIENTS OF AUTOMORPHIC FORMS

In the theory of modular forms (automorphic forms on the modular group PSL(2, $\mathbb{Z}$)), a topic that has absorbed the interest of mathematicians for at least 50 years, is the order of magnitude of the Fourier coefficients. The problem was stimulated by the conjecture of Ramanujan, recently proved by Deligne, that $\tau(n) = 0(n^{11/2+\varepsilon})$, where

$$\Delta(w) = \sum_{1}^{\infty} \tau(n)\, e^{2\pi i n w},$$

the classical "discriminant," is a form in $A_6$(PSL(2, $\mathbb{Z}$)). This profound result was accomplished in a number of steps starting with the first non-trivial estimate of certain exponential sums by Kloosterman in 1927; the more recent work (Weil, Eichler, Deligne, Serre) makes heavy use of algebraic geometry and is particular to the arithmetic structure of the modular group. Much work has been done on other groups of finite invariant area having parabolic cusps; for all of this see the summary article of Selberg [**27**].

In our setting it is more natural to consider the Taylor coefficients of automorphic forms and to allow the group $G$ to have infinite area. The results obtained here, however, are on the surface and do not reflect the deeper and more particular properties of $G$. Better results may perhaps be obtained in the future by the methods of Section 6.

5.1.1 *Definition.* $s_n(f) \equiv s_n := \sum_{k=0}^{n} |a_k|^2$, $f \equiv \sum_{k=0}^{\infty} a_k z^k$.

5.1.2 THEOREM. *Let* $f \in A_q^p(G)$, $1 \leqslant p \leqslant \infty$. *Then*

$$|a_k| \leqslant m \|f\|_p\, k^q.$$

*Moreover if G is CT, $p < \infty$.*

$$a_k = o(k^q). \tag{5.1.3}$$

*Let $\infty \geqslant p \geqslant 2$. Then*

$$s_k \leqslant m\|f\|_p^2 \, k^{2q}, \tag{5.1.4}$$

*and if G is CT, $p < \infty$,*

$$s_k = o(k^{2q}). \tag{5.1.5}$$

*Proof.* We have

$$|a_k|\rho^k \leqslant m \int_0^{2\pi} |f(\rho \, e^{i\theta})| \, d\theta, \tag{5.1.6}$$

$$|a_k| \int_0^r \rho^{k+1} (1 - \rho^2)^{q-2} \, d\rho \leqslant m \int_0^{2\pi} \int_0^r (1 - \rho^2)^{q-2} \, |f(\rho \, e^{i\theta})| \, \rho \, d\rho \, d\theta$$
$$= m \iint_{\Delta_r} \lambda^{-q}(z) \, |f(z)| \, d\omega.$$

For each $k$ choose $r = 1 - 1/k$. Then

$$\int_0^r \geqslant m \int_{1-2/k}^{1-1/k} \rho^{k+1}(1 - \rho)^{q-2} \, d\rho \geqslant mk^{-1}(1 - 2/k)^{k+1} \, k^{-q+2}$$
$$\geqslant mk^{1-q}, \qquad m = m(q). \tag{5.1.7}$$

Let $1 < p < \infty$; we omit the calculations for $p = 1, \infty$. By Tsuji's estimate

$$\iint_{\Delta_r} \lambda^{-q}(z)|f(z)| \, d\omega \leqslant m\left\{\iint_{\Delta_r} \lambda^{-pq}|f|^p \, d\omega\right\}^{1/p} \left\{\iint_{\Delta_r} d\omega\right\}^{1/p'} \tag{5.1.8}$$
$$\leqslant m \left\{\iint_{\Omega} n(r, z) \, \lambda^{-pq}(z)|f(z)|^p \, d\omega\right\}^{1/p} (1 - r)^{-1/p'}$$
$$\leqslant m\|f\|_p (1 - r)^{-1/p-1/p'} \leqslant m\|f\|_p k, \qquad r = 1 - 1/k.$$

This proves 5.1.2.

To establish 5.1.4, we use Parseval's formula. Let $2 < p < \infty$ and let $p^*$ be the conjugate of $p/2$, $1 < p/2 < \infty$. We have

$$2\pi \sum_0^\infty |a_j|^2 \rho^{2j} = \int_0^{2\pi} |f(\rho\, e^{i\theta})|^2 \, d\theta,$$

$$2\pi \sum_0^\infty |a_j|^2 \int_0^r \rho^{2j+1}(1 - \rho^2)^{2q-2} \, d\rho = \iint_{\Delta_r} \lambda^{-2q} |f|^2 \, d\omega$$

$$\leqslant \left\{ \iint_{\Delta_r} \lambda^{-pq} |f|^p \, d\omega \right\}^{2/p} \left\{ \iint_{\Delta_r} d\omega \right\}^{1/p^*}.$$

By 5.1.7 with $r = 1 - 1/k$ the left member exceeds $m \sum_0^k |a_j|^2 \, . \, k^{1-2q}$, and by a calculation analogous to 5.1.8 the right member is less than $m\|f\|_p^2 \, k$. This gives 5.1.4. The proof for $p = 2, \infty$ is simpler. We have not been able to get this result for $1 \leqslant p < 2$.

Next, suppose $G$ is CT, $p < \infty$. The $o$-results would follow from

$$n(r, z) = o(1/(r - 1)). \tag{5.1.9}$$

for the particular group $G$. In fact, we have, for integral $k$, $k^{-1}\, n(1 - k^{-1}, z) \to 0$ as $k \to \infty$ for each $z$, and by Tsuji's estimate $k^{-1} n(1 - k^{-1}, z) \leqslant m(G)$ for all $k$. By Lebesgue's dominated convergence theorem

$$\iint_\Omega k^{-1} n(1 - k^{-1}, z) \lambda^{-pq}(z) |f(z)|^p \, d\omega = o(1)$$

as $k \to \infty$; hence

$$\iint_{\Delta_r} \lambda^{-q} |f| \, d\omega = (o(k))^{1/p} k^{1/p'} = o(k), \qquad r = 1 - k^{-1}.$$

The $o$-results follow from this. Thus we have proved:

(5.1.10) 5.1.9 implies 5.1.3 and 5.1.5.

So we have to prove 5.1.9 for $G$ of CT. But this follows easily from 2.2.5. Take $s = 1$; the right member converges, and since the integral in the left member increases and hence approaches a limit, it follows that the limit $L(z) = \lim_{r \to 1} n(r, z)(1 - r) \geqslant 0$ exists. Suppose $L(z) > 0$ for some $z$. Then

$$\int_{r_0}^{r} n(t, z)\, dt > 2^{-1} L(z) \int_{r_0}^{r} (1 - t)^{-1}\, dt$$

for $r_0$ near enough to 1, and the integral $\to \infty$ as $r \to 1$. Hence $L(z) = 0$ for all $z$, proving 5.1.9.

When $G = \{\mathrm{id}\}$ the above results hold since $G$ is CT. However by direct methods one obtains the better result that 5.1.3 is true with $q$ replaced by $q - 1$ and 5.1.5 is true with $2q$ replaced by $2q - 1$.

**5.2** Metzger [16] has obtained lower bounds on the Taylor coëfficients.

5.2.1 THEOREM. *If $f = \sum a_k z^k$ is a holomorphic automorphic form on $G$ of weight $q$, then for $s_0 > 0$ and $\varepsilon > 0$,*

$$s_k \neq O(k^{2q + s_0 - 1 - \varepsilon})$$

*and*

$$a_k \neq O(k^{q + s_0/2 - 1 - \varepsilon}).$$

## 6. THE ORBITAL COUNTING FUNCTION

**6.1** The function $n(r, z)$ is a fundamental concept for the group $G$ and has been used extensively in previous sections. We have developed upper and lower bounds in the case of certain groups, limit relations in other cases, but no asymptotic formulas. There is such a formula when $G$ is a group whose quotient space mod $\mathscr{U}$ is closed, i.e. compact without boundary. It was proved by Huber [11] and his paper contains other noteworthy results. The formula reads

$$n(r, z) \sim \frac{\sigma(\Delta_r)}{\sigma(\Omega)} \sim \frac{1}{2(g - 1)} \frac{1}{1 - r}, \qquad r \to 1 \tag{6.1.1}$$

a result that accords exactly with intuition.

Huber's investigation is part of a much larger theory that started with

Maass's introduction in 1949 of "automorphic wave forms", which are nonanalytic automorphic functions that satisfy a certain partial differential equation. Contributions have been made by many writers, of whom Selberg [**28**], Faddeev, Roelcke and Elstrodt are a few. These papers considered groups of finite area with parabolic elements; Roelcke and Elstrodt treated more extensive classes of groups. Huber confined himself to compact groups. Fricker proved sharpened forms of Huber's result for groups acting on three-dimensional hyperbolic half-space. McKean proved some of Huber's results by using the "trace formula" in [**28**]. Randol has contributed new results in the compact case. Very recently Patterson has obtained a more precise form of 6.1.1 by a different method. For references to the above work see the bibliographies in [**8**] and [ **20**].

Here we shall follow Huber [**11**].

**6.2.** In what follows our setting will be $\mathscr{H} = \{z = x + iy : y > 0\}$, the upper half-plane, with Poincaré metric $d\omega(z) \equiv d\omega = \bar{y}^2\, dx\, dy$. The Möbius group acting on $\mathscr{H}$ consists of the transformations $T: z \to (\alpha z + \beta)/(\gamma z' + \delta)$ with real $\alpha, \beta, \gamma, \delta$ and $\alpha\delta - \beta\gamma = 1$. We suppose that $G$ is a Fuchsian group acting without fixed points on $\mathscr{H}$ and that $\mathscr{H}/G$ is closed, i.e. compact with no boundary. It is well known that $G$ consists of 1 and hyperbolic elements, each of which acts without fixed points in $\mathscr{H}$ but has two distinct fixed points on $\partial\mathscr{H}$. Moreover, $G$ is horocyclic, its genus $g > 1$, and $\Omega$ (= Dirichlet region with centre $i$) consists of $4g$ sides and lies compact in $\mathscr{H}$. The $H$-area of $\Omega$ is $4\pi(g - 1)$. For details the reader should see Chapter 7.

Let $\Delta$ be the Laplace–Beltrami operator acting on $\mathscr{U}/G$. In the coordinates of $\mathscr{H}$, $\Delta$ is defined by

$$\Delta f \equiv \Delta_z f(z) := y^2(f_{xx} + f_{yy}) = 4y^2 f_{z\bar{z}}(z),$$

where as usual $2f_z = f_x - if_y$, etc. We say $\phi: \mathscr{H} \to \mathbb{C}$ is an *eigenfunction* of $\Delta$ with eigenvalue $\lambda$ if

$$\text{(6.2.1)} \qquad \phi \in C^2(\mathscr{H}), \qquad \phi \not\equiv 0, \qquad \phi(Az) = \phi(z) \qquad \text{for} \quad A \in G$$

and

$$\text{(6.2.2)} \qquad \Delta_z \phi(z) + \lambda\phi(z) = 0, \qquad z \in \mathscr{H}.$$

It is known [10], p. 64 that the operator $\Delta$ has a discrete point spectrum

$$0 = \lambda_0' < \lambda_1' \leqslant \lambda_2' \leqslant \ldots \to +\infty$$

where each eigenvalue is written as many times as its (finite) multiplicity; we have anticipated the result that $\lambda_0' = 0$ has multiplicity 1. There is a system of real eigenfunctions $\phi_j$ corresponding to $\lambda_j'$, orthonormal with respect to $d\omega$, i.e. $\iint_\Omega \phi_j(z)\,\phi_k(z)\,d\omega = \delta_{jk}$. Since $\mathscr{U}/G$ is compact, $\phi_0$ is harmonic everywhere and so constant, and by the normalization,

$$\phi_0(z) \equiv (4\pi(g-1))^{-1/2}. \tag{6.2.3}$$

Thus $\lambda_0' = 0$ has multiplicity 1, as stated above. Furthermore,

$$\sum_{j=1}^{\infty} (1/\lambda_j')^2 < \infty, \quad \sum_{j=1}^{\infty} (\phi_j(z)/\lambda_j')^2 < \infty \tag{6.2.4}$$

the convergence of the second series being uniform in $z \in \mathscr{H}$. Finally, every $f$ satisfying 6.2.1 can be expanded as

$$f(z) = \sum_{n=0}^{\infty} a_n \phi_n(z), \qquad a_n = \iint_\Omega f(z)\,\phi_n(z)\,d\omega. \tag{6.2.5}$$

With $d$ the $H$-distance in $\mathscr{H}$ we abbreviate $d(z, i) = d(z)$. There is an invariant of the metric:

$$\cosh d(z, w) = 1 + \frac{|z - w|^2}{2\operatorname{Im} z \operatorname{Im} w},$$

as can be seen from the formula (3.6) in Chapter 2. Define the Dirichlet series

$$J(z, s) = \sum_{A \in G} \cosh^{-s} d(Az), \qquad s = \sigma + it. \tag{6.2.6}$$

We shall see that (6.2.6) converges absolutely and uniformly for $\sigma \geqslant \sigma_0 > 1$, $z \in \mathscr{H}$ and satisfies a functional equation

$$\Delta_z J(z, s) + s(1 - s)\, J(z, s) + s(1 + s)\, J(z, s + 2) = 0, \sigma > 1. \tag{6.2.7}$$

For fixed $s$ the function $J(z, s)$ can be expanded in a Fourier series in the eigenfunctions $\phi_j(z)$, as in (6.2.5), and by the use of (6.2.7) we shall show, after a long discussion, that

$$(6.2.8) \qquad J(z, s) = \frac{\alpha}{s-1} + \frac{\pi^{\frac{1}{2}} 2^{s-1}}{\Gamma(s)} H(z, s), \qquad \alpha = \frac{1}{2(g-1)}$$

with

$$(6.2.9) \qquad H(z, s) = \sum_{1}^{\infty} \Gamma\left(\frac{s - s^{+}(\lambda_n')}{2}\right) \Gamma\left(\frac{s - s^{-}(\lambda_n')}{2}\right) \phi_n(z)\, \phi_n(i).$$

Here

$$(6.2.10) \qquad s^{+}(\lambda) = \tfrac{1}{2}(1 + \sqrt{1 - 4\lambda}), \qquad s^{-}(\lambda) = \tfrac{1}{2}(1 - \sqrt{1 - 4\lambda})$$

and $H(z, s)$ converges absolutely and uniformly in every closed set that avoids the points

$$(6.2.11) \qquad P = \{s^{+}(\lambda_n') - 2l, s^{-}(\lambda_n') - 2l \colon n \geqslant 1, l = \text{integer} \geqslant 0\}.$$

This remarkable formula provides the analytic continuation of the Dirichlet series for $J$ and shows $J$ to be meromorphic in the whole $s$-plane with poles in the set $P$.

If $\lambda_n'$ is one of the finite number of values in $(0, \frac{1}{4})$, the points $s^{+}(\lambda_n')$, $s^{-}(\lambda_n')$ lie in $(0, 1)$; otherwise $s^{+}, s^{-}$ are complex conjugates on $\sigma = \frac{1}{2}$. Hence $J(z, s) - \alpha(s-1)$ is a holomorphic function of $s$ in $\sigma > 1 - \delta$ for some $\delta > 0$. We can now apply the Wiener–Ikehara Tauberian theorem, which can be stated in the following form ([**32**], p. 241): let $\psi(x) \geqslant 0$ be non-decreasing in $0 \leqslant x < \infty$ and let $\psi(x) = 0(x)$, $x \to \infty$. If

$$(*) \qquad \phi(s) := \int_{1}^{\infty} x^{-s}\, d\psi(x) = \frac{\alpha}{s-1} + R(s), \qquad \sigma > 1, \qquad \alpha \neq 0$$

is analytic for $\sigma > 1$ and $R(s)$ is continuous on $\sigma = 1$, then

$$\psi(x) \sim \alpha x, \qquad x \to \infty.$$

To use this for our purpose, define

$$N(z, u) = \operatorname{card}\{A \in G \colon d(Az) \leqslant u\}, \tilde{N}(x) = N(z, \operatorname{arc\,cosh} x),$$

and choose $\psi(x) = \tilde{N}(x)$. We shall see (Lemma 6.3.1) that $N(z,u) < m\,e^u$, so that $|\psi(x)| \leqslant m \exp(\operatorname{arc}\cosh x) \leqslant mx$. And

$$\phi(s) = \int_1^\infty x^{-s}\,dN(z, \operatorname{arccosh} x) = \int_0^\infty \cosh^{-s} u\,dN(z,u)$$
$$= \sum_{A \in G} \cosh^{-s} d(Az) = J(z,s),$$

so that (*) is fulfilled with $\alpha = \frac{1}{2}(g-1)$, in view of 6.2.8 and the following lines. Hence

$$\psi(\cosh t) = \tilde{N}(\cosh t) = N(z,t) \sim \alpha \cosh t \sim \alpha\, e^{t/2}.$$

6.2.12 Theorem. $N(z,t) \sim \alpha\, e^t/2, \alpha = \frac{1}{2}(g-1)$.

We can interpret this result in our old situation in which the group acts on $\mathscr{U}$. Map $\mathscr{H}$ to $\mathscr{U}$ by a Möbius transformation $\xi = Tz$ that carries $i$ to 0. $T$ carries the Poincaré metric on $\mathscr{H}$ to a metric on $\mathscr{U}$ given by $\lambda_1|d\xi|$ with $\lambda_1 = 2/(1-|z|^2)$, differing from our former metric by a factor of 2. In the new metric $d(0,\xi) = \log(1+|\zeta|)/(1-|\zeta|)$, $\sigma(\Delta_r) = 4\pi r^2/(1-r^2)$; moreover, for a group of genus $g > 1$, $\sigma(\Omega) = 4\pi(g-1)$. Let $G_1 = TGT^{-1}$, $A_1 = TAT^{-1}$, $A \in G$. Then

$$\text{(6.2.13)}\quad N(z,t) = \operatorname{card}\{A \in G : d(z,i) \leqslant t\} = \operatorname{card}\{A_1 \in G_1 : d(\zeta,0) \leqslant t\}$$
$$= \operatorname{card}\{A_1 : |\zeta| \leqslant \tanh(t/2)\} = n_0(\tanh(t/2), \zeta; G_1).$$

Since $G$ has no elliptic elements, $n_0(r,\zeta) = n(r,\zeta)$. Hence by the theorem,

$$\text{(6.2.14)}\qquad n(r,\zeta) \sim \frac{\alpha}{2}\frac{1+r}{1-r} \sim \frac{1}{2(g-1)}\frac{1}{1-r} \sim \frac{\sigma(\Delta_r)}{\sigma(\Omega)}, \qquad r \to 1$$

as announced in 6.1.1.

**6.3** In order to carry out this programme we note first

6.3.1 Lemma $\qquad N(z,t) < me^t, \qquad t > 0, \qquad z \in \mathscr{H}$

By 6.2.13 this reduces to Tsuji's estimate, 1.2.4.

6.3.2 LEMMA. *The series for $J(z, s)$ and the series obtained by term-by-term differentiation up to derivatives of the second order converge absolutely and uniformly for $\sigma \geqslant \sigma_0 > 1$, $z$ in a compact subset of $\mathscr{H}$. Moreover, $J(z, s) \in C^2(\mathscr{H})$ and is G-automorphic. The partial derivatives of J up to second order may be obtained by term-by-term differentiation of the series for J. Finally,*

$$|J(z, s)| \leqslant m\sigma/(\sigma - 1), \qquad z \in \mathscr{H}, \qquad \sigma > 1. \tag{6.3.3}$$

The lemma is proved by utilizing Lemma 6.3.1 and a formula analogous to 2.2.5, namely,

$$\sum_{t_1 \leqslant d(Az) < t_2} \cosh^{-\sigma} d(Az) = N(z, t_2) \cosh^{-\sigma} t_2 + \sigma \int_{t_1}^{t_2} N(z, u) \cosh^{-\sigma - 1} u \sinh u \, du. \tag{6.3.4}$$

■

6.3.5 LEMMA. *Equation 6.2.7 holds.*

For the proof we first note that $\Delta_z$ commutes with every $T \in \mathbf{M\ddot{o}b}\, \mathscr{H}$, i.e.

$$\Delta_z(f \circ T)(z) = (\Delta_u f(u))_{u=Tz}. \tag{6.3.6}$$

Now

$$\cosh d(z) = 1 + \frac{|z - i|^2}{2y}, \tag{6.3.7}$$

as already noted in the lines following 6.2.5. Using 6.3.6 and 6.3.7 we verify that

$$\Delta_z \cosh^{-s} d(Az) = s(s - 1) \cosh^{-s} d(Az) - s(s + 1) \cosh^{-(s+2)} d(Az).$$

Finally apply Lemma 6.3.2 to get

$$\Delta_z J(z, s) = \sum_{A \in G} \Delta_z \cosh^{-s} d(Az).$$

Next, consider the orthogonal expansion 6.2.5

$$J(z, s) = \sum_{n=0}^{\infty} a_n(s) \phi_n(z),$$

where, because of $\phi_n(Az) = \phi_n(z)$, we get

$$a_n(s) = \iint_{\Omega} J(z, s)\, \phi_n(z)\, d\omega = \sum_{A \in G} \iint_{\Omega} \cosh^{-s} d(Az) \,.\, \phi_n(z)\, d\omega$$

$$= \sum_{A} \iint_{A\Omega} \cosh^{-s} d(z)\, \phi_n(z)\, d\omega,$$

(6.3.8)
$$a_n(a) = \iint_{\mathcal{U}} \cosh^{-s} d(z)\, \phi_n(z)\, d\omega.$$

We introduce new coordinates $\theta = \theta(z)$, $\tau = \tau(z)$, where $\theta$ is the positive angle between the geodesic form $i$ to $z$ and a fixed geodesic through $i$, and $\tau(z) = \log\cosh d(z)$. We then find that

(6.3.9)
$$a_n(s) = \int_0^\infty e^{-(s-1)\tau} \Phi_n(\tau)\, d\tau, \qquad \Phi_n(\tau) = \int_0^{2\pi} \phi_n(z(\tau, \theta))\, d\theta.$$

Since $\Phi_n(\tau)$ is continuous and bounded for $\tau \geqslant 0$, $\Phi_n(0) = 2\pi\phi_n(i)$, and $a_n(s)$ is the Laplace transform of $\Phi_n$, we get from known theorems,

(6.3.10)
$$\lim_{\sigma \to \infty} \sigma a_n(\sigma) = 2\pi\phi_n(i).$$

Let

(6.3.11)
$$f_n(s) = a_n(s) 2^{-s} \Gamma(s) / \Gamma\left(\frac{s - s_n^+}{2}\right) \Gamma\left(\frac{s - s_n^-}{2}\right),$$

where we are writing

(6.3.12)
$$s_n^+ = s^+(\lambda_n'), \qquad s_n^- = s^-(\lambda_n').$$

6.3.13 LEMMA. *$f_n(s)$ is holomorphic in $\sigma > 1$ and satisfies*

$$f_n(s + 2) = f_n(s).$$

*Proof.* The holomorphicity is clear from 6.3.11 and properties of the gamma function. Multiply the functional equation 6.2.7 by $\phi_n(z)$, integrate over $\Omega$, and use 6.2.1, 6.2.6 and 6.3.8 to get

$$\iint_{\Omega} \phi_n(z)\, \Delta_z J(z, s)\, d\omega + s(1 - s)\, a_n(s) + s(1 + s)\, a_n(s + 2) = 0. \tag{6.3.14}$$

We elaborate the integral by Green's formula. If $\mathscr{L}$ is the ordinary Laplacian, $\Delta_z = y^2 \mathscr{L}$ and $y^2\, d\omega = dx\, dy$, so

$$(*) \qquad \iint_{\Omega} \phi_n \Delta_z J\, d\omega = \iint_{\Omega} \phi_n \mathscr{L} J\, dx\, dy =$$

$$\iint_{\Omega} J \mathscr{L} \phi_n\, dx\, dy + \int_{\partial\Omega} \left( \phi_n y \frac{\partial J}{\partial \nu} - J y \frac{\partial \phi_n}{\partial \nu} \right) \frac{|dz|}{y},$$

$\nu$ being the outer normal. Now $\partial\Omega$ is the union of a finite number of pairs of arcs $(\gamma, A\gamma)$ and the contribution

$$\int_{\gamma \cup A\gamma} \phi_n y \frac{\partial J}{\partial \nu} \frac{|dz|}{y} = 0$$

if we take into account that $\phi_n$, $|dz|/y$, and $|\partial J/\partial(\nu/y)|$ are $G$-automorphic, while the map $\gamma \to A\gamma$ reverses the sense of the normal. Similarly

$$\int_{\gamma \cup A\gamma} J y \frac{\partial \phi_n}{\partial \nu} \frac{|dz|}{y} = 0.$$

Hence (*) yields

$$\iint_{\Omega} \phi_n(z)\, \Delta_z J(z, s)\, d\omega = \iint_{\Omega} J(z, s)\, \mathscr{L} \phi_n(z)\, dx\, dy \tag{6.3.15}$$

$$= \iint_{\Omega} J(z, s)\, \Delta_z \phi_n(z)\, d\omega = -\lambda_n' \iint_{\Omega} J(z, s)\, \phi_n(z)\, d\omega = -\lambda_n' a_n(s),$$

E

where we used 6.2.2 for the eigenfunction $\phi_n$. Combining 6.3.14 and 6.3.15, we get

$$s(s+1)\,a_n(s+2) = (s^2 - s + \lambda_n')\,a_n(s) = (s - s_n^+)(s - s_n^-)a_n$$

by 6.2.10 and 6.3.13. The lemma now follows from the functional equation for the gamma function.

By employing Stirling's formula for $\Gamma(s)$ we get from 6.3.10 that $f_n(\sigma) \to 2^{-1}\pi^{-\frac{1}{2}}\phi_n(i)$, $\sigma \to \infty$, and the last lemma now yields

$$f_n(s) = \text{const} = 2^{-1}\,\pi^{-\frac{1}{2}}\,\phi_n(\mathrm{i}).$$

Hence from 6.3.11 we have

6.3.16 LEMMA. *Let*

$$F(s, \lambda_n') = \Gamma\left(\frac{s - s^+(\lambda_n')}{2}\right)\Gamma\left(\frac{s - s^-(\lambda_n')}{2}\right), \qquad \lambda_n' \geqslant 0.$$

*Then*

$$a_n(s) = \pi^{\frac{1}{2}}2^{s-1}\,F(s, \lambda_n')\,\phi_n(i)/\Gamma(s), \qquad \sigma > 1.$$

**6.4** Introduce 6.3.16 into the Fourier series for $J(z, s)$:

$$J(z, s) = \frac{\pi^{\frac{1}{2}}2^{s-1}}{\Gamma(s)}\sum_{n=0}^{\infty} F(s, \lambda_n')\,\phi_n(z)\,\phi_n(i).$$

We separate the term $n = 0$: it is $\pi^{\frac{1}{2}}\,2^{s-1}\,\phi_0(z)\,\phi_0(i)/\Gamma(s)$ times

$$F(s, \lambda_0') = F(s, 0) = \Gamma\left(\frac{s-1}{2}\right)\Gamma\left(\frac{s}{2}\right) = \frac{2}{s-1}\Gamma\left(\frac{s}{2}\right)\Gamma\left(\frac{s+1}{2}\right)$$

$$= \frac{2^{2-s}\pi^{\frac{1}{2}}}{s-1}\Gamma(s),$$

by virtue of the "duplication formula"

$$2^{s-1}\Gamma\left(\frac{s}{2}\right)\Gamma\left(\frac{s+1}{2}\right) = \pi^{\frac{1}{2}}\,\Gamma(s).$$

Hence, noting the value of $\phi_0(z)$ in 6.2.3, we get

$$(6.4.1)\quad J(z, s) = \frac{\alpha}{s-1} + \frac{\pi^{\frac{1}{2}} 2^{s-1}}{\Gamma(s)} \sum_{n=1}^{\infty} F(s, \lambda'_n)\, \phi_n(z)\, \phi_n(i)$$

$$= \frac{\alpha}{s-1} + \frac{\pi^{\frac{1}{2}} 2^{s-1}}{\Gamma(s)} H(z, s), \qquad \alpha = \frac{1}{2(g-1)}$$

with $H$ defined as in 6.2.9.

What remains to be proved is that $H$ is meromorphic in the whole $s$-plane with poles in the set $P$ of 6.2.11. This would prove 6.2.8 and therefore Theorem 6.2.12.

Let $h \geqslant 2$ be an integer and let

$$(6.4.2)\quad Q_h = \{s : |\sigma| \leqslant h,\ |t| \leqslant h\}, \qquad N_0(h) = \max\{n : \lambda'_n \leqslant 4h^2 + \tfrac{1}{4}\}.$$

The desired properties of $H$ follow from the estimate

$$(6.4.3)\qquad \sum_{n=k}^{l} |F(s, \lambda'_n)\phi_n(z)\phi_n(i)| \leqslant m(4h^{-2})^{h-1}\Gamma(3h) \left\{ \sum_{n=k}^{l} \left( \frac{\phi_n(z)}{\lambda'_n} \right)^2 \right\}^{\frac{1}{2}},$$

where

$$l \geqslant k \geqslant N_0(h), \qquad s \in Q_h, \qquad z \in \mathscr{H}.$$

In fact if $s_0 \in P$, say $s_0 = s^+(\lambda'_n) - 2l$ for given $n$ and $l$, then we can take $h > (4\lambda'_n - 1)^{\frac{1}{2}}/4$ and the series $\sum_{j>n} F(s, \lambda'_j)\, \phi_j(z)\, \phi_j(i)$ by 6.4.3 and 6.2.4 will converge uniformly and absolutely on $Q_h$ and therefore on every compact set in the $s$-plane. But for each $h$ this series differs from $H(z, s)$ by a finite number of terms, all of which are holomorphic at $s_0$ except for a finite number of the form $F(s, \lambda'_{n+j})\, \phi_{n+j}(z)\, \phi_{n+j}(i)$, $1 \leqslant j \leqslant j_0$, corresponding to an eigenvalue $\lambda'_n$ of multiplicity $j_0$.

The proof of 6.4.3 requires a number of steps. Substitute the formula 6.3.16 for $a_n(s)$ into Bessel's inequality:

$$\sum_{1}^{\infty} |F(s, \lambda'_n)\, \phi_n(i)|^2 \leqslant \frac{|\Gamma(s)|^2}{\pi 2^{2\sigma-2}} \iint_{\Omega} |J(z, s)|^2\, d\omega.$$

Now $|\Gamma(s)| \leqslant \Gamma(\sigma)$ for $\sigma > 0$ and $\Gamma(\sigma)$ is increasing for $\sigma \geqslant 2$. Also $|J(z, s)| \leqslant m\sigma/(\sigma - 1)$ by 6.3.3. Hence

$$\sum_1^\infty |F(s, \lambda_n') \phi_n(i)|^2 \leqslant m\Gamma^2(\sigma_0), \qquad 2 \leqslant \sigma \leqslant \sigma_0. \tag{6.4.4}$$

Next we must estimate $F(s, \lambda)$. Using the functional equation for $\Gamma(s)$ we find

$$F(s, \lambda) = \frac{4}{(s - s^+(\lambda))(s - s^-(\lambda))} \prod_{k=1}^{h-1} (s + 2k - s^+(\lambda))^{-1}$$
$$\times (s + 2k - s^-(\lambda))^{-1} F(s + 2h, \lambda).$$

The rational function of $s$ can be estimated and we obtain:

$$|F(s, \lambda)| \leqslant m\lambda^{-1}(4h^{-2})^{h-1} |F(s + 2h, \lambda)|, \tag{6.4.5}$$
$$s \in Q_h, \qquad \lambda \geqslant 4h^2 + \tfrac{1}{4}, \qquad h \geqslant 2.$$

We can now prove 6.4.3. From 6.4.5 we have

$$\sum_{n=k}^{l} |F(s, \lambda_n') \phi_n(z) \phi_n(i)| \leqslant m(4h^{-2})^{h-1} \sum_{n=k}^{l} |F(s + 2h, \lambda_n')\phi_n(i)| \frac{|\phi_n(z)|}{\lambda_n'}$$
$$\leqslant m(4h^{-2})^{h-1} \left\{\sum_h^l |F)s + 2h, \lambda_n') \phi_n(i)|^2\right\}^{\frac{1}{2}} \left\{\sum_k^l \left(\frac{\phi_n(z)}{\lambda_n'}\right)^2\right\}^{\frac{1}{2}}.$$

When $s \in Q_h$, $h \geqslant 2$, we have $2 \leqslant h \leqslant \sigma + 2h \leqslant 3h$; hence by 6.4.4 the first sum in the right member is bounded by $m\Gamma(3h)$. 6.4.3 follows.

This completes the proof of Theorem 6.2.12. ■

**6.5** Lehner (unpublished) has shown

6.5.1 Theorem. *There exist groups G of* DT *with* $\sigma(G) = \infty$ *such that*

$$n(r, z) = o(1/(1 - r)), \qquad r \to 1.$$

The example uses the group $G = \{W \in F: exp_{a_1} W = 0\}$; where $F$ is a surface group of genus $g > 1$ and $\exp_{a_1} W$ is the algebraic sum of the exponents of $a_1$ in the expression for $W$ as a word in the generators. Moreover, Nicholls (unpublished) has proved that $\liminf n(r, z)(1-r) = 0,\ r \to 1$, if $G$ is DT and $\sigma(G) = \infty$. Thus 1.2.7 is characteristic for groups of finite area.

## REFERENCES


1. A. F. Beardon, Inequalities for certain Fuchsian groups, *Acta Math.* **127** (1971), 221–258.
2. A. F. Beardon, The exponent of convergence of Poincaré series, *Proc. Lond. math. Soc.* **18** (1968), 461–483.
3. L. Bers, *in* "Proceedings of the International Symposium on Linear Spaces, Jerusalem". Academic Press, London and New York, 1961, pp. 88–100.
4. L. Bers, Automorphic forms and Poincaré series for infinitely generated Fuchsian groups, *Amer. J. Math.* **87** (1965), 196–214.
5. D. Drasin, Cusp forms and Poincaré series, *Amer. J. Math.* **90** (1968), 356–365.
6. C. J. Earle, A reproducing formula for integrable automorphic forms, *Amer. J. Math.* **88** (1966), 867–870.
7. C. J. Earle, Some remarks on Poincaré series, *Compositio math.* **21** (1969), 167–176.
8. J. Elstrodt, Die Resolvente zum Eigenwertproblem der automorphen Formen in der hyperbolischen Ebene, I, *Math. Ann.* **203** (1973), 295–330.
9. L. R. Ford, "Automorphic Functions". Chelsea, New York, 1951.
10. D. Hilbert, "Grundzuge einer allgameinen Theorie der linearen Integralglcichen", Teubner, Berlin, 1912.
11. H. Huber, Zur analytischen Theorie hyperbolischen Raumformen und Bewegungsgruppen, *Math. Ann.* **138** (1959), 1–26.
12. I. Kra, "Automorphic Forms and Kleinian Groups". Benjamin, Reading, Mass., 1972.
13. J. Lehner, On the boundedness of integrable automorphic forms, *Illinois J. Math.* **18** (1974), 575–584.
14. J. Lehner, "Discontinuous Groups and Automorphic Functions". American Mathematical Society, Providence, 1964.
15. A. Marden, Universal properties of Fuchsian groups in the Poincaré metric, *Ann. Math. Stud.* **79** (1974), 315–339.
16. T. A. Metzger, On the growth of the Taylor coefficients of automorphic forms, *Proc. Amer. math. Soc.* **39** (1971), 321–328.
17. R. Nevanlinna, "Uniformisierung", Springer, 1953.
18. J. Nielsen and K. Fenchel, "Groups of Linear Transformations", in preparation.
19. S. J. Patterson, The limit set of a Fuchsian group, II, *Acta Math.* **136** (1976) 241–273.
20. S. J. Patterson, A lattice point problem in hyperbolic space, *Mathematika*, **22** (1975) 81–88.
21. H. Petersson, Über eine Metrisierung der automorphen Formen und die Theorie der Poincaréschen Reihen, *Math. Ann.* **117** (1940), 453–537.
22. H. Petersson, Theorie der automorphen Formen beliebigen reeller Dimension und ihre Darstellung eine neue Art Poincaréschen Reihen, *Math. Ann.* **103** (1930), 369–436.
23. Ch. Pommerenke, On inclusion relations for spaces of automorphic forms, *in* "Advances in Complex Function Theory". Springer Lecture Notes No. 505, pp. 92–100.

24. Ch. Pommerenke, On Bloch functions, *J. London Math. Soc.* **2** (1970), 689–695.
25. R. V. Rajeswara Rao, On the boundedness of $p$-integrable automorphic forms, *Proc. Amer. math. Soc.* **44** (1974), 278–282.
26. R. V. Rajeswara Rao, Fuchsian groups of convergence type and Poincaré series of dimension $-2$, *J. Math. Mech.* **18** (1969), 629–644.
27. A. Selberg, On the estimation of Fourier coefficients of modular forms, *in* "Symposium on Number Theory". American Mathematical Society, 1965.
28. A. Selberg, Harmonic analysis and discontinuous groups on weakly symmetric spaces with applications to Dirichlet series, *J. Indian Math. Soc.* **20** (1956), 47–87.
29. G. Shimura, "Introduction to the Arithmetic Theory of Automorphic Functions".
30. C. L. Siegel, "Topics in Complex Function Theory", Vol. II. Wiley–Interscience, New York, 1969.
31. M. Tsuji, "Modern Potential Theory". Maruzen, Tokyo, 1959.
32. W. A. Veech, "A Second Course in Complex Analysis". W. A. Benjamin, New York, 1967.

# 4. Quasiconformal Homeomorphisms and Beltrami Equations

OLLI LEHTO

*University of Helsinki, Finland*



This is a quick survey of the properties of quasiconformal mappings in the plane, with some historical remarks and with applications to the theory of Teichmüller spaces in mind. Corresponding more or less to the lectures given, the representation is necessarily concise. For detailed proofs and less standard real analysis background, the readers can consult the monograph [**22**], to which we refer throughout the text.

## 1. QUASICONFORMAL DIFFEOMORPHISMS

Historically, there have been two approaches to quasiconformal mappings. One is based on a natural generalization of the local behaviour of conformal mappings, while the other is via the Beltrami equation encountered in surface theory. Our starting point is the first one; the connection with the Beltrami equation will be subsequently elucidated.

Let $f$ be a diffeomorphism between finite domains $A$ and $A'$ of the plane, i.e. $f: A \to A'$ is one-to-one and onto and $f$ and its inverse $f^{-1}$ are continuously differentiable in $A$ and $A'$, respectively. A diffeomorphism maps infinitesimal circles onto infinitesimal ellipses, and if the image ellipses have a uniformly bounded ratio of axes, the mapping is quasiconformal:

1.1 *Definition.* A diffeomorphism $f: A \to A'$ is $K$-quasiconformal if $f$ is sense-preserving and

$$\max_{\alpha} |\partial_{\alpha} f(z)| \leqslant K \min_{\alpha} |\partial_{\alpha} f(z)|$$

for every $z \in A$, where $\partial_{\alpha} f$ denotes the derivative of $f$ in the direction $\alpha$.

Using auxiliary Möbius transformations, we can extend the definition to the cases where $\infty$ lies in $A$ or $A'$. In the following, we shall often disregard the special point $\infty$, leaving it to the reader to decide what modifications are possibly needed to take account of it.

From the definition it is clear that a diffeomorphism is 1-quasiconformal if and only if it is conformal.

The dilatation condition 1.1 is equivalent to a global condition which can be expressed in very general terms. In order to formulate this, let us consider a family $\Gamma$ of Jordan arcs or curves lying in the domain $A$. Let $P(\Gamma)$ be the family of all non-negative Borel functions $\rho$ with the property $\int_{\gamma} \rho |dz| \geqslant 1$ for every locally rectifiable $\gamma \in \Gamma$. The greatest lower bound

$$M(\Gamma) = \inf_{\rho \in P(\Gamma)} \iint_{A} \rho^2 \, dx \, dy$$

is called the *module* and $1/M(\Gamma)$ the *extremal length* of $\Gamma$.

The module, which is a conformal invariant, remains quasi-invariant under quasiconformal mappings:

1.2 THEOREM. *If $f$ is a $K$-quasiconformal diffeomorphism, then*

$$M(f(\Gamma)) \leqslant K M(\Gamma)$$

*for every family $\Gamma$ of $A$.*

The validity of 1.2 follows easily from the definition of $M$ and the inequality 1.1 ([**22**], p. 18). ■

In fact, conditions 1.1 and 1.2 are equivalent for diffeomorphisms. The converse to Theorem 1.2 can be expressed under apparently weaker hypotheses, using special families associated with quadrilaterals or ring domains.

Let $Q(z_1, z_2, z_3, z_4)$ be a quadrilateral, i.e. a Jordan domain $Q$ and a sequence $z_1, z_2, z_3, z_4$ of points on the boundary of $Q$. Suppose $Q$ is mapped conformally onto a rectangle $R$ so that the points $z_1, z_2, z_3, z_4$ correspond to the vertices of $R$. Let $a$ and $b$ be the lengths of the sides of $R$, the side of length $a$ corresponding to the arc $(z_1, z_2)$ with endpoints $z_1$ and $z_2$. Then

$$\operatorname{mod} Q(z_1, z_2, z_3, z_4) = a/b$$

is the module of $Q(z_1, z_2, z_3, z_4)$. It is well defined and conformally invariant.

Let $\Gamma$ be the family of locally rectifiable arcs joining in $Q$ the sides $(z_1, z_2)$ and $(z_3, z_4)$. Then

$$\operatorname{mod} Q(z_1, z_2, z_3, z_4) = M(\Gamma). \tag{1.3}$$

Because of the conformal invariance of module, we may assume in proving 1.3 that $Q$ is a rectangle, with $z_1 = 0$, $z_2 = M > 0$, $z_3 = M + i$, $z_4 = i$. Then $\operatorname{mod} Q(z_1, z_2, z_3, z_4) = M$. On the other hand, for any $p \in P(\Gamma)$,

$$1 \leqslant \left( \int_0^1 \rho(x + iy)\, dy \right)^2 \leqslant \int_0^1 \rho^2(x + iy)\, dy,$$

and so

$$M \leqslant \iint_Q \rho^2\, dx\, dy.$$

Equality holds for the constant function 1, which is in $P(\Gamma)$, and 1.3 follows.

We conclude from 1.2 and 1.3 that a $K$-quasiconformal diffeomorphism of $A$ satisfies the inequality

$$\operatorname{mod} f(Q) \leqslant K \operatorname{mod} Q \tag{1.4}$$

for every quadrilateral $Q$ (if no confusion is possible we let the symbol $Q$ alone stand for a quadrilateral) whose closure $\bar{Q}$ lies in $A$.

There is the following connection between 1.4 and 1.1 ([**22**], p. 50).

1.5 THEOREM. *Let $f$ be a homeomorphism of $A$ satisfying* 1.4 *for every quadrilateral $Q$, $\bar{Q} \subset A$. Then* 1.1 *holds at every point $z \in A$ at which $f$ is differentiable.*

It follows, in particular, that 1.2 holds for a sense-preserving diffeomorphism $f$ if and only if $f$ is $K$-quasiconformal.

The characterization 1.3 of mod $Q$ makes it possible to estimate mod $Q$ in geometric terms. For instance, let $s_a$ and $s_b$ denote the Euclidean distances in $Q$ between the sides $(z_1, z_2)$, $(z_3, z_4)$ and $(z_2, z_3)$, $(z_4, z_1)$, respectively, and $m(Q)$ the area of $Q$. Choosing $\rho = 1/s_a$ and $\rho = 1/s_b$, we obtain from 1.3 the following estimates (*Rengel's inequality*).

The module of the quadrilateral satisfies the double inequality

$$s_b^2/m(Q) \leqslant \operatorname{mod} Q(z_1, z_2, z_3, z_4) \leqslant m(Q)/s_a^2.$$

Equality holds if and only if $Q$ is a rectangle ([**22**], p. 22).

Another important concept in the theory of quasiconformal mappings is the module of a ring domain. If a ring domain B is conformally equivalent to an annulus $\{z : r_1 < |z| < r_2\}$, then the module of $B$ is defined by $\operatorname{mod} B = \log(r_2/r_1)$. Let $\Gamma$ be the family of rectifiable Jordan curves separating the boundary components of $B$. Then

$$\operatorname{mod} B = 2\pi M(\Gamma).$$

This leads to estimates analogous to Rengel's inequality ([**22**], pp. 30–33).

## Notes

Definition 1.1 of a quasiconformal diffeomorphism was given by Grötzsch [**18**] in 1928. He proved that 1.1 and 1.4 are equivalent for diffeomorphisms. Ahlfors [**1**] pointed out the importance of these Grötzsch mappings in his studies on covering surfaces in 1935 and suggested the name quasiconformal. At about the same time, Lavrentiev [**20**] introduced a class of homeomorphisms, also by generalizing the local behaviour of conformal mappings. They are somewhat more general than the Grötzsch mappings, as will be explained in the Notes of Section 6. A few years later, Teichmüller published a number of papers ([**32, 33, 34**], and others) in which he made fundamental discoveries of the role of quasiconformal mappings in the function theory on Riemann surfaces. His work has greatly inspired subsequent research: an extensive theory has been developed around the notion of Teichmüller space.

Ahlfors and Beurling [**7**] introduced extremal length in 1950. A brief historical account is in Ahlfors [**5**], in which reference is made to earlier underlying ideas of Beurling. The use of module rather than extremal length was advocated by Fuglede [**14**] whose generalizations have turned out to be important in the higher-dimensional theory of quasiconformal mappings.

## 2. QUASICONFORMAL HOMEOMORPHISMS

The requirement that a quasiconformal mapping be continuously differentiable is too restrictive in many applications. An elegant generalized definition of quasiconformality can be given in terms of the module of curve families. There are some practical advantages, however, to base the definition on the modules of quadrilaterals, as did Ahlfors and Pfluger in the early fifties in generalizing Grötzsch mappings.

2.1 *Definition.* A homeomorphism $f: A \to A'$ is $K$-quasiconformal if $f$ is sense-preserving and

$$\operatorname{mod} f(Q) \leqslant K \operatorname{mod} Q$$

for every quadrilateral $Q$ with $\overline{Q} \subset A$.

A mapping is quasiconformal if it is $K$-quasiconformal for some $K$, $1 \leqslant K < \infty$. The infimum of the numbers $K$ satisfying 2.1 is called the maximal dilatation of $f$. If $f$ is $K$-quasiconformal, then clearly its inverse is also $K$-quasiconformal.

From Theorem 1.2 it follows that a $K$-quasiconformal diffeomorphism is $K$-quasiconformal in the sense of this more general definition. The converse is not true, but we shall soon see that a quasiconformal homeomorphism is differentiable almost everywhere. Hence, by Theorem 1.5, the dilatation condition 1.1 holds at almost all points.

It is easy to prove that a $K$-quasiconformal mapping of $A$ satisfies the inequality

$$\operatorname{mod} f(B) \leqslant K \operatorname{mod} B$$

for every ring domain $B$ with $\overline{B} \subset A$. Actually even the general inequality 1.2 remains valid, but the proof requires deeper insight into the properties of quasiconformal mappings ([**22**], p. 171). In particular, we could replace 2.1 by 1.2 in the above definition.

As in the case of diffeomorphisms, a 1-quasiconformal mapping is conformal; this follows from Rengel's inequality without difficulty ([**22**], p. 28). An important gain is the following compactness property, not enjoyed by quasiconformal diffeomorphisms. Let $f_n$ be a sequence of $K$-quasiconformal mappings of $A$ which converges to a non-constant limit function $f$, uniformly on every compact subset of $A$. Then $f$ is $K$-quasiconformal ([**22**], p. 74).

*Definition.* Let $R = \{x + iy : a \leqslant x \leqslant b, c \leqslant y \leqslant d\}$ be a rectangle, $I_x$ the vertical segment with abscissa $x$ joining the horizontal sides of $R$, and $I'_y$ the horizontal segment with ordinate $y$ joining the vertical sides of $R$. A complex-valued function $f$ continuous in a domain $A$ is said to be *absolutely continuous on lines* (ACL) in $A$ if the following condition is satisfied: For every $R$ lying in $A$, the restrictions $f|I_x$ and $f|I'_y$ are absolutely continuous, the former for almost all $x \in [a, b]$, the latter for almost all $y \in [c, d]$.

It follows from well known theorems of real analysis that a function which is ACL has finite partial derivatives a.e. In the general case, the existence of partial derivatives does not tell us much about the differentiability of the function. For homeomorphisms, which we shall be dealing with, the situation is different.

2.2 Lemma. *Let $f$ be a continuous open mapping of a plane domain into the plane. If $f$ has finite partial derivatives almost everywhere then $f$ is differentiable almost everywhere.*

The proof is based on the maximum principle which is combined with standard density theorems; for the details we refer to [**22**], p. 128.

By means of Rengel's inequality it can be proved that a quasiconformal mapping is ACL ([**22**], p. 162). Using Lemma 2.2, we can thus deduce a number of important properties of quasiconformal mappings.

2.3 Theorem. *Let $f$ be a $K$-quasiconformal mapping. Then $f$ is differentiable a.e., satisfies the dilatation condition* 1.1 *a.e., and its derivatives are locally $L^2$-integrable.*

*Proof.* Since $f$ is ACL, its partials exist a.e. and differentiability a.e. is implied by Lemma 2.2. The validity of 1.1 a.e. then follows from Theorem 1.5. In order to establish the third statement, we first conclude from the equation

$$\partial_\alpha f = f_z + f_{\bar{z}}\, e^{-2i\alpha} \tag{2.4}$$

that

$$\max |\partial_\alpha f(z)| = |f_z(z)| + |f_{\bar{z}}(z)|, \tag{2.5}$$

$$\min |\partial_\alpha f(z)| = |f_z(z)| - |f_{\bar{z}}(z)|.$$

The Jacobian $J$ of $f$ is

$$J = |f_z|^2 - |f_{\bar{z}}|^2. \tag{2.6}$$

Multiplying both sides of 1.1 by $\max |\partial_\alpha f(z)|$ we thus obtain

$$\max |\partial_\alpha f(z)|^2 \leqslant KJ(z).$$

Since the Jacobian of an a.e. differentiable homeomorphism is locally integrable (see inequality 3.4 in the following section), we see that the partial derivatives of $f$ are locally in $L^2$.

We shall see later that the derivatives are in fact locally in $L^p$ for some $p > 2$ (Theorem 6.7).

**Notes**

Definition 2.1 was suggested by Pfluger [**27**]. Ahlfors was the first to use these quasiconformal mappings in [**2**]. This paper revived interest in Teichmüller's work and gave impetus to rapidly expanding new research,

That 2.1 implies the seemingly much more general inequality of Theorem 1.2 was proved by Väisälä [**35**]; the proof makes use of Theorem 2.3,

Strebel [**31**] and Mori [**24**] proved that a quasiconformal mapping is ACL; the particularly simple proof to which reference is made in the text is due to Pfluger [**28**]. Differentiability a.e. was first established by Mori [**24**], whose proof was based on a distortion inequality for quasiconformal mappings and Rademacher-Stepanoff's theorem. Lemma 2.2, which will be applied in Section 3 also, is due to Gehring and Lehto [**16**].

## 3. GENERALIZED DERIVATIVES

Let $f$ be a complex-valued function defined in a domain $A$ of the plane. The function $f$ is said to have (generalized) $L^p$-derivatives, $p \geqslant 1$, in $A$ if the following two conditions are fulfilled in $A$: (i) $f$ is ACL (ii) the derivatives $f_z$ and $f_{\bar{z}}$ belong locally to $L^p$.

An equivalent definition is as follows: $f$ has $L^p$-derivatives in $A$ if and only if $f$ is continuous in $A$ and there are functions $g, h \in L^p$ locally in $A$ such that

$$\iint_A (f\phi_z + g\phi)\,dx\,dy = 0, \quad \iint_A (f\phi_{\bar{z}} + h\phi)\,dx\,dy = 0$$

for every $\phi \in C_0^1(A)$. If this condition holds, then $g = f_z$, $h = f_{\bar{z}}$ a.e.

A third characterization depends on the fact that functions with $L^p$-derivatives admit a particular $C_0^\infty$-approximation. For convenience of later reference we express this result as a lemma ([**22**], p. 147).

3.1 LEMMA. *Let $f, g, h$ be functions in $A$, $g$ and $h$ belonging locally to $L^p$. Further, let $f_n \in C_0^\infty(A)$, $n = 1, 2, \ldots$, be functions such that for every compact set $F \subset A$,*

(i) $\lim_{n\to\infty} f_n(z) = f(z)$ *uniformly on $F$,*

(ii) $\lim_{n\to\infty} \iint |(f_n)_z - g|^p\,dx\,dy = 0$, $\lim_{n\to\infty} \iint |(f_n)_{\bar{z}} - h|^p\,dx\,dy = 0$.

*Then $f$ has $L^p$-derivatives in $A$. Conversely, if $f$ has $L^p$-derivatives in $A$, there are functions $f_n \in C_0^\infty(A)$ satisfying conditions* (i) *and* (ii), *with $g = f_z$, $h = f_{\bar{z}}$.*

The classical Green's formula can be generalized with the aid of the latter part of this theorem ([**22**], p. 148). Let $f$ and $g$ be functions having respectively $L^p$- and $L^q$-derivatives in $A$, where $1/p + 1/q = 1$. If $D$, $\bar{D} \subset A$, is a Jordan domain with a rectifiable boundary on which $g$ is of bounded variation, then

$$\int_{\partial D} f\,dg = -2i \iint_D (f_z g_{\bar{z}} - f_{\bar{z}} g_z)\,dx\,dy;$$

here $\partial D$ is the positively oriented boundary of $D$. This relation also holds if $f$ has $L^1$-derivatives and $g$ belongs, say, to class $C^2$.

The choices $g(z) = z$ and $g(z) = \bar{z}$ yield the important formulae

$$\int_{\partial D} f\,dz = 2i \iint_D f_{\bar{z}}\,dx\,dy, \quad \int_{\partial D} f\,d\bar{z} = -2i \iint_D f_z\,dx\,dy. \tag{3.2}$$

From these we obtain immediately the following generalization of a well-known classical result.

3.3 LEMMA. *Let $g$ and $h$ be functions in a simply connected domain $A$ which have $L^1$-derivatives and satisfy $g_{\bar{z}} = h_z$ a.e. Then there exists a function $f \in C^1(A)$ such that $f_z = g, f_{\bar{z}} = h$ a.e.*

*Proof.* Applying 3.2 we conclude that

$$\int_{\partial D} g \, dz + h \, d\bar{z} = 0.$$

After this, the proof is as in the classical case.

Let $f: A \to A'$ be a homeomorphism and $B \subset A$ a Borel set. The mapping $f$ is said to be (locally) *absolutely continuous* in $A$ if the set function $B \mapsto m(f(B))$ is (locally) absolutely continuous in $A$, where $m$ denotes the two-dimensional Lebesgue measure. The following result is an easy consequence of standard theorems of measure and integration theory ([**22**], p. 131). If a sense-preserving homeomorphism $f: A \to A'$ is differentiable a.e. in $A$, then

$$\iint_B J \, dx \, dy \leqslant m(f(B)) \tag{3.4}$$

for every Borel set $B \subset A$. Equality holds for every $B$ if and only if $f$ is locally absolutely continuous in $A$. Then

$$\iint_E J \, dx \, dy = m(f(E)) \tag{3.5}$$

for every measurable set $E \subset A$,

This criterion for absolute continuity, in conjunction with the generalized Green's formula, yields the following result ([**22**], p. 150).

3.6 LEMMA. *A homeomorphism having $L^2$-derivatives in locally absolutely continuous.*

We can now give applications to quasiconformal mappings.

3.7 THEOREM. *A quasiconformal mapping $f$ of $A$ is locally absolutely continuous in $A$, with $J(z) > 0$ and $f_z(z) \neq 0$ a.e. in $A$.*

*Proof.* By Theorem 2.3, $f$ has $L^2$-derivatives. Consequently, by Lemma 3.6, $f$ is locally absolutely continuous. Thus 3.5 holds. Since the inverse of a quasiconformal mapping is quasiconformal, it follows from 3.5 that $J(z) > 0$ a.e. From 2.6 we then deduce that $f_z(z) \neq 0$ a.e. ■

For functions with $L^p$-derivatives the following composition theorem is true ([22], p. 151).

*Let $w: A \to A'$ and its inverse $w^{-1}: A' \to A$ be homeomorphisms with $L^p$-derivatives, $p \geqslant 2$, and $f$ a function with $L^q$-derivatives in $A'$, $1/p + 1/q = 1$. Then $f \circ w$ has $L^1$-derivatives in $A$, and its derivatives are calculated by the customary chain rule.*

With the aid of this result, a new characterization can be given for quasiconformality ("Analytic definition").

3.8 THEOREM. *Let $f$ be a sense-preserving homeomorphism of a domain $A$ which is ACL in $A$ and satisfies 1.1 a.e. in $A$. Then $f$ is $K$-quasiconformal in $A$.*

*Proof.* By Lemma 2.2, $f$ is differentiable a.e. Repeating the reasoning used in the proof of Theorem 2.3, we conclude that $f$ has $L^2$-derivatives. Let $Q$ be an arbitrary quadrilateral, $\bar{Q} \subset A$, and $g: Q \to R$, $h: f(Q) \to R'$, conformal maps onto rectangles $R$ and $R'$ whose vertices correspond to the distinguished boundary points of $Q$ and $f(Q)$. The theorem follows if we prove that

$$\text{mod}\, R' \leqslant K \,\text{mod}\, R \tag{3.9}$$

The decisive step in proving 3.9 is to deduce from the above composition theorem that the mapping $h \circ f \circ g^{-1}: R \to R'$ satisfies the hypotheses of Theorem 3.8 in $R$. After this, 3.9 can be established by the standard length-area method. (For details see [22], p. 168). ■

The counter-example $x + iy \mapsto x + i(y + \phi(x))$, $A = \{x + iy: 0 < x < 1, 0 < y < 1\}$, where $\phi$ is the Cantor function, shows that ACL is indispensable in Theorem 3.8.

## Notes

The notion of $L^p$-derivatives, fundamental in the modern theory of partial differential equations, is attributed to Friedrichs [**13**] and Sobolev [**29**]. Morrey [**25**] in fact used $L^2$-derivatives in 1938.

The history of Theorem 3.7 will be told in the Notes of Section 4. Theorem 3.8, under the additional hypotheses that $f$ is differentiable a.e. and has $L^1$-derivatives, was first established by Yûjôbô [**39**]. Later Bers [**9**] and Pfluger [**28**] relaxed *a priori* requirements. Under the above minimal conditions, the theorem was proved by Gehring and Lehto [**16**].

## 4. THE BELTRAMI EQUATION

Let $f$ be a $K$-quasiconformal mapping of a domain $A$ and $z \in A$ a point at which $f$ is differentiable. From 2.5 it follows that the dilatation condition 1.1 is equivalent to

$$(4.1) \qquad |f_{\bar{z}}(z)| \leqslant \frac{K-1}{K+1}|f_z(z)|.$$

We say that $z \in A$ is a *regular point* of $f$ if $f$ is differentiable at $z$ and $J(z) > 0$. At a regular point $f_z(z) \neq 0$, and so the quotient

$$(4.2) \qquad \mu(z) = f_{\bar{z}}(z)/f_z(z)$$

can be defined. By Theorems 2.3 and 3.7, $\mu(z)$ exists a.e. in $A$. The function $\mu$ is called the *complex dilatation* of the mapping $f$. It is a Borel measurable function, and by 4.1

$$(4.3) \qquad |\mu(z)| \leqslant \frac{K-1}{K+1} < 1.$$

Complex dilatation has a simple geometric interpretation: from 2.5 we deduce that

$$\max_{\alpha} |\partial_\alpha f(z)| / \min_{\alpha} |\partial_\alpha f(z)| = (1 + |\mu(z)|)/(1 - |\mu(z)|),$$

and from 2.4 that $|\partial_\alpha f(z)|$ assumes its maximum when $\alpha = \frac{1}{2}\arg \mu(z)$.

The definition 4.2 leads us to consider the *Beltrami equation*

$$(4.4) \qquad w_{\bar{z}} = \mu w_z,$$

where $\mu$ is a measurable function with $\|\mu\|_\infty < 1$. If $\mu$ is identically zero, 4.4 reduces to the Cauchy-Riemann equation $w_{\bar{z}} = 0$. We say that $w$ is an $L^p$-solution of 4.4 if $w$ has $L^p$-derivatives and 4.4 holds a.e.

Reinterpreting the results of Theorems 2.3 and 3.8, we obtain the following new characterization of quasiconformality.

4.5 THEOREM. *A K-quasiconformal mapping is an $L^2$-solution of* 4.4 *where* $\mu$ *satisfies* 4.3. *Conversely, if* $\mu$ *is a measurable function satisfying* 4.3 *a.e. then every homeomorphic $L^2$-solution of* 4.4 *is a K-quasiconformal mapping.*

With the introduction of the Beltrami equation, a number of important new problems arise, such as the existence and uniqueness of homeomorphic $L^2$-solutions of 4.4, representation of the mapping in terms of $\mu$, dependence of the mapping on $\mu$, etc.

It is not difficult to show that complex dilatation determines a quasiconformal mapping up to conformal transformations. ("The Uniqueness theorem").

4.6 THEOREM. *Let $f$ and $g$ be quasiconformal mappings of $A$ with complex dilatations $\mu_f$ and $\mu_g$. If $\mu_f(z) = \mu_g(z)$ a.e. in $A$, then $g \circ f^{-1}$ is conformal.*

*Proof.* Formal computation yields

$$\mu_{g\circ f^{-1}}(\zeta) = \frac{\mu_g(z) - \mu_f(z)}{1 - \bar{\mu}_f(z)\mu_g(z)} e^{2i\arg f_z(z)},$$

where $\zeta = f(z)$. Since $f$ preserves null-sets, condition $\mu_g(z) = \mu_f(z)$ a.e. in $A$ thus implies $\mu_{g\circ f^{-1}}(\zeta) = 0$ a.e. in $f(A)$. It follows from Theorem 3.8 that $g \circ f^{-1}$ is 1-quasiconformal and hence conformal. ■

The problems regarding the existence and representation of the solutions of 4.4 are more difficult to handle, and the main part of Sections 5 and 6 will be devoted to these questions. In view of a later application, we mention here the following approximation theorem.

4.7 THEOREM. *Let $f_n$ be a sequence of K-quasiconformal mappings of $A$ which converges to a quasiconformal mapping $f$ with complex dilatation $\mu$, uniformly on compact subsets of $A$. If the complex dilatations $\mu_n$ of $f_n$ tend to a limit $\mu_\infty$ a.e. in $A$, then $\mu_\infty(z) = \mu(z)$ a.e. in $A$.*

For the proof we refer to **[22]**, p. 187.

## Notes

Homeomorphic $L^2$-solutions of Beltrami equations were systematically studied by Morrey [**25**] in 1938. So, in view of Theorem 4.5, it was actually he who introduced general quasiconformal mappings. But it took almost twenty years before Bers [**9**] observed the connection of Morrey's work with quasiconformal mappings and thus united two apparently different theories.

Mori [24] posed the problem of whether a quasiconformal mapping maps null-sets onto null-sets (Theorem 3.7). Morrey [**25**] had proved that this is so for homeomorphic $L^2$-solutions of Beltrami equations. Hence, Mori's question was answered in the affirmative by the observation of Bers. Later Gehring and Väisälä [**17**] gave a purely "geometric" proof for Theorem 3.7, based on module of quadrilaterals.

Theorem 4.7 is due to Bers [**9**].

## 5. REPRESENTATION THEOREM

By Theorem 4.6, a complex dilatation $\mu$ determines quasiconformal mappings $f$ of the plane up to Möbius transformations. For suitably normalized mappings $f$ we shall now derive a representation formula which gives $f(z)$ in terms of $\mu$.

The starting point is a representation formula for functions with $L^1$-derivatives, which coincides with Cauchy's integral formula in the case of an analytic function. Let $f$ be a function having $L^1$-derivatives in a domain $A$. $D$ be a Jordan domain with rectifiable boundary, $\bar{D} \subset A$, and $D_r = \{\zeta: |\zeta - z| < r\}$ with $\bar{D}_r \subset D$. Set $\psi(\zeta) = f(\zeta)(\zeta - z)^{-1}$ for $\zeta \in A - D_r$, $\psi(\zeta) = r^{-2} f(\zeta)(\bar{\zeta} - \bar{z})$ for $\zeta \in D_r$, and apply the first formula 3.2 to $\psi$ in $D$ and $D_r$. Subtracting and letting $r \to 0$, we then obtain (generalized Pompeiu formula)

$$(5.1) \quad f(z) = \frac{1}{2\pi i} \int_{\partial D} \frac{f(\zeta)}{\zeta - z} d\zeta - \frac{1}{\pi} \iint_D \frac{f_{\bar{\zeta}}(\zeta)}{\zeta - z} d\xi \, d\eta, \qquad z \in D, \zeta = \xi + i\eta.$$

Here the first integral defines a holomorphic function and the second exists as a Cauchy principal value.

Suppose that $f$ has $L^1$-derivatives in the whole plane and $f(z) \to 0$ as $z \to \infty$. With the notation

$$T\omega(z) = -\frac{1}{\pi} \iint \frac{\omega(\zeta)}{\zeta - z} d\xi \, d\eta$$

where integration is over the whole plane, it follows from 5.1 that

$$f = Tf_{\bar{z}}. \tag{5.2}$$

Let $\omega$ be a function which has $L^p$-derivatives in the plane, $p > 2$, and vanishes for $|z| > R$. We can then apply Hölder's inequality to estimate $|T\omega(z)|$ and conclude that

$$|T\omega(z)| \leqslant C \|\omega\|_p, \tag{5.3}$$

$$|T\omega(z_1) - T\omega(z_2)| \leqslant C' \|\omega\|_p |z_1 - z_2|^{1-2/p}, \tag{5.4}$$

where $C$ and $C'$ depend on $p$ and $R$ only.

If $\omega \in C_0^\infty$ in the plane, we can differentiate $T\omega$ and obtain

$$(T\omega)_{\bar{z}} = \omega, \ (T\omega)_z = S\omega, \tag{5.5}$$

where

$$S\omega(z) = -\frac{1}{\pi} \iint \frac{\omega(\zeta)}{(\zeta - z)^2} \, d\xi \, d\eta$$

exists as a Cauchy principal value. The linear operator $S$ is called the *Hilbert transformation.* A simple application of the second formula 3.2 shows that $T\omega$ and $S\omega$ belong to the class $C^\infty$ ([**22**], p. 155–157).

With the help of 3.2 we can also prove that $S$ preserves the $L^2$-norms of $C_0^\infty$-functions: $\|S\omega\|_2 = \|\omega\|_2$ ([**22**], p. 157). A much deeper result is the *Calderón-Zygmund inequality* which says that $S$ is bounded in $L^p \cap C_0^\infty$ for every $p > 1$. In other words, there is a constant $c$ depending only on $p$ such that

$$\|S\omega\|_p \leqslant c \|\omega\|_p. \tag{5.6}$$

([**4**], p. 106, [**30**], p. 35).

Since $C_0^\infty$ is dense in the space $L^p$ and $L^p$ is complete, 5.6 can be used to extend $S$ to the whole space $L^p$ by continuity ([**22**], p. 159). Let $c_p$ denote the $p$-norm of $S$, i.e. the infimum of the numbers $c$ for which 5.6 is valid. By means of Riesz's convexity theorem, it can be proved that $c_p$ is continuous in $p$ ([**4**], p. 113).

Use of the Hilbert transformation yields the desired representation formula for quasiconformal mappings.

5.7 THEOREM. *Let $f$ be a quasiconformal mapping of the plane with complex dilatation $\mu$. If $\mu$ has bounded support and $f(z) - z \to 0$ as $z \to \infty$, then*

$$f(z) = z + \sum_{n=1}^{\infty} T\phi_n(z),$$

*where the functions $\phi_n$ are defined by*

$$\phi_1 = \mu, \ \phi_n = \mu S\phi_{n-1}, \ n = 2, 3, \ldots . \tag{5.8}$$

*The series is absolutely and uniformly convergent in the whole plane.*

*Proof.* We first remark that, owing to the normalization at infinity, $\mu$ determines the mapping $f$ uniquely.

Applying 5.2 to the function $z \mapsto f(z) - z$ we obtain

$$f(z) = z + Tf_{\bar{z}}(z). \tag{5.9}$$

Differentiation with respect to $z$ yields

$$f_z = 1 + Sf_{\bar{z}} \ \text{a.e.} \tag{5.10}$$

If $f_{\bar{z}} \in C_0^\infty$ this is seen from 5.5; in the general case we can apply Lemma 3.1, since $f$ has $L^2$-derivatives. Because $f_{\bar{z}} = \mu f_z$ a.e. we conclude from 5.10 that $f_{\bar{z}}$ satisfies a.e. the equation

$$\omega = \mu + \mu S\omega. \tag{5.11}$$

This can be solved by iteration. If $k = \|\mu\|_\infty$, the Calderón-Zygmund inequality 5.6 gives for the function $\phi_n$, defined by 5.8,

$$\|\phi_{n+1}\|_p \leqslant kc_p\|\phi_n\|_p \leqslant \ldots \leqslant (kc_p)^n \|\mu\|_p.$$

Now choose $p > 2$ so that $kc_p < 1$; since $p \mapsto c_p$ is continuous and $c_2 = 1$, this is possible. Then

$$\omega_n = \sum_{i=1}^{n} \phi_i$$

is a convergent sequence in $L^p$. Because $\omega_{n+1} = \mu + \mu S\omega_n$, its limit $\omega$ satisfies 5.11. If $\psi$ is another solution of 5.11, then $\| \omega - \psi \|_p \leqslant kc_p \| \omega - \psi \|_p$, and hence $\psi = \omega$ a.e. We conclude that $\omega_n \to f_{\bar{z}}$ in $L^p$. Considering 5.9 and 5.3, we obtain 5.7 and see that the series is absolutely and uniformly convergent. ■

## Notes

Vekua [**37**] and Ahlfors [**3**] seem to have been the first to use the two-dimensional Hilbert-transformation in connection with Beltrami equations. The importance of the inequality 5.6, established by Calderón and Zygmund [**12**] in 1952, was noticed by Bojarski [**10**], whose reasoning we have followed in proving Theorem 5.7. Another method to derive the Calderón-Zygmund inequality is presented in Vekua's monograph [**36**]; [**4**]. Further progress in the theory of singular integrals is reflected in Stein's monograph [**30**], which contains a lucid proof for a generalization of Calderón-Zygmund's inequality.

Bojarski used 5.7 to prove the Existence theorem and the local $L^p$-integrability of the derivatives of quasiconformal mappings for some $p > 2$ (Theorems 6.1 and 6.7, in the following Section 1).

## 6. EXISTENCE THEOREM

In this section we shall prove that the complex dilation of a quasiconformal mapping can be prescribed almost everywhere ("Existence theorem").

6.1 THEOREM. *Let $\mu$ be a measurable function in a domain $A$ with $\| \mu \|_\infty < 1$. Then there exists a quasiconformal mapping of $A$ whose complex dilatation agrees with $\mu$ a.e.*

*Proof.* We may clearly assume that $A$ is the finite plane.

The proof is in three steps. We first show that if $\mu \in C_0^\infty$, the equation

$$w_{\bar{z}} = \mu w_z \tag{6.2}$$

has a locally injective solution. Next, a topological argument shows that this solution is globally injective. Finally, approximating an arbitrary $\mu$ by $C_0^\infty$-functions we obtain the desired general solution.

*Step 1.* If $\mu \in C_0^\infty$, there is a locally injective $C^1$-function $f$ which satisfies 6.2.

*Proof.* Consider the non-homogeneous Beltrami equation

$$w_{\bar{z}} = \mu w_z + \mu_z. \tag{6.3}$$

Imitating the method used in the previous section in deriving the representation formula, we consider the equation

$$\omega = S\mu_z + S(\mu\omega). \tag{6.4}$$

Set $\phi_1 = S\mu_z$, $\phi_n = S(\mu\phi_{n-1})$, $n = 2, 3, \ldots$, and choose $p > 2$ so that $c_p\|\mu\|_\infty < 1$. Then

$$\|\phi_{n+1}\|_p \leqslant (c_p\|\mu\|_\infty)^n \|\mu_z\|_p.$$

Consequently,

$$\omega_n = \sum_1^n \phi_i$$

converges in $L^p$ to a limit, and this is a solution of 6.4.

The function

$$w = T(\mu_z + \mu\omega) \tag{6.5}$$

is an $L^p$-solution of 6.3. In order to prove this, we choose functions $\omega_n \in C_0^\infty$ so that $\|\omega - \omega_n\|_p \to 0$. For $f_n = T(\mu_z + \mu\omega_n)$ it follows from 5.5 that $(f_n)_z = S(\mu_z + \mu\omega_n)$, $(f_n)_{\bar{z}} = \mu_z + \mu\omega_n$. Applying Lemma 3.1 to $f = w$, $f_n$, $g = S(\mu_z + \mu\omega)$, $h = \mu_z + \mu\omega$, we conclude that $w$ has $L^p$-derivatives and that $w_z = S(\mu_z + \mu\omega)$, $w_{\bar{z}} = \mu_z + \mu\omega$. By 6.4, $w_z = \omega$, and so 6.3 holds.

Next we apply Lemma 3.3 to $g = e^w$, $h = \mu e^w$. This can be done because $g$ and $h$ have $L^p$-derivatives and $h_z = e^w(\mu w_z + \mu_z) = e^w w_{\bar{z}} = g_{\bar{z}}$. Hence, the $C^1$-function $f$,

$$f(z) = \int_0^z e^{w(t)}\,(dt + \mu(t)\,d\bar{t}), \tag{6.6}$$

satisfies $f_z = e^w$, $f_{\bar{z}} = \mu e^w$, and is thus a solution of 6.2. For its Jacobian we have

$$J(z) = |e^{w(z)}|^2\,(1 - |\mu(z)|^2) > 0,$$

which implies local injectiveness.

*Step 2.* The function $f$ of Step 1 is a homeomorphism of the plane.

*Proof.* Since $\mu_z + \mu\omega$ has bounded support, we conclude from 6.5 that $w(z) \to 0$ as $z \to \infty$. Thus $f_z(z) = e^{w(z)} \to 1$, and it follows from 6.6 that $f(z)$ has the limit $\infty$ at $\infty$. This implies that $f$ maps the plane $A$ onto itself. Otherwise there are points $z_n$ such that $f(z_n)$ converges to a finite boundary point of the image. Since the points $z_n$ cannot accumulate at $\infty$, a contradiction follows. The same argument shows that no point can have infinitely many pre-images. It follows that every point has an open neighbourhood $V$ onto which each component of $f^{-1}V$ maps topologically. Thus the pair $(f, A)$ is a universal covering surface of $A$. Since $A$ is simply-connected, we conclude from the Monodromy theorem that $f$ is a homeomorphism of $A$.

*Step 3.* Let $f_n$, $n = 1, 2, \ldots$, be a $\mu_n$-quasiconformal self-mapping of the finite plane, $f_n(0) = 0$, $f_n(1) = 1$, $\|\mu_n\| \leqslant k < 1$. If $\mu_n \to \mu$ a.e. then there exists a subsequence of $f_n$ which converges to a $\mu$-quasiconformal mapping of the plane.

*Proof.* The mappings $f_n$ constitute a normal family ([**22**], p. 73). Thus a subsequence exists which is uniformly convergent in every compact subset of the plane. The normalization guarantees that the limit function $f$ is a quasiconformal mapping of the plane. From $\lim \mu_n = \mu$ a.e. it follows by Theorem 4.7 that $f$ is $\mu$-quasiconformal.

The proof of Theorem 6.1 can now be completed. Let $\mu$ be an arbitrary measurable function in the plane with $\|\mu\|_\infty < 1$. Then there are $C_0^\infty$-functions $\mu_n$ so that $\mu(z) = \lim \mu_n(z)$ a.e. and $\|\mu_n\|_\infty \leqslant \|\mu\|_\infty$. By Step 2, there exist $\mu_n$-quasiconformal mappings $f_n$ of the plane, and we can normalize them so that the conditions of Step 3 are fulfilled. Applying the result of Step 3, we obtain the desired mapping. ■

Using Theorem 6.1 we obtain from our previous results about singular integrals an improvement to the result that the derivatives of a quasiconformal mapping belong locally to $L^2$.

6.7 THEOREM. *The derivatives of a quasiconformal mapping are locally in a class $L^p$ for some $p > 2$.*

*Proof.* Let $f$ be a quasiconformal mapping of a domain $A$ and $E$ a compact subset of $A$. By Theorem 6.1, there is a quasiconformal mapping $\phi$ of the plane whose complex dilatation coincides with that of $f$ in a bounded domain $D$, with $E \subset D, \bar{D} \subset A$, which is conformal outside $D$ and for which $\phi(z) - z \to 0$ as $z \to \infty$. By Theorem 4.6, $f = h \circ \phi$, where $h$ is conformal in $\phi(D)$. Hence we may assume that $f = \phi$.

By 5.11 we then have

$$f_{\bar{z}} = \mu + \mu S f_{\bar{z}}.$$

From this we see that, for every $p > 2$ for which $c_p \|\mu\|_\infty < 1$,

$$\|f_{\bar{z}}\|_p \leqslant \|\mu\|_p (1 - c_p \|\mu\|_\infty)^{-1}.$$

From 5.10 and 5.6 we deduce that $f_z$ also belongs to this $L^p$ in $E$. ■

Let $p(K)$ be the supremum of the numbers $p$ for which all $K$-quasiconformal mappings have $L^p$-derivatives. By the above theorem $p(K) > 2$, while the example $z \mapsto z|z|^{1/K-1}$ shows that

$$p(K) \leqslant 2K/(K - 1). \tag{6.8}$$

Hence $p(K) \to 2$ as $K \to \infty$. From the result of Calderón and Zygmund [**12**] that $c_p/p$ is bounded it follows that

$$\liminf_{K \to 1} (K - 1)\, p(K) > 0.$$

It is an interesting open problem to determine $p(K)$; the conjecture has been expressed that equality holds in 6.8.

Let $\Sigma_k$ be the class of all quasiconformal mappings $f$ of the plane whose complex dilatations are bounded by $k$ in absolute value and vanish outside the unit disc and which satisfy $f(z) - z \to 0$ as $z \to \infty$. Let $B_k$ denote the subset of $L^\infty$ whose functions $\mu$ satisfy $\|\mu\|_\infty < k$ and vanish outside the unit disc. By Theorems 6.1 and 4.6 there is a one-to-one correspondence between the elements of $\Sigma_k$ and $B_k$. Equation 5.7, which expresses this correspondence, is explicit enough to admit conclusions about the dependence of $f(z)$ on $\mu$. For instance, if $\mu$ depends holomorphically on a complex parameter $\zeta$, then $f(z)$ is an analytic function of $\zeta$, for every fixed $z$. This result has important applications in the theory of Teichmüller spaces (see Chapter 5, Sections 10–12). It can also be used in the study of univalent functions with quasiconformal extensions, which has gained a lot of popularity in recent years [**21**].

## Notes

Historically, the Beltrami equation was introduced in connection with the problem of finding isothermal coordinates for a given surface. The equation was considered already in 1822 by Gauss [**15**], who proved the existence of a locally injective solution if $\mu$ is real-analytic. The equation also played an important role in the studies of Beltrami [**8**] on surface theory. Generalizing the result of Gauss, Korn [**19**] and Lichtenstein [**23**] showed the existence of an injective solution if $\mu$ is Hölder-continuous. In the above generality, the Existence theorem was first proved by Morrey [**25**].

For an arbitrary continuous $\mu$, the Beltrami equation need not have a $C^1$-solution. Somewhat similar to the observation of Bers, clarifying the position of Beltrami equations in the theory of quasiconformal mappings (Notes of Section 4), was the discovery of Bojarski [**11**] in 1957 that homeomorphic $L^2$-solutions corresponding to continuous $\mu$ coincide with Lavrentiev's class of mappings with continuous characteristics (c.f. notes of Section 1). For these mappings, Lavrentiev [**20**] had already proved an Existence Theorem in 1935. A systematic exposition of Lavrentiev mappings is in Volkovyskij's monograph [**38**]; their position in the hierarchy of quasiconformal mappings has been studied by Näätänen [**26**].

A brief account of the many ways to prove Theorem 6.1 is given in [**22**], p. 194. The above arrangement of proof is essentially from Ahlfors-Bers [**6**]. In Step 1 we could assume that $\mu$ is a polynomial, in which case the considerations become completely elementary (see [**22**], p. 208). In Step 2, the purely topological argument based on the Monodromy theorem could be replaced by an application of the uniformization theorem.

## REFERENCES

1. L. Ahlfors. Zur Theorie der Überlagerungsflächen. *Acta Math.* **65** (1935) 157–194.
2. L. Ahlfors. On quasiconformal mappings. *J. Analyse Math.* **3** (1954) 1–58, 207–208.
3. L. Ahlfors. Conformality with respect to Riemannian metrics. *Ann. Acad. Sci. Fenn. Ser. AI* **206** (1955).
4. L. Ahlfors. "Lectures on Quasiconformal Mappings". Van Nostrand, Princeton, N.J. 1966.
5. L. Ahlfors. "Conformal Invariants: Topics in Geometric Function Theory." McGraw-Hill, New York, 1973.
6. L. Ahlfors and L. Bers. Riemann's mapping theorem for variable metrics. *Ann. of Math.* **72** (1960) 385–404.
7. L. Ahlfors and A. Beurling. Conformal invariants and function-theoretic null-sets. *Acta Math.* **83** (1950) 101–129.

8. E. Beltrami. Ricerce di analisis applicata alla geometria. G. Mat. Battaglini **2** (1864).
9. L. Bers. On a theorem of Mori and the definition of quasiconformality. *Trans. Amer. Math. Soc.* **84** (1957) 75–*b*4.
10. B. V. Bojarski. Gomeomorfnye rešenija sistem Bel'trami. *Dokl. Akad. Nauk SSSR.* **102** (1955) 661–664.
11. B. V. Bojarski. Obobščennye rešenija sistemy differencial'nyh uravneniĭ pervogo porjadka èllipticeskogo tipa s razryvnymi koefficentami. *Mat. USSR-Sb.* **43** (1957) 451–503.
12. A. P. Calderón and A. Zygmund. On the existence of certain singular integrals. *Acta Math.* **88** (1952) 85–139.
13. K. O. Friedrichs. The identity of weak and strong extensions of differential operators. *Trans. Amer. Math. Soc.* **55** (1944) 132–151.
14. B. Fuglede. Extremal length and functional completion. *Acta Math.* **98** (1967) 171–219.
15. C. F. Gauss. "Astronomische Abhandlungen" Vol 3. (H. C. Schumacher ed.) Altona 1825. Or "Carl Friedrich Gauss Werke", Vol. IV, pp. 189–216. Dieterichsche Universitäts-Druckerei, Göttingen 1880.
16. F. W. Gehring and O. Lehto. On the total differentiability of functions of a complex variable. *Ann. Acad. Sci. Fenn. Ser. A I* **272** (1959).
17. F. W. Gehring and J. Vaisälä. On the geometric definition of quasiconformal mappings. *Comment. Math. Helv.* **36** (1962) 19–32.
18. H. Grötzsch. Uber die Verzerrung bei schlichten nichtkonformen Abbildungen und über eine damit zusgammenhängende Erweiterung des Picardschen Satzes. *Ber. Verh. Sächs. Akad. Wiss.* **80** (1928) 503–507.
19. A. Korn. Zwei Anwendungen der Methode der sukzessiven Annäherungen. Mathematische Abhandlungen Hermann Amandus Schwarz zu seinem fünfzigjährigen Doktorjubiläum, pp. 215–229. Springer Berlin, 1914.
20. M. A. Lavrentieff. Sur une classe de reprèsentations continues *Mat. USSR-Sb.* **48** (1935) 407–423.
21. O. Lehto. Quasiconformal mappings and singular integrals. Symposia Mathematica XVIII, Istituto Nazionale di Alta Matematica, Roma 1976.
22. O. Lehto and K. I. Virtanen. Quasiconformal mappings in the plane. Springer-Verlag, Berlin 1973.
23. L. Lichtenstein. Zur Theorie der konformen Abbildungen; Konforme Abbildungen nicht-analytischer singularitätenfreier Flächenstücke auf ebene Gebiete. *Bull. Acad. Sci. Cracovie,* (1916) 192–217.
24. A. Mori. On quasi-conformality and pseudo-analyticity. *Trans. Amer. Math. Soc.* **84** (1957) 56–77.
25. C. B. Morrey. On the solution of quasilinear elliptic partial differential equations. *Trans. Amer. Soc.* **43** (1938) 126–166.
26. M. Näätänen. Maps with continuous characteristics as a subclass of quasiconformal maps. *Ann. Acad. Sci. Fenn. Ser. A I* **410** (1967).
27. A. Pfluger. Quasikonforme Abbildungen und logarithmische Kapazität. *Ann. Inst. Fourier (Grenoble).* **2** (1951) 69–80.
28. A. Pfluger. Über die Äquivalenz der geometrischen und der analytischen Definition quasikonformer Abbildungen. *Comment. Math. Helv.* **33** (1959) 23–33.
29. S. L. Sobolev. "Applications of Functional Analysis in Mathematical Physics." American Mathematical Society, Providence, R.I. 1963.

30. E. M. Stein. "Singular Integrals and Differentiability Properties of Functions." Princeton University Press, Princeton, N.J. 1970.
31. K. Strebel. On the maximal dilatation of quasiconformal mappings. *Proc. Amer. Math. Soc.* **6** (1955) 903–909.
32. O. Teichmüller. Untersuchungen über konforme und quasikonforme Abbildung, *Deutch. Math.* **3** (1938) 621–678.
33. O. Teichmüller. Extremale quasikonformen Abbildungen und quadratische Differentiale. *Abh. preuss. Akad. Wiss.* **22** (1940) 1–197.
34. Bestimmung der extremalen quasikonforme Abbildungen bei geschlossenen orientierten Riemannschen Flächen. *Abh. Preuss. Akad. Wiss.* Kl **4** (1944).
35. J. Väslälä. On quasiconformal mappings in space. *Ann. Acad. Sci. Fenn. Ser. A I* **298** (1961).
36. I. N. Vekua. Generalized Analytic Functions. Pergamon Press, Oxford. 1962.
37. I. N. Vekua. Ob odnoĭ sisteme singuljarnyh integralo-differencial-nyh uravneniĭ. *Trudy Tbliss. Mat. Inst.* **24** (1957) 136–147.
38. L. I. Volkovyskii. "Kvazikonformnye otobrazenija". L'vov: Izdatel'stvo L'vov-skogo Universiteta, 1954.
39. Z. Yûjôbô. On absolutely continuous functions of two or more variables in the Tonelli sense and quasiconformal mappings in the A. Mori sense. *Comment. math. Univ. S. Paul*, **4**, (1955) 67–92.

# 5. Teichmüller Theory

CLIFFORD J. EARLE‡

*Cornell University, Ithaca, New York, U.S.A.*



## 1. INTRODUCTION

Since these lectures were intended for non-specialists, they are rather elementary, concentrating on the basic facts and definitions of Teichmüller theory. Fortunately there are already a number of more advanced surveys of the theory. The reader is encouraged to consult the articles [**11**], [**17**], or [**28**] as well as other chapters in this volume. One important subject that went unmentioned for lack of time is the generalization of Teichmüller space from Fuchsian groups to finitely generated Kleinian groups. A good introduction to that subject is given in "A Crash Course on Kleinian Groups",

‡ The author is grateful to the National Science Foundation for financial support.

(L. Bers and I. Kra eds), Lecture Notes in Mathematics No. 400, Springer, Berlin, 1974.

## 2. CONFORMAL STRUCTURES IN THE PLANE

Since Teichmüller theory has to do with putting different conformal structures onto the same surface, we want to begin by describing these structures as conveniently as we can. Any Riemannian metric on a plane region $D$ can be written in the form

$$ds^2 = E\,dx^2 + 2F\,dx\,dy + G\,dy^2$$

where $E$, $F$, and $G$ are real valued functions in $D$ such that $E$ and $EF - G^2$ are positive everywhere. However it is better for our purposes to use complex notation and write

$$ds = \lambda(z)|dz + \mu(z)\,d\bar{z}| \tag{2.1}$$

where $\lambda(z)$ is a positive function in $D$, and $\mu(z)$ is complex valued with $|\mu(z)| < 1$. It is well known and easy to verify that every metric can be written uniquely in the form 2.1.

Now suppose that $w: D \to D^*$ is a sense preserving diffeomorphism of $D$ onto a region $D^*$, that $D$ has the metric 2.1, and that

$$ds^* = \lambda^*(\zeta)|d\zeta + \mu^*(\zeta)\,d\bar{\zeta}|$$

is a metric on $D^*$. Then $w$ is a conformal map (with respect to the given metrics) if and only if $|dz + \mu(z)\,d\bar{z}|$ is proportional to

$$|d\zeta + \mu^*(\zeta)\,d\bar{\zeta}| = |dw(z) + \mu^*(w(z))\overline{dw(z)}|$$

at all $z$ in $D$. For example, if $ds^* = |d\zeta|$ is the Euclidean metric in $D^*$, then $w$ is conformal if and only if it is a solution of the Beltrami equation

$$w_{\bar{z}} = \mu(z)w_z \tag{2.2}$$

in $D$. Here of course $w_z$ and $w_{\bar{z}}$ are the complex partial derivatives

$$w_z = \frac{1}{2}\left(\frac{\partial w}{\partial x} - \mathrm{i}\frac{\partial w}{\partial y}\right),$$

$$w_{\bar{z}} = \frac{1}{2}\left(\frac{\partial w}{\partial x} + \mathrm{i}\frac{\partial w}{\partial y}\right).$$

We call the metrics $\mathrm{d}s$ and $\mathrm{d}s^*$ on $D$ *conformally equivalent* if and only if $\mathrm{d}s$ and $\mathrm{d}s^*$ are proportional at every point of $D$, or in other words the identity map of $D$ onto itself is conformal (with respect to the given metrics $\mathrm{d}s$ and $\mathrm{d}s^*$). A conformal equivalence class of metrics is called a *conformal structure*. It is clear from the representation 2.1 that the conformal structures on $D$ are in one-to-one correspondence with the complex valued functions $\mu(z)$ in $D$ with $|\mu(z)| < 1$. In fact each such function corresponds to the conformal structure determined by the metric

$$\mathrm{d}s = |\mathrm{d}z + \mu(z)\,\mathrm{d}\bar{z}|.$$

## 3. CONFORMAL STRUCTURES ON A SURFACE

Let $S$ be any smooth (class $C^\infty$) surface. Once again we call two smooth Riemannian metrics on $S$ conformally equivalent if they are proportional at every point, and we call an equivalence class of metrics a *conformal structure on* $S$. We shall denote the set of conformal structures on $S$ by $\mathscr{M}(S)$.

An oriented surface $S$ with a given conformal structure is called a *Riemann surface*. A Riemann surface is often defined to be a connected one-dimensional complex manifold, but the above definition is readily seen to be equivalent. In fact if $S$ is oriented and has a given conformal structure, then the sense-preserving conformal maps from open sets of $S$ into the complex plane $\mathbb{C}$ (with its natural Euclidean metric) form a complex analytic atlas on $S$. Conversely, every connected one-dimensional complex manifold $S$ has a natural orientation, and it is a classical fact, following easily from the uniformization theorem, that $S$ has a Riemannian metric such that the complex coordinate functions on $S$ are sense preserving conformal maps into $\mathbb{C}$.

A conformal map between Riemann surfaces is a sense-preserving diffeomorphism which is conformal with respect to the given conformal structures. The Riemann surfaces $S$ and $S'$ are called equivalent if there is a conformal map of $S$ onto $S'$. Notice that this is a much weaker equivalence relation than conformal equivalence of metrics. Two Riemannian metrics on $S$ are conformally equivalent if the identity map on $S$ is conformal (with respect to the given metrics). They define equivalent Riemann surfaces if there is some sense preserving diffeomorphism $f: S \to S$ which is conformal.

## 4. THE TEICHMÜLLER SPACE T($p$, 0)

Let $S$ be a smooth oriented closed surface of genus $p \geqslant 0$; we fix $S$ once and for all. If $W$ is any closed Riemann surface of genus $p$, choose a Riemannian metric $ds^*$ on $W$ belonging to its given conformal structure, and choose a sense-preserving diffeomorphism $f: S \to W$. Using $f$, pull back the metric $ds^*$ to a metric $ds = f^*(ds^*)$ on $S$. ($ds$ is defined by requiring the length of any smooth path $\gamma$ in $S$ to be the length of $f \circ \gamma$ in $W$.) The Riemann surface defined by the metric $ds$ on $S$ is equivalent to $W$, since $f: S \to W$ is a conformal map. We see therefore that every closed Riemann surface of genus $p$ is equivalent to one defined by some conformal structure on $S$. However, the set $\mathcal{M}(S)$ of conformal structures on $S$ is much too big; too many structures in $\mathcal{M}(S)$ determine equivalent Riemann surfaces. We want to factor $\mathcal{M}(S)$ by a suitable equivalence relation.

*Definition.* Let $\mu$ in $\mathcal{M}(S)$ be any conformal structure, and $f: S \to S$ any sense-preserving diffeomorphism. By $f^*\mu$ we mean the unique conformal structure $\nu$ on $S$ such that $f$ maps $S_\nu$ conformally onto $S_\mu$. Here of course $S_\nu$ and $S_\mu$ denote $S$ with the conformal structures $\nu$ and $\mu$ respectively.

The structure $f^*\mu$ is obtained by choosing a Riemannian metric $ds^*$ in the conformal equivalence class $\mu$ and pulling back $ds^*$ as we did before.

It is clear that if $\nu = f^*\mu$, then $\mu$ and $\nu$ determine equivalent Riemann surfaces. Conversely, if $S_\nu$ and $S_\mu$ are equivalent Riemann surfaces, there is a conformal map $f: S_\nu \to S_\mu$, and $\nu = f^*(\mu)$. That proves the following.

PROPOSITION. *Let the group* $\mathit{Diff}^+(S)$ *of sense preserving diffeomorphisms of* $S$ *act on* $\mathcal{M}(S)$ *by* $\mu \cdot f = f^*(\mu)$. *Then the quotient space* $\mathcal{M}(S)/\mathit{Diff}^+(S)$ *is (in canonical one-to-one correspondence with) the set of equivalence classes of closed Riemann surfaces of genus* $p$.

The above quotient space is called the *Riemann space* or the *space of moduli*. Direct study of the Riemann space is difficult, and following Teichmüller we divide $\mathscr{M}(S)$ by a normal subgroup of $\mathrm{Diff}^+(S)$ rather than the full group.

*Definition.* $\mathrm{Diff}_0(S)$ is the group of diffeomorphisms of $S$ which are homotopic to the identity map.

*Definition.* For $p \geqslant 2$, the *Teichmüller space* $\mathbf{T}(p, 0)$ is the quotient space $\mathbf{T}(p, 0) = \mathscr{M}(S)/\mathrm{Diff}_0(S)$.

COROLLARY. *The mapping class group* $\mathit{Diff}^+(S)/\mathit{Diff}_0(S)$ *acts on* $\mathbf{T}(p, 0)$, *producing the space of moduli as quotient.* (See chapters 1 and 6.)

In fact since $\mathrm{Diff}_0(S)$ is a normal subgroup of $\mathrm{Diff}^+(S)$, the action of $\mathrm{Diff}^+(S)$ on $\mathscr{M}(S)$ automatically induces an action of $\mathrm{Diff}^+(S)/\mathrm{Diff}_0(S)$ on $\mathbf{T}(p, 0)$.

## 5. THE TEICHMÜLLER SPACE $\mathbf{T}(p, n)$

A Riemann surface $W$ is said to have *finite type* if there exist a closed Riemann surface $\overline{W}$ and a conformal map $i$: $W \to \overline{W}$ so that $\overline{W}\backslash i(W)$ is a finite point set. We say that $W$ has type $(p, n)$ if $\overline{W}$ has genus $p$ and $\overline{W}\backslash i(W)$ consists of exactly $n$ points. It is customary to identify $W$ and $i(W)$ so that $i$ becomes the inclusion map. The points of $\overline{W}\backslash W$ are called the puncture or distinguished points.

Choose any set $\{x_1, \ldots, x_n\}$ of $n$ points on our fixed closed surface $S$ of genus $p$. Put $S' = S\backslash\{x_1, \ldots, x_n\}$. Every conformal structure $\mu$ in $\mathscr{M}(S)$ makes $S$ a Riemann surface of type $(p, n)$. $\mu$ and $\nu$ will determine equivalent Riemann surfaces $S'$ if and only if $\nu = f^*(\mu)$ for some sense preserving diffeomorphism $f: S \to S$ which fixes the point set $\{x_1, \ldots, x_n\}$.

*Definition.* $\mathrm{Diff}^+(S, n)$ is the group of $f$ in $\mathrm{Diff}^+(S)$ which map the set $\{x_1, \ldots, x_n\}$ onto itself. $\mathrm{Diff}_0(S, n)$ is the group of $f$ in $\mathrm{Diff}^+(S, n)$ whose restriction to $S'$ is homotopic to the identity in $S'$.

It is important to notice that $\mathrm{Diff}_0(S, n)$ is a subgroup of $\mathrm{Diff}_0(S)$. Moreover, replacing the set $\{x_1, \ldots, x_n\}$ by $\{y_1, \ldots, y_n\}$ has the effect of replacing $\mathrm{Diff}_0(S, n)$ by a conjugate subgroup of $\mathrm{Diff}_0(S)$. That is true because there is a diffeomorphism $f$ in $\mathrm{Diff}_0(S)$ such that $f(x_j) = y_j$, $1 \leqslant j \leqslant n$.

*Definition.* For $2p - 2 + n > 0$, the *Teichmüller space* $\mathbf{T}(p, n)$ is the quotient space $\mathscr{M}(S)/\mathrm{Diff}_0(S, n)$.

Of course the mapping class group $\mathrm{Diff}^+(S, n)/\mathrm{Diff}_0(S, n)$ of type $(p, n)$ acts on $\mathbf{T}(p, n)$. The inequality $2p - 2 + n > 0$, which reduces to the inequality $p \geqslant 2$ of Section 4 if $n = 0$, means that the surface $S'$ has negative Euler characteristic and that every Riemann surface of type $(p, n)$ has the unit disc $\mathscr{U}$ for its universal covering space.

Since $\mathrm{Diff}_0(S, n)$ is a subgroup of $\mathrm{Diff}_0(S)$, there is a natural projection of $\mathbf{T}(p, n) = \mathscr{M}(S)/\mathrm{Diff}_0(S, n)$ onto $\mathbf{T}(p, 0) = \mathscr{M}(S)/\mathrm{Diff}_0(S)$. This projection is a fibration, with the homogeneous space $\mathrm{Diff}_0(S)/\mathrm{Diff}_0(S, n)$ as fibre.

## 6. BELTRAMI DIFFERENTIALS FOR A FUCHSIAN GROUP

Next we want to obtain a more concrete model for $\mathbf{T}(p, 0)$. Choose some conformal structure on $S$. That structure lifts to a conformal structure on the universal covering space $\tilde{S}$. Mapping $\tilde{S}$ conformally onto $\mathscr{U}$ we get a smooth covering map $\pi\colon \mathscr{U} \to S$ so that the group $G$ of cover transformations consists of conformal maps. $G$ is a Fuchsian group of the first kind, acting freely in $\mathscr{U}$, and isomorphic to the fundamental group $\pi_1(S)$. The orbit space $\mathscr{U}/G$ is a Riemann surface, equivalent to $S$ with its given conformal structure. Since $\pi\colon \mathscr{U} \to S$ is a local diffeomorphism, every conformal structure on $S$ can be lifted (pulled back) to a conformal structure on $\mathscr{U}$, and every cover transformation $g$ in $G$ will be conformal with respect to this lifted structure. In this way we get a bijective map between $\mathscr{M}(S)$ and

$$\mathscr{M}(G) = \{\mu \in \mathscr{M}(\mathscr{U}); g\colon \mathscr{U}_\mu \to \mathscr{U}_\mu \text{ is conformal for all } g \text{ in } G\}.$$

LEMMA. *Let $\mu \in \mathscr{M}(\mathscr{U})$. Then $\mu \in \mathscr{M}(G)$ if and only if*

$$\mu(gz)\overline{g'(z)}/g'(z) = \mu(z) \text{ for all } z \in \mathscr{U},\ g \in G. \tag{6.1}$$

*Proof.* $\mathscr{U}_\mu$ ($= \mathscr{U}$ with the conformal structure $\mu$) is a simply connected Riemann surface, so there is a conformal map $f$ from $\mathscr{U}_\mu$ onto a region $D$ in $\mathbb{C}$ (with its standard conformal structure). The map $f\colon \mathscr{U} \to D$ is a solution of the Beltrami equation, 2.2, as explained in Chapter 4, Theorem 4.5.

Now by definition $\mu \in \mathscr{M}(G)$ if and only if for every $g$ in $G$, the map $h = f \circ g \circ f^{-1}$ of $D$ onto itself is conformal in the usual sense (with respect to the standard conformal structure on $D$). But $h$ is conformal if and only if $h \circ f$ ($= f \circ g$) is a solution of 2.2. Direct computation shows that $f \circ g$ solves 2.2 if and only if 6.1 holds. ∎

COROLLARY. *If* $\mu \in \mathcal{M}(G)$, *then* $\|\mu\|_\infty < 1$.

*Proof.* By 6.1, $|\mu(gz)| = |\mu(z)|$ if $g \in G$. Therefore $|\mu|$ defines a continuous function on the compact surface $\mathcal{U}/G$, so $|\mu|$ attains its maximum, which must be less than one. ■

*Definition.* The space of *Beltrami differentials* for $G$ consists of the $L^\infty$ functions $\mu(z)$ on $\mathcal{U}$ satisfying 6.1 a.e. The space $\mathbf{M}(G)$ consists of the Beltrami differentials $\mu$ with $\|\mu\|_\infty < 1$.

*Remark.* For each $\mu$ in $\mathbf{M}(G)$ let $w_\mu : \mathcal{U} \to \mathcal{U}$ be the quasiconformal ($qc$) map that solves 2.2 and fixes the three boundary points $\pm 1$ and $i$. Then $w_\mu \circ g \circ (w_\mu)^{-1}$ is a Möbius transformation for each $g$ in $G$.

## 7. THE GROUP $D_0(G)$

With $G$ and $\pi : \mathcal{U} \to \mathcal{U}/G = S$ as above we define $\mathbf{D}_0(G)$ to be the group consisting of those $f$ in $\mathrm{Diff}^+(\mathcal{U})$ such that $f \circ g \circ f^{-1} = g$ for all $g$ in $G$.

7.1 LEMMA. *Every $f$ in $\mathbf{D}_0(G)$ extends to a homeomorphism of $\bar{\mathcal{U}} = \mathrm{Cl}(\mathcal{U})$ onto itself, fixing $\partial\mathcal{U}$ pointwise.*

*Proof.* Since $f \circ g \circ f^{-1}$ is a conformal map for all $g$ in $G$, $\mu = f_{\bar{z}}/f_z$ belongs to $\mathcal{M}(G)$. Therefore $f: \mathcal{U} \to \mathcal{U}$ is quasiconformal and extends to $\bar{\mathcal{U}}$ as a homeomorphism. If $\zeta$ in $\partial\mathcal{U}$ is the attractive hyperbolic fixed point of $g$ in $G$, then we have

$$\zeta = \lim_{n\to\infty} g^n(f(0)) = \lim_{n\to\infty} f(g^n(0)) = f(\zeta).$$

Therefore $f$ fixes the limit set $L(G)$ pointwise. But $L(G) = \partial\mathcal{U}$. ■

7.2 LEMMA. *For $f$ in $\mathbf{D}_0(G)$ define $\hat{f}: S \to S$ by $\hat{f} \circ \pi = \pi \circ f$. The map $f \mapsto \hat{f}$ is a group isomorphism of $\mathbf{D}_0(G)$ onto* $\mathrm{Diff}_0(S)$.

*Proof.* Note that $\hat{f}$ is a well defined member of $\mathrm{Diff}^+(S)$. We must prove that

(i) $f \mapsto \hat{f}$ is one-to-one,

(ii) $f \mapsto \hat{f}$ is onto $\mathrm{Diff}_0(S)$,

(iii) $\hat{f} \in \mathrm{Diff}_0(S)$.

The proof of (i) is easy. If $\hat{f}$ is the identity map on $S$, then $f$ belongs to $G$, so $f$ is the identity because $f$ fixes $\partial\mathscr{U}$ pointwise. The proof of (ii) and (iii) is well known. We sketch the proof here, following the methods of Ahlfors [**1**] and Bers [**6**]. Let $h: S \to S$ belong to $\mathrm{Diff}_0(S)$. Choose a homotopy from the identity to $h$, and lift the homotopy to $\mathscr{U}$. This gives a continuous family of maps $f_t: \mathscr{U} \to \mathscr{U}$, $0 \leqslant t \leqslant 1$, such that $f_0$ is the identity, $f_1$ covers $h$, and each $f_t$ covers a map from $S$ to $S$. For each $g$ in $G$ and each $t$, $0 \leqslant t \leqslant 1$, there is a $g_t$ in $G$ so that $f_t \circ g = g_t \circ f_t$. Clearly $g_0 = g$, and $g_t$ depends continuously on $t$. Since $G$ is a discrete group, $g_t = g$ for all $t$. It follows that $f_1 \in \mathbf{D}_0(G)$ and $h = \hat{f}_1$. That proves (ii).

Finally, given $f$ in $\mathbf{D}_0(G)$, $z$ in $\mathscr{U}$, and $t$, $0 \leqslant t \leqslant 1$, let $f_t(z)$ be the point on the hyperbolic geodesic segment from $z$ to $f(z)$ such that the (hyperbolic) distance from $z$ to $f_t(z)$ is $t$ times the distance from $z$ to $f(z)$. Then $f_t \circ g = g \circ f_t$ for all $g$ in $G$, and the maps $f_t$ cover a homotopy on $S$ from the identity to $\hat{f}$. ∎

## 8. THE TEICHMÜLLER SPACE T(G)

We continue with $G$ and $\pi: \mathscr{U} \to S$ as above. We have lifted $\mathscr{M}(S)$ and $\mathrm{Diff}_0(S)$ to $\mathscr{M}(G)$ and $\mathbf{D}_0(G)$. The action of $\mathrm{Diff}_0(S)$ on $\mathscr{M}(S)$ lifts to an action of $\mathbf{D}_0(G)$ on $\mathscr{M}(G)$. Our first lemma says the lifted action is the obvious one.

8.1 LEMMA. *If $\mu \in \mathscr{M}(G)$ and $f \in \mathbf{D}_0(G)$, then*

$$\mu \cdot f = f^*(\mu).$$

*Proof.* Let $\nu = f^*(\mu)$. Then $\nu \in \mathscr{M}(G)$ because $f^{-1} \circ g \circ f = g$ for all $g$ in $G$. Let $\nu'$ and $\mu'$ in $\mathscr{M}(S)$ correspond to $\nu$ and $\mu$ respectively. Then all maps in the commutative diagram below are locally conformal, so $\nu' = \hat{f}^*(\mu') = \mu' \cdot \hat{f}$. Hence, by definition, $\nu = \mu \cdot f$.

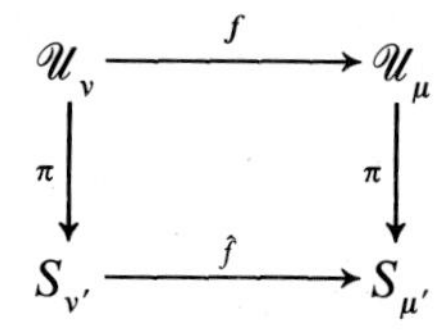


∎

*Definition.* For any $\mu$ in $L^\infty(\mathscr{U})$ with $\|\mu\|_\infty < 1$, we denote by $w_\mu$ the unique solution of 2.2 that maps $\mathscr{U}$ homeomorphically onto itself fixing $\pm 1$ and $i$.

8.2 LEMMA. *Let $\mu, \nu \in \mathcal{M}(G)$. Then $\nu = f^*(\mu)$ for some $f$ in $\mathbf{D}_0(G)$ if and only if $w_\mu = w_\nu$ on $\partial\mathcal{U}$.*

*Proof.* Let $\nu = f^*(\mu)$. Then $w_\mu \circ f$ is a solution of $w_{\bar{z}} = \nu w_z$. Since $f$ fixes $\partial\mathcal{U}$ pointwise we have $w_\nu = w_\mu \circ f$, and $w_\nu = w_\mu$ on $\partial\mathcal{U}$.

Conversely, let $w_\mu = w_\nu$ on $\partial\mathcal{U}$. Then

$$w_\nu \circ g \circ (w_\nu)^{-1} = w_\mu \circ g \circ (w_\mu)^{-1}$$

on $\partial\mathcal{U}$, hence in all of $\mathcal{U}$, for any $g$ in $G$. (Recall that both sides of the above equation are Möbius transformations.) But then $f = (w_\mu)^{-1} \circ w_\nu$ belongs to $\mathbf{D}_0(G)$, and $w_\nu = w_\mu \circ f$, so $\nu = f^*(\mu)$. ■

COROLLARY. *The covering map $\pi: \mathcal{U} \to S = \mathcal{U}/G$ induces a canonical map of $\mathbf{T}(p, 0)$ onto the set of boundary maps $w_\mu: \partial U \to \partial U$, $\mu \in \mathcal{M}(G)$.*

*Definition.* For any Fuchsian group $G$, we define the *Teichmüller space* $\mathbf{T}(G)$ to be the set of boundary maps $w_\mu: \partial\mathcal{U} \to \partial\mathcal{U}$, $\mu \in \mathbf{M}(G)$. Recall that $\mathbf{M}(G)$ consists of the $\mu$ in $L^\infty(U)$ such that $\|\mu\|_\infty < 1$ and 6.1 holds a.e. in $\mathcal{U}$.

We shall see later that, for the special groups $G$ we have been considering, replacing $\mathcal{M}(G)$ by $\mathbf{M}(G)$ has no effect on the size of $\mathbf{T}(G)$. In general, each boundary map $w_\mu: \partial\mathcal{U} \to \partial\mathcal{U}$, $\mu \in \mathbf{M}(G)$, can be obtained from some real analytic function $\mu$ in $\mathcal{M}(G)$.

*Remark.* Each boundary map $w_\mu: \partial\mathcal{U} \to \partial\mathcal{U}$ in $\mathbf{T}(G)$ determines an isomorphism $\theta_\mu$ of $G$ onto a Fuchsian group $G_\mu$ by the rule $\theta_\mu(g) = w_\mu \circ g \circ (w_\mu)^{-1}$. If $G$ is of the first kind, the isomorphism $\theta_\mu$ determines the boundary map $w_\mu$. Indeed if $g$ in $G$ is hyperbolic, $w_\mu$ must map the attractive fixed point of $g$ onto the attractive fixed point of $\theta_\mu(g)$, and these points are dense in $L(G)$.

## 9. SOME HISTORICAL REMARKS

For finitely generated Fuchsian groups the study of $\mathbf{T}(G)$, interpreted as a set of isomorphisms $\theta$, goes back at least as far as the work of Fricke and Klein and the later work of Fenchel and Nielsen. Quasiconformal maps were introduced into the subject by Teichmüller, who studied extremal quasiconformal maps between closed Riemann surfaces. Ahlfors [1] introduced the more flexible geometric definition of quasiconformal mappings and proved

Teichmüller's theorem. Bers [5] placed the Beltrami differentials and the Beltrami equation in their present central position. The definition of $\mathbf{T}(G)$ for arbitrary Fuchsian groups is due to Bers [9]. Our discussion of $\mathbf{T}(G)$ in the next few sections closely follows his lectures [8]. See also the lecture notes of Ahlfors [3]. The definitions of the spaces $\mathbf{T}(p, n)$ given here in Sections 4 and 5 are from the articles [15] and [16] of Eells and myself. One advantage of using smooth conformal structures is that with the $C^\infty$ topology the groups $\mathrm{Diff}_0(S, n)$ become topological groups and the projection maps from $\mathscr{M}(S)$ to $\mathbf{T}(p, n)$ become principal fibre bundles with the groups $\mathrm{Diff}_0(S, n)$ as fibre. It follows that the groups $\mathrm{Diff}_0(S, n)$ are contractible, provided that $2p - 2 + n > 0$. We shall not pursue this matter here, but refer to [16] for details.

## 10. QUASI–FUCHSIAN GROUPS

First we state a general principle.

10.1 PROPOSITION. *Let $\Gamma$ be any Kleinian group and let $w: \mathbb{C} \to \mathbb{C}$ be a quasiconformal map. Suppose $\mu = w_{\bar{z}}/w_z$ satisfies*

$$\mu(gz)\overline{g'(z)}/g'(z) = \mu(z) \text{ a.e. in } \mathbb{C} \tag{10.1}$$

*for all $g$ in $\Gamma$. Then $w\Gamma w^{-1}$ is a Kleinian group, and its limit set is $w(L(\Gamma))$.*

*Proof.* If $g \in \Gamma$, then $w$ and $w \circ g$ are both quasiconformal mappings of the extended plane onto itself. Equation 10.1 implies that $w$ and $w \circ g$ solve the same Beltrami equation, so $w \circ g \circ w^{-1}$ is a Möbius transformation and $w\Gamma w^{-1}$ is a group of Möbius transformations. Clearly its limit set is $w(L(\Gamma))$. ■

A Kleinian group obtained from $\Gamma$ by this method is called a *quasiconformal deformation* of $\Gamma$. A quasiconformal deformation of a Fuchsian group is called a *quasi-Fuchsian* group.

*Definition.* Let $G$ be any Fuchsian group acting in $\mathscr{U}$. For any $\mu \in \mathbf{M}(G)$ we denote by $w^\mu$ the unique quasiconformal map of $\mathbb{C}$ onto itself such that $w_{\bar{z}} = \mu w_z$ in $\mathscr{U}$, $w_{\bar{z}} = 0$ in $\mathscr{U}^*$, the exterior of $\mathscr{U}$, and $w(z) - z \to 0$ as $|z| \to \infty$. We also denote by $G^\mu$ the quasi-Fuchsian group $w^\mu G (w^\mu)^{-1}$ and we write $g^\mu = w^\mu \circ g \circ (w^\mu)^{-1}$ for all $g$ in $G$.

LEMMA (BERS). *$w_\mu = w_\nu$ on $\partial\mathscr{U}$ if and only if $w^\mu = w^\nu$ on $\partial\mathscr{U}$, hence in $\mathscr{U}^*$.*

*Proof.* Suppose $w_\mu = w_\nu$ on $\partial\mathcal{U}$. Define the quasiconformal map $h: \mathbb{C} \to \mathbb{C}$ by putting $h = (w_\nu)^{-1} \circ w_\mu$ in $\mathcal{U}$ and $h(z) = z$ in $\mathcal{U}^*$. Put $w = w^\mu \circ h^{-1}$. Then $w$ is quasiconformal in $\mathbb{C}$, $w_{\bar{z}} = \nu w_z$ in $\mathcal{U}$, and $w_{\bar{z}} = 0$ in $\mathcal{U}^*$. Further, in $\mathcal{U}^*$, $w(z) - z = w^\mu(z) - z \to 0$ as $|z| \to \infty$. Therefore $w = w^\nu$. But $w = w^\mu$ in $\mathcal{U}^*$.

Conversely, suppose $w^\mu - w^\nu$ in $\mathcal{U}^*$. Then $w^\nu \circ (w_\nu)^{-1}$ and $w^\mu \circ (w_\mu)^{-1}$ both map $\mathcal{U}$ conformally onto $w^\nu(\mathcal{U}) = w^\mu(\mathcal{U})$. Hence there is a conformal map $f: \mathcal{U} \to \mathcal{U}$ such that $w^\nu \circ (w_\nu)^{-1} = w^\mu \circ (w_\mu)^{-1} \circ f$ in $\mathcal{U}$. Thus $(w^\mu)^{-1} \circ w^\nu = (w_\mu)^{-1} \circ f \circ w_\nu$ in $\overline{\mathcal{U}}$, and on $\partial\mathcal{U}$ we have $\mathrm{id} = (w_\mu)^{-1} \circ f \circ w_\nu$ or $w_\mu = f \circ w_\nu$. But $w_\mu$ and $w_\nu$ fix the points $\pm 1$ and $i$. Therefore so does the Möbius transformation $f$, and $f$ is the identity. ∎

COROLLARY. $\mathbf{T}(G)$ *is in natural one-to-one correspondence with the set of conformal maps* $w^\mu$, $\mu \in \mathbf{M}(G)$.

## 11. THE BERS EMBEDDING

*Definition.* $\mathbf{B}(G)$ is the Banach space of holomorphic functions $\varphi(z)$ in $\mathcal{U}^*$ such that

$$\varphi(gz)g'(z)^2 = \varphi(z) \text{ for all } z \in \mathcal{U}^*, g \in G, \tag{11.1}$$

$$\|\varphi\|_* = \sup\{|\varphi(z)|(1 - |z|^2)^2; z \in \mathcal{U}^*\} < \infty. \tag{11.2}$$

The functions $\varphi$ in $\mathbf{B}(G)$ are called *bounded quadratic differentials.*

Recall that the *Schwarzian derivative* $[f]$ of the locally one-to-one meromorphic function $f$ is defined by

$$[f] = (f''/f')' - \tfrac{1}{2}(f''/f')^2.$$

We need the following two properties of Schwarzian derivatives. Both are classical and easy to verify.

$$[f \circ g] = ([f] \circ g)(g')^2 + [g]. \tag{11.3}$$

$$[f] = 0 \text{ if and only if } f \text{ is a Möbius transformation.} \tag{11.4}$$

*Definition.* For each $\mu$ in $\mathbf{M}(G)$ we denote by $\varphi^\mu$ the Schwarzian derivative of the conformal map $w^\mu$ in $\mathcal{U}^*$.

THEOREM (BERS). *Suppose* $\mu \in \mathbf{M}(G)$. *Then* $\varphi^\mu \in \mathbf{B}(G)$ *and* $\|\varphi^\mu\|_* \leqslant 6$. *Moreover* $\varphi^\mu = \varphi^\nu$ *if and only if* $w^\mu = w^\nu$ *in* $\mathscr{U}^*$.

*Proof.* Let $g \in G$. Using 11.3, 11.4, and Proposition 10.1 we find that

$$\begin{aligned} \varphi^\mu &= [w^\mu] = [g^\mu \circ w^\mu] = [w^\mu \circ g] \\ &= ([w^\mu] \circ g)(g')^2 = (\varphi^\mu \circ g)(g')^2. \end{aligned}$$

Therefore $\varphi^\mu$ satisfies 11.1. As for 11.2, the fact that $w^\mu$ is one-to-one in $\mathscr{U}^*$ implies that $\|\varphi^\mu\|_* \leqslant 6$, see [**26**], so $\varphi^\mu \in \mathbf{B}(G)$. If $w^\mu = w^\nu$ in $\mathscr{U}^*$ then of course $\varphi^\mu = \varphi^\nu$. Conversely, suppose $\varphi^\mu = \varphi^\nu$. Put $h = w^\mu \circ (w^\nu)^{-1}$ in $w^\nu(\mathscr{U}^*)$. Then in $\mathscr{U}^*$

$$\varphi^\mu = [w^\mu] = [h \circ w^\nu] = ([h] \circ w^\nu)((w^\nu)')^2 + \varphi^\nu,$$

so $[h] = 0$ in $w^\nu(\mathscr{U}^*)$ and $h$ is a Möbius transformation. But $h(z) - z \to 0$ as $|z| \to \infty$, so $h$ is the identity. ■

COROLLARY. *The map* $\mu \mapsto \varphi^\mu$ *of* $\mathbf{M}(G)$ *into* $\mathbf{B}(G)$ *induces an injective map of* $\mathbf{T}(G)$ *into* $\mathbf{B}(G)$.

From now on we identify $\mathbf{T}(G)$ with the set $\{\varphi^\mu; \mu \in \mathbf{M}(G)\}$ in $\mathbf{B}(G)$, and we define a map $\Phi: \mathbf{M}(G) \to \mathbf{T}(G) \subset \mathbf{B}(G)$ by $\Phi(\mu) = \varphi^\mu$. The next theorem lists some properties of $\Phi$. The first two statements were established by Bers [**9**]; the last by Ahlfors and Weill [**4**].

THEOREM. $\Phi: \mathbf{M}(G) \to \mathbf{B}(G)$ *is an open holomorphic map. Its differential* $\Phi'(\mu)$ *is surjective and has a bounded right inverse at every* $\mu$ *in* $\mathbf{M}(G)$. *For* $\varphi \in \mathbf{B}(G)$ *put*

$$\text{(11.5)} \qquad \mathscr{S}\varphi(z) = -\tfrac{1}{2}\bar{z}^{-4}\varphi(1/\bar{z})(1 - |z|^2)^2, \; z \in \mathscr{U}.$$

*If* $\|\varphi\|_* < 2$, *then* $\mathscr{S}\varphi \in \mathbf{M}(G)$ *and* $\Phi(\mathscr{S}\varphi) = \varphi$.

COROLLARY. $\mathbf{T}(G)$ *is a bounded region in* $\mathbf{B}(G)$, *contained in the open ball of radius* 6 *and containing the open ball of radius* 2 *about the origin.*

*Remarks.* (1) $\mathbf{M}(G)$ and $\mathbf{T}(G)$ are open sets in complex Banach spaces. The map $\Phi$ is holomorphic in the sense that its Fréchet derivative $\Phi'(\mu)$ exists at each

$\mu$ in $\mathbf{M}(G)$ and is a bounded $\mathbb{C}$-linear map (see for example Hille and Phillips [**22**]).

(2) Since it is an open set in $\mathbf{B}(G)$, $\mathbf{T}(G)$ is a complex manifold modelled on $\mathbf{B}(G)$. The space $\mathbf{B}(G)$, and therefore $\mathbf{T}(G)$, is infinite dimensional unless $G$ is finitely generated and of the first kind. In the finite dimensional case, let $\mathscr{U}'$ be the complement in $\mathscr{U}$ of the discrete set of elliptic fixed points of $G$. Then $\mathscr{U}'/G$ is a finite Riemann surface of some finite type $(p, n)$ (see Chapter 7, Section 1.5), and $\mathbf{T}(G)$ is naturally isomorphic to $\mathbf{T}(p, n)$. The fact that up to isomorphism $\mathbf{T}(G)$ depends only on the type of $\mathscr{U}'/G$ was proved first by Bers and Greenberg [**14**].

(3) Formula 11.5 shows that $\mathscr{S}\varphi$ is a real analytic function if $\varphi \in \mathbf{B}(G)$. Therefore the set of points $\{\Phi(\mu) : \mu \in \mathbf{M}(G)$ and $\mu$ real analytic$\}$ covers a neighbourhood of the origin in $\mathbf{T}(G)$. An easy connectedness argument shows that it equals $\mathbf{T}(G)$, as we asserted in Section 8.

(4) The complex structure on $\mathbf{T}(p, 0)$ was first obtained by Ahlfors [**2**]. Partial results had been obtained by Rauch [**27**]. Bers obtained the embedding as a bounded region, first for the finite dimensional spaces [**7**], then for all spaces $\mathbf{T}(G)$ in [**9**].

## 12. THE COTANGENT SPACE TO T(G)

The Bers' map $\Phi$ displays $\mathbf{T}(G)$ as an open region in $\mathbf{B}(G)$ and therefore as a complex manifold whose tangent space at each point is $\mathbf{B}(G)$. However this description of the tangent space to $\mathbf{T}(G)$ is not completely intrinsic since it depends on choosing a particular realization of $\mathbf{T}(G)$ as a region in a complex Banach space. We want a more intrinsic description of the tangent space and its dual the cotangent space.

*Definition.* $\mathscr{A}(G)$, the space of integrable quadratic differentials on $\mathscr{U}$, consists of the holomorphic functions $\varphi(z)$ in $\mathscr{U}$ satisfying

$$\varphi(gz)g'(z)^2 = \varphi(z) \text{ for all } z \in \mathscr{U}, g \in G,$$

and

$$\|\varphi\| = \iint_{\mathscr{U}/G} |\varphi(z)| \, dx \, dy < \infty.$$

THEOREM. *Let $\mu$ be a Beltrami differential. Then* $\Phi'(0)\mu \; (= \lim_{t \to 0} \Phi(t\mu)/t) = 0$

*if and only if*

$$\iint_{\mathscr{U}/G} \mu(z)\varphi(z)\,dx\,dy = 0 \text{ for all } \varphi \in \mathscr{A}(G). \tag{12.1}$$

*Proof.* We use the representation theorem for $w^{t\mu}(z)$ from Chapter 4, Theorem 5.7. Recall that if $\|t\mu\|_\infty < 1$, then

$$w_t(z) = w^{t\mu}(z) = z + \sum_{n=1}^{\infty} t^n \sigma_n(z),$$

where

$$\sigma_1(z) = -\frac{1}{\pi}\iint_{\mathscr{U}} \frac{\mu(\zeta)}{\zeta - z}\,d\xi\,d\eta.$$

Now for $z$ in $\mathscr{U}^*$ we have

$$w_t'(z) = 1 + t\sigma_1'(z) + o(t),$$

$$w_t''(z) = t\sigma_1''(z) + o(t),$$

$$w_t''(z)/w_t'(z) = t\sigma_1''(z) + o(t),$$

and

$$\Phi(t\mu)(z) = (w_t''(z)/w_t'(z))' - \tfrac{1}{2}(w_t''(z)/w_t'(z))^2$$

$$= t\sigma_1'''(z) + o(t).$$

Therefore, for all $z$ in $\mathscr{U}^*$,

$$\Phi'(0)\mu(z) = \lim_{t\to 0}\frac{1}{t}\Phi(t\mu)(z) = \sigma_1'''(z)$$

$$= -\frac{6}{\pi}\iint_{\mathscr{U}} \frac{\mu(\zeta)}{(\zeta - z)^4}\,d\xi\,d\eta$$

$$= -\frac{6}{\pi}\sum_{n=3}^{\infty} c_n z^{-(n+1)},$$

where

$$c_n = n(n-1)(n-2) \iint_{\mathscr{U}} \zeta^{n-3} \mu(\zeta) \, d\xi \, d\eta.$$

Now $\Phi'(0)\mu = 0$ if and only if $c_n = 0$ for all $n \geqslant 3$, and this holds if and only if

$$\iint_{\mathscr{U}} \mu(\zeta) f(\zeta) \, d\xi \, d\eta = 0$$

for all polynomials $f$, and therefore for all integrable holomorphic functions $f(\zeta)$ in $\mathscr{U}$. But that condition is equivalent to 12.1. In fact

$$\iint_{\mathscr{U}} \mu(z) f(z) \, dx \, dy = \iint_{\mathscr{U}/G} \mu(z) \, [\boldsymbol{\Theta} f(z)] \, dx \, dy$$

where $\boldsymbol{\Theta} f(z)$ is the Poincaré theta series, and we know from Chapter 3, Theorem 3.1 that every $\varphi$ in $\mathscr{A}(G)$ can be written in the form $\boldsymbol{\Theta} f$. ■

Let us denote by $L^\infty(G)$ the Banach space of all Beltrami differentials, with the $L^\infty$ norm, and by $L^1(G)$ the space of measurable functions $f$ on $\mathscr{U}$ such that

$$f(gz) g'(z)^2 = f(z) \quad \text{a.e.}$$

for all $g$ in $G$, and

$$\|f\| = \iint_{\mathscr{U}/G} |f(z)| \, dx \, dy < \infty.$$

Then $\mathscr{A}(G)$ is a closed subspace of $L^1(G)$, and $L^\infty(G) = L^1(G)^*$ by the natural pairing

$$l_\mu(f) = \iint_{\mathscr{U}/G} \mu(z) f(z) \, dx \, dy.$$

Our theorem tells us that the kernel of $\Phi'(0)$ is $\mathscr{A}(G)^\perp$.

COROLLARY. *If* $\mathbf{T}(G)$ *is finite dimensional its cotangent space at zero is* $\mathscr{A}(G)$.

*Proof.* $\Phi'(0)$ defines an isomorphism between the tangent space to $\mathbf{T}(G)$ at zero and the space $L^\infty(G)/\ker(\Phi'(0)) = L^1(G)^*/\mathscr{A}(G)^\perp = \mathscr{A}(G)^*$. The cotangent space is $\mathscr{A}(G)^{**}$, which equals $\mathscr{A}(G)$ in the finite dimensional case. ∎

*Remark.* A Beltrami differential $\mu$ in $\mathbf{M}(G)$ is called *trivial* if $\Phi(\mu) = 0$ or equivalently $w_\mu(z) = z$ on $\partial\mathscr{U}$. The set of trivial Beltrami differentials is denoted by $\mathbf{M}_0(G)$. The theorem in Section 11 implies that $\mathbf{M}_0(G)$ is a closed complex submanifold of $\mathbf{M}(G)$, and the theorem above tells us that the tangent space to $\mathbf{M}_0(G)$ at $\mu = 0$ is $\mathscr{A}(G)^\perp$. That fact, for finite dimensional Teichmüller spaces, was known long before the Bers embedding was discovered, and is known as Teichmüller's lemma (see Bers [5]).

## 13. GROUPS ON THE BOUNDARY OF T(G)

For each $\varphi$ in $\mathbf{B}(G)$ there is a unique locally one-to-one meromorphic function $w_\varphi$ in $\mathscr{U}^*$ whose Schwarzian derivative is $\varphi$ and whose Laurent expansion in $\mathscr{U}^*$ has the form

$$w_\varphi(z) = z + \sum_{n=1}^{\infty} b_n z^{-n}, \qquad z \in \mathscr{U}^*. \tag{13.1}$$

The function $w_\varphi$ depends continuously on $\varphi$ in the sense that if a sequence converges in $\mathbf{B}(G)$ then the corresponding sequence of functions $w_\varphi$ converges uniformly on compact sets in $\mathscr{U}^*$, with respect to the spherical metric on $\mathbb{C} \cup \{\infty\}$. If $\varphi = \Phi(\mu)$ belongs to $\mathbf{T}(G)$, then $w_\varphi = w^\mu$ is a conformal map in $\mathscr{U}^*$.

PROPOSITION (Bers [10]). *For each $\varphi \in \partial\mathbf{T}(G)$, $w_\varphi$ is a conformal map of $\mathscr{U}^*$ into $\mathbb{C}$, and $G_\varphi = w_\varphi G(w_\varphi)^{-1}$ is a Kleinian group whose regular set contains the simply connected region $w_\varphi(\mathscr{U}^*) \cup \{\infty\}$.*

*Proof.* Choose a sequence $\varphi_n = \Phi(\mu_n)$ in $\mathbf{T}(G)$ converging to $\varphi$. The corresponding conformal maps converge to $w_\varphi$, so $w_\varphi$ is either conformal or constant. The normalization 13.1 prevents $w_\varphi$ from being constant, so $w_\varphi$ is conformal. For each $g$ in $G$ the Möbius transformations $g^{\mu_n}$ converge to $w_\varphi \circ g \circ (w_\varphi)^{-1}$, so $G_\varphi$ is a group of Möbius transformations. It is clear that $w_\varphi(\mathscr{U}^*) \cup \{\infty\}$ contains no limit points of $G_\varphi$. ∎

*Remark.* If $\mathbf{T}(G)$ is finite dimensional, then $w_\varphi(\mathscr{U}^*) \cup \{\infty\}$ is a simply connected component of the regular set of $G_\varphi$, invariant under $G_\varphi$. Finitely generated Kleinian groups whose regular set has a simply connected invariant component are called $B$-groups. They have interesting and important properties. See the papers of Bers [**10**] and Maskit [**25**] and Chapters 8 and 10.

## 14. THE BERS FIBRE SPACE

For every $\varphi = \Phi(\mu)$ in $\mathbf{T}(G)$ there exists a well defined Jordan region $D_\varphi = w^\mu(\mathscr{U})$ and a well defined conformal map $w_\varphi = w^\mu$ of $\mathscr{U}^*$ onto the complement of $D_\varphi$. The quasi-Fuchsian group $G_\varphi = w^\mu G(w^\mu)^{-1}$ is also independent of $\mu$ (provided that $\Phi(\mu) = \varphi$), since $g_\varphi = w_\varphi \circ g \circ (w_\varphi)^{-1}$ in the complement of $D_\varphi$. The quotient space $D_\varphi/G_\varphi$ is equivalent to the Riemann surface represented by $\varphi$ in $\mathbf{T}(G)$. With Bers we define

$$\mathbf{F}(G) = \{(\varphi, z) \in \mathbf{T}(G) \times \mathbb{C}; z \in D_\varphi\}$$

and we define an action of $G$ on $\mathbf{F}(G)$ by

$$g(\varphi, z) = (\varphi, g_\varphi(z)). \tag{14.1}$$

THEOREM (BERS [**12**]). *$\mathbf{F}(G)$ is a bounded region in $\mathbf{B}(G) \times \mathbb{C}$. $G$ is a discontinuous group of biholomorphic maps of $\mathbf{F}(G)$ onto itself. The quotient space $\mathbf{V}(G) = \mathbf{F}(G)/G$ is a complex manifold. The obvious projection $(\varphi, z) \mapsto \varphi$ of $\mathbf{F}(G)$ onto $\mathbf{T}(G)$ induces a holomorphic projection $\pi: \mathbf{V}(G) \to \mathbf{T}(G)$ so that $\pi^{-1}(\varphi)$ is the Riemann surface $D_\varphi/G_\varphi$ for each $\varphi$ in $\mathbf{T}(G)$.*

*Remark.* By definition, a *complex analytic family of closed Riemann surfaces* consists of a pair of complex manifolds $\mathbf{V}$ and $\mathbf{B}$ and a holomorphic map $\pi$ of $\mathbf{V}$ onto $\mathbf{B}$ such that:

(i) $\pi$ is a submersion (its differential is surjective everywhere),
(ii) $\pi$ is proper (the inverse image of every compact set is compact),
(iii) $\pi^{-1}(t)$ is a closed Riemann surface for every $t$ in $\mathbf{B}$.

If the quotient space $\mathscr{U}/G$ is compact, then the above projection $\pi: \mathbf{V}(G) \to \mathbf{T}(G)$ defines a holomorphic family of closed Riemann surfaces of genus $p$ over the Teichmüller space $\mathbf{T}(G) = \mathbf{T}(p, n)$. For $n = 0$, Grothendieck [**21**] called attention to a universal property of this family and showed how this property makes possible an axiomatic description and construction of **T**$(p, 0)$. For $n > 0$, a corresponding universal property was observed by Engber [**20**] and by me [**15**]. Engber uses this property to give a construction of **T**$(p, n)$. Some of these matters are also discussed by Hubbard [**23**].

## 15. FURTHER PROPERTIES OF T(*p*, *n*)

In closing we want to mention a few of the more delicate analytic properties of the spaces **T**$(p, n)$. First of all, Teichmüller's theorem asserts that **T**$(p, n)$ is homeomorphic to $\mathbb{R}^{6p-6+2n}$. Proofs were given by Ahlfors [**1**] and Bers [**6**]. Earlier proofs, without quasiconformal maps, were given by Fricke and Klein and by Fenchel and Nielsen. See also Chapter 9, Section 1.

Secondly, **T**$(p, n)$ is a domain of holomorphy. This was proved by Bers and Ehrenpreis [**13**].

Finally, we want to describe the hyperbolic metric of **T**$(p, n)$. Recall that the *hyperbolic* or *Kobayashi* (*pseudo*) *metric* on a complex manifold $X$ is defined as the largest pseudometric $d$ on $X$ such that

$$d(f(z_1), f(z_2)) \leqslant \rho(z_1, z_2)$$

for all holomorphic maps $f$ of $\mathcal{U}$ into $X$ and for all $z_1$, $z_2$ in $\mathcal{U}$. Here $\rho$ is the Poincaré metric in $\mathcal{U}$:

$$\rho(z_1, z_2) = \tanh^{-1}|(z_1 - z_2)(1 - \bar{z}_1 z_2)^{-1}|.$$

See Kobayashi's book [**24**] for further discussion. The *Teichmüller metric* on **T**$(G)$ is defined as follows. First define the Teichmüller metric on **M**$(G)$ by

$$d'(\mu, \nu) = \tfrac{1}{2}\log K,$$

where $K$ is the maximal dilatation of the quasiconformal map $w_\mu \circ w_\nu^{-1}$ as defined in Chapter 4, p. 131. The Teichmüller metric on **T**$(G)$ is the quotient metric of $d'$ under the map $\Phi$:**M**$(G) \to$ **T**$(G)$. It was proved by Royden [**29**] that the Teichmüller metric on **T**$(p, n)$ is the hyperbolic metric. Since the Teichmüller metric is rather explicit this theorem is very useful. Royden [**29**] used it to determine all the biholomorphic maps of **T**$(p, 0)$ onto itself; they are given by members of the mapping class group, acting on **T**$(p, 0)$ as we described in Section 4. More recently Hubbard [**23**] and Kra and I, [**18**] and

[19] have used Royden's theorem to study holomorphic sections of the families $\mathbf{V}(G) \to \mathbf{T}(G)$ over $\mathbf{T}(G) = \mathbf{T}(p, n)$.

## REFERENCES


1. L. V. Ahlfors, On quasiconformal mappings, *J. Analyse Math.* **3** (1954), 1–58.
2. L. V. Ahlfors, The complex analytic structure of the space of closed Riemann surfaces, *in* "Analytic Functions" (R. Nevanlinna *et al.* eds) pp. 45–66. Princeton University Press, Princeton, N.J. 1960.
3. L. V. Ahlfors, "Lectures on Quasiconformal Mappings". Van Nostrand-Reinhold, Princeton, N.J. 1966.
4. L. V. Ahlfors and G. Weill, A uniqueness theorem for Beltrami equations, *Proc. Amer. Math. Soc.* **13** (1962), 975–978.
5. L. Bers, Spaces of Riemann surfaces, *in* Proceedings of the International Congress of Mathematicians (Edinburgh, 1958), pp. 349–361. Cambridge University Press, N.Y. 1960.
6. L. Bers, Quasiconformal mappings and Teichmüller's theorem, *in* "Analytic Functions" (R. Nevanlinna *et al.* eds) pp. 89–119, Princeton University Press, Princeton, N.J. 1960.
7. L. Bers, Correction to "Spaces of Riemann surfaces as bounded domains". *Bull. Am. math. Soc.* **67** (1961), 465–466.
8. L. Bers, "On Moduli of Riemann Surfaces". Lecture notes from E.T.H. Zürich, 1964.
9. L. Bers, Automorphic forms and general Teichmüller spaces, *in* "Proceedings of the Conference on Complex Analysis (Minneapolis, 1964)", pp. 109–113, Springer, Berlin, 1965.
10. L. Bers, On boundaries of Teichmüller spaces and Kleinian groups I, *Ann. of Math.* (2) **91** (1970), 570–600.
11. L. Bers, Uniformization, moduli, and Kleinian groups, *Bull. London Math. Soc.* **4** (1972) 257–300.
12. L. Bers, Fiber spaces over Teichmüller spaces, *Acta Math.* **130** (1973) 89–126.
13. L. Bers and L. Ehrenpreis, Holomorphic convexity of Teichmüller spaces, *Bull. Amer. Math. Soc.* **70** (1964) 761–764.
14. L. Bers and L. Greenberg, Isomorphisms between Teichmüller spaces, *in* "Advances in the Theory of Riemann surfaces" (L. Ahlfors *et al.* eds) Ann. Math. Studies, No. 66, pp. 53–79. Princeton University Press, Princeton, N.J. 1971.
15. C. J. Earle, On holomorphic families of pointed Riemann surfaces, *Bull. Amer. Math. Soc.* **79** (1973), 163–166.
16. C. J. Earle and J. Eells, A fibre bundle description of Teichmüller theory. *J. Differential Geometry* **3**(1969) 19–43.
17. C. J. Earle and J. Eells, Deformations of Riemann surfaces, in "Lectures in Modern Analysis and Applications I" (C. T. Taam ed) pp. 122–149. Springer, Berlin, 1969.
18. C. J. Earle and I. Kra, On holomorphic mappings between Teichmüller spaces *in* "Contributions to Analysis" (L. V. Ahlfors *et al.* eds) pp. 107–124. Academic Press, New York and London, 1974.
19. C. J. Earle and I. Kra, On sections of some holomorphic families of closed Riemann surfaces, *Acta Math.* **137** (1976) 49–79.

20. M. Engber, Teichmüller spaces and representability of functors, *Trans. Amer. Math. Soc.* **201** (1975) 213–226.
21. A. Grothendieck, Techniques de construction en géométrie analytique, Séminaire H. Cartan, 13 ème année: 1960/61, Exp. 7, 9–17.
22. E. Hille and R. S. Phillips, "Functional Analysis and Semigroups" (American Mathematical Society, Colloquoquium Publications, vol. 31). American Mathematical Society, Providence, R.I. 1957.
23. J. Hubbard, Sur les sections analytiques de la courbe universelle de Teichmüller, *Mem. Amer. Math. Soc.* **26** (1976).
24. S. Kobayashi, "Hyperbolic Manifolds and Holomorphic Mappings". Dekker, New York, 1970.
25. B. Maskit, On boundaries of Teichmüller spaces and on Kleinian groups II, *Ann. of Math.* (2) **91** (1970) 607–639.
26. Z. Nehari, The Schwarzian derivative and schlicht functions, *Bull. Am. Math. Soc.* **55** (1949) 545–551.
27. H. E. Rauch, On the transcendental moduli of algebraic Riemann surfaces, *Proc. Nat. Acad. Sci. USA* **41** (1955), 42–49.
28. H. E. Rauch, A transcendental view of the space of algebraic Riemann surfaces, *Bull. Amer. Math. Soc.* **71** (1965) 1–39.
29. H. L. Royden, Automorphisms and isometries of Teichmüller space, *in* "Advances in the theory of Riemann surfaces" (L. Ahlfors *et al.* eds) Ann. of Math. Studies No. 66, pp. 369–383. Princeton University Press, Princeton, N.J. 1971.

# 6. The Algebraic Structure of Surface Mapping Class Groups

JOAN S. BIRMAN*

*Department of Mathematics, Columbia University, New York, U.S.A.*



## 1. INTRODUCTION

Let $T_{g,0}$ denote a closed orientable surface of genus $g$, and let $T_{g,n}$ denote that same surface after $n$ distinct points $z_1^0, \ldots, z_n^0$ have been removed. The *mapping class group* $\mathcal{M}_{g,n}$ of $T_{g,n}$ is the group of all orientation-preserving piecewise-linear homeomorphisms of $T_{g,n} \to T_{g,n}$, modulo the subgroup of those homeomorphisms which are isotopic to the identity map. Our focus in this chapter will be on the algebraic structure of the groups $\mathcal{M}_{g,n}$, $g \geqslant 0$, $n \geqslant 0$.

In this introductory section we will review some classical properties of surface mappings. We will also try to show why this particular class of groups is of interest. Note that we are investigating a very special and concrete situation, rather than a general mathematical concept which can be applied to diverse situations. Our special class of groups does, however, have a surprising universality in its own right, because it plays an important role in such widely diverse areas as classical analysis (Riemann surface theory), infinite group theory (the structure of the automorphism group of a free

* Supported in part by the Alfred P. Sloane Foundation and by NSF Grant #MPS-71-03442.

group) and geometric topology (the study of closed, orientable 3-manifolds, with recent hints at connection with 4-manifold topology). The connection with analysis will be developed in other chapters, notably Chapter 5. In this introductory section we will try to indicate some of the reasons why the groups $\mathcal{M}_{g,n}$ are of interest to algebraists and topologists.

Section 2 treats certain special types of surface mappings, "twists", "spins" and "interchanges." Section 3 contains an investigation of the relationship between mapping class groups of closed surfaces and surfaces with a finite set of punctures. In Section 4 we discuss generators for the groups $\mathcal{M}_{g,0}$ and in Section 5, defining relations in these groups to the extent that these are known. Section 6 sketches a proof that the groups $\mathcal{M}_{g,0}$ are residually finite.

We begin by recording two nice properties of surfaces which will be assumed in our investigations. The theorems stated below were originally established by Baer [**1, 2**], and more recently reinvestigated and developed in full generality by Epstein [**10**].

Let $z^0_{n+1}$ be the base points for $\pi_1 T_{g,n}$.

1.1 THEOREM. *Let $f_0, f_1 : (S^1, s_0) \to (T_{g,n}, z^0_{n+1})$ be piecewise-linear embeddings of $S^1$ which are homotopic via a homotopy which keeps the base point fixed. Assume that $f_0(S^1)$ does not bound a disc in $T_{g,n}$. Then $f_0, f_1$ are piecewise-linear ambient isotopic via an isotopy which keeps the base point fixed.*

1.2 THEOREM. *Let $h:(T_{g,n}, z^0_{n+1}) \to (T_{g,n}, z^0_{n+1})$ be a piecewise-linear orientation-preserving homeomorphism which is homotopic to the identity map via a homotopy which keeps the base point fixed. Then h is piecewise-linear isotopic to the identity map via an isotopy which fixed the base point.*

The reader is referred to Epstein's excellent paper [**10**] for a careful study of these and related questions about curves and isotopies on 2-manifolds.

In our definition of $\mathcal{M}_{g,n}$ we regarded $T_{g,n}$ as having been obtained from $T_{g,0}$ by removing $n$ points. Alternatively, one may regard $T_{g,n}$ as being obtained from $T_{g,0}$ by removing $n$ discs $D_1, \ldots, D_n$, where $z^0_i$ is an interior point of $D_i$, $1 \leqslant i \leqslant n$. From this point of view, admissible maps and isotopies are required to keep $\{\partial D_1 \cup \ldots \cup \partial D_n\}$ fixed setwise.

1.3 PROPOSITION. *Each orientation-preserving homeomorphism $\alpha:(T_{g,0} - D_1 \cup \ldots \cup D_n) \to (T_{g,0} - D_1 \cup \ldots \cup D_n)$ extends to a homeomorphism $\alpha'$: $T_{g,n} \to T_{g,n}$. Moreover, $\alpha$ is isotopic to the identity map (via an isotopy which keeps each $\partial D_i$ fixed setwise) if and only if $\alpha'$ is isotopic to the identity map.*

This theorem is easily proved using techniques in J. F. P. Hudson's book "Piecewise linear topology", Benjamin, 1969.

In view of Proposition 1.3, the group $\mathscr{M}_{g,n}$ admits a second interpretation as the group of mapping classes of $(T_{g,0} - D_1 \cup \ldots \cup D_n)$, where admissible maps and isotopies keep $\{\partial D_1 \cup \ldots \cup \partial D_n\}$ fixed setwise.

We will also have occasion to treat the subgroup $\mathscr{P}_{g,n}$ of index $n!$ in $\mathscr{M}_{g,n}$ consisting of mapping classes which fix $z_1^0, \ldots, z_n^0$ individually. Note that, if $\mathbf{G}_{g,n}$ is the group of all orientation-preserving homeomorphisms of $T_{g,n} \to T_{g,n}$ which fix $z_1^0, \ldots, z_n^0$ individually, and if $\mathbf{G}_{g,n}$ is topologized by the compact-open topology, then the group $\mathscr{P}_{g,n}$ may be interpreted as the set of arc components in $\mathbf{G}_{g,n}$; this set carries a natural group structure because $\mathbf{G}_{g,n}$ is a topological group.

Isotopic deformations of $T_{g,n}$ obviously induce inner automorphisms of $\pi_1 T_{g,n}$, but a striking property of surface mappings is the converse: if $\alpha: T_{g,n} \to T_{g,n}$ is a homeomorphism which induces $\alpha_* \in \mathrm{Inn}(\pi_1 T_{g,n})$, then $\alpha$ is isotopic to the identity map. (See Chapter 1, Section 3.12, also [**21**] for the classical proof for the case $n = 0$, and p. 133 of [**31**] for a different proof, applicable in the general case $n \geqslant 0$.) To see that this pleasant phenomenon is not universally true, one need only increase the dimension by one, and consider a solid handlebody† $X_g$ of genus $g \geqslant 1$, and the homeomorphism of $X_g$ defined by slicing $X_g$ along a meridinal disc, twisting 360°, and glueing together again; this homeomorphism is clearly not isotopic to the identity map, yet the induced automorphism of $\pi_1 X_g$, which is a free group of rank $g$, is trivial.

If $n = 0$, a classical result of Nielsen, proved in [**25**], asserts that each automorphism of $\pi_1 T_{g,0}$ is induced by a topological mapping. See also Chapter 1, Section 3.12.1 and Chapter 9, Section 1.9 where the general Nielsen–Fenchel theorem (for $n > 0$) is discussed. Orientation-preserving homeomorphisms of $T_{g,0}$ induce a well-defined subgroup $\mathrm{Aut}^+ \pi_1 T_{g,0}$ of index 2 in $\mathrm{Aut}\, \pi_1 T_{g,0}$ and $\mathscr{M}_{g,0} = \mathscr{P}_{g,0}$ is naturally isomorphic to $\mathrm{Aut}^+ \pi_1 T_{g,0} / \mathrm{Inn}\, \pi_1 T_{g,0}$. If $n \geqslant 1$ the corresponding assertions are somewhat more complicated. Choose a system of generators for $\pi_1 T_{g,n}$ as indicated in Fig. 1. Then $\pi_1 T_{g,n}$ admits the two presentations:

$$(1)\ \langle w_1, \ldots, w_{2g}, u_1, \ldots, u_n;\ \prod_{i=1}^{g} [w_i, w_{i+g}]\, u_1 u_2, \ldots, u_n \rangle$$

$$(2)\ \langle w_1, \ldots, w_{2g}, u_1, \ldots, u_{n-1} \rangle.$$

† A handlebody is the closure of a regular neighbourhood of a graph in $E^3$.

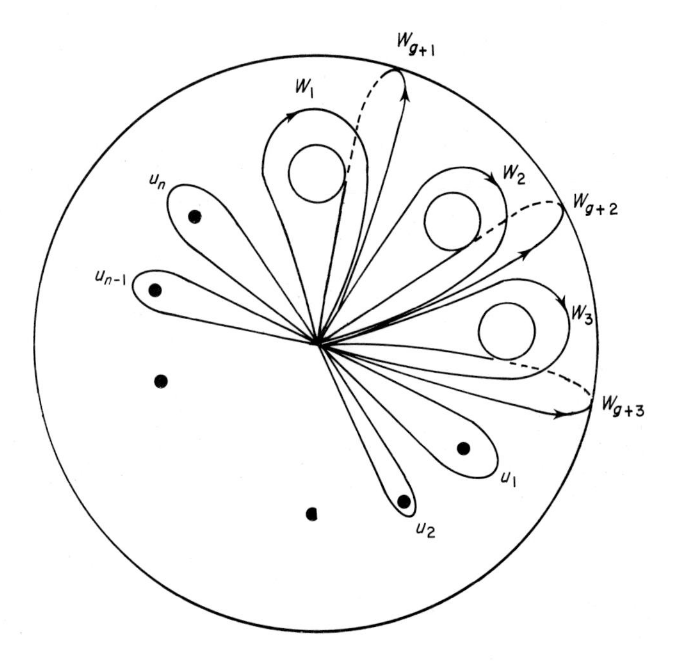


*Figure 1.*

From (1), it is immediate that topologically induced automorphisms in $\mathrm{Aut}^+\pi_1\, T_{g,0}$ must map

$$\Pi_n = \prod_{i=1}^{g} [w_i, w_{i+g}]\, u_1, \ldots, u_n$$

to a conjugate of itself. In each outer automorphism class we may then choose a representative which stabilizes $\Pi_n$.

An automorphism $\alpha$ of $\pi_1 T_{g,n}$ represents an element of the group $\mathscr{P}_{g,n}$ only if $\alpha(u_i)$ is a conjugate of $u_i$ ($i = 1, \ldots, n$), because a loop around a single boundary point must be mapped to a loop about that same point. Let $F_{2g+n-1}$ be the free group of the presentation (2). Let $A^*$ denote the subgroup of $\mathrm{Aut}\, F_{2g+n-1}$ consisting of the automorphisms which stabilize the conjugacy classes of $u_1, u_2, \ldots, u_{n-1}$, and fix the element

$$\Pi_{n-1} \prod_{i=1}^{g} [w_i, w_{i+g}] \, u_1 u_2, \ldots, u_{n-1}.$$

Then from the discussion above it is clear that $A^*$ maps homomorphically onto $\mathscr{P}_{g,n}$. Let $\langle \tilde{\pi} \rangle$ denote the cyclic subgroup of $A^*$ generated by the inner automorphism "conjugation by $\Pi_{n-1}$."
Then $\mathscr{P}_{g,n}$ is isomorphic to $A^*/\langle \tilde{\pi} \rangle$. (For the neatest proof, see Theorem 6 of [**20**]. Other proofs are in [**25**], [**21**], [**30**], [**31**], or Theorem 1 of [**12**].)

These assertions about algebraic characterizations of the groups $\mathscr{M}_{g,n}$ and $\mathscr{P}_{g,n}$ are summarized by:

1.4 Theorem. *The group $\mathscr{M}_{g,0}$ is canonically isomorphic to $Aut^+ \pi_1 T_{g,0}/Inn\ \pi_1 T_{g,0}$. The group $\mathscr{P}_{g,n}$, $n \geqslant 1$, is canonically isomorphic to $A^*/\langle \tilde{\pi} \rangle$, where $A^* = \{\alpha \in Aut\ F_{2g+n-1}/\alpha(u_i) = t_i u_i t_i^{-1}, i = 1, \ldots, n-1, \alpha(\pi_{n-1}) = t_{n-1} \pi_{n-1} t_n^{-1}$,*

$$\pi_{n-1} = \prod_{i=1}^{g} [w_i, w_{i+g}] \, u_1 u_2, \ldots, u_{n-1}\},$$

*and $\tilde{\pi}$ is the inner automorphism "conjugation by $\pi_{n-1}$". The group $\mathscr{M}_{g,n}$ is an extension of index $n!$ of $\mathscr{P}_{g,n}$, the quotient group being the full symmetric group on $n$ letters.*

One might hope that these very beautiful correspondences between algebra and topology would lead to constructive solutions to many questions about the structure of the group $\mathscr{M}_{g,n}$, but unfortunately that has not been the case, as the algebraic problems which are involved in the study of $\mathrm{Aut}^+ \pi_1 T_{g,0}/\mathrm{Inn}\ \pi_1 T_{g,0}$ and $A^*$ appear to be extremely difficult. Very recently, the work of McCool [**24**], who uses powerful techniques involving Whitehead transformations in the free group $F_{2g+n-1}$, $n \geqslant 1$, have begun to show promise of a successful algebraic attack on some of the problems we will investigate. We will, however, restrict ourselves to a primarily geometric approach, because McCool's techniques have not yet been sharpened sufficiently to yield concrete results, although it is clear that they have that potential.

Before ending this introductory section, I wish to justify the assertion that the groups $\mathscr{M}_{g,n}$ are of interest in several distinct areas of mathematics. The fact that the groups $\mathscr{M}_{g,n}$, $n \geqslant 0$, are of interest to algebraists should be clear from Theorem 1.4. The role of the group $\mathcal{M}_{g,n}$ in Teichmüller theory will be discussed in Chapters 5 and 9. Connections between groups of hyperbolic 3-space and 3-manifolds will be explored in Chapter 8. We now add a few examples which illustrate why the group $\mathcal{M}_{g,0}$ is of direct interest in 3-manifold topology.

We consider, first, the special class of 3-manifolds which are obtained as the quotient space $T_{g,0} \times I/\sim$, where the equivalence relation $\sim$ is defined by choosing a representative $\alpha$ of $[\alpha] \in \mathcal{M}_{g,0}$, and identifying $(p, 0)$ with $(\alpha(p), 1)$, $p \in T_{g,0}$. Let $\mathbf{M}(\alpha)$ denote the 3-manifold so obtained. It is a known result that if $\alpha, \alpha'$ represent distinct elements $[\alpha]$, $[\alpha'] \in \mathcal{M}_{g,0}$, then the 3-manifolds $\mathbf{M}(\alpha)$ and $\mathbf{M}(\alpha')$ will be homeomorphic if and only if $[\alpha]$ and $[\alpha']$ represent conjugate elements of $\mathcal{M}_{g,0}$. Thus the homeomorphism problem for this special class of 3-manifolds is equivalent to the conjugacy problem in the group $\mathcal{M}_{g,0}$. At this moment, that problem is unsolved; however useful topological invariants of $\mathbf{M}(\alpha)$ can be obtained very easily by examining the group of topologically induced automorphisms of $H_1(T_{g,0})$ i.e. the *symplectic modular group* $\mathrm{Sp}(2g, \mathbb{Z})$, which is a natural homomorphic image of $\mathcal{M}_{g,0}$. Let $\mu: \mathcal{M}_{g,0} \to \mathrm{Sp}(2g, \mathbb{Z})$ denote the homomorphism defined by $\mu([\alpha]) = \alpha_*: H_1(T_{g,0}) \to H_1(T_{g,0})$. Then the $2g$ distinct coefficients of the characteristic polynomial of $\alpha_*$ will be topological invariants of $\mathbf{M}(\alpha)$. Also, if $\rho_p$ is the natural homomorphism from $\mathrm{Sp}(2g, \mathbf{Z})$ to $\mathrm{Sp}(2g, \mathbb{Z}_p)$ then the order of the conjugacy class of $\rho_p \alpha_*$ will be a topological invariant of $\mathbf{M}(\alpha)$.

A more general construction involves the notion of a *Heegaard splitting*. Let $X_g$, $X'_g$ be handlebodies of genus $g$, which are obtained by embedding a surface $T_{g,0}$ in the 3-sphere $S_3$ that it divdes $S^3$ into two homeomorphic components, $X_g$ and $X'_g$. Let i: $\partial X_g \to \partial X'_g$ be the identity map. Let $[\alpha] \in \mathcal{M}_{g,0}$ be represented by $\alpha$: $\partial X_g \to \partial X_g$, so that $i\alpha$: $\partial X_g \to \partial X'_g$. Then we may define a 3-manifold $M_\alpha$ as $X_g \cup_{i\alpha} X'_g$, where $\partial X_g$ and $\partial X'_g$ are identified by the rule $i\alpha(p) = p$, $p \in \partial X_g$. The manifold $M_\alpha$ is represented as a "Heegaard splitting." It will be closed (i.e. compact and without boundary) and orientable. It is not difficult to see that *every* closed orientable 3-manifold may be so obtained, if we accept the fact that 3-manifolds may be triangulated, for we may find $X_g$ as the closure of a regular neighbourhood of the 1-skeleton of the triangulation, and $X'_g$ as the closure of a regular neighbourhood of the 1-skeleton of the dual triangulation, where both of these neighbourhoods have been increased so that their union is the entire 3-manifold.

As an example of a way in which knowledge about the algebraic structure of $\mathcal{M}_{g,0}$ has been applied to Heegaard splittings to answer questions about 3-manifolds, see [**16**], where knowledge of a system of generators for $\mathcal{M}_{g,0}$

(Section 3 below) was used to (i) obtain a new representation theorem for closed orientable 3-manifolds, and (ii) give a proof of the triviality of the combinatorial cobordism group for 3-manifolds.

We remark that, using the notion of a Heegaard splitting, it is possible to reduce many classical problems in 3-manifolds (e.g. the homeomorphism problem, the Poincaré conjecture, the Smith Conjecture) to explicit questions about $\mathscr{M}_{g,0}$. It seems reasonable to expect that as our knowledge of the group $\mathscr{M}_{g,0}$ increases some of these problems will be solved.

## 2. TWISTS, SPINS AND INTERCHANGES

In this section we describe several special types of surface mappings which will be needed later. We begin by defining the fundamental concept of a *twist* about a simple closed curve.

Let $N_0$ be an oriented cylindrical surface, parametrized by cylindrical coordinates $(y, \theta)$, $-1 \leqslant y \leqslant +1, 0 \leqslant \theta \leqslant 2\pi$. We define a map $h: N_0 \to N_0$ by the rule $h(y, \theta) = (y, \theta + \pi(y + 1))$. Observe that $h$ restricted to $\partial N_0$ is the identity.

Now let $c$ be a simple closed curve on (oriented) $T_{g,n}$ with $N$ a neighbourhood of $c$. Then we may find an orientation-preserving embedding $e: N_0 \to T_{g,n}$, with $e(N_0) = N$, $e(\{(0, \theta)\}) = c$. The map $ehe^{-1}$ is then a homeomorphism of $N \to N$, which is the identity on the two boundary curves of $N$, and so may be extended by the identity map to a self-homeomorphism of $T_{g,n}$. The resulting map $\tau_c$ and its inverse are known as *Dehn twists*. Intuitively, $\tau_c$ is defined by cutting open $T_{g,n}$ along one of the boundaries of $N$, twisting the free edge of $N$ through 360°, and gluing together again.

Note that if $p$ is a path which crosses the curve $c$ at a finite number of points, say $a_1, \ldots, a_r$, then the effect on $p$ of a twist $\tau_c$ will be to break $p$ at each point $a_i$ and insert a copy of $c$. The manner in which these copies of $c$ are joined to $p$ may be decided by the following rule. Let $\Delta p \subset p$ be a small segment of $p$, which has $a_i$ as one of its end points ($i = 1, \ldots, r$). Give $\Delta p$ a local orientation toward $c$. Then, an observer who is sitting on $\Delta p$, facing in the direction of the arrow, must move to the right (respectively left) when $\tau_c$ (respectively $\tau_c^{-1}$) is applied. This rule will be valid for directed segments on either side of $c$, as long as the local orientation is always toward $c$. (Thus the sense of $\tau_c$ does not depend on the assignment of an orientation to either $p$

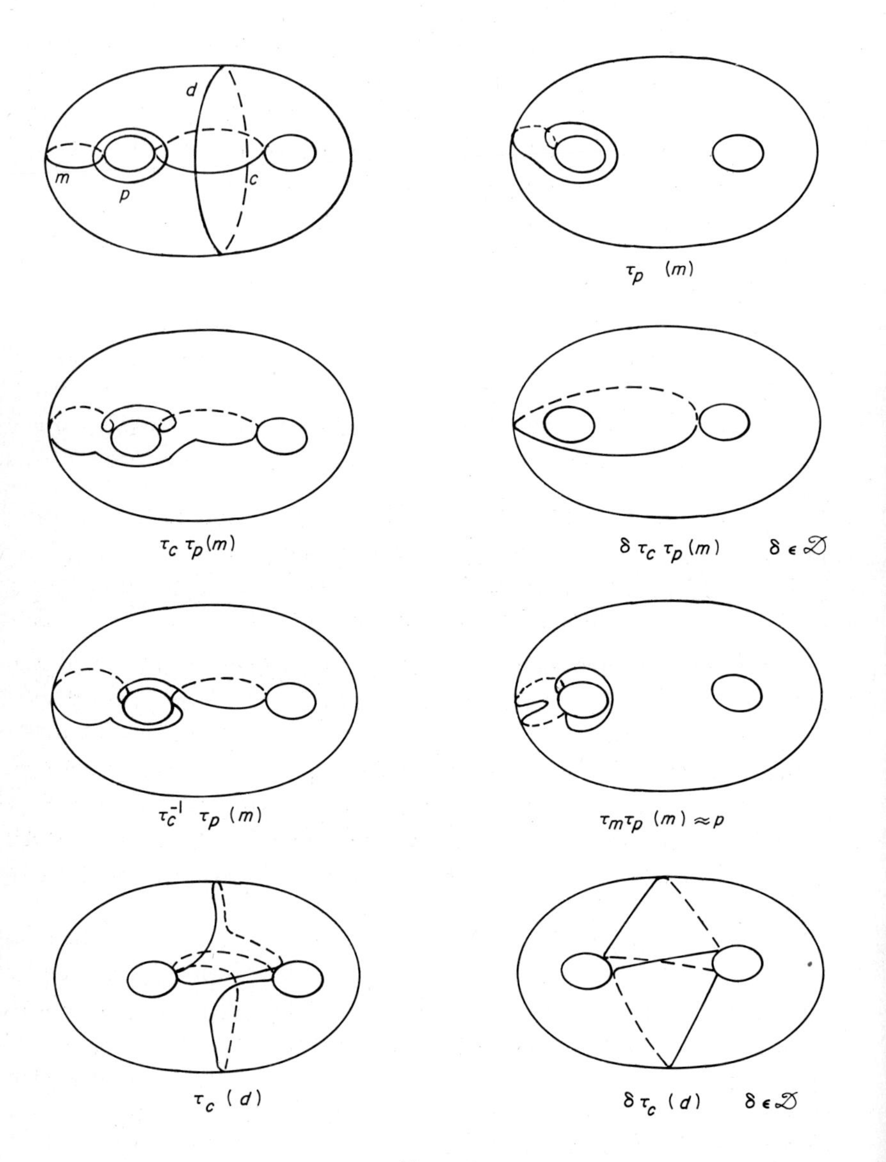


*Figure 2.*

or $c$.) Figure 2 gives examples which illustrate the effect of twist maps on simple closed curves on $T_{g,n}$. In Fig. 2, $\mathscr{D}$ denotes the subgroup of $\mathbf{G}(g, n)$ consisting of maps isotopic to the identity.

The isotopy class of $\tau_c$ is trivial on $T_{g,n}$ if $c$ bounds a disc, or a disc which contains one boundary point; otherwise the isotopy class of $\tau_c$ has infinite order. We will prove in Section 3 that the isotopy classes of a finite set of twist maps generate $\ker(i_*: \mathscr{M}_{g,n} \to \mathscr{M}_{g,0})$, and in Section 4 that the isotopy classes of a finite set of twist maps generate $\mathscr{M}_{g,0}$ for every $g \geqslant 0$.

Twist maps have several important properties.

2.1. PROPOSITION.

(i) *If* $\alpha: T_{g,0} \to T_{g,0}$ *is a homeomorphism, then* $\tau_{\alpha(c)} = \alpha\tau_c\alpha^{-1}$.

(ii) *If* $c$ *and* $p$ *are isotopic simple closed curves on* $T_{g,n}$, *then* $\tau_c$ *is isotopic to* $\tau_p$.

(iii) *Let* $z_1^0, \ldots, z_n^0$ *be an arbitrary set of* $n$ *distinguished points of* $T_{g,0}$. *Then any twist* $\tau_c$ *is isotopic to a twist which fixes* $z_1^0, \ldots, z_n^0$. *Thus any twist on* $T_{g,0}$ *is isotopic to a twist on* $T_{g,n}$.

*Proof.* The proofs of (i) and (iii) follow directly from the definition. To establish (ii), let $\alpha_s$ be an isotopic deformation of $T_{g,0}$ with $\alpha_0$ = identity, $\alpha_1(p) = c$. Then $\alpha_s\tau_p\alpha_s^{-1}$ is an isotopy between $\tau_p$ and $\tau_c$. Note that, by Theorem 1.1, the word "isotopic" in assertion (ii) can be replaced by "homotopic." ■

A second type of homeomorphism of $T_{g,n} \to T_{g,n}$ $(n \geqslant 1)$ which will be useful and important to us is defined by a product of twists, and will be called a "*spin*." Spin maps are maps of surfaces with punctures, which move the punctures non-trivially. Hence in defining a spin as a product of twists, we will have to be careful to choose our embeddings of $N_0$ in $T_{g,n}$ so that the distinguished set meets $e(N_0)$ in a particular manner. Let $z_n^0$ be one of the punctures, and let $c$ be a simple closed curve on $T_{g,n}$ which contains $z_n^0$. Let $c'$, $c''$ be the boundaries of an annular neighbourhood of $c$ on $T_{g,n}$, and let $N' = e'(N_0)$, $N'' = e''(N_0)$ be annular neighbourhoods of $c'$ and $c''$ respectively which are disjlint from $z_j^0, j \neq n$, and which are chosen so that $c \subset \partial N', \partial N''$. See Fig. 3. Then the twist product $\tau_{c'}\tau_{c''}^{-1}$ is a *spin of* $z_n^0$ *about* $c$. In Fig. 3 the effect of $\tau_{c'}\tau_{c''}^{-1}$ on the ray $e'(y, 0) \cup e''(y, 0)$ is pictured. Our spin is seen to drag the point $z_n^0$ about the path $c$. The map $\tau_{c'}\tau_{c''}^{-1}$ represents a non-trivial element of $\mathscr{P}_{g,n}$, but it represents the identity element in the group $\mathscr{P}_{g,n-1}$, i.e. it is in $\ker(i_*: \mathscr{P}_{g,n} \to \mathscr{P}_{g,n-1})$.

A slightly different map, which is closely related to a spin, may be defined on closed surfaces. Let $m$ be any non-separating simple closed curve on $T_{g,0}$ and let $c'$ be any other, separating or non-separating, which is disjoint from

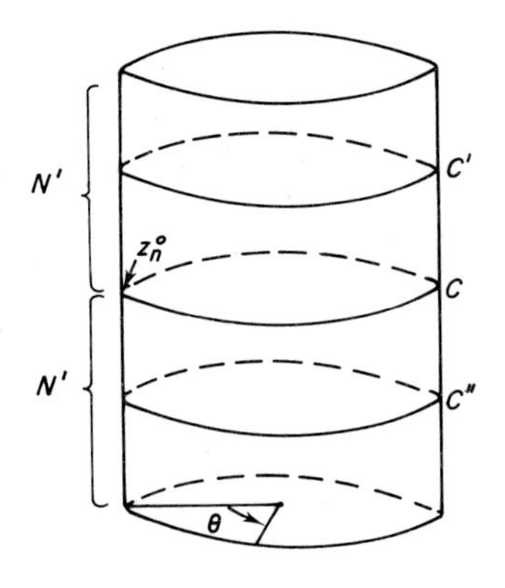


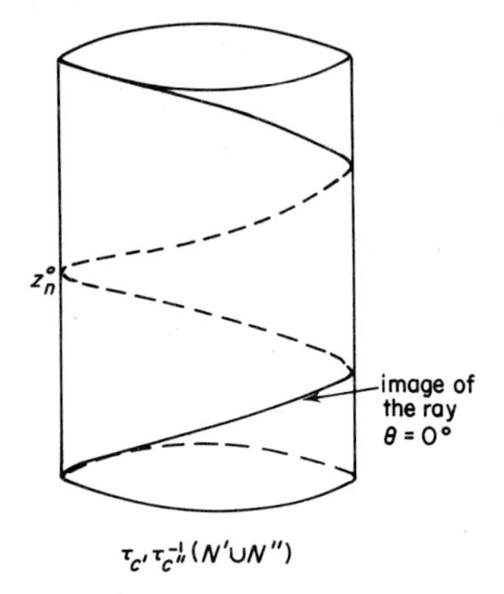


*Figure 3.*

$m$. Choose an arc $a$ which joins $c'$ to $m$, and construct a curve $c''$ which is disjoint from $c'$ and bounds a neighbourhood of $m \cup a \cup c'$, in the manner illustrated in Fig. 4.1. Now consider the twist pair $\tau_{c'}\tau_{c''}^{-1}$. In order to understand this map, let us cut $T_{g,0}$ open along $m$ (Fig. 4.2) to obtain a surface of genus $g - 1$ with two boundary components $m_1$ and $m_2$. Let $D$ be a closed disc that caps $m_2$, and let $z^0$ be interior to that disc. Note that, on the surface $T' = (T_{g,0} - m + D)$ the curves $c'$ and $c''$ are homotopic curves, which are separated by the point $z^0$. We may then find a curve $c$ which (i) contains $z^0$, (ii) is disjoint from each of the curves $c'$ and $c''$, and (iii) is homotopic in $T'$ to $c'$ and $c''$. It is now clear that $\tau_{c'}\tau_{c''}^{-1}: T' \to T'$ may be interpreted as a spin of $z^0$ about $c$.

Returning to our original surface $T_{g,0}$, we now see that $\tau_{c'}\tau_{c''}^{-1}$ has the effect of dragging the curve $m$ along the arc $a$, then around the curve $c'$, and finally back to its initial position via the arc $a$ traversed in the opposite sense. This interpretation of $\tau_{c'}\tau_{c''}^{-1}$ is easily visualized when the surface $T_{g,0}$ is embedded in 3-space in the manner illustrated in Fig. 4.3 (instead of as in Fig. 4.1). Our map simply "walks" the foot of the handle along $a$, about $c'$, and back where it started along $a^{-1}$. Our map would have been a bit more difficult to visualize if we had replaced the curve $c'$ with the curve $d'$ in Fig. 4.3. In that case the surface $T_{g,0}$ would appear to be "knotted" after executing $\tau_{d'}\tau_{d''}^{-1}$. (This knotting is, of course, a property of the *embedding* of $T_{g,0}$ in 3-space; it is not a 2-dimensional phenomenon.) A still more complicated case occurs if $m$ does not bound a disc in the inside handlebody $X_g$, e.g.

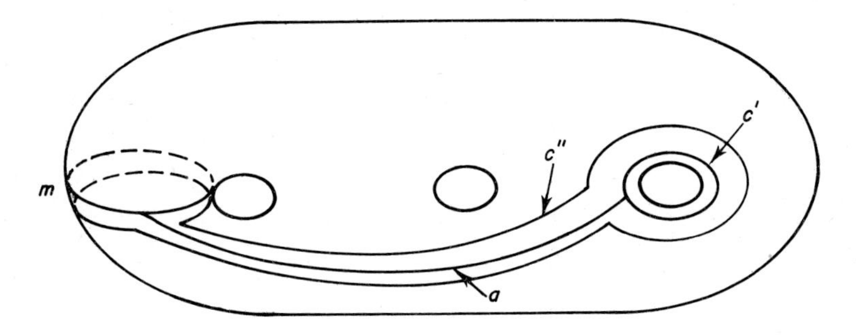


*Figure 4.1.*

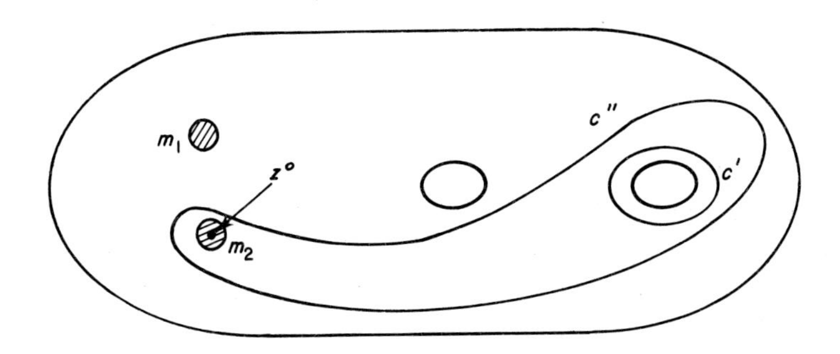


*Figure 4.2.*

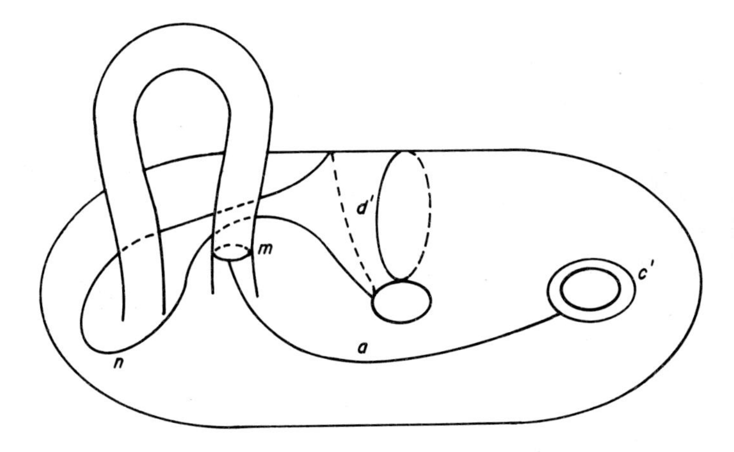


*Figure 4.3.*

replace $m$ by $n$ in Fig. 4.3. One cannot "see" any map which spins $n$ about $c'$ because such a map cannot be realized in 3-space.

Having at our disposal twists and spins, we next define an *interchange.* Let $a$, $c$ be non-separating simple closed curves on $T_{g,0}$. Suppose that there exists a curve $b$ on $T_{g,0}$ which meets each of the curves $a$ and $c$ transversally in one point. Then we assert that the twist product $\tau_c\tau_b\tau_c\tau_a\tau_b\tau_c$ interchanges the curves $a$ and $c$, and maps $b$ to itself. (The proof of this assertion will follow easily from Lemma 4.2 in Section 4.) Note that an interchange is supported in a neighborhood of the three defining curves $a$, $b$, $c$. Special cases of interchanges are:

*Case* (*i*) (Hilden's interchange). The surface $T_{g,0}$ is regarded as the boundary of a handlebody $X_g$. The curves $a$ and $c$ are chosen to be curves that bound discs in $X_g$. Then an interchange of $a$ and $c$ extends to $X_g$.

*Case* (*ii*) (*Powell's interchange*). If $a$ and $c$ are homologous simple closed curves, then an interchange of $a$ and $c$ (composed with a map which reverses orientation on $a$ and $c$) induces the identity on homology.

Curve interchanges are the principle tools in Hilden's (new) solution to the problem of finding generators for the subgroup of $\mathscr{M}_{g,0}$ consisting of classes of maps which have representatives that extend to $X_g$ [**13**].

## 3. FIBRATIONS

In this section we will study the natural homomorphism $i_*: \mathscr{P}_{g,n} \to \mathscr{P}_{g,n-1}$, defined by considering a homeomorphism $f$: $T_{g,n} \to T_{g,n}$, and extending it to a homeomorphism of $T_{g,n-1} \to T_{g,n-1}$ by defining $f(z_n^0) = z_n^0$. We will show that ker $i_*$ is generated by appropriate spins $\tau_{c'}\tau_{c''}^{-1}$. We will also investigate the group-theoretical structure of the kernel of the homomorphism from $\mathscr{M}_{g,n} \to \mathscr{M}_{g,0}$ induced by $i_*$. Our principle tools will be certain elementary properties of fibrations, hence we begin by reviewing the basic properties of fibre spaces which will be needed. For a general reference on fibre spaces, see [**28**, **14**].

A sequence of groups $\{G_k\}$ and homomorphisms $\{\Phi_k\}$, with

$$\xrightarrow{\Phi_{k+1}} G_{k+1} \xrightarrow{\Phi_k} G_k \xrightarrow{\Phi_{k-1}} G_{k-1} \longrightarrow$$

is *exact* if $\ker \Phi_k = \operatorname{im} \Phi_{k+1}$ for each $k$. Wè will use the following notation:

$X$ a topological space
$F$ a subspace of $X$
$x_0$ an arbitrary but fixed point of $F$
$I^k$ = unit $k$-cube = $\{(t_1, t_2, \ldots, t_k) \in \mathbb{R}^k : t_i \in [0, 1], 1 \leqslant i \leqslant k\}$
$I^{k-1} = \{(t_1, \ldots, t_{k-1}, 0) \in I^k\}$
$\dot{I}^k = \{(t_1, \ldots, t_k) \in I^k$: at least one $t_j = 0$ or 1, for $1 \leqslant j \leqslant k\}$
$J^{k-1} = \dot{I}^k - I^{k-1}$

The $k^{th}$ *relative homotopy group* $\pi_k(X, F, x_0)$ is the group† whose elements are homotopy classes of maps $f: (I^k, I^{k-1}, J^{k-1}) \to (X, F, x_0)$ with an appropriate rule for composing maps (see [**28**]). In the special case where $F = x_0$ this group is the $n^{th}$ *homotopy group* $\pi_k(X, x_0) = \pi_k(X, x_0, x_0)$. Define the maps:

$i$: $(F, x_0) \to (X, x_0)$ by inclusion
$j$: $(X, x_0, x_0) \to (X, F, x_0)$ by inclusion
$\partial$: $\pi_n(X, F, x_0) \to \pi_{n-1}(F, x_0)$ by $\partial[f] = [f|I_{n-1}]$.

3.1. Theorem (See [**28**].) *The sequence of groups and homomorphisms*

$$\cdots \xrightarrow{j_*} \pi_k(X, F, x_0) \xrightarrow{\partial} \pi_{k-1}(F, x_0) \xrightarrow{i_*} \pi_{k-1}(X, x_0) \xrightarrow{j_*} \pi_{k-1}(X, F, x_0) \xrightarrow{\partial} \cdots$$

$$\xrightarrow{\partial} \pi_0(F, x_0) \xrightarrow{i_*} \pi_0(X, x_0)$$

*is exact.*‡

Let $X$, $B$ be topological spaces, with $B$ arcwise-connected, and let $p$: $(X, x_0) \to (B, b_0)$ be a continuous map. Let $F$ be a fixed topological space. Then $X$ is a *fibre space* over the *base* $B$ with *fibre* $F$ and *projection* $p$ if

(i) for all $b \in B$: $p^{-1}(b)$ is homeomorphic to $F$,
(ii) for all $b \in B$ there exists a neighborhood $U$ of $b$ in $B$ and a homeomorphism $\Phi$: $U \times F \to p^{-1}(U)$ such that $p\Phi(u, f) = u$ for each $u \in U$, $f \in F$.

† If $k = 0$ we regard $\pi_k(X, F, x_0)$ as a set, which may or may not have a group structure. Similarly, $\pi_k(X, x_0)$ may not be a group if $k = 0$.

‡ If the sets near the right end of the sequence do *not* have a group structure, then one can only say that the indicated maps take the neutral element to the neutral element and that that kernel of such a transformation is the inverse image of the neutral element.

The fundamental property of fibre spaces which will be important in our situation is the existence of an isomorphism

$$p_*: \pi_k(X, F, x_0) \to \pi_k(B, b_0), \qquad k \geqslant 1.$$

This fact allows one to define a homomorphism $\bar{\partial}: \pi_k(B, b_0) \to \pi_{k-1}(F, x_0)$ by $\bar{\partial} = p_*^{-1}\partial$. The exact sequence of homotopy groups for the triplet $(X, F, x_0)$ then goes over to an exact sequence for the fibration $p: X \to B$:

$$(3) \cdots \xrightarrow{p_* j_*} \pi_k(B, x_0) \xrightarrow{\bar{\partial}} \pi_{k-1}(F, x_0) \xrightarrow{i_*} \pi_{k-1}(X, x_0) \xrightarrow{p_* j_*} \pi_{k-1}(B, x_0)$$
$$\cdots \xrightarrow{\bar{\partial}} \pi_0(F, x_0) \xrightarrow{i_*} \pi_0(X, x_0) \to 1$$

We will make use of this latter exact sequence to relate the groups $\mathscr{P}_{g,n}$ and $\mathscr{P}_{g,n-1}$, $n \geqslant 1$.

Recall that $\mathbf{G}_{g,n}$ is the group of orientation-preserving homeomorphisms of $T_{g,n} \to T_{g,n}$ where if $\alpha \in \mathbf{G}_{g,n}$ then $\alpha(z_i^0) = z_i^0$, $1 \leqslant i \leqslant n$. We topologize the group $\mathbf{G}_{g,n}$ with the compact-open topology, that is, take as a subbase† for the open sets of $\mathbf{G}_{g,n}$ the totality of sets $W(C, U)$, where if $C \subset T_{g,n}$ is an arbitrary compact subset and $U \subset T_{g,n}$ is an arbitrary open subset, then $W(C, U) = \{g \in \mathbf{G}_{g,n} | g(C) \subset U\}$. This makes $\mathbf{G}_{g,n}$ into a topological group, and we recognize that $\mathscr{P}_{g,n}$ is precisely the space of arc components $\pi_0 G_{g,n}$.

Define an *evaluation map* $p: \mathbf{G}_{g,n-1} \to T_{g,n-1}$, $n \geqslant 1$, by $p(\alpha) = \alpha(z_n^0)$. Observe that $p$ is continuous with respect to the given topologies on $\mathbf{G}_{g,n-1}$ and $T_{g,n-1}$. We assert:

3.2 THEOREM **[4, 26]**. *The space* $\mathbf{G}_{g,n-1}$ *is a fibre space over the base* $T_{g,n-1}$ *with projection p and fibre* $\mathbf{G}_{g,n}$.

*Proof.* We first show that property (i) for fibre spaces is satisfied, i.e. for each point $z \in T_{g,n-1} = (T_{g,0} - z_1^0 \cup \ldots \cup z_{n-1}^0)$ the preimage $p^{-1}(z)$ is homeomorphic to $\mathbf{G}_{g,n}$. Note that $\mathbf{G}_{g,n}$ is a closed subgroup of the topological group $\mathbf{G}_{g,n-1}$, and that two elements $\alpha$ and $\beta \in \mathbf{G}_{g,n-1}$ have the same image under $p$ if and only if they are in the same left coset of $\mathbf{G}_{g,n}$ in $\mathbf{G}_{g,n-1}$. Thus the set of points $p^{-1}(z)$ is in one to one correspondence with the elements of $\mathbf{G}_{g,n}$, and the correspondence is easily seen to be a homeomorphism.

To establish property (ii) for fibre spaces, consider the point $z_n^0 \in T_{g,n-1}$, and let $U$ be an open neighbourhood of $z_n^0$ in $T_{g,n-1}$. Construct a family of

† Open sets in $\mathbf{G}_{g,n}$ are thus arbitrary unions of finite intersections of sets $W(C, U)$.

homeomorphisms $\{\alpha: T_{g,n-1} \to T_{g,n-1}\}$, depending continuously on $u \in U$, such that each $\alpha_u$ is supported on $T_{g,n-1} - U$ and satisfies $\alpha_u(z_n^0) = u$. Define $\Phi: U \times \mathbf{G}_{g,n} \to p^{-1}(U)$ by $\Phi(u, \alpha) = \alpha_u$. Then $p\Phi(u, \alpha) = \alpha_u(z_n^0) = u$, as required. Since the space $T_{g,n-1}$ is homogeneous, a similar construction holds for any $z \in T_{g,n-1}$, hence our theorem is established. ■

As an immediate consequence of Theorem 3.2 we have:

3.3 COROLLARY. *The sequence of groups and homomorphisms*

$$\cdots \xrightarrow{i_*} \pi_1 \mathbf{G}_{g,n-1} \xrightarrow{p_*} \pi_1 T_{g,n-1} \xrightarrow{\bar{\partial}} \pi_0 \mathbf{G}_{g,n} \xrightarrow{i_*} \pi_0 \mathbf{G}_{g,n-1} \xrightarrow{p_*} \pi_0 T_{g,n-1} = 1$$

*is exact.*

Let us interpret the homomorphism $\bar{\partial}$ explicitly in our situation. Choose $z_n^0$ to be the base point for $\pi_1 T_{g,n-1}$. Let $[c] \in \pi_1 T_{g,n-1}$ be the homotopy class of a simple $z_n^0$-based loop $c$ on the surface $T_{g,n-1}$ which represents one of the generators of $\pi_1 T_{g,n-1}$. Let $c', c''$ be the boundaries of an annular neighbourhood of $c$ on $T_{g,n-1}$. Note that $c'$ and $c''$ are homotropic on $T_{g,n-1}$, but are *not* homotopic on $T_{g,n}$ because they are separated by the point $z_n^0$. A careful examination of the definition of $\partial$ and $p_*$ shows that $\bar{\partial}[c]$ is precisely the twist product $\tau_{c'}\tau_{c''}^{-1}$, which we discussed earlier and dubbed a spin of $z_n^0$ about $c$. By Proposition 2.1 the isotopy class of $\tau_{c'}\tau_{c''}^{-1}$ depends only on the isotopy class of $c$, and by Theorem 1.1 the isotopy class of $c$ on $T_{g,n-1}$ depends only on the homotopy class of $c$, hence the map $\bar{\partial}$ is indeed well-defined. The sequence in Corollary 3.3 is exact, hence we expect that image $\bar{\partial} = \ker i_*$, and indeed each of the maps $\tau_{c'}\tau_{c''}^{-1}$ (where $c'$ represents a generator of $\pi_1 T_{g,n-1}$) is non-trivial as an element of $G_{g,n}$, but its extension to an element of $G_{g,n-1}$ is trivial. Finally, choose any map $\alpha \in \ker i_* \subset G_{g,n}$. Then $\alpha$ is *not* isotopic to the identity as a map of $T_{g,n} \to T_{g,n}$, but $\alpha$ *is* isotopic to the identity as a map of $T_{g,n-1} \to T_{g,n-1}$. The isotopy $\alpha_t$ taking $\alpha$ to the identity then defines a loop $\alpha_t(z_n^0)$ through the base point $z_n^0$ of $\pi_1 T_{g,n-1}$, hence it determines an element $[\alpha_t(z_n^0)]$ in $\bar{\partial}^{-1}(\alpha)$. Thus, all appears to be correct and reasonable. Since $\mathscr{P}_{g,n} = \pi_p \mathbf{G}_{g,n}$ Corollary 3.3 implies

3.4 COROLLARY. *The kernel of* $i_*: \mathscr{P}_{g,n} \to \mathscr{P}_{g,n-1}$ *is generated by* $2g + n - 2$ *spins of the base point* $z_n^0$ *about curves* $c_1, \ldots, c_{2g+n-2}$ *whose homotopy classes generate* $\pi_1 T_{g,n-1}$.

*Proof.* By exactness of the sequence in Corollary 3.2, we know that $\ker i_* =$ image $\bar{\partial}$, hence we need at most one spin map for each generator of $\pi_1 T_{g,n-1}$. ■

3.5. COROLLARY. *The kernel of the natural homomorphism from* $\mathscr{M}_{g,n} \to \mathscr{M}_{g,0}$ *is generated by spins and twists.*

*Proof.* The group $\mathscr{P}_{g,n}$ may be built up from $\mathscr{P}_{g,0}$ by repeated extension, using the exact sequence of Corollary 3.3. Each twist on $T_{g,k}$ is isotopic to one on $T_{g,k+1}$ by Proposition 2.1, part (iii). Thus $\ker \mathscr{P}_{g,n} \to \mathscr{P}_{g,0}$ is generated by spins. The group $\mathscr{M}_{g,n}$ is an extension of index $n!$ of $\mathscr{P}_{g,n}$, and is clearly generated by any set of mapping classes which induce all possible permutations of the boundary points $z_1^0, \ldots, z_n^0$. A twist about a curve $c$ which contains $z_i^0$ and $z_{i+1}^0$ and is supported in a disc which includes $z_i^0$ and $z_{i+1}^0$ but no other $z_j^0$, will interchange $z_i^0$ and $z_{i+1}^0$, and $n - 1$ such transpositions generate the full group of permutations of $z_1^0, \ldots, z_n^0$. ■

We may now continue, and uncover the algebraic structure of the kernel from $\mathscr{M}_{g,n} \to \mathscr{M}_{g,0}$.

3.6. LEMMA. *Ker* $\bar{\partial} \subseteq$ *centre* $\pi_1(T_{g,n-1})$.

*Proof.* Suppose $x \in \ker \bar{\partial} =$ image $p_*$. Let $H \in \pi_1 \mathbf{G}_{g,n-1}$ be such that $p_* H = x$. The element $H$ is represented by a loop $h = \{h_t : 0 \leqslant t \leqslant 1\}$, where each $h_t$ is in $\mathbf{G}_{g,n-1}$ and $h_0 = h_1 =$ identity. Then $p(h_t) = h_t(z_n^0)$, $0 \leqslant t \leqslant 1$, represents $x$. Let $y \in \pi_1 T_{g,n-1}$ be represented by $y(s)$, $0 \leqslant s \leqslant 1$. Define $G$: $I \times I \to T_{g,n-1}$ by $G(t, s) = h_t(y(s))$, $(t, s) \in I \times I$. Then $G$ is continuous, and $G$ restricted to $\dot{I}_2$ represents the homotopy class of $xyx^{-1}y^{-1}$. Since $G$ is defined on the entire disc $I \times I$, it follows that $xyx^{-1}y^{-1} = 1$. As $y$ was an arbitrary element of $\pi_1 T_{g,n-1}$, we may conclude that $x \in$ centre $\pi_1 T_{g,n-1}$.

3.7 THEOREM [4]. *Let* $i_* : \mathscr{P}_{g,n} \to \mathscr{P}_{g,n-1}$ *be the natural homomorphism. Then ker* $i_*$ *is isomorphic to* $\pi_1 T_{g,n-1}$, *under the embedding* $\bar{\partial} : \pi_1 T_{g,n-1} \to \pi_0 \mathbf{G}_{g,n} = \mathscr{P}_{g,n}$ *for every* $g, n$ *except* $g = n = 1$ *and* $g = 0, n = 1, 2$. *If* $g = n = 1$ *or* $g = 0, n = 1, 2$ *then ker* $i_* = 1$.

*Proof.* By Lemma 3.6, we know $\ker \bar{\partial} \subset$ centre $\pi_1 T_{g,n-1}$, and except for the special cases cited above this implies $\ker \bar{\partial} = 1$, because $\pi_1 T_{g,n-1}$ is centreless. The groups $\mathscr{P}_{0,0}$ and $\mathscr{P}_{0,1}$ are each trivial groups, hence if $g = 0, n = 1$ then $\ker i_* = 1$. If $g = n = 1$ or $g = 0, n = 2$, then $\pi_1 T_{g,n}$ is abelian, hence centre $\pi_1 T_{g,n} = \pi_1 T_{g,n}$. It is easy to verify that in each of these cases $\bar{\partial}\pi_1 T_{g,n} \subset \ker \bar{\partial}$, hence by exactness $\ker i_* = 1$. ■

*Remark.* It is possible to determine the kernel of the homomorphism from

$\mathscr{M}_{g,n} \to \mathscr{M}_{g,0}$ quite explicitly for each pair $(g, n)$, because Theorem 3.7 contains all the information needed to determine it uniquely. It is, of course, constructed inductively by a sequence of extensions, where in each case the kernel is free. The group so obtained is the *Fox n-string braid group of the surface* $T_{g,0}$. For an introduction to the theory of braid groups, and a survey of the literature, see [**6**]. For explicit computations of these braid groups, see [**26**, **3**].

## 4. GENERATORS FOR $\mathscr{M}_{g,0}$

We focus our attention next on the proposition

4.1 THEOREM. [**9**, **16**, **17**, **18**]. *Every piecewise-linear orientation-preserving homeomorphism of a closed orientable surface of genus $g \geqslant 0$ is isotopic to a product of Dehn twists about non-separating simple closed curves.*

Later in this section we will prove that $\mathscr{M}_{g,0}$ may also be generated by elements of finite order (Corollary 4.6), and finally that $\mathscr{M}_{g,0}$ may be generated by a *finite* set of twists about canonical non-separating curves (Theorem 4.7).

*Proof of Theorem* 4.1. The proof which we will give is new, but it owes its inspiration and some of its details to the very elegant proof of Lickorish [**16**, **17**, **18**].

Suppose that $\alpha\colon T_{g,0} \to T_{g,0}$ is an orientation-preserving homeomorphism. Our plan is to examine the image $p = \alpha(m)$ of a standard non-separating curve $m$ on $T_{g,0}$ under $\alpha$ and to find a sequence of twists $\{\tau_{c_i};\, i = 1, \ldots, r\}$ about curves $c_1, \ldots, c_r$ and an isotopy $\delta$ such that $\alpha' = \delta\tau_{c_r}^{\varepsilon_r} \ldots \tau_{c_1}^{\varepsilon_1}\alpha$, $\varepsilon_i = \pm 1$, keeps $m$ fixed pointwise. It will then follow that the restriction of $\alpha'$ to $T_{g,0} - m$ is an orientation-preserving homeomorphism of a surface of genus $g - 1$ with two boundary components. Induction of the genus, together with the knowledge which we have already acquired about the kernel of the natural homomorphism from $\mathscr{P}_{g-1,2} \to \mathscr{P}_{g-1,0}$ will then complete the argument.

Note that the theorem is trivially true if $g = 0$. We assume, inductively, that our theorem is true for closed surfaces of genus $g - 1$.

Let us fix notation. Greek letters $\alpha, \beta, \tau, \ldots$ will be used to denote elements of the group $\mathbf{G}_{g,0}$ of orientation-preserving homeomorphisms of $T_{g,0}$, or of its subgroup $\mathscr{D}$ of homeomorphisms which are isotopic to the identity map. Lower case roman letters $m, p, c, \ldots$ will be used to denote simple closed curves on $T_{g0}$. The symbols $\tau_m, \tau_p, \tau_c, \ldots$ denote twists about $m, p, c, \ldots$. We write;

$\alpha \cong \beta$ if there exists $\delta \in \mathscr{D}$ such that $\alpha = \delta\beta$.
$p \approx m$ if there exists $\delta \in \mathscr{D}$ such that $\delta(p) = m$.
$p \underset{\tau}{\sim} m$ if there exists non-separating curves $c_1, \ldots, c_r$ such that

$$\tau_{c_r}^{\varepsilon_r} \ldots \tau_{c_1}^{\varepsilon_1}(p) \approx m.$$

$|p \cap m|$ = cardinality of $p \cap m$, when $p$ and $m$ are in general position.

4.2 LEMMA. *Let $r$, $s$ be simple closed curves on $T_{g,0}$. Suppose that $|r \cap s| = 1$. Then $r \underset{\tau}{\sim} s$.*

*Proof*: Note that, since $|r \cap s| = 1$, it is implicit that $r$ and $s$ are non-separating. In Figs 5.1, 5.2 and 5.3 the solid lines show the curves $s$, $\tau_r(s)$ and $\tau_s\tau_r(s)$. The "loop" in $\tau_s\tau_r(s)$ in Fig. 5.3 may be pulled back by an isotopy which is supported in a neighbourhood of $p$ to deform $\tau_s\tau_r(s)$ to $r$. Then $r \underset{\tau}{\sim} s$.

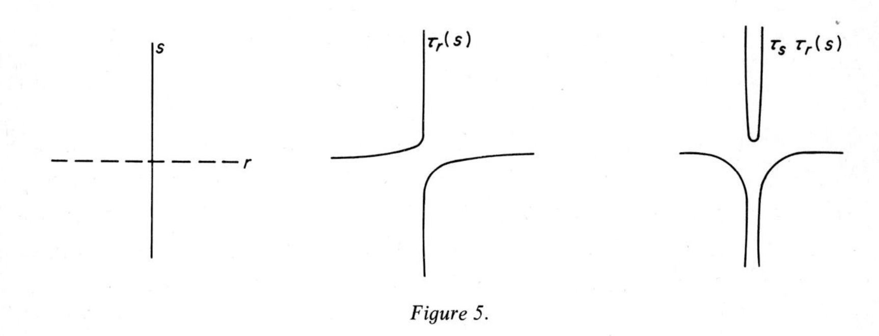


*Figure 5.*

4.3 LEMMA. *Let $m$ be a non-separating oriented simple closed curve on $T_{g,0}$. Then there exists a sequence of twists about non-separating curves, with product $\alpha$, such that $\alpha(m) \approx m$, but the induced orientation of $\alpha(m)$ is opposite to that of $m$.*

*Proof.* Find a non-separating simple closed curve $p$ such that $|p \cap m| = 1$. Then by Lemma 4.2, we have $\tau_m\tau_p(m) \approx p$. A second application of the construction used in Lemma 4.2 yields $\tau_p\tau_m(p) \approx m$, hence $\tau_p\tau_m^2\tau_p(m) \approx m$, and one may verify easily that the induced orientation of $\tau_p\tau_m^2\tau_p(m)$ is opposite to that of $m$.

4.4 LEMMA. *Let r, s be non-separating disjoint simple closed curves on* $T_{g,0}$. *Then* $r \underset{\tau}{\sim} s$.

*Proof.* The result is trivial if $r \approx s$. If $r \not\approx s$, we assert we can find a simple closed curve $c$ with $|c \cap r| = |c \cap s| = 1$. If such a curve $c$ exists it must be non-separating (because it meets $r$ once), hence by Lemma 4.2 we have $r \underset{\tau}{\sim} c$ and $c \underset{\tau}{\sim} s$, hence $r \underset{\tau}{\sim} s$. To see that $c$ exists, choose points $A_r$ and $A_s$ on $r$ and $s$ respectively. Suppose first that $r \cup s$ separates. Since $r$ and $s$ are each non-separating, it follows that $T_{g,0} - \{r \cup s\}$ has two components, hence we may join $A_r$ to $A_s$ by arcs $c_1$ and $c_2$ in each component to obtain $c = c_1 \cup c_2$. If $r \cup s$ does not separate, choose small arcs $c_r = \overline{A_{r1}A_{r2}}$ and $c_2 = \overline{A_{s1}A_{s2}}$ which cut $r$ and $s$ transversally at $A_r$ and $A_s$ respectively. Join $A_{r_1}$ and $A_{s_1}$ by an arc $c_1$ on $T_{g,0} - \{r \cup s \cup c_r \cup c_s\}$. Then $T_{g,0} - \{r \cup s \cup c_r \cup c_s \cup c_1\}$ is still connected, hence we may find an arc $c_2$ in it joining $A_{r_2}$ to $A_{s_2}$, to obtain $c = c_1 \cup c_r \cup c_2 \cup c_s$.

4.5 LEMMA. *Let p, m be arbitrary non-separating curves on* $T_{g,0}$. *Then* $p \underset{\tau}{\sim} m$.

*Proof.* By Lemmas 4.2 and 4.4 the result is true if $|p \cap m| = 0$ or 1. Suppose that $|p \cap m| > 1$. We consider two cases.

*Case* 1. The set $p \cap m$ contains two adjacent points (on $m$), say $A$ and $B$, which are such that $p$ is oriented on the same direction at $A$ as it is at $B$ with respect to an assigned orientation on $m$. See Fig. 6. Choose a point $A'$ close to $A$, and not on $p$ or $m$. Let $c$ be a simple path which starts at $A'$ and proceeds, close to $p$ but without crossing $p$ until it reaches $B'$ near $B \in p \cap m$, and then crosses each of the curves $p$ and $m$ once to return to $A'$. Since $|c \cap p| = 1$, the curve $c$ must be non-separating. By Lemma 4.2, we have $p \underset{\tau}{\sim} \subseteq$. Note that

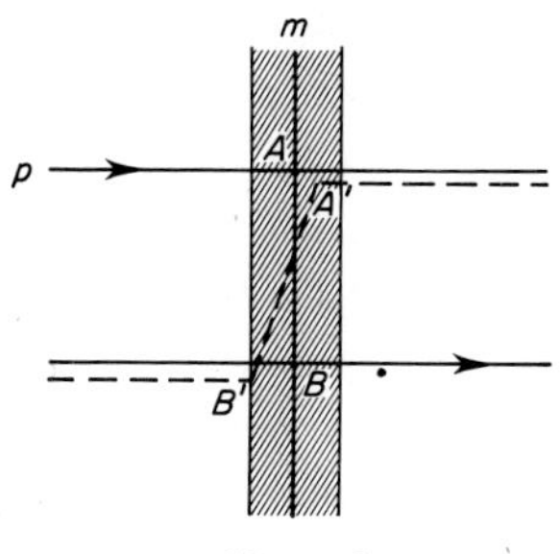


*Figure 6.*

$|c \cap m| < |p \cap m|$, hence by induction on $|p \cap m|$ we have $c \underset{\tau}{\sim} m$. Therefore $p \underset{\tau}{\sim} m$.

*Case* 2. The set $p \cap m$ contains no pair of adjacent points (on $m$) which satisfy the condition of Case 1. Then the intersections of $p$ with $m$ alternate as in Fig. 7.1. Let $N_p$ be a regular neighbourhood of $p$ on $T_{g,0}$, and let $c$ be one of the boundary curves of $N_p$. Then $c$ is a non-separating simple closed curve on $T_{g,0}$ and the points of $c \cap m$ alternate with those of $p \cap m$ in the manner indicated in Fig. 7.1 (because if $p$ is oriented, then $c$ will always appear to the right of $p$ to an observer walking along $p$ in the assigned direction). Choose two points $A$ and $B$ in $c \cap m$ which are adjacent on $m$, and such that one of the arcs $AB$ of $m$ contains no points of $p \cap m$. Let us now examine the intersections of a neighbourhood $N_m$ of $m$ with $c$ in the vicinity of $A$ and $B$. See Fig. 7.2. The boundary of $N_m$ will meet $c$ in four points, $A_1, A_2, B_1, B_2$, thus dividing $c$ into four arcs, denoted $b = \overline{B_2 B_1}$ (containing $B$), $d_1 = \overline{B_1 A_1}$, $a = \overline{A\ \ A_2}$ (containing $A$) and $d_2 = \overline{A_2 B_2}$. Let $e_1 = \overline{A_1 B_1}$ and $e_2 = \overline{A_2 B_2}$ be the boundaries of $N_m$ in the region of interest. Then we may construct simple closed curves $c_1 = d_1 \cup e_1$, $c_2 = d_2 \cup e_2$, each disjoint from $p$, and each meeting $m$ fewer times than $p$.

We assert that *one* of these curves $c_1$ or $c_2$ must be non-separating. For, suppose that *both* were separating. Then $T_{g,0} - c_1$ contains two components, and $c_2$ must be in one of them. Similarly, $T_{g,0} - c_2$ contains two components and $c_1$ must be in one of them. Therefore $T_{g,0} - c_1 - c_2$ has three components. Hence $T_{g,0} - c_1 - c_2 - a - b$ has four components (since $a + e_2 + b + e_1$ bounds a disc). Then $T_{g,0} - c_1 - c_2 - a - b + e_2$ has three components (since we have glued together two of the four components of $T_{g,0} - c_1 - c_2 - a - b$) and $T_{g,0} - c_1 - c_2 - a - b + e_2 + e_1 = T_{g,0} - c$ has two components. This gives the sought for contradiction, because $c$ is non-separating.

Suppose that $c_i$ is non-separating ($i = 1$ or $2$). By construction, $|c_i \cap p| = 0$ and $|c_i \cap m| < |p \cap m|$. Then by Lemma 4.4, we know that $p \underset{\tau}{\sim} c_i$ and by induction on $|p \cap m|$ we know that $c_i \underset{\tau}{\sim} m$, hence $p \underset{\tau}{\sim} m$.

We resume the proof of Theorem 4.1. By Lemma 4.5 we may find a sequence of twists $\tau_1, \ldots, \tau_r$ about non-separating curves $c_1, \ldots, c_r$ such that $\tau_\tau^{\varepsilon_r} \ldots \tau_1^{\varepsilon_1}\alpha(m) \approx m$. By Lemma 4.3 we may do this in such a way that if $m$ is oriented, then $\tau_\tau^{\varepsilon_r} \ldots \tau_1^{\varepsilon_1}\,\alpha(m)$ has an induced orientation which agrees with that of $m$. Hence there exists some $\delta \in \mathscr{D}$ such that $\alpha' = \delta\tau_r^{\varepsilon_r} \ldots \tau_1^{\varepsilon_1}\,\alpha(m)$ leaves $m$ fixed pointwise.

By cutting the surface $T_{g,0}$ along $m$, we may now regard $\alpha'$ as a homeo-

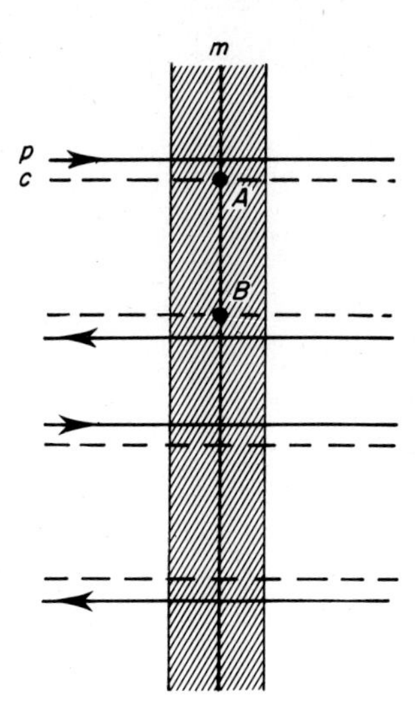


*Figure* 7(*i*).

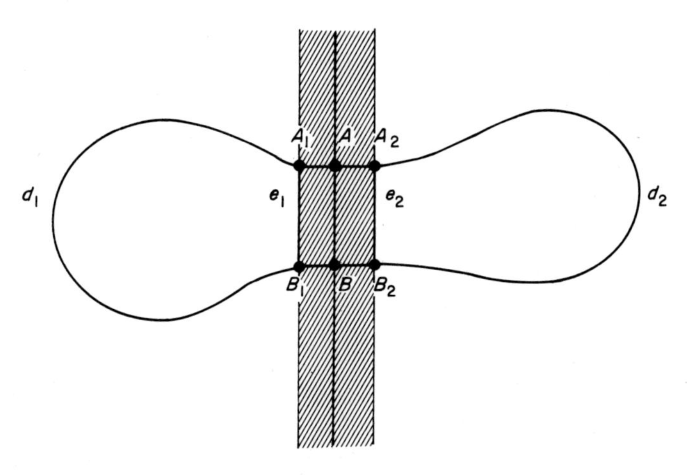


*Figure* 7(*ii*).

morphism of $T_{g-1}$ minus two discs, $D_1$ and $D_2$, where $\alpha'|\partial D_i$ – identity, $i = 1, 2$. Let $z_1^0$, $z_2^0$ be interior points of $D_1$ and $D_2$. Then $\alpha'$ extends to a map $\alpha'': T_{g-1,2} \to T_{g-1,2}$ which fixes $z_1^0$ and $z_2^0$. Moreover, $\alpha'$ will be isotopic to the identity map only if $\alpha''$ is isotopic to the identity map.

By the induction hypothesis on genus, the group $\mathscr{M}_{g-1,0} = \mathscr{P}_{g-1,0}$ is generated by the isotopy classes of twists about non-separating curves. By

Proposition 2.1, part (iii) each twist on $T_{g-1,0}$ is isotopic to a twist on $T_{g-1,2}$. Hence our theorem will be true if the kernel of the homomorphism from $\mathscr{P}_{g-1,2} \to \mathscr{P}_{g-1,0}$ is generated by twists about non-separating simple closed curves. This latter fact was established in Corollary 3.4. Hence Theorem 4.1 is true. ■

As observed earlier, the isotopy class of a twist about a non-separating simple closed curve on $T_{g,0}$ always has infinite order. Surprisingly, the group $\mathscr{M}_{g,0}$ can also be generated by elements of finite order.

4.6 COROLLARY. [19]. *Let $m_1, \ldots, m_{2g+1}$ be the simple closed curves on $T_{g,0}$ which are illustrated in Fig. 8. Let $\tau_i$ denote the isotopy classes of the twist about $m_i$, $1 \leqslant i \leqslant 2g+1$. Define*

$$\alpha = \tau_1 \tau_2 \ldots \tau_{2g+1}, \beta = \tau_1^{-1} \alpha.$$

*Then $\mathscr{M}_{g,0}$ is generated by the conjugacy classes of $\alpha$ and $\beta$.*

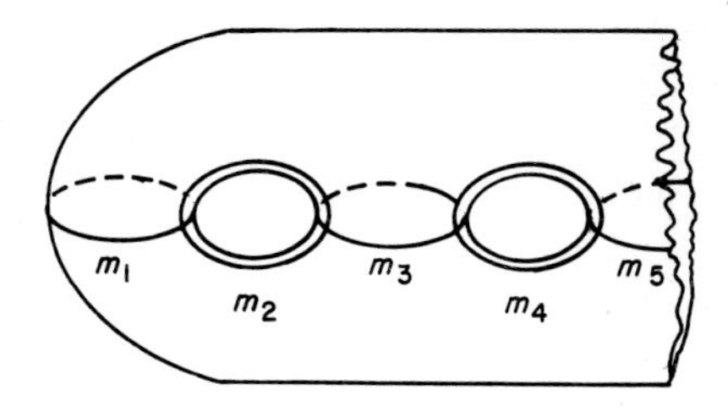

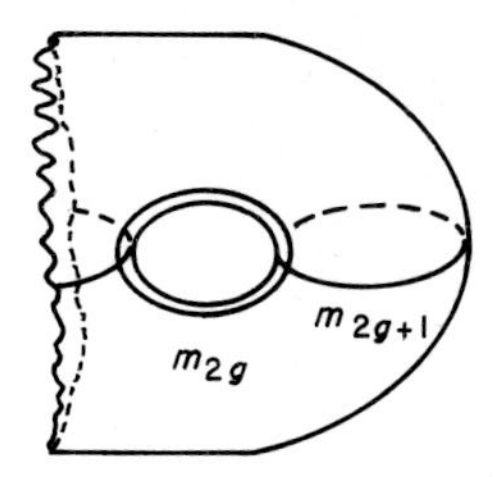


*Figure 8.*

*Moreover, $\alpha$ has order $2g+2$ and $\beta$ has order $4g+2$ in $\mathscr{M}_{g,0}$.*

*Proof.* To prove the second, assertion, it is adequate to show that the automorphisms of $\pi_1 T_{g,0}$ induced by $\alpha^{2g+2}$ and $\beta^{4g+2}$ are inner. This may be established by an ugly but routine computation, composing automorphisms induced by the twists $\tau_i$ on $\pi_1 T_{g,0}$. The latter are given explicitly in equations (12)–(14), p. 109, of [7].

Let $N$ be the normal subgroup of $\mathscr{M}_{g,0}$ generated by all conjugates of $\alpha$ and $\beta$. Then $\tau_1 = \alpha\beta^{-1} \in N$. Note that, if $p$ is any non-separating simple closed curve on $T_{g,0}$, then there exists a homeomorphism $\alpha$: $T_{g,0} \to T_{g,0}$ such that $\alpha(p) = m_1$. Hence, by Proposition 2.1, part (i), the isotopy class of the twist $\tau_p$ is in $N$. By Theorem 4.1 it then follows that $N = \mathscr{M}_{g,0}$. ■

A much stronger result than Theorem 4.1 is in fact true, namely that a finite set of twists about non-separating curves generate $\mathcal{M}_{g,0}$. We establish this stronger result.

4.7 THEOREM. *The isotopy classes of the twists about the curves* $\{u_i, y_i, z_{ij}, v_{ij}: 1 \leqslant i < j \leqslant g\}$ *in Fig. 9 generate* $\mathcal{M}_{g,0}$.

4.8 COROLLARY [**17**, **18**]. *The isotopy classes of the twists about the curves* $\{u_i, y_i, z_{j,j+1}; 1 \leqslant i \leqslant g, 1 \leqslant j \leqslant g-1\}$ *in Fig. 9 generate* $\mathcal{M}_{g,0}$.†

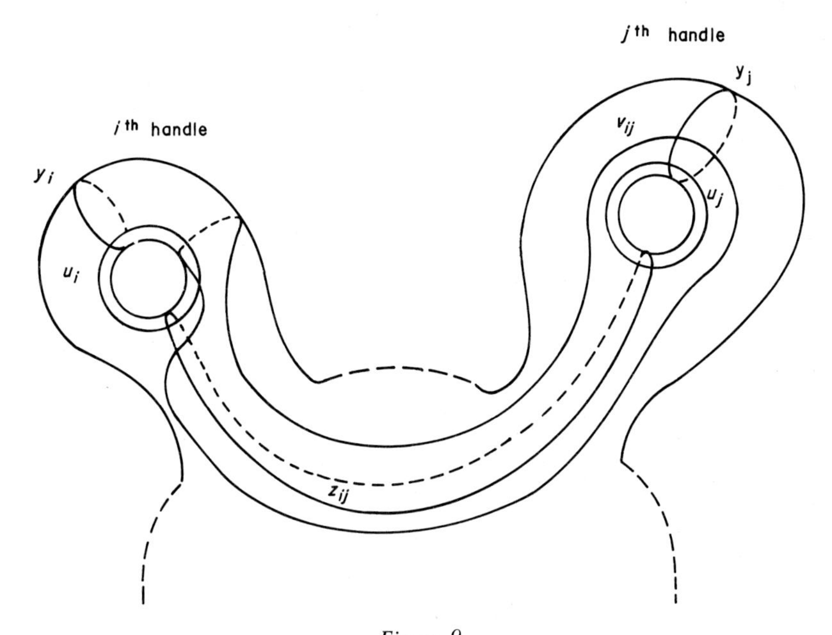


*Figure 9.*

We will prove Theorem 4.7. The proof of Corollary 4.8 will be left to the reader. (It is not difficult to establish if one uses Proposition 2.1, part (i).) Our proof of Theorem 4.7 is modeled on the proof in [**17**] and [**18**], but it is shorter than that proof because the results of Section 3 enables us to induct on genus. In particular we are able to avoid the lengthy enumeration of cases which was required in [**17**] and [**18**].

† *Note added in proof*: S. Humphreys has proved that the $2_{g+1}$ curves $\{u_i, y_1, y_2, z_j; 1\ i\ g, 1\ j\ g-1\}$ suffice, and that at least $2g+1$ twists are needed.

*Proof of Theorem* 4.7. Let $G$ denote the subgroup of $\mathscr{M}_{g,0}$ which is generated by the isotopy classes of the twists described in the statement of the theorem. We wish to prove that $G = \mathscr{M}_{g,0}$.

Since $\mathscr{M}_{0,0}$ is a trivial group the assertion is trivially true if $g = 0$. We assume, then, that it is true for surfaces of genus $< g$. Consider a surface $T_{g,0}$ of genus $g \geqslant 1$.

By Theorem 4.1, it is adequate to establish the assertion for a map which is a twist about a single non-separating curve on $T_{g,0}$. Let $p$ be that curve. Let $m_1$, $m_2$, $m_3$ be the simple closed curves on $T_{g,0}$ which are illustrated in Fig. 10. We wish to induct on $\max\{|p \cap m_i|, i = 1, 2\}$. We begin by investigating the situation when $|p \cap m_1| = 0$ or $2_0$ and $|p \cap m_2| = 0$, 1 or $2_0$, where "$2_0$" means two intersections, but no algebraic intersections. Note that $|p \cap m_1|$ cannot be 1, because $m_1$ is a separating curve. Note that $m_2$ and $m_3$ are members of the canonical set of curves in the statement of the theorem.

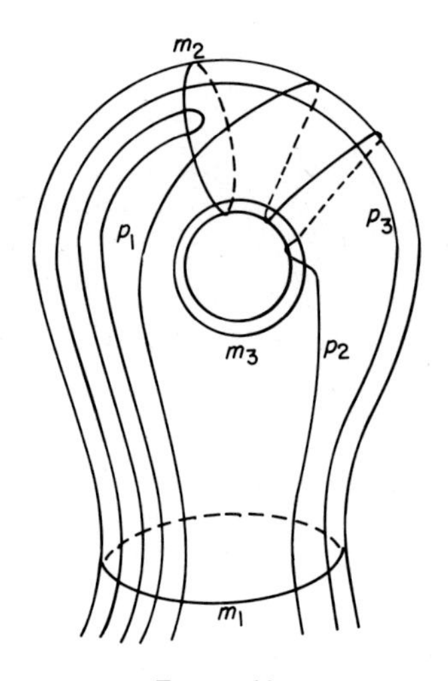


*Figure 10.*

*Case* 1. Suppose $|p \cap m_2| = 0$. Since $\tau_p$ is the identity map outside a neighbourhood of $p$, we may choose that neighbourhood to avoid $m_2$, so that $\tau_p$ restricted to $m_2$ = identity map. Hence, cutting open along $m_2$, our map $\tau_p$ restricts to a homeomorphism of $T_{g,0} - m_2$. By capping the boundary curves of $T_{g,0} - m_2$ with discs, as in the proof of Theorem 4.1, we may then regard $\tau_p$ as an element in the group $\mathscr{P}_{g-1,2}$. Just as in the proof of Theorem 4.1, the group $\mathscr{P}_{g-1,2}$ is generated by the generators of $\mathscr{M}_{g-1,0} = \mathscr{P}_{g-1,0}$

(lifted to $\mathscr{M}_{g,0}$ by reversing the cutting and capping procedure) together with the generators of ker $(i_*: \mathscr{P}_{g-1,2} \to \mathscr{P}_{g-1,0})$, c.f. Corollary 3.4). We leave it to the reader to verify that ker $i_*$ is generated by elements in $G$. (It is only necessary to check that the required spins may be obtained as products of twists in our collection.) The induction hypothesis on genus then allows us to complete the proof.

*Case* 2. Suppose $|p \cap m_1| = 0$, $|p \cap m_2| \neq 0$. Then $\tau_p$ restricts to a map of the "torus part" of the surface $T_{g,0}$. For the mapping class group of $T_{1,1}$ the result is well known (see Section 5 below). We omit details.

*Case* 3. Suppose $|p \cap m_1| = |p \cap m_2| = 2_0$. Then $p$ contains an arc which lies entirely on the torus part of $T_{g,0}$, and since that arc meets $m_2$ twice with zero algebraic intersection it must look like the arc $p_1$ in Fig. 10. But then, an isotopy brings us back to case (1).

*Case* 4. $|p \cap m_1| = 2_0$, $|p \cap m_2| = 1$. We study the arc of $p$ which is on the torus part of $T_{g,0}$. By classification of simple closed curves on a torus we see immediately that our arc must be homotopic to one which looks like either $p_2$ or $p_3$ in Fig. 10. If we are in the situation of $p_2$, then by twisting about $m_2$ we can reduce the number of intersections to arrive at the situation of $p_3$. Apply the twist product $\tau_{m_3}\tau_{m_2}$ to $p_3$ (see Fig. 11). The curve $\tau_{m_3}\tau_{m_2}(p)$ can be isotoped off $m_2$, therefore we are in the situation of case 1, and again we are done.

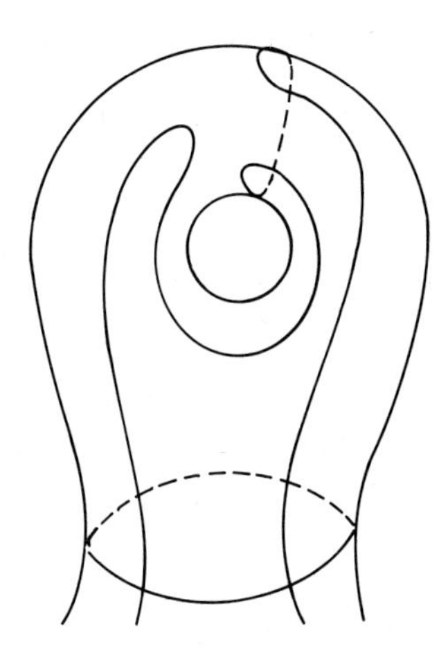

*Figure 11.*

We are now ready to consider the general situation. Let $q = \max\{|p \cap m_1|, |p \cap m_2|\}$. If $q = 0$ or 1 or $2_0$, we are dône. Suppose that $q \geqslant 2$, where if $q = 2$, then $q \neq 2_0$. Suppose also that our theorem is true whenever $\max\{|p \cap m_1|, |p \cap m_2|\} < q$. Let $q_i = |p \cap m_i| \leqslant q$, $i = 1, 2$. At least one of the $q_i$ is equal to $q$, and we may assume without loss of generality that $q_1 = q$ (because our proof does not depend in any way on the separating or non-separating properties of $m_i$). We will treat separately the cases $q_2 < q$ and $q_2 = q$. The proof will also depend on how $p$ meets $m_3$.

*Case* 1.1. Assume $q_2 < q$. Assume also that the set $|p \cap m_1|$ contains 2 adjacent points on $m_1$, say $A$ and $B$, which are such that $p$ is oriented in the same direction at $A$ as it is at $B$ with respect to an assigned orientation on $m_1$. See Fig. 12.1. This is similar to the situation that occurred in the proof of Lemma 4.5. Choose a neighbourhood $N_1$ of $m_1$ which is disjoint from $m_2$,

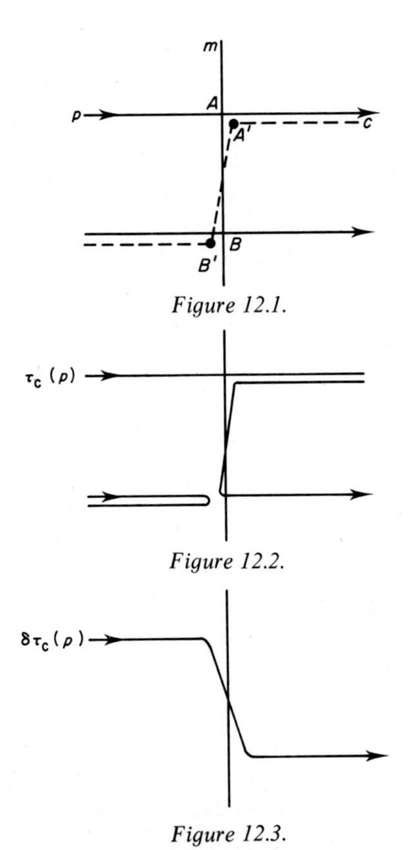


*Figure 12.1.*

*Figure 12.2.*

*Figure 12.3.*

and construct $c$ as in the proof of that Lemma, with $c \subset p \cup N_1$. From the picture, it is clear that $|c \cap m_1| < |p \cap m_1| = q_1 = q$. Also, since $c \subset p \cup N_1$ and $N_1 \cap m_2 = \varnothing$, we must have $|c \cap m_2| \leqslant |p \cap m_2| = q_2 < q$. By the inductive hypothesis, it follows that $\tau_c \in G$.

Fig. 12.2 shows $\tau_c(p)$. By inspection of the picture, it is clear that we may find a $\delta \in \mathscr{D}$ such that $u = \delta\tau_c(p) \subset p \cup N_2$ and $|u \cap m_1| < q$, as illustrated in Figure 12.3. Since $u \subset p \cup N_2$, we also have $|u \cap m_2| \leqslant |p \cap m_2| = q_2 < q$. Hence by induction on $q$ we see that $\tau_u \in G$.

Applying Proposition 2.1, part (i), we find that since $u = \delta\tau_c(p)$, therefore $\tau_p = \tau_c^{-1}\delta^{-1}\tau_u\delta\tau_c$. Since $\tau_c, \tau_u \in G$, it then follows that $\tau_p \in G$.

*Case* 1.2. Let $q_1 = q_2 = q$, and assume that the oriented curves $p$ and $m_1$ intersect exactly as in Case 1.1. Construct $c$ as before. If $|c \cap m_2| < |p \cap m_2| = q$, we may proceed as in Case 1.1 to prove that $\tau_c, \tau_u \in G$ and hence that $\tau_p \in G$. Suppose that $|c \cap m_2| = |p \cap m_2| = q$. This means that all of the points of $|p \cap m_2|$ occurred on the $\overrightarrow{AB}$ segment of $p$. But then we may alter our construction of $c$ so that $c$ runs in the opposite direction, next to $p$, i.e. $c \subset N_1 \cup \overrightarrow{BA}$ instead of $N_1 \cup \overrightarrow{AB}$. Then $|c \cap m_2| = 0$, so that we may clearly proceed as in Case 1.1 to conclude that $\tau_p \in G$.

*Case* 2.1. Assume first that $q_2 < q$. The case where $q_2 = q$ will be treated later (Case 2.2) by an easy modification of the argument given below, just as in Cases 1.1 and 1.2. We may now assume that the set $p \cap m_1$ contains no pair of adjacent points which satisfy the condition of Case 1.1. Then the intersection of $p$ and $m_1$ alternate as in Fig. 7.1. Choose 3 consecutive crossing points $A$, $B$, $D$ on $m_1$, as in Fig. 13.1. One of the $p$-segments $\overrightarrow{AD}$ or $\overrightarrow{DA}$ does not contain $B$; assume that the notation is chosen so that this segment is $\overrightarrow{DA}$. Let $N_1$ be a neighborhood of $m_1$ which is disjoint from $m_2$. Choose points $D'$, $A'$ in $N_1$ close to $D$, $A$ respectively. These points are to be chosen in such a way that they are both on the same side of $p$, with respect to the given orientation. Let $c$ be a simple path which starts at $A'$ and proceeds, close to $p$, to the point $D'$, then crosses $m$ once in $N_1$ to return to $A'$. Note that $|c \cap m_1| < |p \cap m_1| = q_1 = q$. Since $c \subset p \cup N_1$, and $N_1 \cap m_2 = \varnothing$, we must also have $|c \cap m_2| < |p \cap m_2| = q_2 < q$. By the induction hypothesis, it then follows that $\tau_c \in G$.

Figure 13.2 shows $\tau_c(p)$. By inspection of the picture, it is clear that we may find a $\delta \in \mathscr{D}$ such that $u = \delta\tau_c(p) \subset p \cap N_1$ and $|u \cap m_1| < q$, as illustrated in Fig. 13.3. Since $u \subset p \cup N_1$, it is also clear that $|u \cap m_2| \leqslant |p \cap m_2| < q$. Hence by induction on $q$ we see that $\tau_u \in G$.

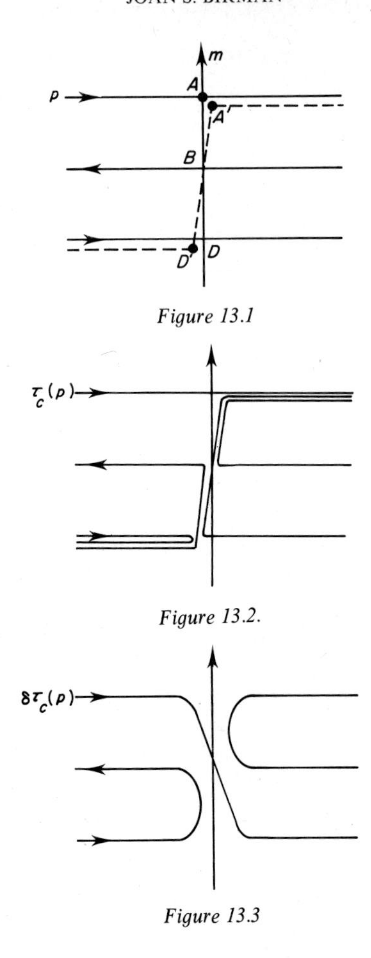


*Figure 13.1*

*Figure 13.2.*

*Figure 13.3*

Applying Proposition 2.1, part (i), we now find that since $u = \delta\tau_c(p)$, therefore $\tau_p = \tau_c^{-1}\delta^{-1}\tau_u\delta\tau_c$. Since $\tau_c, \tau_u \in G$, it then follows that $\tau_p \in G$.

*Case* 2.2. The intersections of $p$ and $m$ occur exactly as in Case 2.1, but now $q_1 = q_2 = q$. Construct $c$ as in Case 2.1. If $|c \cap m_2| < |p \cap m_2| = q_2 = q$ then we may proceed as in case 2.1 to prove that $\tau_c, \tau_u \in G$, and hence that $\tau_p \in G$. Suppose, then, that $|c \cap m_2| = |p \cap m_2| = q$. Denote the points of $p \cap m_1$ by $A, B, D, E, \ldots$ in the order in which they are encountered on $m_1$. (There are at least four points in $p \cap m_1$, because by the hypothesis the intersections occur in pairs, as in Fig. 7.1, and we are assuming that there are at least three points, hence there must be four points.) These points divide $p$ into arcs, and in Case 2.1 we constructed $c$ so that $c \subset N_1 \cup \overrightarrow{AD}$ where $\overrightarrow{AD}$ does not contain $B$.

Since $|c \cap m_2| = |p \cap m_2| = q$, it then follows that all the intersections of $p$ and $m_2$ occurred on the $\overrightarrow{AD}$ segment of $p$, and *none* on the $\overrightarrow{DA}$ section of $p$. We could however just as easily have oriented $p$ in the opposite sense and based our construction on the points $B$, $D$, $E$ (instead of $A$, $B$, $D$). With this new choice we will necessarily obtain a curve $c$ which meets $m_2$ fewer times than $p$ meets $m_2$. We may then proceed as in Case 2.1 to conclude that $\tau_p \in G$. This completes the proof that $\tau_p \in G$.

## 5. DEFINING RELATIONS IN $\mathcal{M}_{g,0}$

In this section we discuss the problem of obtaining presentations for the groups $\mathcal{M}_{g,0}$. Most of the results presented here will be descriptive in nature with reference to the literature replacing detailed proofs.

5.1. THEOREM. *The group $\mathcal{M}_{g,0}$ is finitely presented for each $g \geqslant 0$.*

*Proof.* [**24**] An alternative proof has also been found, it will appear in a future manuscript by Earle and Marden.† It should be noted that McCool's proof [**24**] is constructive, however the presentations which can (in principle) be obtained by his methods are too cumbersome to be of practical interest, at this moment.

In spite of Theorem 5.1, the problem of obtaining defining relations for the group $\mathcal{M}_{g,0}$ is unsolved, at this time, for $g \geqslant 3$.‡ I will next describe known results for the classical cases $g = 0$ and 1, and more recent results which solve the problem for $g = 2$ and at the same time yield information about certain subgroups of $\mathcal{M}_{g,0}$, when $g \geqslant 3$.

The group $\mathcal{M}_{0,0}$ is the trivial group. For an elementary proof of this classical result see Theorem 4.4 of [**6**]. Note that, since $\mathcal{M}_{0,0} = 1$, it follows that the group $\mathcal{M}_{0,n}$ is isomorphic to the kernel of the natural homomorphism from $\mathcal{M}_{0,n}$ to $\mathcal{M}_{0,0}$, studied in Section 3. Since a procedure was given in that section for finding a presentation for the kernel of the homomorphism $(\mathcal{M}_{0,n} \to \mathcal{M}_{0,0})$, it follows that a presentation can be obtained for $\mathcal{M}_{0,n}$. This fact will be important to us later, because the group $\mathcal{M}_{2,0}$ is closely related to the group $\mathcal{M}_{0,6}$. For a derivation of defining relations in $\mathcal{M}_{0,n}$ see Section 4 of [**6**], or see [**22**]. (The methods used in [**22**] are very different from those discussed here.)

† See Chapter 8, Section 7.3.
‡ *Note added in proof*: Hatcher and Thurston have announced a procedure for determining a complete set of defining relations for $\mathcal{M}_{g,0}$.

The group $\mathcal{M}_{1,0}$ is, by Theorem 1.4, canonically isomorphic to $\text{Aut}^+\pi_1 T_{1,0}$, because $\pi_1 T_{1,0}$ is abelian, so that $\text{Inn}\,\pi_1 T_{1,0} = \{1\}$. Since $\pi_1 T_{1,0}$ is a free abelian group of rank 2, it then follows that $\mathcal{M}_{1,0}$ is isomorphic to the group of $2 \times 2$ integer matrices having determinant $+1$, i.e. $SL(2, \mathbb{Z})$. That group is generated by the matrices

$$u_1 = \begin{bmatrix} 1 & 1 \\ 0 & 1 \end{bmatrix}, \quad y_1 = \begin{bmatrix} 1 & 0 \\ -1 & 1 \end{bmatrix}$$

and has defining relations

$$u_1 y_1 u_1 = y_1 u_1 y_1, \qquad (u_1 y_1 u_1)^4 = 1$$

(A proof may be found in [**23**].) In this presentation the matrices $u_1$, $y_1$ represent the automorphisms induced by the isotopy classes of twists about the curves $u_1$, $y_1$ in Fig. 9.

We consider, next, the case $g = 2$. We will treat this problem in a somewhat more general setting, specializing later to the case $g = 2$. Let $[\alpha]$ be an element in the group $\mathcal{M}_{g,0}$ which has finite order $k$, and let $G$ be the cyclic subgroup generated by $[\alpha]$. Let $N_G$ be the normalizer of $G$ in $\mathcal{M}_{g,0}$. We discuss a phenomenon which allows us to relate the structure of the group $N_G$ to that of a special subgroup of $\mathcal{M}_{h,n}$ for some appropriate $h < g, n \geqslant 0$.

A classical theorem of Nielsen asserts that if $[\alpha] \in \mathcal{M}_{g,0}$ has finite order $k$, then one may find a homeomorphism $\alpha: T_{g,0} \to T_{g,0}$ which represents $[\alpha]$ having the property that $\alpha^k$ is the identity map. We are interested in the $k$-fold covering space projection $\pi: T_{g,0} \to T_{g,0}/\alpha$ induced by the collapsing map from $T_{g,0}$ to the orbit space $T_{g,0}/\alpha$ denoted by $S$. A homeomorphism $\delta: T_{g,0} \to T_{g,0}$ will be said to be *fibre-preserving* (with respect to $\alpha$) if $\delta$ belongs to the normalizer of the cyclic subgroup $G$ of order $k$ generated by $\alpha$. (The condition that $\delta$ be fibre-preserving is equivalent to the requirement that $\delta$ project to a homeomorphism $\hat{\delta}$ of the orbit space $S$ defined by $\hat{\delta} = \pi\delta\pi^{-1}$.) A question which is very closely related to the theorem of Nielsen, quoted above, is whether, if $[\delta]$ belongs to the normalizer of $[\alpha]$ in $\mathcal{M}_{g,0}$, it follows that $[\delta]$ has a representative $\delta$ which belongs to the normalizer of $\alpha$? This is settled by Lemma 5.2.

5.2 Lemma. *Let $[\alpha]$, $[\delta] \in \mathcal{M}_{g,0}$, with $[\alpha]$ having finite order $k$ and $[\delta]\,[\alpha]\,[\delta]^{-1} = [\alpha]^p$, $1 \leqslant p < k$. Then $[\alpha]$ may be represented by an element $\alpha$ of order $k$ and $[\delta]$ may be represented by a fibre-preserving homeomorphism $\delta$, with $\delta\alpha\delta^{-1} = \alpha^p$.*

*Proof.* See Theorem 3 of [**8**].

In view of Lemma 5.2, the isotopy classes of fibre-preserving homeomorphisms define the subgroup $N_G$ of $\mathcal{M}_{g,0}$, that is, the normalizer of $G$ in $\mathcal{M}_{g,0}$.

Let $\delta$ be a fibre-preserving map, and let $\hat{\delta} = \pi\delta\pi^{-1}$. The covering-space projection $\pi: T_{g,0} \to S$ will be branched if $\delta$ has fixed points, and in that case the map $\hat{\delta}: S \to S$ will necessarily preserve the branch set. Thus, if $T_{g,0}/\alpha$ has genus $h$, and if $\pi$ is branched over $n$ points of $S$ then $\hat{\delta}$ will represent an element $[\hat{\delta}]$ in the group $\mathcal{M}_{h,n}$. In this situation we have:

5.3 THEOREM. *There is a well-defined homomorphism*

$$\tau: N_G \to \mathcal{M}_{h,n}$$

*given by* $\tau([\delta]) = [\hat{\delta}]$. *The kernel of* $\tau$ *is the cyclic subgroup* $G$ *of order* $k$ *of* $N_G \subseteq \mathcal{M}_{g,0}$.

*Proof.* [**8**] [**12**] [**29**] The difficult part of the proof of Theorem 5.3 is to show that $\tau$ is well-defined. The group $N_G$ is a subgroup of $\mathcal{M}_{g,0}$. Each element in $N_G$ is represented by a fibre-preserving homeomorphism, which thus necessarily preserves points in $\pi^{-1}$ (branch set). However, admissible *isotopies* need not preserve fibres; thus intuition suggests that $\tau$ will not be well-defined. It is a rather remarkable property of surface mappings that, if $\delta$ and $\delta'$ represent the same element $[\delta]$ of $N_G$, then in fact there is a *fibre-preserving isotopy* between $\delta$ and $\delta'$.

*Remark.* Results in [**12**] indicate that an even more general result than Theorem 5.3 is true. In Theorem 5.3 our group $N_G$ is the normalizer of the cyclic subgroup of order $k$ generated by $[\alpha]$. Suppose that we replace this cyclic subgroup by an arbitrary finite subgroup $G$ of $\mathcal{M}_{g,0}$. Let $N_G$ be the normalizer of $G$ in $\mathcal{M}_{g,0}$. Then the results in Section 4 of [**12**] indicate that Theorem 5.3 is still true. We are unable to state the theorem in this generality, however, because the analogue of Lemma 5.2 is not known for the case of a general finite subgroup $G$ of $\mathcal{M}_{g,0}$. (This is the Hurwitz Nielsen realization problem, discussed in Chapter 9, Section 3.2.)

Let us now specialize the situation discussed in Theorem 5.3 to the case $g = 2$. The surface $T_{g,0}$ may be assumed to be embedded in a symmetric manner in $E^3$, e.g. as in Fig. 14.1, so that it is invariant under reflections in the three coordinate planes. Let $\alpha: T_{2,0} \to T_{2,0}$ be defined by $\alpha(x, y, z) = (-x, y, -z)$. Then $[\alpha]$ has order 2 in $\mathcal{M}_{2,0}$. Let $c_1, \ldots, c_5$ be the curves illustrated in Fig. 14.1. By Theorem 4.7 the group $\mathcal{M}_{2,0}$ is generated by the isotopy classes

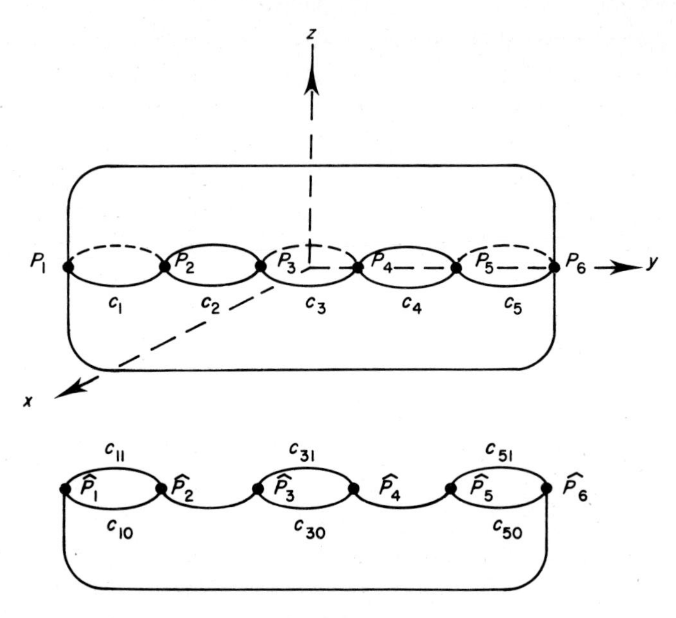


*Figure 14.*

$\{[\tau_{c_i}]: i = 1, \ldots, 5\}$. It is not difficult to see that $[\tau_{c_i}]$ commutes with $[\alpha]$ for each $i = 1, \ldots, 5$, hence the normalizer of $[\alpha]$ in $\mathcal{M}_{2,0}$, for this particular choice of $\alpha$, coincides with the full group $\mathcal{M}_{2,0}$. Thus the possibility arises of using Theorem 5.3 constructively to obtain a presentation for $N_\alpha$, if we are able to show that $\tau(N_\alpha)$ is a known group. With this end in mind, let us examine the orbit space projection $\pi: T_{2,0} \to S$. A fundamental set for $S$ is illustrated in Fig. 14.2. (The arcs $c_{i0}$ and $c_{i1}$ must be identified, for $i =$ 1, 3, 5.) One sees easily that $S$ is topologically equivalent to a sphere, and that the orbit space projection $\pi: T_{2,0} \to S \approx T_{0,6}$ is a branched covering space projection with 6 branch points (labeled $\hat{P}_1, \ldots, \hat{P}_6$ in Fig. 14.2). Thus the homomorphism $\tau$ of Theorem 5.3 maps $\mathcal{M}_{2,0}$ to $\mathcal{M}_{0,6}$.

It was noted earlier in this section that the methods of Section 3 allow one to obtain an explicit presentation for the group $\mathcal{M}_{0,n}$, $n \geqslant 0$. The generators which are obtained by these methods are, in fact, precisely the projections of $\{[\tau_{c_i}]; i = 1, \ldots, 5\}$ to the orbit space $S$, so that $\tau$ is surjective. These observations give the following Corollary to Theorem 5.3.

5.4 COROLLARY. *If $\alpha$ is the involution defined above, then the homomorphism $\tau$ maps $\mathcal{M}_{2,0}$ onto $\mathcal{M}_{0,6}$, and has as its kernel the cyclic group of order 2 generated by $[\alpha]$.*

*Remarks.* See [6] for a detailed proof of Theorem 5.3, specialized to the case considered in Corollary 5.4, and for the application of these ideas to obtain an explicit presentation for the group $\mathcal{M}_{2,0}$. The proof of Theorem 4.7 in [6] is essentially that of [8], specialized to the explicit case considered in Corollary 5.4. As noted earlier, other proofs are in [8], [12] and [29], and also in [7]. (The proof in [7] is however of limited interest, because it is much longer than that in [8] and applies to a much more restricted geometric situation.)

Before ending this section, we note one more result about the group $\mathcal{M}_{g,0}$, concerning its abelianized quotient group.

5.5. THEOREM. *The group* $\mathcal{M}_{g,0}/\mathcal{M}'_{g,0}$ *has order* 1 *or* 2 *if* $g \geqslant 3$.† (*If* $g = 2$ *it has order* 10, *and if* $g = 1$ *it has order* 12.)

*Proof.* [5] We will prove Theorem 5.5 for $g \geqslant 3$.

By Theorem 4.1, the group $\mathcal{M}_{g,0}$ is generated by twists about non-separating curves on $T_{g,0}$. By elementary results on classification of surfaces any two such curves are equivalent under a homeomorphism of $T_{g,0}$, hence by Proposition 2.1, part (i), the group $\mathcal{M}_{g,0}$ is generated by a single conjugacy class. Therefore $\mathcal{M}_{g,0}/\mathcal{M}'_{g,0}$ is cyclic, and is generated by the image in $\mathcal{M}_{g,0}/\mathcal{M}'_{g,0}$ of the isotopy class of a single twist about any non-separating curve. Call that generator $t$. In [5] explicit maps are given on $T_{g,0}$ which have finite order. These maps are given as products of twists about non-separating simple closed curves, and they imply by an easy calculation that $t^{56} = 1$ if $g = 3$, that $t^{10} = 1$ if $g \geqslant 3$ and that $t^{28} = 1$ if $g \geqslant 4$. Together these imply that $t^2 = 1$ if $g \geqslant 3$, establishing Theorem 5.5.

## 6. RESIDUAL FINITENESS OF $\mathcal{M}_{g,0}$

If one works in infinite group theory, one tends to adopt the attitude that everything is known about finite groups! Thus it is natural to ask how rich a supply of finite quotients is available for the study of a particular infinite group. Note that our group $\mathcal{M}_{g,0}$ is (by Powell's new results) a perfect group if $g \geqslant 3$. Therefore the class of finite groups which are homomorphic images

† *Note added in proof*: Powell has proved that the order is 1 if $g \geqslant 3$. (J. Powell, Two theorems on the mapping class group, *Proc. Amer. Math. Soc.* to appear).

of $\mathcal{M}_{g,0}$ (if $g \geqslant 3$) is a subclass of the class of finite perfect groups. In spite of this limitation, we will show in this section that the group $\mathcal{M}_{g,0}$ has a very "large" supply of finite quotients.

A group $G$ is said to be *residually finite* if for every $x \in G$ there is a normal subgroup $N_x$ of finite index in $G$ such that $x \notin N_x$. Thus there is a canonical homomorphism $\Phi$ from $G$ onto a finite group $G/N_x$, and the image of $x$ under $\Phi$ is non-trivial. Every finite group is, of course, residually finite. Our object in this section is to sketch Grossman's proof of Theorem 6.1.

6.1 THEOREM [**11**] *The mapping class group $\mathcal{M}_{g,0}$ is residually finite for every $g \geqslant 0$.*

*Remarks.* Our proof is totally non-constructive! We will not succeed in producing a single explicit finite quotient, and indeed at this writing the only finite quotients of $\mathcal{M}_{g,0}$ which have been investigated in any detail are those which factor through the symplectic group $\mathrm{Sp}(2g, \mathbb{Z})$.

*Proof of Theorem* 6.1. Recall that $\mathcal{M}_{g,0}$ is naturally isomorphic to $\mathrm{Aut}^+ \pi_1 T_{g,0}/\mathrm{Inn}\, \pi_1 T_{g,0}$, and is thus a subgroup of index 2 in the group $\mathrm{Out}\, \pi_1 T_{g,0}$ of outer automorphisms of $\pi_1 T_{g,0}$. Hence $\mathcal{M}_{g,0}$ will be residually finite if $\mathrm{Out}\, \pi_1 T_{g,0}$ is residually finite. To prove that $\mathrm{Out}\, \pi_1 T_{g,0}$ is residually finite, we will use two properties of $\pi_1 T_{g,0}$.

*Property* 1. The group $\pi_1 T_{g,0}$ is *conjugacy separable*, i.e. for all $k, h \in \pi_1 T_{g,0}$ either $k$ is conjugate to $h$ or else there exists a finite group $\Omega$ and a homomorphism $\Phi: \pi_1 T_{g,0} \to \Omega$ such that $\Phi(k)$ is not conjugate to $\Phi(h)$ in $\Omega$. (For a proof that $\pi_1 T_{g,0}$ has property (1), see Stebe ,[**27**].)

*Property* 2. Every class-preserving automorphism of $\pi_1 T_{g,0}$ is inner. For a proof, see [**11**]. The proof depends on a "word" type of argument.

We will now show that properties (1) and (2) imply that $\mathrm{Out}\, \pi_1 T_{g,0}$ is residually finite. We must show for all $k \in \pi_1 T_{g,0}$ there exists a homomorphism $\Phi: \mathrm{Out}\, \pi_1 T_{g,0} \to \Omega$, where $\Omega$ is finite, such that $\Phi(k) \neq 1$. This is equivalent to the existence of a collection of homomorphisms $\{\overline{\Phi}_\nu\}$, $\overline{\Phi}_\nu: \mathrm{Out}\, \pi_1 T_{g,0} \to \overline{\Omega}_\nu$, where $\overline{\Omega}_\nu$ is finite, such that $\cap_\nu(\ker \overline{\Phi}_\nu) = 1$. This, in turn, is equivalent to the assertion that there exists a collection of homomorphisms $\{\Phi_\nu\}$, $\Phi_\nu: \mathrm{Aut}\, \pi_1 T_{g,0} \to \Omega_\nu$ where $\Omega_\nu$ is finite, such that

$$\bigcap_\nu (\ker \Phi_\nu) = \mathrm{Inn}\, \pi_1 T_{g,0}.$$

Consider the collection $\{C_v\}$ of characteristic subgroups of finite index in $\pi_1 T_{g,0}$. Let $i(v)$ denote the number of distinct conjugacy classes in the finite group $(\pi_1 T_{g,0}/C_v)$, and let $\Omega_v$ denote the group of permutations of the $i(v)$ conjugacy classes. Each automorphism $a$ of $\pi_1 T_{g,0}$ induces an automorphism $a_*$ of $(\pi_1 T_{g,0}/C_v)$, which permutes conjugacy classes in $(\pi_1 T_{g,0}/C_v)$. Thus we have a homomorphism $\Phi_v$: $\text{Aut}\, \pi_1 T_{g,0} \to \Omega_v$.

We claim that $\cap_v(\ker \Phi_v) = \text{Inn}\, \pi_1 T_{g,0}$. To see this, suppose first that $a \in \text{Inn}\, \pi_1 T_{g,0}$. Then $a_*$ will also be inner, hence $\Phi_v(a)$ will be the identity permutation. Therefore $\text{Inn}\, \pi_1 T_{g,0} \subseteq \cap_v(\ker \Phi_v)$. To prove the reverse inclusion, assume that $\beta \in \bigcap_v (\ker \Phi_v)$, and suppose that there exists some $x \in \pi_1 T_{g,0}$ such that $\beta(x)$ is not conjugate to $x$. Since $\pi_1 T_{g,0}$ has property (1), we may find a normal subgroup $N$ of $\pi_1 T_{g,0}$ which has finite index $m$, such that $\beta(x)$ and $x$ map onto non-conjugate elements of $(\pi_1 T_{g,0}/N)$. The intersection of all normal subgroups of index $m$ in $\pi_1 T_{g,0}$ is a characteristic subgroup of $\pi_1 T_{g,0}$ which has finite index because $\pi_1 T_{g,0}$ is finitely generated. hence $N \supseteq C_v$ for some $v$. But then $\beta(x)$ and $x$ map onto non-conjugate elements of $\pi_1 T_{g,0}/C_v$, contradicting the fact that $\beta \in \ker \Phi_v$. Therefore $\beta$ must be class-preserving, and since $\pi_1 T_{g,0}$ has property (2) it follows that $\beta \in \text{Inn}\, \pi_1 T_{g,0}$. This completes the proof.

## REFERENCES


1. R. Baer, Kurventypen auf Flächen. *J. reine angew, Math.* **156** (1927) 231–246.
2. R. Baer, Isotopie von kurven auf orientierbaren, geschlossenen Flächen und ihr Zusammenhang mit der topologischen Deformation der Flächen, *J. Reine Angew Math.* **159** (1928) 101–111.
3. J. S. Birman, On braid groups, *Comm. Pure Appl. Math.* **22** (1969) 41–72.
4. J. S. Birman, Mapping class groups and their relationship to braid groups. *Comm. Pure Appl. Math.* **22** (1969) 213–238.
5. J. S. Birman, Abelian quotients of the mapping class group of a 2-manifold. *Bull. Amer. Math. Soc.* **76** (1970) 147–150.
6. J. S. Birman, Braids, Links and Mapping Class Groups, Ann. of Math. Stud. No. 82, 1975.
7. J. S. Birman and H. Hilden, Mapping class groups of closed surfaces and covering spaces. Ann. of Math. Stud. No. 66, pp. 81–115, 1971.
8. J. S. Birman and H. Hilden. On isotopies of homeomorphisms of Riemann surfaces, *Ann. of Math.* **97** (1973) 424–439.
9. M. Dehn, Die Gruppe der Abbildungsklassen *Acta Math.* **69** (1938) 135–206.
10. D. B. A. Epstein, Curves on 2-manifolds and isotopies. *Acta Math.* **115** (1966) 83–107.
11. Edna Grossman, On the residual finiteness of certain mapping class groups. *J. London Math. Soc.* **9**, Part 1 (1974), 160–164.

12. W. J. Harvey and C. Maclachlan, On mapping class groups and Teichmüller spaces, *Proc. London Math. Soc.* (3). XXX (1985) 496–512.
13. Hugh M. Hilden, A finite set of generators for the mapping class group of a 3-dimensional handlebody, to appear.
14. P. Hilton, "Homotopy Theory." Cambridge University Press, 1966.
15. M. Kupferwasser and P. Shalen, "Endomorphisms of Free Groups and Surface Mappings." In preparation.
16. W. B. R. Lickorish, A representation of orientable, combinatorial 3-manifolds. *Ann. of Math.* **76** (1962) 531–540.
17. W. B. R. Lickorish. A finite set of generators for the homeotopy group of a surface, *Proc. Camb. Philos. Soc.* **60** (1964) 769–778.
18. W. B. R. Lickorish. Corrigendum: a finite set of generators for the homeotopy group of a surface. *Proc. Camb. Philos. Soc.* **62** (1966) 679–681.
19. C. Maclachlan. Modulus space is simply connected, *Proc. Amer. Math. Soc.* **29** (1971) 85–86.
20. C. Maclachan, Modular groups and fibre spaces over Teichmüller spaces. *In* "Discontinuous Groups and Riemann Surfaces," (L. Greenburg, ed.) Ann. of Math. Stud. No. 79, 1974 pp. 297–313.
21. W. Mangler, Die Klassen topologischer Abbildungen einer geschlossenen Fläche auf sich. *Math. Z.* **44** (1939) 541–554.
22. W. Magnus, Uber Automorphismen von Fundamentalgruppen Berandeter Flächen, *Math. Ann.* **109** (1934) 617–646.
23. W. Magnus, A. Karass and D. Solitar, "Combinatorial Group Theory," John Wiley–Interscience, N.Y. 1966.
24. J. McCool, Some finitely presented subgroups of the automorphism group of a free group, *J. Algebra* **35** (1975) 205–213.
25. J. Nielsen, Untersuchungen zur Topologie der geschlossen zweiseitigen Flächen. *Acta Math.* **50** (1927) 189–358.
26. G. P. Scott, Braid groups and the group of homeomorphisms of a surface. *Proc. Camb. Philos. Soc.* **68** (1970) 605–617.
27. P. F. Stebe, Conjugacy separability of certain Fuchsian groups. *Trans. Amer. Math. Soc.* **163** (1972) 173–188.
28. N. Steenrod, "Topology of Fibre Bundles." Princeton University Press, N.J. 1951.
29. H. Zieschang, On the homeotopy groups of surfaces. *Math. Ann.* **206** L (1973) 1–21.
30. H. Zieschang, Über Automorphismen ebener discontinuerlicher Gruppen. *Math. Ann.* **166** (1966) 148–167.
31. H. Zieschang, E. Vogt and H.-D. Coldewey, "Flächen und ebene diskontinuierliche Gruppen." Lecture Notes in Math. No. 122, Springer-Verlag, Berlin, 1970.

# 7. Finiteness Theorems for Fuchsian and Kleinian Groups

L. GREENBERG

*University of Maryland, U.S.A.*



## INTRODUCTION

A basic theorem concerning Fuchsian groups states that such a group is finitely generated if and only if it has a fundamental polygon with finitely many sides. This fact has many geometric and algebraic consequences. For example, if $\Gamma$ is a finitely generated Fuchsian group, operating in the unit disc $\mathscr{D}$, then.

(i) The quotient surface $S = \mathscr{D}/\Gamma$ has finite topological type.
(ii) The covering map $\mathscr{D} \rightarrow S$ is branched over finitely many points in $S$ (if at all).
(iii) $\Gamma$ has finitely many conjugacy classes of elements of finite order.
(iv) There are finitely many $\Gamma$-orbits of parabolic fixed points.
(v) The translation lengths of (conjugacy classes of) hyperbolic elements in $\Gamma$ do not accumulate. (Equivalently, the traces of hyperbolic elements do not accumulate).
(vi) The intersection of two finitely generated subgroups is again finitely generated.

Unfortunately, the basic theorem fails for Kleinian groups. That is to say, there exist finitely generated Kleinian groups which have no finite sided fundamental polyhedron in the upper half-space $\mathscr{H}^3$. At first such groups (which we call geometrically infinite) were shown to exist, but recently, concrete examples have been given by Jørgensen.

The situation is partly salvaged by Ahlfors' finiteness theorem, which must be considered the deepest theorem presently known about Kleinian groups. This theorem states that if $\Gamma$ is a finitely generated Kleinian group whose regular set is $\Omega \subset \mathbb{C}$, then

(i) $\Omega/\Gamma$ has finitely many components;
(ii) each component is a closed Riemann surface with a finite number of punctures;
(iii) there are at most a finite number of points in $\Omega/\Gamma$ over which the covering map $\Omega \rightarrow \Omega/\Gamma$ is branched.

It is curious that, although the theorem is geometric, the only known proof is heavily analytic.* The theorem can be used to show that $\Gamma$ has a finite sided fundamental polygon in $\Omega$. In fact the Ford polygon, and the boundary of any Dirichlet region are finite sided.

If we restrict our attention to groups with finite sided polyhedra (which we call geometrically finite groups) then most of the theorems for Fuchsian groups remain true. Some unpleasant things can still happen, however. For example, a geometrically finite group can contain geometrically infinite subgroups.

In this article we shall give most of the proofs of the above facts. At the end, we shall discuss some examples of geometrically infinite groups. We have also included a section on automorphism groups of Riemann surfaces.

* New proof: Dennis Sullivan, "Quasiconformal Homeomorphisms + Dynamics I: the solution of the Fatou-Julia problem on wandering d

This has been included because our exposition of Fuchsian groups gives all the necessary preliminaries, and this pretty chapter of Riemann surface theory does not seem to be widely known in the mathematical community.

## 1. FUCHSIAN GROUPS

### 1.1 Notation

The Fuchsian groups we shall discuss will act in the disc

$$\mathscr{D} = \{z \in \mathbb{C} : |z| < 1\}.$$

The boundary $\partial\mathscr{D} = \{z : |z| = 1\}$ will be denoted by $\mathscr{E}$. We shall use the Poincaré metric

$$ds^2 = \frac{4(dx^2 + dy)^2}{(1 - x^2 - y^2)^2}$$

in $\mathscr{D}$. The curvature is $k = -1$, the element of area is

$$dA = \frac{4dx\, dy}{(1 - x^2 - y^2)^2}$$

and the geodesics are circular arcs orthogonal to $\mathscr{E}$. The Poincaré distance will be denoted $d(z, w)$. With this metric, $\mathscr{D}$ is a model of the non-Euclidean (hyperbolic) plane. A *horocycle* at $p \in \mathscr{E}$ is a circle in $\mathscr{D}$, tangent to $\mathscr{E}$ at $p$. It is invariant under a parabolic transformation leaving $p$ fixed. A *hypercycle* is a circular arc in $\mathscr{D}$ with end points $p_1, p_2 \in \mathscr{E}$. It is invariant under a hyperbolic transformation with fixed points $p_1, p_2$. $L(\Gamma)$ denotes the *limit set* of a Fuchsian group $\Gamma$. $L(\Gamma)$ is the set of accumulation points of any orbit $\Gamma(z)$, $z \in \mathscr{D}$. A connected component of $\mathscr{E} \setminus L(\Gamma)$ is called an *interval of discontinuity*. For details of elementary facts the reader is referred to Chapter 2.

### 1.2. Fundamental polygons

A *fundamental polygon* for a Fuchsian group $\Gamma$ is a non-Euclidean polygon $\mathscr{P}$ (relatively closed in $\mathscr{D}$) such that:

(i) $\Gamma\mathscr{P} = \mathscr{D}$.
(ii) The interior $\mathring{\mathscr{P}}$ does not intersect any of its $\Gamma$-images.
(iii) For each side $s$ of $\mathscr{P}$, there is another side $\bar{s}$ and an element $\gamma \in \Gamma$, such that $\gamma(s) = \bar{s}$, and $\gamma(\mathscr{P})$ is a polygon adjacent to $\mathscr{P}$ along $\bar{s}$.

(iv) A compact set $K \subset \mathscr{D}$ intersects only finitely many $\Gamma$-images of $\mathscr{P}$. (c.f. Chapter 2, Section 6).

1.2.1 PROPOSITION. *Property* (iv) *is a consequence of the other properties.*

*Proof.* It suffices to show that each point $p \in \mathscr{D}$ has a neighbourhood which intersects only a finite number of $\Gamma$-images of $\mathscr{P}$. This is evident if $p \in \breve{\mathscr{P}}$, or $p$ is equivalent (under $\Gamma$) to a point in $\mathring{\mathscr{P}}$, since $\mathring{\mathscr{P}}$ (or a $\Gamma$-image of $\mathring{\mathscr{P}}$) is such a neighbourhood. The conclusion is also clear if $p$ has a $\Gamma$-image which lies on a side of $\mathscr{P}$, but is not a vertex. In this case, two adjacent $\Gamma$-images of $\mathring{\mathscr{P}}$, together with their common side, constitute the required neighbourhood. It suffices to show that each vertex $p$ of $\mathscr{P}$ has the required neighbourhood.

If only finitely many vertices of $\mathscr{P}$ were equivalent to $p$ under $\Gamma$, then we could form the required neighbourhood from finitely many $\Gamma$-images of $\mathring{\mathscr{P}}$ (together with their common sides). In particular, if $\mathscr{P}$ were compact, this would be true. Thus we may suppose that $\mathscr{P}$ is not compact.

Let $\mathscr{D}_r = \{z \in \mathbb{C} : d(z, p) < r\}$ and $\mathscr{E}_r = \partial\mathscr{D}_r = \{z \in \mathbb{C} : d(z, p) = r\}$. Since $\mathscr{P}$ is not compact, any $\Gamma$-image of $\mathscr{P}$ which meets $\mathscr{D}_r$ must also intserct $\mathscr{E}_r$. $\mathscr{P}$ has at most a countable number of vertices (since by definition of polygon, the vertices of a polygon are not allowed to accumulate in $\mathscr{D}$). Since $\Gamma$ is also countable, there are only a countable number of points in $\mathscr{D}$ which are $\Gamma$-images of vertices of $\mathscr{P}$. Therefore, there is a value of $r$, such that no $\Gamma$-image of a vertex belongs to $\mathscr{E}_r$. Each point of $\mathscr{E}_r$ has a neighbourhood which meets only finitely many $\Gamma$-images of $\mathscr{P}$. Therefore the same is true for $\mathscr{E}_r$. Since each $\Gamma$-image of $\mathscr{P}$ which meets $\mathscr{D}_r$ must also intersect $\mathscr{E}_r$, we see that $\mathscr{D}_r$ is the required neighbourhood of $p$. ■

1.2.2 PROPOSITION. *The elements $\gamma$ from property* (iii) *generate* $\Gamma$.

*Proof.* Let $A$ be the set of elements in $\Gamma$ which identify sides of $\mathscr{P}$, according to property (iii). We want to show that $A$ generates $\Gamma$.

Let $\gamma \in \Gamma$. Since $\mathscr{D}$ is connected and $\Gamma\mathscr{P} = \mathscr{D}$, there is a finite sequence of $\Gamma$-images of $\mathscr{P}: \gamma_0\mathscr{P}, \gamma_1\mathscr{P}, \ldots, \gamma_n\mathscr{P}$ such that $\gamma_0 = 1$, $\gamma_n = \gamma$ and $\gamma_k\mathscr{P}$ is adjacent to $\gamma_{k+1}\mathscr{P}$ along a common side for $k = 1, \ldots, n-1$. Therefore $\mathscr{P}$ is adjacent to $\gamma_k^{-1}\gamma_{k+1}\mathscr{P}$ along a common side, and so $\gamma_k^{-1}\gamma_{k+1} = \alpha_{k+1} \in A$. Thus $\gamma = \gamma_n = (\gamma_0^{-1}\gamma_1)(\gamma_1^{-1}\gamma_2)\ldots(\gamma_{n-1}^{-1}\gamma_n) = \alpha_1\alpha_2\ldots\alpha_n$. ■

Let $z_0$ be a point in $\mathscr{D}$ which is not fixed by any elliptic element in $\Gamma$. The

*Dirichlet region* for $\Gamma$ (also called the *normal polygon*), centred at $z_0$, is

$$\mathscr{P} = \{z \in \mathscr{D} : d(z, z_0) \leqslant d(z, \gamma(z_0)), \quad \text{for } \gamma \in \Gamma\}.$$

A Dirichlet region is a fundamental polygon for $\Gamma$ (see [**47**]). For each $\gamma \in \Gamma$ ($\gamma \neq 1$) the set $S_\gamma = \{z \in \mathscr{D} : d(z, z_0) = d(z, \gamma(z_0))\}$ is the geodesic which is the perpendicular bisector of the segment $[z_0, \gamma(z_0)]$. The sides of $\mathscr{P}$ are segments contained in some of the $S_\gamma$.

1.2.3 LEMMA. *Let $\mathscr{P}$ be a Dirichlet region for $\Gamma$, centred at $z_0$. Let $\gamma \in \Gamma$ ($\gamma \neq 1$) and suppose that $z$ and $\gamma(z)$ both belong to $\mathscr{D} \cap \partial\mathscr{P}$. Then $d(z_0, z) = d(z_0, \gamma(z))$.*

*Proof.* For $\gamma \in \Gamma$ ($\gamma \neq 1$), let

$$R_\gamma = \{z \in \mathscr{D} : \ d(z, z_0) \leqslant d(z, \gamma(z_0))\},$$
$$R_{\gamma^{-1}} = \{z \in \mathscr{D} : \ d(z, z_0) \leqslant d(z, \gamma^{-1}(z_0))\},$$

and $R = R_\gamma \cap R_{\gamma^{-1}}$. Note that $\mathscr{P} \subset R$. We claim that if $z$ and $\gamma(z)$ both belong to $R$, then $d(z_0, z) = d(z_0, \gamma(z))$. For, since $z \in R_{\gamma^{-1}}$, $d(z, z_0) \leqslant d(z, \gamma^{-1}(z_0)) = d(\gamma(z), z_0)$. Also, since $\gamma(z) \in R_\gamma$, $d(\gamma(z), z_0) \leqslant d(\gamma(z), \gamma(z_0)) = d(z, z_0)$. Therefore $d(z_0, z) = d(z_0, \gamma(z))$.

Now suppose that $z$ and $\gamma(z)$ both belong to $\mathscr{D} \cap \partial\mathscr{P}$. Since $\mathscr{P} \subset R$, $z$ and $\gamma(z)$ belong to $R$, so that $d(z_0, z) = d(z_0, \gamma(z))$. ∎

We shall use the following non-standard terminology.

*Definition.* A *periodic parabolic region* for $\Gamma$ is a disc, bounded by a horocycle, which is invariant under a parabolic element in $\Gamma$. A *periodic hyperbolic region* for $\Gamma$ is a region in $\mathscr{D}$, bounded by a hypercycle and an interval of discontinuity in $\mathscr{E}$, which is invariant under a hyperbolic element in $\Gamma$.

1.2.4. LEMMA. *Let $\mathscr{P}$ be a Dirichlet region and $\mathscr{Q}$ a periodic parabolic or hyperbolic region for $\Gamma$. Then only a finite number of sides of $\mathscr{P}$ meet $\mathscr{Q}$.*

*Proof.* We may assume that $\mathscr{P}$ has infinitely many sides. We shall first consider the parabolic case. Let $H$ be the horocycle boundary of $\mathscr{Q}$, and $\gamma \in \Gamma$ a parabolic element leaving $\mathscr{Q}$ invariant. Let $\mathscr{S} = \{z \in \mathscr{D} : d(z, z_0) = d(z, \gamma(z_0))\}$ and $\mathscr{S}' = \{z \in \mathscr{D} : d(z, z_0) = d(z, \gamma^{-1}(z_0))\}$. $\mathscr{S}$ and $\mathscr{S}'$ meet at the parabolic fixed point $p$, and $\mathscr{P} \cap \mathscr{Q}$ is contained in the cuspidal region bounded by

$H$, $\mathcal{S}$ and $\mathcal{S}'$. Let $H_0$ be the closed interval on $H$ which lies between $\mathcal{S}$ and $\mathcal{S}'$.

For each side $t$ of $\mathcal{P}$, let $T$ denote the complete geodesic which contains $t$ as a subinterval. The geodesics $T$ accumulate only at the limit set of $\Gamma$, which lies on $\mathcal{E}$. Therefore only a finite number of the extended sides $T$ meet a compact set $K \subset \mathcal{D}$. In particular this is true for $K = H_0$, Suppose $T$ meets $\mathcal{Q}$, but not $H_0$. Then it meets the cuspidal region bounded by $H_0$, $\mathcal{S}$ and $\mathcal{S}'$, while it does not intersect $H_0$. Therefore $T$ intersects both $\mathcal{S}$ and $\mathcal{S}'$, and consequently separates $\mathcal{P}$ from the parabolic fixed point $p$ at the vertex of the cusp. In this case $\mathcal{P} \cap \mathcal{Q}$ is contained in the compact region bounded by $H_0$, $\mathcal{S}$, $\mathcal{S}'$ and $T$. Only a finite number of sides may meet this region, since it is compact.

The hyperbolic case is similar, with the following modifications. $\mathcal{Q}$ is bounded by a hypercycle $H$ and an interval of discontinuity $I$ on $\mathcal{E}$. There is a hyperbolic element $\gamma \in \Gamma$ leaving $\mathcal{Q}$ invariant. The geodesics $\mathcal{S} = \{z \in \mathcal{D} : d(z, z_0) = d(z, \gamma(z_0))\}$ and $\mathcal{S}' = \{z \in \mathcal{D} : d(z, z_0) = d(z, \gamma^{-1}(z_0))\}$ intersect $H$ and they each have an endpoint in $I$. Let $H_0$ and $I_0$ be the closed subintervals of $H$ and $I$ which lie between $\mathcal{S}$ and $\mathcal{S}'$. $\mathcal{P} \cap \mathcal{Q}$ is contained in the region bounded by $H_0$, $I_0$, $\mathcal{S}$ and $\mathcal{S}'$. Only a finite number of the extended sides $T$ can meet $H_0$ and $I_0$, since the latter contain no limit points. If $T$ meets $\mathcal{Q}$, but not $H_0$ or $I_0$, then it separates $\mathcal{P}$ from $I_0$, and the argument concludes as in the parabolic case. ■

1.2.5 THEOREM. *Let $\mathcal{P}$ be a Dirichlet region for the Fuchsian group $\Gamma$. Then $\Gamma$ is finitely generated if and only if $\mathcal{P}$ has a finite number of sides.*

*Proof.* If $\mathcal{P}$ is finite-sided, then $\Gamma$ is finitely generated, by Proposition 1.2.2.

Suppose $\Gamma$ is finitely generated. Let $s_1, s_2, \ldots$ be the sides of $\mathcal{P}$, and $\gamma_1 . \gamma_2, \ldots$ the corresponding generators. Since $\Gamma$ is finitely generated, it is generated by finitely many of the $\gamma_i$, say $\gamma_1, \gamma_2, \ldots, \gamma_n$. Choose $r$ sufficiently large so that the disc $\mathcal{D}_0 = \{z \in \mathcal{D} : d(z, z_0) < r\}$ intersects the sides $s_1, s_2, \ldots, s_n$ (where $z_0$ is the centre of the Dirichlet region $\mathcal{P}$), and let $R_0 = \overline{\mathcal{D}}_0 \cap \mathcal{P}$. For each $s_i$, there is a side $\bar{s}_i$, such that $\gamma_i(s_i) = \bar{s}_i$. By Lemma 1.2.3, $d(z_0, z) = d(z_0, \gamma_i(z))$ for $z \in s_i$. Therefore the sides $\bar{s}_i$ intersect $R_0$ also, and $\gamma_i(R_0 \cap s_i) = R_0 \cap \bar{s}_i$. It is easily seen that this fact implies that the region $R = \Gamma R_0$ is connected.

Since $R_0$ is compact, it intersects only a finite number of sides of $\mathcal{P}$, which we may suppose to be the sides $s_i$ and $\bar{s}_i$ $(1 \leqslant i \leqslant n)$. Besides these sides, the boundary $\partial R_0$ consists of a finite number of arcs of the circle $\partial \mathcal{D}_0$, which we denote by $t_1, t_2, \ldots, t_m$.

In order to show that $\mathcal{P}$ has only a finite number of sides, we shall show that only finitely many sides meet each of the sets $R$ and $\mathscr{D} \setminus R$. The first statement is clear, since $\mathscr{P} \cap R = R_0$.

Let $C$ be a connected component of $\mathscr{D} \setminus R$, and let $\partial C$ be its relative boundary in $\mathscr{D}$. $\partial C$ is composed of certain $\Gamma$-images of the arcs $t_i$. If $\mathscr{P}$ meets $C$, then (since it intersects $R \subset \mathscr{D} \setminus C$) it also meets $\partial C$. Thus $\mathscr{P}$ intersects only those components $C$ which contain the arcs $t_i$ in their boundaries. Consequently $\mathscr{P}$ intersects only a finite number of the components $C$. It now suffices to show that only finitely many sides of $\mathscr{P}$ meet a given component $C$.

If $\partial C$ is not connected, then $C$ separates $R$. Since $R$ is connected, $\partial C$ is also. Either $\partial C$ is composed of finitely many or infinitely many $\Gamma$-images of the arcs $t_i$. In the first case, $\partial C$ is a closed curve, and $C$ is relatively compact. Therefore $C$ meets only finitely many $\Gamma$-images of $\mathscr{P}$ and consequently only finitely many sides.

Suppose $\partial C$ contains infinitely many $\Gamma$-images of the arcs $t_i$. Then two of the arcs in $\partial C$ are $\Gamma$-equivalent. Thus there is an element $\gamma \in \Gamma$ ($\gamma \neq 1$) such that $\gamma(\partial C) = \partial C$ (and therefore $\gamma(C) = C$). If $\gamma$ is elliptic, then $\partial C$ is composed of a finite number of arcs. Therefore $\gamma$ is hyperbolic or parabolic. There is a finite subinterval $\partial_0$ of $\partial C$, so that $\partial C = \bigcup_{n \in \mathbb{Z}} \gamma^n(\partial_0)$. Thus, in the hyperbolic case, $\partial C$ has bounded distance from the axis of $\gamma$, and in the parabolic case, $\partial C$ has bounded distance from a horocycle invariant by $\gamma$. It follows that $C$ is contained in a periodic hyperbolic or parabolic region for $\Gamma$, and the proof concludes with the help of Lemma 1.2.4. ■

*Remark.* In [**18**] and [**35**] it is shown that the theorem is true for any fundamental polygon, provided an element $\gamma \in \Gamma$ can identify only a finite number of pairs of sides.

### 1.3 The Nielsen convex region

We shall now discuss another property which is equivalent to finite generation. A subset $S \subset \overline{\mathscr{D}}$ is convex (in the non-Euclidean sense) if for any pair of points $s_1, s_2 \in S$, the geodesic segment $[s_1, s_2] \subset S$. The *convex hull* of a set $S \subset \overline{\mathscr{D}}$ is

$$[S] = \cap\{T \subset \overline{\mathscr{D}} : T \supset S \quad \text{and } T \text{ convex}\}.$$

Let $L(\Gamma)$ denote the limit set of $\Gamma$. The *Nielsen convex region* is $\mathscr{K}(\Gamma) = \mathscr{D} \cap [L(\Gamma)]$. The boundary $\partial\mathscr{K}(\Gamma)$ is composed of geodesics which span the intervals of discontinuity in $\mathscr{E}$. We shall truncate $\mathscr{K}(\Gamma)$ by removing certain parabolic regions. For this purpose, we need the following fact.

1.3.1 LEMMA. *At each parabolic fixed point p, there is a horocycle $H_p$, with the following properties:*

(i) $\gamma H_p = H_{\gamma(p)}$, *for* $\gamma \in \Gamma$;
(ii) $H_p \cap H_q = \varnothing$, *if* $p \neq q$;
(iii) $H_p \subset \mathscr{K}(\Gamma)$.

*Proof*. Let $H$ be a horocycle tangent to $\mathscr{E}$ at the parabolic fixed point $p$. Let $\Gamma_p = \{\gamma \in \Gamma : \gamma(p) = p\}$, and choose an element $\delta \in \Gamma_p$ $(\delta \neq 1)$. The distance $d = d(z, \delta(z))$ is constant for $z \in H$. Also $d(\gamma(z), (\gamma\delta\gamma^{-1})(\gamma z)) = d(z, \delta(z))$, so that any point on a horocycle $\gamma H$ (for $\gamma \in \Gamma$) is moved distance $d$ by some element in $\Gamma$. This implies that the horocycles $\{\gamma H\}$ do not accumulate in $\mathscr{D}$.

Let $H_0$ be a fundamental interval on $H$ for $\Gamma_p$ (i.e. $H_0$ is a compact interval such that $H = \Gamma_p H_0$). Since the $\Gamma$-images of $H$ do not accumulate, only a finite number of them, say $\gamma_1 H, \ldots, \gamma_n H$ intersect $H_0$. Then the $\Gamma$-images of $H$ which intersect $H$ are precisely $\gamma\gamma_i H$, where $\gamma \in \Gamma_p$, $1 \leqslant i \leqslant n$. We can choose a smaller horocycle $H'$ at $p$, which does not intersect any $\gamma_i H$. The corresponding images $\gamma_i H'$ lie inside $\gamma_i H$, so these do not intersect $H'$. Thus $H'$ does not intersect any of its $\Gamma$-images.

For each interval of discontinuity $I$ in $\mathscr{E}$, let $G_I$ denote the geodesic which spans $I$. Since the intervals $I$ are disjoint, the $G_I$ do not accumulate in $\mathscr{D}$. Therefore, only a finite number of the $G_I$ intersect a fundamental interval $H'_0$ in $H'$. We may therefore find a smaller horocycle $H_p$ at $p$, so that $H_p$ does not intersect any of the $G_I$ or any horocycle $\gamma H_p$ $(\gamma \in \Gamma, \gamma \neq 1)$. Since the $G_I$ constitute the boundary of $\mathscr{K}(\Gamma)$, and $p \notin I$ for any interval of discontinuity $I$, it follows that $H_p \subset \mathscr{K}(\Gamma)$. For $\gamma \in \Gamma$, we define $H_{\gamma(p)} = \gamma H_p$.

We may now work on the remaining equivalence classes inductively. Suppose that we have found horocycles $H_{p_1}, H_{p_2}, \ldots, H_{p_{n-1}}$ which satisfy the lemma. Suppose $p_n$ is a parabolic fixed point which is not $\Gamma$-equivalent to $p_1, p_2, \ldots, p_{n-1}$. As above, we may find a horocycle $H$ at $p_n$, such that $H \subset \mathscr{K}(\Gamma)$ and $H$ does not intersect any of its $\Gamma$-images. Choose a fundamental interval $H_0$ for $H$ modulo $\Gamma_{p_n}$. Only finitely many horocycles $\gamma H_{p_k}$ $(\gamma \in \Gamma$, $1 \leqslant k \leqslant n-1)$ can intersect $H_0$. Choose a smaller horocycle $H_{p_n}$ (at $p_n$) which does not intersect any $\gamma H_{p_i}$. Define $H_{\gamma(p_n)} = \gamma H_{p_n}$. ■

For each horocycle $H_p$ of the previous lemma, let $D_p$ be the open disc bounded by $H_p$. We define the *truncated Nielsen convex region* to be

$$\mathscr{K}^*(\Gamma) = \mathscr{K}(\Gamma) \setminus \bigcup_p \mathscr{D}_p.$$

$\mathscr{K}^*(\Gamma)$ is not unique, since a choice of horocycles is required. Note that $\mathscr{K}^*(\Gamma)$ is invariant under the action of $\Gamma$.

1.3.2 THEOREM. *$\Gamma$ is finitely generated if and only if $\mathscr{K}^*(\Gamma)/\Gamma$ is compact.*

*Proof.* Suppose that $\Gamma$ is finitely generated. Let $\mathscr{P}$ be a Dirichlet region for $\Gamma$. By Theorem 1.2.5, $\mathscr{P}$ has a finite number of sides. It is easily seen that this implies that $\mathscr{E} \cap \partial\mathscr{P}$ has finitely many connected components, and each component is either a parabolic fixed point (which is a vertex of $\mathscr{P}$) or a subinterval of discontinuity (possibly degenerating to a single point). These are all cut off in passing to $\mathscr{P} \cap \mathscr{K}^*(\Gamma)$. Thus $\mathscr{P} \cap \mathscr{K}^*(\Gamma)$ is compact. Since this is a fundamental region for $\mathscr{K}^*(\Gamma)$ mod $\Gamma$, $\mathscr{K}^*(\Gamma)/\Gamma$ is also compact.

Suppose that $\mathscr{K}^*(\Gamma)/\Gamma$ is compact. Let $\phi: \mathscr{K}^*(\Gamma) \to \mathscr{K}^*(\Gamma)/\Gamma$ denote the canonical projection, and let $p_0 \in \mathscr{K}^*(\Gamma)/\Gamma$. The distance $d(p_0, p)$ has an upper bound for $p \in \mathscr{K}^*(\Gamma)/\Gamma$ (since the latter is compact), say $d(p_0, p) < r$. Let $z_0$ be a point in $\mathscr{K}^*(\Gamma)$, such that $\phi(z_0) = p_0$, and let $B = \{z \in \mathscr{D}: d(z_0, z) < r\}$. Then $\phi$ maps $B \cap \mathscr{K}^*(\Gamma)$ onto $\mathscr{K}^*(\Gamma)/\Gamma$, and therefore $\mathscr{K}^*(\Gamma) \subset \Gamma B$.

Since $B$ is relatively compact in $\mathscr{D}$, there are only a finite number of elements $\gamma \in \Gamma$ such that $B \cap \gamma(B) \neq \varnothing$. We claim that these elements $\gamma_1, \gamma_2, \ldots, \gamma_n$, generate $\Gamma$. The argument, which we omit here, is similar to the one used to prove Proposition 1.2.2. ■

## 1.4 Conjugacy classes

In this section and the remaining ones of this chapter, we shall present some consequences of the previous theorems.

The boundary of $\mathscr{K}^*(\Gamma)$ consists of curves $H_p$ (horocycles at parabolic fixed points $p$) and $G_I$ (geodesics which span intervals of discontinuity $I \subset \mathscr{E}$). The elements ($\neq 1$) in $\Gamma$ which leave these boundary curves invariant are called the *boundary elements*. These are all of the parabolic elements and any hyperbolic elements which leave invariant an interval of discontinuity. If $I$ is an interval of discontinuity which is invariant by a non-trivial element, we shall call $I$ *periodic*. An element $\gamma \in \Gamma$ is *primary* if it is not a power of any element in $\Gamma$ (except $\gamma$ and $\gamma^{-1}$).

1.4.1 THEOREM. *Let $\Gamma$ be a finitely generated Fuchsian group. Then $\Gamma$ has only a* *finite number of conjugacy classes of elliptic elements, primary parabolic elements and primary hyperbolic boundary elements. Furthermore, every interval of discontinuity is periodic.*

*Proof.* Let $\mathscr{P}$ be a Dirichlet region for $\Gamma$. Any elliptic element is conjugate to one whose fixed point is a vertex of $\mathscr{P}$. Since $\mathscr{P}$ has finitely many vertices, and each vertex is fixed by only a finite number of elements, $\Gamma$ has finitely many conjugacy classes of elliptic elements.

Since $\mathscr{K}^*(\Gamma)/\Gamma$ is a compact surface, it has a finite number of boundary curves, each of which is a closed curve. Each boundary curve corresponds to an orbit of horocycles $H_p$, or an orbit of geodesics $G_I$. Each such orbit corresponds to two conjugacy classes of primary boundary elements. Therefore $\Gamma$ has only a finite number of conjugacy classes of primary boundary elements. Since $\mathscr{K}^*(\Gamma)/\Gamma$ is compact, each $G_I$ projects to a closed curve, and so each interval of discontinuity $I$ is periodic. ■

Let $\gamma$ be a hyperbolic transformation with fixed points $z_1, z_2 \in \mathscr{E}$. The geodesic which connects $z_1$ and $z_2$ is called the *axis* of $\gamma$, and is denoted $A_\gamma$. Every point $z \in A_\gamma$ is moved by $\gamma$ the same distance $\lambda(\gamma) = d(z, \gamma(z))$, and $\lambda(\gamma)$ is called the *translation length* of $\gamma$. Another characterization of $\lambda(\gamma)$ is: $\lambda(\gamma) = \min_{z \in \mathscr{D}} d(z, \gamma(z))$. The trace $t = \operatorname{tr}(\gamma)$ (where $\gamma$ is considered as a $2 \times 2$ matrix) is related to $\lambda = \lambda(\gamma)$ by $|t| = 2\cosh(\lambda/2)$. (This is easily seen by transforming $\mathscr{D}$ to $\mathscr{H}$, the upper half plane, and the fixed points of $\gamma$ to $0, \infty$, and computing the translation length in the upper half-plane. (cf. Chapter 2) It is clear that $\lambda(\gamma)$ depends only on the conjugacy class of $\gamma$. However, different conjugacy classes of hyperbolic elements may have the same translation length. In the following theorem, we consider each translation length to be listed as many times as there are conjugacy classes corresponding to it.

1.4.2 THEOREM. *Let $\Gamma$ be a finitely generated Fuchsian group. Then the translation lengths of conjugacy classes of hyperbolic elements do not accumulate in the set of real numbers. In particular, there is a positive minimum. Equivalently, the traces do not accumulate, and in particular, are bounded away from $\pm 2$.*

*Proof.* The axis $A_\gamma$ of a hyperbolic element $\gamma$ is contained in the Nielsen convex region $\mathscr{K}(\Gamma)$. Since $A_\gamma$ cannot be contained inside a horocycle, it meets $\mathscr{K}^*(\Gamma)$. The compactness of $\mathscr{K}^*(\Gamma)/\Gamma$ implies (as we saw in the proof of Theorem 1.3.2) that there is a disc $B = \{z \in \mathscr{D} : d(z_0, z) < r\}$ such that

$\mathscr{K}^*(\Gamma) \subset \Gamma B$. Therefore, there is an element $\delta \in \Gamma$, so that $\delta A_\gamma$ meets $B$. Noting that $\delta(A_\gamma) = A_{\delta\gamma\delta^{-1}}$, we see that each hyperbolic element has a conjugate whose axis meets $B$. Since $B$ is relatively compact and $\Gamma$ is discontinuous, for any $\lambda_0 > 0$ there can be only a finite number of elements $\gamma \in \Gamma$, so that $d(z, \gamma(z)) < \lambda_0$ for some point $z \in B$. Therefore there are only finitely many conjugacy classes of hyperbolic elements $\gamma \in \Gamma$ with $\lambda(\gamma) < \lambda_0$. ■

## 1.5 Signature, presentation and area

If $\Gamma$ is a finitely generated Fuchsian group, then $S^* = \mathscr{K}^*(\Gamma)/\Gamma$ is a compact surface with boundary. Let $S^*$ have genus $g$ and $t$ boundary components. $\Gamma$ has a finite number of orbits of elliptic fixed points: $\Gamma z_1, \Gamma z_2, \ldots, \Gamma z_n$. Suppose that the stabilizer of $z_k$ (which is a finite, cyclic, elliptic subgroup) has order $v_k$. The ordered set $\sigma = (g; v_1, v_2, \ldots, v_n; t)$ is called the *signature* of $\Gamma$.

From a slightly different point of view, let $S = \mathscr{D}/\Gamma$ (considered as a Riemann surface). Since a Dirichlet region for $\Gamma$ has a finite number of sides, $S$ is a surface of finite topological type. In fact $S^* \subset S$ and $S \setminus S^*$ has $n$ components, each of which is conformally equivalent to an annulus or a punctured disc. Thus $S$ is homeomorphic to the interior of $S^*$, and $S^*$ is a deformation retract of $S$. The canonical projection $\phi : \mathscr{D} \to S$ is a ramified covering. There are $n$ points $s_1, s_2, \ldots, s_n \in S$ over which $\phi$ is branched, and the branching order over $s_k$ is $v_k$. The signature $\sigma$ contains the essential topological data concerning $S$ and $\phi$, but no conformal data.

1.5.1 THEOREM. *A Fuchsian group $\Gamma$ with signature $(g; v_1, \ldots, v_n; t)$ has the following group presentation:*

$$\Gamma = \left\langle a_1, b_1, \ldots, a_g, b_g, e_1, \ldots, e_n, f_1, \ldots, f_t : \prod_{k=1}^{g} (a_k b_k a_k^{-1} b_k^{-1}) e_1 \ldots e_n f_1 \ldots f_t = e_1^{v_1} = \ldots = e_n^{v_n} = 1 \right\rangle$$

*Proof.* Let $\mathscr{D}_0$ denote $\mathscr{D}$ minus all elliptic fixed points, and $S_0 = \phi(\mathscr{D}_0)$. Then $\phi : \mathscr{D}_0 \to S_0$ is an unramified, normal covering, and $\Gamma$ is the group of covering transformations. Choose base points $s_0 \in S_0$ and $z_0 \in \phi^{-1}(s_0) \subset \mathscr{D}_0$, and consider the fundamental groups $F = \pi_1(S_0, s_0)$ and $N = \phi^*(\pi_1(\mathscr{D}_0, z_0))$. By a well-known theorem about covering spaces (cf. Chapter 1 pp. 26–27), the group of covering transformations $\Gamma$ is isomorphic to $F/N$ (see [**40**]). In particular, if $\Gamma$ contains no elliptic transformations, then $\mathscr{D}_0 = \mathscr{D}$, $S_0 = S$, $N = \{1\}$ and we are finished. For the remainder of the proof, we shall assume that $\Gamma$ contains elliptic transformations.

In $S_0$ there exist closed curves $\alpha_1, \beta_1, \ldots, \alpha_g, \beta_g$ going around the handles, $\varepsilon_1, \varepsilon_2, \ldots, \varepsilon_n$ going around the removed branch points, and $\eta_1, \eta_2, \ldots, \eta_t$ going around the ideal boundaries, whose homotopy classes $a_i = [\alpha_i]$, $b_i = [\beta_i]$, $e_j = [\varepsilon_j]$, $f_k = [\eta_k]$ generate $F$, with the defining relation $\prod_{i=1}^{g} (a_i b_i a_i^{-1} b_i^{-1}) e_1 \ldots e_n f_1 \ldots f_t = 1$. If we show that $N$ is the smallest normal subgroup containing $e_1^{\nu_1}, e_2^{\nu_2}, \ldots, e_n^{\nu_n}$, then the theorem will be proved.

Let $z_1, z_2, \ldots$ denote the (infinite) collection of elliptic fixed points in $\mathscr{D}$, so that $\mathscr{D}_0 = \mathscr{D} \setminus \{z_1, z_2, \ldots\}$. There exist closed curves $\delta_j$ which go once around $z_j$, whose homotopy classes $d_j = [\delta_j]$ freely generate $\pi_1(\mathscr{D}_0, z_0)$. If $z_j$ lies over the branch point $s_k$, then the branching order is $\nu_k$, and $\phi(\delta_j)$ goes $\nu_k$ times around $s_k$. Therefore $\phi^*(d_j) = [\phi(\delta_j)]$ is conjugate to $e_k^{\nu_k}$. Thus $e_k^{\nu_k} \in N$ (for $k = 1, 2, \ldots, n$).

Now suppose $u \in N$. Then $u = \phi^*(v)$, where $v \in \pi_1(\mathscr{D}_0, z_0)$. Since the $d_j$ generate $\pi_1(\mathscr{D}_0, z_0)$, $v$ may be expressed $v = d_{j_1}, d_{j_2}, \ldots, d_{j_m}$. Then $u = \phi^*(v) = \phi^*(d_{j_1}), \ldots, \phi^*(d_{j_m})$. But, as we saw above, each $\phi^*(d_j)$ is conjugate to some $e_k^{\nu_k}$. Therefore $N$ is the smallest normal subgroup containing the elements $e_k^{\nu_k}$.† ■

*Remark.* The same argument shows that an infinitely generated Fuchsian group is a free product of cyclic groups.

We shall now associate a finite area to each finitely generated Fuchsian group $\Gamma$. This will not be the (non-Euclidean) area of $\mathscr{D}/\Gamma$, which may be infinite, but the area of $\mathscr{K}(\Gamma)/\Gamma$. We shall use the Gauss–Bonnet formula (see Chapter 2). Since the curvature is $-1$, this formula states that the area of a geodesic triangle with angles $\alpha, \beta, \gamma$ is $A = \pi - (\alpha + \beta + \gamma)$. More generally, a geodesic $n$-gon with angles $\alpha_1, \alpha_2, \ldots, \alpha_n$ has area $A = (n-2)\pi - \sum_{i=1}^{n} \alpha_i$.

1.5.2 THEOREM. *A Fuchsian group $\Gamma$ is finitely generated if and only if $\mathscr{K}(\Gamma)/\Gamma$ has finite area.*

*Proof.* Suppose that $\Gamma$ is finitely generated. Then $\mathscr{K}^*(\Gamma)/\Gamma$ is compact, so it has finite area. $\mathscr{K}(\Gamma)/\Gamma$ differs from $\mathscr{K}^*(\Gamma)/\Gamma$ by finitely many parabolic cusps (i.e. quotients of periodic parabolic regions). Since a parabolic cusp has finite area, so does $\mathscr{K}(\Gamma)/\Gamma$.

† This result also follows from theorem 2.7.1 of Chapter 1.

Now suppose that $\mathscr{K}(\Gamma)/\Gamma$ has finite area. Let $\mathscr{P}$ be a Dirichlet region for $\Gamma$, centred at $z_0 \in \mathscr{K}(\Gamma)$. $Q = \mathscr{P} \cap \mathscr{K}(\Gamma)$ is a convex fundamental polygon for $\mathscr{K}(\Gamma) \bmod \Gamma$. Thus $\text{Area}(Q) = \text{Area}(\mathscr{K}(\Gamma)/\Gamma)$. We shall show that $Q$ has finitely many sides. The proof of Proposition 1.2.2 shows that the elements in $\Gamma$ which identify sides of $Q$ generate $\Gamma$. Therefore, $\Gamma$ is finitely generated if $Q$ has finitely many sides.

We shall begin by showing that $\partial Q$ has finitely many points (if any) on $\mathscr{E}$. Suppose $\partial Q$ has infinitely many points on $\mathscr{E}$. Let $z_1, z_2, \ldots$ be a countable subset of these, and let $Q_n$ be the convex hull of $\{z_1, z_2, \ldots, z_n\}$. Since $Q$ is convex, $Q_n \subset Q$. $Q_n$ has exactly $n$ vertices (namely $z_1, z_2, \ldots, z_n$) and all angles are zero. Therefore $\text{Area}(Q_n) = (n - 2)\pi$. Since $n$ can be chosen arbitrarily large, this contradicts the finiteness of $\text{Area}(Q)$. Therefore $\partial Q$ has only a finite number of points on $\mathscr{E}$.

Now suppose $Q$ has infinitely many vertices in $\mathscr{D}$. Each cycle of vertices (i.e. $\Gamma$-equivalence class of vertices) is finite, and so there are infinitely many vertex cycles. The sum of the angles at all of the vertices of a cycle is $2\pi$ (if the vertices are not elliptic fixed points, and lie in the interior of $\mathscr{K}(\Gamma)$), $\pi$ (if the vertices lie on $\partial\mathscr{K}(\Gamma)$) or $2\pi/v$ (if the vertices are elliptic fixed points of order $v$). In all cases, the sum of these angles is of the form $2\pi/v$, for some positive integer $v$. If there are any elliptic fixed points of order 2 whose cycles consist of one vertex, then the angle of this vertex is $\pi$. We shall consider the two sides meeting at such a vertex to be a single side, and we shall ignore such vertices. With this understanding about sides, it still suffices to show that there are a finite number of sides (since each new side consists of at most two old sides). This implies that any cycle of elliptic vertices of order 2 may be supposed to contain at least two vertices.

Let $\mathscr{D}_r = \{z \in \mathscr{D} : d(z_0, z) \leqslant r\}$ and $\hat{Q}_r = Q \cap D_r$. (Note that Lemma 1.2.3 shows that if a vertex of $Q$ occurs in $\hat{Q}_r$, then its entire vertex cycle also occurs there.) Part of $\partial\hat{Q}_r$ may consist of intervals of the circle $\partial\mathscr{D}_r$. Cut off these circular arcs by geodesic segments with the same endpoints, and denote this smaller polygon by $Q_r$. We shall only consider values of $r$ such that $\partial\mathscr{D}_r$ does not pass through any vertex of $Q$. We may choose $r$ large enough so that any given finite set of vertices of $Q$ occur in $Q_r$. We want to estimate the area of $Q_r$.

Suppose that exactly $n$ vertex cycles $C_1, C_2, \ldots, C_n$ occur in $Q_r$, and the angle sum in $C_i$ is $2\pi/v_i$. We may suppose that the first $h$ cycles (possibly $h = 0$) have angle sum $2\pi$, the next $k$ cycles (possibly $k = 0$) have angle sum $\pi$, and the remaining cycles $C_i$ have angle sum $2\pi/v_i$, where $v_i \geqslant 3$. Each of the first $h$ cycles must contain at least three vertices, since each angle is $<\pi$

(because $Q_r$ is convex). Each of the next $k$ cycles must contain at least two vertices (because of our convention about 1-vertex cycles of elliptic fixed points of order 2). Thus the union of all cycles in $Q_r$ contains at least $3h + 2k + (n - h - k) = 2h + k + n$ vertices. In addition, $Q_r$ has $m$ accidental sides, which cut off the circular intervals of $\partial D_r$. Each of these gives rise to two more vertices, whose sum is $< 2\pi$. Thus $Q_r$ has at least $2h + k + n + 2m$ vertices (and sides), and the sum of the angles is at most

$$2\pi h + \pi k + \sum_{i=h+k+1}^{n} \frac{2\pi}{v_i} + 2\pi m,$$

where each $v_i \geqslant 3$. Applying the Gauss–Bonnet formula, we now have

$$\begin{aligned}
\text{Area}(Q_r) &\geqslant (2h + k + n + 2m - 2)\pi - \left(2\pi h + \pi k + \sum_{i=h+k+1}^{n} \frac{2\pi}{v_i} + 2\pi m\right) \\
&= (n - 2)\pi - \sum_{i=h+k+1}^{n} \frac{2\pi}{v_i} \\
&\geqslant (n - 2)\pi - \frac{2}{3} n\pi \\
&= \frac{n\pi}{3} - 2\pi.
\end{aligned}$$

Since $n$ may be arbitrarily large, this contradicts the finiteness of Area($Q$).

We have now shown that $Q$ has a finite number of sides, and therefore $\Gamma$ is finitely generated. ■

We now sum up the facts from Theorems 1.2.5, 1.3.2 and 1.5.2.

1.5.3 THEOREM. *Let $\Gamma$ be a Fuchsian group, and let $\mathscr{P}$ be a Dirichlet region for $\Gamma$. The following are equivalent.*

(i) *$\Gamma$ is finitely generated.*
(ii) *$\mathscr{P}$ has a finite number of sides.*
(iii) *$\mathscr{H}^*(\Gamma)/\Gamma$ is compact.*
(iv) *$\mathscr{H}(\Gamma)/\Gamma$ has a finite area.* ■

We shall now compute the area of $\mathscr{H}(\Gamma)/\Gamma$.

1.5.4 THEOREM. *Let $\Gamma$ be a Fuchsian group with signature $(g; v_1, v_2, \ldots, v_n; t)$. Then*

$$\text{Area}(\mathscr{K}(\Gamma)/\Gamma) = 2\pi\left[2g - 2 + t + \sum_{i=1}^{n}\left(1 - \frac{1}{v_i}\right)\right].$$

*Proof.* $S = \mathscr{K}(\Gamma)/\Gamma$ is a surface of genus $g$ with $p$ punctures and $b$ boundary curves, where $p + b = t$. Let $\bar{S}$ be the compactification of $S$, obtained by adding a point at each puncture. The total angle at each such point is 0. $\text{Area}(\mathscr{K}(\Gamma)/\Gamma) = \text{Area}(S) = \text{Area}(\bar{S})$.

Triangulate $\bar{S}$ by geodesic triangles, in such a way that each point in $\bar{S} \setminus S$ and each point corresponding to an elliptic fixed point is a vertex. If these are $V$ vertices, $E$ edges and $F$ triangles, then $V - E + F = 2 - 2g - b$.

Let $V_k$ be the number of vertices in the interior of $\bar{S}$ at which the total angle is $2\pi/k$. (In particular there are $p = V_\infty$ vertices in $\bar{S} \setminus S$.) Let $V_b$ be the number of boundary vertices, $E_i$ the number of edges which meet the interior of $\bar{S}$, and $E_b$ the number of boundary edges. Since the boundary edges and vertices occur on closed curves, $V_b = E_b$. Since each triangle contains three edges, each inner edge belongs to two triangles and each boundary edge belongs to one triangle, it follows that $3F = 2E_i + E_b$.

Let the triangles be denoted $T_1, T_2, \ldots$, and suppose that $T_q$ has angles $\alpha_q, \beta_q, \gamma_q$. Then

$$\begin{aligned}\text{Area}(\mathscr{K}(\Gamma)/\Gamma) = \text{Area}(\bar{S}) &= \sum_{q=1}^{F} \text{Area}(T_q) = \sum_{q=1}^{F} (\pi - \alpha_q - \beta_q - \gamma_q)\\ &= \pi F - (\text{sum of all angles})\\ &= \pi F - \sum_{k=1}^{\infty} V_k \frac{2\pi}{k} - V_b \pi\end{aligned}$$

Now using $3F = 2E_i + E_b$, or $F = -2F + 2E_i + E_b$, we have

$$\text{Area}(\mathscr{K}(\Gamma)/\Gamma) = -2\pi F + 2\pi E_i + \pi E_b - \sum_{k=1}^{\infty} V_k \frac{2\pi}{k} - V_b \pi.$$

Next, using the relations $V = V_b + \sum_{1 \leqslant k < \infty} V_k + V_\infty$, $E = E_i + E_b$, $V_b = E_b$, $p + b = t$, $p = V_\infty$ and $V - E + F = 2 - 2g - b$, we obtain

$$\begin{aligned}\text{Area}(\mathscr{K}(\Gamma)/\Gamma) &= 2\pi(-V + E - F) + 2\pi \sum_{1 \leqslant k < \infty}^{\infty} V_k\left(1 - \frac{1}{k}\right)\\ &\quad + 2\pi V_\infty + \pi V_b - \pi E_b\\ &= 2\pi\left[2g - 2 + t + \sum_{i=1}^{n}\left(1 - \frac{1}{v_i}\right)\right].\end{aligned}$$

■

1.5.5 THEOREM (Siegel). *Let* $\Gamma$ *be a finitely generated Fuchsian group with more than two limit points. Then* $\text{Area}(\mathscr{K}(\Gamma)/\Gamma) \geqslant \pi/21$. *Equality occurs only in the case where* $\Gamma$ *is the* (2, 3, 7) *triangle group (i.e. the group of signature* (0; 2, 3, 7; 0)).

*Proof.* The assumption that the limit set $L(\Gamma)$ contains more than two points guarantees that $\text{Area}(\mathscr{K}(\Gamma)/\Gamma) > 0$. If $\Gamma$ has signature $(g; v_1, v_2, \ldots, v_n; t)$ then by Theorem 1.5.4,

$$\text{Area}(\mathscr{K}(\Gamma)/\Gamma) = 2\pi\left[2g - 2 + t + \sum_{i=1}^{n}\left(1 - \frac{1}{v_i}\right)\right].$$

We shall use the abbreviated notation $A(\Gamma) = \text{Area}(\mathscr{K}(\Gamma)/\Gamma)$. Further, to make the proof more concise, we shall include the term $t$ in the sum $\sum(1 - (1/v_i))$ by including $t$ more terms $v_i$, with value $v_i = \infty$. With this notation,

$$A(\Gamma) = 2\pi\left[2g - 2 + \sum_{i=1}^{s}\left(1 - \frac{1}{v_i}\right)\right],$$

where $s = t + n$ and $2 \leqslant v_1 \leqslant v_2 \leqslant \ldots \leqslant v_s \leqslant \infty$. (Of course, $s$ may be zero, in which case the sum is zero, by definition.) We shall consider several cases to determine the minimum positive value of the right-hand side of the above equation.

*Case 1.* $g \geqslant 2$.

$$A(\Gamma) \geqslant 2\pi\left[2 + \sum_{i=1}^{s}\left(1 - \frac{1}{v_i}\right)\right]$$
$$\geqslant 4\pi.$$

*Case 2.* $g = 1$.

$$A(\Gamma) = 2\pi\sum_{i=1}^{s}\left(1 - \frac{1}{v_i}\right) \geqslant \pi.$$

*Case 3.* $g = 0$.

$$A(\Gamma) = 2\pi\left[-2 + \sum_{i=1}^{s}\left(1 - \frac{1}{v_i}\right)\right].$$

(i) $s \geqslant 5$.

$$A(\Gamma) \geqslant 2\pi\left[-2 + \frac{1}{2} + \frac{1}{2} + \frac{1}{2} + \frac{1}{2} + \frac{1}{2}\right] = \pi.$$

(ii) $s = 4$.

$$A(\Gamma) \geqslant 2\pi\left[-2 + \frac{1}{2} + \frac{1}{2} + \frac{1}{2} + \frac{2}{3}\right] = \frac{\pi}{3}.$$

(iii) $s = 3$, *all* $v_i \geqslant 3$.

$$A(\Gamma) \geqslant 2\pi\left[-2 + \frac{2}{3} + \frac{2}{3} + \frac{3}{4}\right] = \frac{\pi}{6}.$$

(iv) $s = 3, v_1 = 2, v_2 \geqslant 4$. In this case $v_3 \geqslant 5$, and

$$A(\Gamma) \geqslant 2\pi\left[-2 + \frac{1}{2} + \frac{3}{4} + \frac{4}{5}\right] = \frac{\pi}{10}.$$

(v) $s = 3, v_1 = 2, v_2 = 3$. In this case $v_3 \geqslant 7$, and

$$A(\Gamma) \geqslant 2\pi\left[-2 + \frac{1}{2} + \frac{2}{3} + \frac{6}{7}\right] = \frac{\pi}{21}.$$

Thus the minimal area is $\pi/21$, and this occurs only for signature $(0;2,3,7;0)$.■

## 1.6 Some group theoretic results

In this section we shall prove two purely group theoretic theorems about Fuchsian groups, using the preceding geometric methods. In particular, these theorems are valid for free groups.

1.6.1 LEMMA. *Let $S$ be a closed subset of $\mathscr{E}$ which contains at least two points. If $S$ is invariant under a Fuchsian group $\Gamma$, then $S$ contains the limit set $L(\Gamma)$.*

*Proof.* See Chapter 2, Section 6.

1.6.2 LEMMA. *Let $\Gamma$ be a Fuchsian group and $\Delta$ a non-trivial normal subgroup. Then $L(\Delta) = L(\Gamma)$.*

*Proof.* Since $\Delta \subset \Gamma$, $L(\Delta) \subset L(\Gamma)$. We shall prove the reverse inclusion. First we show that $L(\Delta)$ is invariant under $\Gamma$.

Let $z_0 \in L(\Delta)$. There is a sequence $\{\delta_n\} \subset \Delta$, so that $\lim_{n\to\infty} \delta_n(z) = z_0$ for all $z \in \mathscr{D}$. Let $\gamma \in \Gamma$. Then $\lim_{n\to\infty} \delta_n\gamma^{-1}(z) = z_0$, and $\lim_{n\to\infty} \gamma\delta_n\gamma^{-1}(z) = \gamma(z_0)$.

But since $\Delta$ is normal, $\gamma\delta_n\gamma^{-1} \in \Delta$, and $\lim_{n\to\infty} \gamma\delta_n\gamma^{-1}(z) \in L(\Delta)$. This shows that $\gamma(z_0) \in L(\Delta)$, and $L(\Delta)$ is invariant under $\Gamma$.

If $L(\Delta)$ contains at least two points, then Lemma 1.6.1 implies that $L(\Delta) \supset L(\Gamma)$. Otherwise $L(\Delta)$ is empty or consists of one point.

If $L(\Delta) = \varnothing$, then $\Delta$ is finite, so it has a fixed point $z_0 \in \mathscr{D}$. Thus $\Delta$ is a finite, cyclic, elliptic group. Since $\Delta$ is normal, $\Gamma$ leaves invariant the set of fixed points of $\Delta$. But $z_0$ is the only fixed point, so $\Gamma$ leaves $z_0$ fixed. Therefore $\Gamma$ is finite, and $L(\Gamma) = \varnothing = L(\Delta)$.

Finally, suppose $L(\Delta) = \{z_0\}$. Then $z_0$ is fixed by $\Delta$. The elements of $\Delta$ must be parabolic, since hyperbolic elements would give rise to more limit points. Thus $\Delta$ is a cyclic parabolic group, which leaves $z_0$ fixed. For the same reason as in the previous case, $\Gamma$ leaves $z_0$ fixed also. $\Gamma$ must not contain hyperbolic elements, since a discontinuous group cannot contain hyperbolic and parabolic elements with a common fixed point. Therefore $\Gamma$ is a cyclic, parabolic group, which leaves $z_0$ fixed. Thus $L(\Gamma) = \{z_0\} = L(\Delta)$. ■

Recall that a subgroup $\Delta \subset \Gamma$ is *subnormal*, if there is a finite sequence of groups $\Gamma_1 \subset \Gamma_2 \subset \ldots \subset \Gamma_n$ such that $\Gamma_i$ is normal in $\Gamma_{i+1}$, $\Gamma_1 = \Delta$ and $\Gamma_n = \Gamma$.

1.6.3 THEOREM. *Let $\Gamma$ be a finitely generated Fuchsian group, and $\Delta$ a finitely generated subgroup which contains a non-trivial, subnormal subgroup of $\Gamma$. Then the index $[\Gamma:\Delta]$ is finite.*

*Proof.* Since $\Delta$ contains a non-trivial, subnormal subgroup of $\Gamma$, Lemma 1.6.2 implies that $L(\Delta) = L(\Gamma)$. Therefore $\mathscr{K}(\Delta) = \mathscr{K}(\Gamma)$, and we denote this common region by $\mathscr{K}$. Both $\Gamma$ and $\Delta$ operate on $\mathscr{K}$, and we have a (possibly ramified) covering $\phi: \mathscr{K}/\Delta \to \mathscr{K}/\Gamma$. The number of sheets of this covering is equal to the index $[\Gamma:\Delta]$.

Since $\Gamma$ and $\Delta$ are finitely generated, Theorem 1.5.2 implies that $\mathscr{K}/\Delta$ and $\mathscr{K}/\Gamma$ have finite area. Therefore the covering $\phi: \mathscr{K}/\Delta \to \mathscr{K}/\Gamma$ is finite, as is the index $[\Gamma:\Delta]$. Note that Area$(\mathscr{K}/\Gamma) = 0$ only if $\Gamma$ is cyclic or infinite dihedral (i.e. the group of signature $(0; 2, 2; 1)$). The theorem is easily verified in these two cases. ■

1.6.4 THEOREM. *If $\Gamma$ and $\Delta$ are finitely generated subgroups of a Fuchsian group, then $\Gamma \cap \Delta$ is also finitely generated.*

*Proof.* Let $\Phi = \Gamma \cap \Delta$ and let $\Psi$ be the finitely generated Fuchsian group generated by $\Gamma$ and $\Delta$. We may suppose that $\Psi$ contains no parabolic elements, for there are Fuchsian groups which are isomorphic to $\Psi$ and which have no parabolic elements. This has the advantage that $\mathscr{K}^*(\Theta) = \mathscr{K}(\Theta)$ for every subgroup $\Theta \subset \Psi$.

Since $\Gamma$ and $\Delta$ are finitely generated, and without parabolic elements, $\mathscr{K}(\Gamma)/\Gamma$ and $\mathscr{K}(\Delta)/\Delta$ are compact. Therefore there is a disc

$$B = \{z \in \mathscr{D}: |z| < r < 1\}$$

such that $\mathscr{K}(\Delta) \subset \Delta B$ and $\mathscr{K}(\Gamma) \subset \Gamma B$.

Choose coset representatives $\{\gamma_i\}$ and $\{\delta_j\}$ so that

$$\Gamma = \bigcup_i \Phi\gamma_i \quad \text{and} \quad \Delta = \bigcup_j \Phi\delta_j.$$

Then

$$\mathscr{K}(\Gamma) \subset \Gamma B = \Phi \bigcup_i \gamma_i B,$$

$$\mathscr{K}(\Delta) \subset \Delta B = \Phi \bigcup_j \delta_j B.$$

Also

$$\mathscr{K}(\Phi) \subset \mathscr{K}(\Gamma) \cap \mathscr{K}(\Delta),$$

since

$$L(\Phi) \subset L(\Gamma) \cap L(\Delta).$$

We shall now show that $\gamma_i B \cap \mathscr{K}(\Delta) \neq \varnothing$ for only a finite number of representatives $\gamma_i$. For $\phi \in \Phi$, $\gamma_i B \cap \phi\delta_j B \neq \varnothing$ if and only if

$$B \cap \gamma_i^{-1}\phi\delta_j B \neq \varnothing.$$

Since $B$ is relatively compact, there are only a finite number of elements $\psi \in \Psi$, such that $B \cap \psi B \neq \varnothing$. Also note that if $\gamma_i^{-1}\phi_1\delta_j = \gamma_k^{-1}\phi_2\delta_l$, then $\gamma_k\gamma_i^{-1}\phi_1 = \phi_2\delta_l\delta_j^{-1} \in \Gamma \cap \Delta = \Phi$. Therefore $\gamma_k\gamma_i^{-1}$ and $\delta_l\delta_j^{-1} \in \Phi$, so $\gamma_i = \gamma_k$, $\delta_j = \delta_l$ and $\phi_1 = \phi_2$. It follows that there are only a finite number of the $\gamma_i$, $\phi$, $\delta_j$ for which $\gamma_i B \cap \phi\delta_j B \neq \varnothing$. Since $\mathscr{K}(\Delta)$ is contained in the union of the sets $\phi\delta_j B$, there are only finitely many $\gamma_i$ so that $\gamma_i B \cap \mathscr{K}(\Delta) \neq \varnothing$. Note that if $\phi \in \Phi$, then $\phi\mathscr{K}(\Delta) = \mathscr{K}(\Delta)$, so that $\phi\gamma_i B \cap \mathscr{K}(\Delta) \neq \varnothing$ if and only if $\gamma_i B \cap \phi^{-1}\mathscr{K}(\Delta) = \gamma_i B \cap \mathscr{K}(\Delta) \neq \varnothing$. Thus $\gamma_i B \cap \mathscr{K}(\Delta) \neq \varnothing$ if and only if $(\Phi\gamma_i B) \cap \mathscr{K}(\Delta) \neq \varnothing$.

Let $\gamma_1, \gamma_2, \ldots, \gamma_n$ be the coset representatives for which $\gamma_i B \cap \mathscr{K}(\Delta) \neq \varnothing$. Recalling that $\mathscr{K}(\Phi) \subset \Phi \bigcup \gamma_i B$, we now see that $\mathscr{K}(\Phi) \subset \Phi \bigcup_{k=1}^{n} \gamma_k B$. Since $\bigcup_{k=1}^{n} \gamma_k B$ is relatively compact, $\mathscr{K}(\Phi)/\Phi$ is compact, and therefore $\Phi = \Gamma \cap \Delta$ is finitely generated. ∎

### 1.7 Automorphism groups of Riemann surfaces

In this section we shall study the group Aut($S$) of all holomorphic self-homeomorphisms of a Riemann surface $S$. Some examples are the following: Aut($\mathbb{C}$) is the group of affine transformations $f(x) = az + b$ $(a \neq 0)$. Aut($\mathscr{D}$) is the group of linear fractional transformations which leave $\mathscr{D}$ invariant. This group is isomorphic to PSU(1, 1) (see Chapter 2 p. 56) (i.e. the projective, special unitary group which leaves invariant the form $|z_1|^2 - |z_2|^2$). Denoting the Riemann sphere $\mathbb{C} \cup \{\infty\}$ by $\hat{\mathbb{C}}$, Aut($\hat{\mathbb{C}}$) is the group of all linear fractional transformations. Thus Aut($\hat{\mathbb{C}}$) $\cong$ PSL(2, $\mathbb{C}$). In these examples, Aut($S$) is a continuous group. However, the examples are atypical in this respect. In most cases, Aut($S$) is countable and discontinuous, and for most surfaces of finite topological type, Aut($S$) is finite.

We shall need the following theorem, which can be found in [**4**].

UNIFORMIZATION THEOREM. *Let $S$ be a simply connected Riemann surface. Then $S$ is holomorphically equivalent to $\hat{\mathbb{C}}$, $\mathbb{C}$ or $\mathscr{D}$.*

This theorem can be regarded as a strong version of the Riemann mapping theorem. Now let $S$ be an arbitrary Riemann surface, and let $\tilde{S}$ be its universal covering surface. The complex structure on $S$ can be lifted to $\tilde{S}$ so that the projection $\phi: \tilde{S} \to S$ is holomorphic. Let $\Gamma$ denote the group of covering transformations. Then $\Gamma$ is a discontinuous subgroup of Aut($\tilde{S}$) and $S \cong \tilde{S}/\Gamma$. The following table indicates which Riemann surfaces $S$ arise in this manner from the cases $\tilde{S} = \hat{\mathbb{C}}$, $\mathbb{C}$ or $\mathscr{D}$.

| $\tilde{S}$ | $S$ | Genus of Compact $S$ |
|---|---|---|
| $\hat{\mathbb{C}}$ | $\hat{\mathbb{C}}$ | 0 |
| $\mathbb{C}$ | $\mathbb{C}$, $\mathbb{C} - \{0\}$, all tori | 1 |
| $\mathscr{D}$ | all other surfaces | $g \geqslant 2$ |

If $\tilde{S} = \hat{\mathbb{C}}$ or $\mathbb{C}$, then Aut($S$) is a continuous group. This is also true if $S$ is a disc, an annulus, or a punctured disc. In the latter two cases, $S \cong \mathscr{D}/\Gamma$, where $\Gamma$ is a cyclic hyperbolic or parabolic group. In all other cases, the fundamental group $\pi_1(S)$ is non-abelian, and as the following theorem indicates, Aut($S$) is discontinuous on $S$ (i.e. the Aut($S$)-orbits do not accumulate in $S$).

1.7.1 THEOREM. *Let $S$ be a Riemann surface whose fundamental group $\pi_1(S)$ is non-abelian. Then* $\mathrm{Aut}(S)$ *is discontinuous on $S$. (Consequently, this group is countable.) If $\pi_1(S)$ is finitely generated (and non-abelian) then* $\mathrm{Aut}(S)$ *is finite.*

*Proof.* Since $\pi_1(S)$ is non-abelian, the universal covering surface $\tilde{S} \cong \mathscr{D}$ (and we may suppose $\tilde{S} = \mathscr{D}$). The group $\Gamma$ of covering transformations is a Fuchsian group. Let $N(\Gamma)$ denote the normalizer of $\Gamma$ in $\mathrm{Aut}(\mathscr{D})$.

We shall first show that $N(\Gamma)$ is a discrete subgroup of $\mathrm{Aut}(\mathscr{D})$. It suffices to show that if $\{n_k\}$ is a sequence in $N(\Gamma)$ which converges to 1 (the identity transformation), then $n_k = 1$ for large $k$. For such a sequence $\{n_k\}$, and for any $\gamma \in \Gamma$, $\lim_{k\to\infty} n_k \gamma n_k^{-1} = \gamma$. Since $n_k \gamma n_k^{-1} \in \Gamma$, and $\Gamma$ is discrete, it follows that $n_k \gamma n_k^{-1} = \gamma$ for large $k$. But two transformations in $\mathrm{Aut}(\mathscr{D})$ (different from 1) commute if and only if they have the same fixed points. Since $\Gamma$ is non-abelian, it contains elements with different fixed points. If $n_k \neq 1$, then it could not commute with two such elements. Therefore $n_k = 1$ for large $k$, and $N(\Gamma)$ is discrete.

We shall now show that $\mathrm{Aut}(S) \cong N(\Gamma)/\Gamma$. As before, we denote the universal covering map of $S$ by $\phi\colon \mathscr{D} \to S$. For $s \in S$, $\phi^{-1}(s)$ is a $\Gamma$-orbit $\Gamma(z)$. If $n \in N(\Gamma)$, then $n\Gamma(z) = \Gamma n(z)$, so that $n$ maps $\Gamma$-orbits to $\Gamma$-orbits. Thus $n$ induces an automorphism $\bar{n}$ of $S$. The map $\bar{\phi}\colon N(\Gamma) \to \mathrm{Aut}(S)$, defined by $\bar{\phi}(n) = \bar{n}$, is a homomorphism. Since $\phi\colon \mathscr{D} \to S$ is the universal covering, any automorphism of $S$ may be lifted to an automorphism of $\mathscr{D}$. This shows that $\bar{\phi}\colon N(\Gamma) \to \mathrm{Aut}(S)$ is surjective. Since $\mathrm{Ker}(\bar{\phi}) = \Gamma$, $\mathrm{Aut}(S) \cong N(\Gamma)/\Gamma$.

Since $N(\Gamma)$ is discontinuous in $\mathscr{D}$, $\mathrm{Aut}(S)$ is discontinuous in $S$. If $\pi_1(S)$ is finitely generated, then so is $\Gamma$, since $\Gamma \cong \pi_1(S)$. In this case, Theorem 1.5.2 implies that $\mathrm{Area}(\mathscr{K}(\Gamma)/\Gamma)$ is finite. Since $\Gamma$ is normal in $N(\Gamma)$, Lemma 1.6.2 implies that $\mathscr{K}(N(\Gamma)) = \mathscr{K}(\Gamma) = \mathscr{K}$. Thus $\mathrm{Area}(\mathscr{K}/N(\Gamma)) \leqslant \mathrm{Area}(\mathscr{K}/\Gamma) < \infty$, and Theorem 1.5.2 implies that $N(\Gamma)$ is finitely generated. Now Theorem 1.6.3 implies that $[N(\Gamma):\Gamma] < \infty$, so $\mathrm{Aut}(S)$ is finite. ∎

1.7.2 THEOREM (Hurwitz). *Let $S$ be a closed Riemann surface of genus $g \geqslant 2$. Then* $\mathrm{Aut}(S)$ *has order* $|\mathrm{Aut}(S)| \leqslant 84(g-1)$. *Equality occurs if and only if $S = \mathscr{D}/\Gamma$, where $\Gamma$ is a proper normal subgroup of finite index in the* (2, 3, 7) *triangle group.*

*Proof.* Let $\phi\colon \mathscr{D} \to S$ be the universal covering, $\Gamma$ the group of covering transformations and $N(\Gamma)$ the normalizer of $\Gamma$ in $\mathrm{Aut}(\mathscr{D})$. In the proof of the previous theorem, we showed that $N(\Gamma)$ is Fuchsian, and $\mathrm{Aut}(S) \cong N(\Gamma)/\Gamma$.

Therefore

$$|\mathrm{Aut}(S)| = [N(\Gamma):\Gamma] = \frac{\mathrm{Area}(\mathscr{D}/\Gamma)}{\mathrm{Area}(\mathscr{D}/N(\Gamma))}.$$

By the area formula of Theorem 1.5.4, Area $(\mathscr{D}/\Gamma) = 4\pi(g - 1)$. By Theorem 1.5.5, Area$(\mathscr{D}/N(\Gamma)) \geqslant \pi/21$, and equality occurs if and only if $N(\Gamma)$ is the (2, 3, 7) triangle group, which we denote $T(2, 3, 7)$. Thus

$$|\mathrm{Aut}(S)| = \frac{\mathrm{Area}(\mathscr{D}/\Gamma)}{\mathrm{Area}(\mathscr{D}/N(\Gamma))} \leqslant \frac{84(g-1)}{(\pi/21)} = 84(g-1),$$

and equality occurs if and only if $N(\Gamma) = T(2, 3, 7)$. We have proved the first part of the theorem (the inequality). We now consider the case of equality in more detail.

The above argument shows that if $\mathrm{Aut}(S) = 84(g - 1)$, then $S = \mathscr{D}/\Gamma$ where $\Gamma$ is a normal subgroup of $T(2, 3, 7)$. (Here $\Gamma$ is defined to be the group of covering transformations of the universal covering $\phi: \mathscr{D} \to S$.) Conversely, suppose that $\Gamma$ is a proper normal subgroup of finite index in $T(2, 3, 7)$, and $S = \mathscr{D}/\Gamma$. In order to show that $|\mathrm{Aut}(S)| = 84(g - 1)$, we must verify two things:

(i) $\Gamma$ has no elements of finite order. This will ensure that the projection $\mathscr{D} \to \mathscr{D}/\Gamma$ is the universal covering, so that elements in Aut($S$) can be lifted to lie in $N(\Gamma)$.
(ii) $N(\Gamma) = T(2, 3, 7)$.

To prove (i), let $T = T(2, 3, 7)$, $G = T/\Gamma$ and consider the projection $f: T \to G$. $T$ is generated by elements $x$, $y$, $z$ which satisfy the relations $xyz = 1$, $x^2 = y^3 = z^7 = 1$. The elements $\bar{x} = f(x)$, $\bar{y} = f(y)$, $\bar{z} = f(z)$ in $G$ satisfy the same relations. Any element of finite order in $T$ is conjugate (in $T$) to a power of $x$, $y$ or $z$. Therefore, if $\Gamma$ contains an element of finite order, it contains a power of $x$, $y$ or $z$. Since these elements have prime order, $\Gamma$ contains one of the elements $x$, $y$, $z$. Therefore, one of the elements $\bar{x}$, $\bar{y}$, $\bar{z}$ equals 1. The relations $\bar{x}\bar{y}\bar{z} = 1$, $\bar{x}^2 = \bar{y}^3 = \bar{z}^7 = 1$ now imply that $\bar{x} = \bar{y} = \bar{z} = 1$. Thus $G = \{1\}$ and $\Gamma = T$, which contradicts the assumption that $\Gamma$ is a proper subgroup.

To prove (ii), note that $T$ is a maximal Fuchsian group, since it has minimum area. If $\Gamma$ is normal in $T$, then $T \subset N(\Gamma)$, and therefore $T = N(\Gamma)$. ■

1.7.3 THEOREM (Macbeath).

(i) *There are infinitely many* $g$ *for which the upper bound* $84(g-1)$ *is attained.*
(ii) *There are infinitely many* $g$ *for which* $84(g-1)$ *is not attained.*

*Proof.* (i) As before, let $T = T(2, 3, 7)$ denote the (2, 3, 7) triangle group. By the previous theorem, we must consider finite homomorphic images $G = T/\Gamma$. We shall refer to such groups as Hurwitz groups (or $H$-groups). To prove (i), we must show that there exist $H$-groups of infinitely many different orders. This follows from a general fact that a finitely generated matrix group (such as $T$) is residually finite (see [**48**]). Thus for $t \in T$, $t \neq 1$, there is a normal subgroup $\Gamma$ of finite index in $T$, such that $t \notin \Gamma$. Using this fact, we can construct a chain of normal subgroups of finite index

$$T \underset{\neq}{\supset} \Gamma_1 \underset{\neq}{\supset} \Gamma_2 \underset{\neq}{\supset} \Gamma_3 \underset{\neq}{\supset} \dots$$

and corresponding $H$-groups $G_n = T/\Gamma_n$.

To prove (ii), we first note that there is no $H$-group of order 84. For in such a group $G$, the Sylow 7-group $S_7$ must be normal, and the group $G' = G/S_7$ is an $H$-group of order 12. Such an $H$-group does not exist, since the order must be divisible by 84.

Now let $p$ be a prime $> 84$. There is no $H$-group of order $84p$. For if $G$ is such a group, then the Sylow $p$-group $S_p$ is normal, and $G' = G/S_p$ is an $H$-group of order 84. Thus if $g = p + 1$, where $p$ is a prime $> 84$, then there is no $H$-group of order $84(g-1)$. ■

The following theorem, whose proof is too long to describe here, can be found in [**19**].

1.7.4 THEOREM (Greenberg). *Let* $G$ *be a finite group. Then there is a closed Riemann surface* $S$, *such that* $\mathrm{Aut}(S) \cong G$.

### 1.8 Notes

Proposition 1.2.1 was proved by Beardon [**4**]. Theorem 1.2.5 was proved independently by Fenchel and Nielsen [**11**], Greenberg [**18**], Heins [**21**], and Marden [**35**]. Fenchel and Nielsen proved Theorem 1.3.2 first, then used this to prove Theorem 1.2.5. Marden uses the topology of the quotient surface $\mathscr{D}/\Gamma$. The present proof is adapted from [**18**]. The Nielsen convex region was first introduced by Nielsen [**42**] for the case where $\Gamma$ contains only hyperbolic elements. It is discussed more generally by Fenchel and Nielsen [**11**], where Theorem 1.5.2 is proved. Theorem 1.5.2 was first proved by Siegel [**46**] for the case where $L(\Gamma) = \mathscr{E}$ (i.e. Area$(\mathscr{D}/\Gamma) < \infty$). Higher-dimensional versions of this theorem are due to Garland and Raghunathan [**13**] and Wielenberg [**49**]. Theorem 1.4.1 is classical (but only because it was classically assumed that finitely generated groups have finite-sided fundamental polygons). Theorems 1.4.1 and 1.4.2 can be found in Fenchel and Nielsen [**11**]. Theorem 1.5.1 is another classical theorem. It can be proved without Theorem 1.2.5, by using a lemma of Selberg [**44**], which assures the existence of a subgroup of finite index without elliptic elements. Theorem 1.5.4 is classical for groups of the first kind (where $\mathscr{K}(\Gamma) = \mathscr{D}$). But as far as I know, a finite area for arbitrary finitely generated Fuchsian groups is not discussed anywhere except in Fenchel and Nielsen [**11**]. Siegel proved Theorem 1.5.5 in [**46**]. Theorem 1.6.3 was proved by Schreier [**45**] and Karrass and Solitar [**26**] for normal subgroups of free groups. The present version is due to Greenberg [**14**]. Theorem 1.6.4 was proved by Howson [**22**] for free groups and Greenberg [**14**] for Fuchsian groups. Theorems 1.7.1 and 1.7.2 (Hurwitz' theorem) are classical results which can be found in [**47**] and [**23**] (except for the last statement of 1.7.1, which does not seem to be in print). Macbeath proved Theorem 1.7.3 in [**32**]. In that paper he shows that many groups PSL$(2, p)$ are $H$-groups. The proof of Theorem 1.7.4 relies on Teichmüller space considerations.

## 2. KLEINIAN GROUPS

### 2.1 Ahlfors' finiteness theorem

In this chapter, we shall use the term *Kleinian group* to mean a discrete subgroup $\Gamma$ of PSL$(2, \mathbb{C})$. $\Gamma$ acts as a group of linear fractional transformations in the Riemann sphere $\hat{\mathbb{C}} = \mathbb{C} \cup \{\infty\}$. This action extends to the upper half-space $\mathscr{H}^3$, where $\Gamma$ acts discontinuously (see Chapter 2). The limit set $L = L(\Gamma)$ is the set of accumulation points of any orbit $\Gamma(z)$, $z \in \mathscr{H}^3$. The regular set $\Omega = \Omega(\Gamma)$ is the subset of $\hat{\mathbb{C}}$ where $\Gamma$ acts discontinuously. ($\Omega$ may be empty. However some authors use the term Kleinian to mean that $\Omega \neq \varnothing$).

*Definition.* $\Gamma$ is of the *first kind* if $\Omega = \varnothing$, and of the *second kind* if $\Omega \neq \varnothing$.

Until further notice we shall assume that $\Gamma$ is of the second kind, and we shall concentrate on the action of $\Gamma$ on $\Omega$, ignoring the action on the upper half-space. The quotient $\Omega/\Gamma$ is a (possibly disconnected) Riemann surface, and the projection $p: \Omega \to \Omega/\Gamma$ is a (possibly branched) covering map.

*Definitions*

(i) Riemann surface of *finite type* $(g, t)$ is a connected surface which is conformally equivalent to a closed surface of genus $g$ with $t$ punctures.

(ii) If a connected component $S$ of $\Omega/\Gamma$ is of finite type, and contains only a finite number of points over which $p$ is branched, we shall say that $S$ has *finite signature*. The signature is $\sigma = (g; \nu_1, \ldots, \nu_k; t)$, where $S$ has type $(g, t)$ and the branching orders are $\nu_1, \ldots, \nu_k$.

The first objective of this chapter will be to prove the following.

2.1.1 THEOREM (Ahlfors). *Let $\Gamma$ be a finitely generated Kleinian group. Then*

(i) *$\Omega/\Gamma$ has finitely many connected components;*
(ii) *each component has finite signature.*

The proof of the theorem will occupy the next few sections of this chapter. If $\Gamma$ had a fundamental polyhedron (in the upper half-space) with finitely many sides, then the theorem would be obvious. Unfortunately, there exist finitely generated Kleinian groups which have no finite-sided fundamental polyhedron. (We shall give some examples in 2.7.) In spite of the geometric nature of the theorem, the only known proof is analytic.

*Remarks.* The theorem is easily verified for elementary groups. For the remainder of this chapter, we shall assume that $\Gamma$ is not elementary.

In general, $\Omega/\Gamma = \bigcup S_i$ (possibly an infinite union). Let $\Omega_i$ be a connected component of $p^{-1}(S_i)$ and $\Gamma_i$ the stabilizer of $\Omega_i$ in $\Gamma$ (so $S_i = \Omega_i/\Gamma_i$). Since $\Gamma$ is non-elementary, the boundary of $\Omega_i$ must contain more than 2 points, and the universal covering surface of $\Omega_i$ is conformally equivalent to the unit disc $\mathscr{D}$. Let $q: \mathscr{D} \to \Omega_i$ denote the universal covering. The elements of $\Gamma_i$ may be lifted (in all possible ways) to $\mathscr{D}$, yielding a Fuchsian group $\tilde{\Gamma}_i$ such that $\mathscr{D}/\tilde{\Gamma}_i = S_i$. Note that $\mathscr{D} \to S_i$ is branched over the same points (and with the same branching orders) as $\Omega_i \to S_i$, and $S_i$ has the same signature as $\tilde{\Gamma}_i$. We shall call $\tilde{\Gamma}_i$ the *Fuchsian group corresponding to $S_i$*. All the information about $S_i$ (including the branching in $p: \Omega_i \to S_i$) is contained in the Fuchsian group $\tilde{\Gamma}_i$. In particular, $S_i$ has finite signature if and only if the non-Euclidean area of $S_i$ is finite. Since Area$(S_i) \geqslant \pi/21$, Theorem 2.1.1 is equivalent to the following.

2.1.1′ THEOREM. *If $\Gamma$ is finitely generated, then*

$$\text{Area}(\Omega/\Gamma) < \infty.$$

We shall later obtain an upper bound for this area (the Bers inequality).

## 2.2 Automorphic forms

Each component $\Omega_0$ of $\Omega$ has a universal covering isomorphic to $\mathscr{D}$. Thus $\Omega_0$ inherits a hyperbolic metric, which may be expressed as $\lambda(z)|dz|$. This metric is invariant under conformal transformations. For an integer $q \geqslant 2$, an *automorphic form* of weight $q$ for $\Gamma$ is a function $\phi(z)$, holomorphic in $\Omega$, such that

$$\phi(\gamma z)\,\gamma'(z)^q = \phi(z) \qquad (\text{for } \gamma \in \Gamma).$$

Let $\Omega_0$ be an open subset of $\Omega$, which is invariant under $\Gamma$. We shall be interested in the following spaces of automorphic forms. This is discussed in Chapter 3 for the case when $\Gamma$ is Fuchsian and $\Omega_0 = \mathscr{D}$. Let $q$ be an integer $\geqslant 2$. $A_q(\Gamma, \Omega_0)$ is the space of all automorphic forms $\phi(z)$ such that

$$\|\phi\| \overset{\text{def}}{=} \iint_{\Omega_0/\Gamma} |\phi(z)|\lambda^{2-q}(z)\,dx\,dy < \infty.$$

$B_q(\Gamma, \Omega_0)$ is the space of all automorphic forms $\psi(z)$ such that

$$\|\psi\| \overset{\text{def}}{=} \sup_{z \in \Omega_0} |\psi(z)|\lambda^{-q}(z) < \infty.$$

The invariant set $\Omega_0$ will be chosen according to the following conventions.

(i) If $\Gamma$ is the Kleinian group in question (for which we want to establish Ahlfors' theorem), $\Omega_0 = \Omega$.

(ii) For the stabilizer $\Gamma_i$ of a component $\Omega_i$ of $\Omega$, we shall use $\Omega_0 = \Omega_i$.

(iii) For a Fuchsian group, $\Omega_0 = \mathscr{D}$.

With this understanding, we shall write $A_q(\Gamma) = A_q(\Gamma, \Omega_0)$, $B_q(\Gamma) = B_q(\Gamma, \Omega_0)$.

For $\phi \in A_q(\Gamma)$, $\psi \in B_q(\Gamma)$ we form the Peterson inner product

$$\langle \phi, \psi \rangle = \langle \phi, \psi \rangle_{\Gamma, \Omega_0} \overset{\text{def}}{=} \iint_{\Omega_0/\Gamma} \phi(z)\,\bar{\psi}(z)\,\lambda(z)^{2-2q}\,dx\,dy.$$

If $\Gamma$ is a Fuchsian group, then $\Omega_0 = \mathscr{D}$ and $\lambda(z) = \{2(1 - |z|^2)\}^{-1}$. We shall use the following facts for the case where $\Gamma$ is a Fuchsian group (see Chapter 3, [**30**], [**6**]).

(iii) $B_q(\Gamma)$ is the dual of $A_q(\Gamma)$ with respect to the pairing $\langle \phi, \psi \rangle$.
(ii) $\dim A_2(\Gamma) < \infty$ if and only if $\text{Area}(\mathscr{D}/\Gamma) < \infty$.
(iii) If $\text{Area}(\mathscr{D}/\Gamma) < \infty$, and $\Gamma$ has signature $(g; v, \ldots, v_k; t)$, then

$$\dim A_q(\Gamma) = (2q - 1)(g - 1) + t(q - 1) + \sum_{i=1}^{k} \left[ q\left(1 - \frac{1}{v_i}\right) \right].$$

(Here $[x]$ is the largest integer $\leqslant x$.)
(iv) The Poincaré series $\Theta_q F = \sum_{\gamma \in \Gamma} F(\gamma z)\,\gamma'(z)^q$ defines a surjective linear map $\Theta_q : A_q(1) \to A_q(\Gamma)$ of norm $\leqslant 1$. (Here 1 denotes the trivial group).

2.2.1 THEOREM. *$B_q(\Gamma)$ is the dual of $A_q(\Gamma)$ with respect to the pairing $\langle \phi, \psi \rangle$.*

*Proof.* Let $\Omega/\Gamma = \bigcup S_i$, $\hat{\Omega}_i = p^{-1}(S_i)$, $\Omega_i$ a connected component of $\hat{\Omega}_i$, $\Gamma_i$ the stabilizer of $\Omega_i$ in $\Gamma$. Each $\phi \in A_q(\Gamma)$ may be uniquely expressed as $\phi = \Sigma \hat{\phi}_i$, where $\hat{\phi}_i = 0$ in $\Omega - \hat{\Omega}_i$. Let $\phi_i = \hat{\phi}_i | \Omega_i$. Then $\phi_i \in A_q(\Gamma_i)$ and

$$\|\phi\| = \sum \|\phi_i\| < \infty.$$

Similarly, if $\psi \in B_q(\Gamma)$, then $\psi = \sum \hat{\psi}_i$, and $\|\psi\| = \sup \|\psi_i\| < \infty$, where $\psi_i \in B_q(\Gamma_i)$.

Now consider the universal covering map $\mathscr{D} \to \Omega_i$, and let $\tilde{\Gamma}_i$ be the Fuchsian group obtained by lifting $\Gamma_i$. Then $A_q(\tilde{\Gamma}_i) \cong A_q(\Gamma_i)$ and

$$B_q(\tilde{\Gamma}_i) \cong B_q(\Gamma_i).$$

Furthermore, we know for Fuchsian groups (Chapter 3, Theorem 3.2.8) that $B_q(\tilde{\Gamma}_i)$ is the dual of $A_q(\tilde{\Gamma}_i)$. Putting all this together, we get the theorem. ∎

The following is similarly an easy consequence of the Fuchsian case (Chapter 3, Theorem 3.3.3).

2.2.2 THEOREM. *The Poincaré series* $\Theta_q F(z) = \sum_{\gamma \in \Gamma} F(\gamma z)\,\gamma'(z)^q$ *defines a surjective linear map* $\Theta_q : A_q(1, \Omega) \to A_q(\Gamma)$ *of norm* $\leqslant 1$.

The next two Lemmas show that Ahlfors' theorem will follow if we prove that $\dim A_q(\Gamma) < \infty$ (for $q \geqslant 2$).

2.2.3 LEMMA. *Let* $\tilde{\Gamma}$ *be a Fuchsian group of signature*† $(g; v_1, \ldots v_k; t)$. *Then*

$$\dim A_q(\tilde{\Gamma}) \geqslant \frac{(q-1)}{2\pi} \operatorname{Area}(D/\tilde{\Gamma}) + g - 1.$$

*Proof.*

$$\dim A_q(\tilde{\Gamma}) = (2q-1)(g-1) + (q-1)t + \sum_{i=1}^{k} \left[ q\left(1 - \frac{1}{v_i}\right) \right],$$

and

$$\operatorname{Area}(D/\tilde{\Gamma}) = 2\pi \left[ 2(g-1) + t + \sum_{i=1}^{k} \left(1 - \frac{1}{v_i}\right) \right].$$

Let $q = m_i v_i - \mu_i$, where $m_i$ and $\mu_i$ are non-negative integers, such that $0 \leqslant \mu_i < v_i$. Then

$$q\left(1 - \frac{1}{v_i}\right) = q - m_i + \mu_i / v_i,$$

and

$$\left[ q\left(1 - \frac{1}{v_i}\right) \right] = q - m_i = q\left(1 - \frac{1}{v_i}\right) - \frac{\mu_i}{v_i}$$

$$\geqslant q\left(1 - \frac{1}{v_i}\right) - \frac{v_i - 1}{v_i} = (q-1)\left(1 - \frac{1}{v_i}\right).$$

The result follows immediately. ■

† Throughout this section, when we say that a Fuchsian group $\tilde{\Gamma}$ has signature $(g; v, \ldots, v_k; t)$, we shall always assume that $\mathscr{D}/\tilde{\Gamma}$ has finite area. This differs from the usage in section 1.

2.2.4 LEMMA. *Let $\Gamma$ be a Kleinian group.*

(i) *If* $\dim A_2(\Gamma) < \infty$, *then each component $S_i$ of $\Omega/\Gamma$ has finite signature, and there are only a finite number of components which do not correspond to triangle groups. [Recall that a triangle group is a Fuchsian group of signature $(g; v_1, \ldots, v_k; t)$ such that $g = 0$ and $k + t = 3$.]*

(ii) *If, in addition to* (i), $\dim A_q(\Gamma) < \infty$ *for some $q \geqslant 44$, then Ahlfors' theorem is valid.*

*Proof.* (i) Let $\tilde{\Gamma}$ be the Fuchsian group corresponding to $S_i$ (so that $S_i = \mathscr{D}/\tilde{\Gamma}_i$). For Fuchsian groups, we know that $\dim A_2(\tilde{\Gamma}_i) < \infty$ *if and only if Area*$(\mathscr{D}/\tilde{\Gamma}_i) < \infty$, and this is true if and only if $S_i$ has finite signature. Since $A_2(\tilde{\Gamma}_i)$ is isomorphic to a subspace of $A_2(\Gamma)$, $\dim A_2(\tilde{\Gamma}_i) \leqslant \dim A_2(\Gamma) < \infty$. This shows that every component has finite signature.

As in the proof of Theorem 2.2.1, $A_2(\tilde{\Gamma}_i) \cong A_2(\Gamma_i)$ (where $\Gamma_i$ is the stabilizer of $\Omega_i$) and

$$\bigoplus_{i=1}^{N} A_2(\Gamma_i) \subset A_2(\Gamma).$$

Thus,

$$\sum_{i=1}^{\infty} \dim A_2(\tilde{\Gamma}_i) \leqslant \dim A_2(\Gamma) < \infty$$

or

$$\sum \left\{ 3(g_i - 1) + t_i + \sum_{j=1}^{k_i} \left[ 2\left(1 - \frac{1}{v_{ij}}\right) \right] \right\} < \infty,$$

where $\tilde{\Gamma}_i$ has signature $(g_i; v_{il}, \ldots, v_{ik_i}; t_i)$. However, $[2(1 - (1/v))] = 1$, so

$$\sum_{i=1}^{\infty} \{3(g_i - 1) + t_i + k_i\} < \infty.$$

All terms in the series are $\geqslant 0$, and $3(g_i - 1) + t_i + k_i = 0$ only if $\tilde{\Gamma}_i$ is a triangle group. This proves the second statement of part (i).

(ii) We need only show that finitely many components $S_i$ correspond to triangle groups. As in (i), we can show that

$$\sum_{i=1}^{\infty} \dim A_q(\tilde{\Gamma}_i) \leqslant \dim A_q(\Gamma) < \infty.$$

We shall show that each group among the $\tilde{\Gamma}_i$ gives a positive contribution ($\geqslant 1$) to the sum.

Suppose $\tilde{\Gamma}_i$ is a Fuchsian group (of genus $g_i \geqslant 0$). By Lemma 2.2.3,

$$\dim A_q(\tilde{\Gamma}_i) \geqslant \frac{(q-1)}{2\pi} \operatorname{Area}(\mathscr{D}/\tilde{\Gamma}_i) + g_i - 1$$

$$\geqslant \frac{43}{2\pi} \frac{\pi}{21} - 1 > 0.$$

(Of course, since $\dim A_q(\tilde{\Gamma}_i)$ is an integer, it is $\geqslant 1$.) ■

### 2.3 Cohomology

Let $\Pi_{2q-2}$ denote the vector space of polynomials $p(z)$ (with complex coefficients) such that $\deg p(z) \leqslant 2q - 2$. If $\gamma \in \Gamma, p \in \Pi_{2q-2}$, let

$$(p\gamma)(z) := p(\gamma z)\, \gamma'(z)^{1-q}.$$

Thus, if

$$p(z) = \sum_{k=0}^{2q-2} \alpha_k z^k,$$

and

$$\gamma(z) = \frac{az+b}{cz+d},$$

then

$$(p\gamma)(z) = \sum_{k=0}^{2q-2} \alpha_k \left(\frac{az+b}{cz+d}\right)^k (cz+d)^{2q-2},$$

so that $p\gamma \in \Pi_{2q-2}$, and $\Pi_{2q-2}$ is a $\Gamma$-module.

A mapping $\chi: \Gamma \to \Pi_{2q-2}$ is a *cocycle* if $\chi_{\gamma_1 \circ \gamma_2} = \chi_{\gamma_1}\gamma_2 + \chi_{\gamma_2}$, where $\gamma_1, \gamma_2 \in \Gamma$, and $\chi_\gamma = \chi(\gamma)$. $\chi: \Gamma \to \Pi_{2q-2}$ is a *coboundary* if $\chi_\gamma = p\gamma - p$ for some $p \in \Pi_{2q-2}$. The sets $Z^1(\Gamma, \Pi_{2q-2})$ of cocycles and $B^1(\Gamma, \Pi_{2q-2})$ of coboundaries are complex vector spaces. The (first) *cohomology space* is

$$H^1(\Gamma, \Pi_{2q-2}) = Z^1(\Gamma, \Pi_{2q-2})/B^1(\Gamma, \Pi_{2q-2}).$$

2.3.1 LEMMA. *If $\Gamma$ is generated by $N$ elements, then* $\dim H^1(\Gamma, \Pi_{2q-2}) \leqslant (2q-1)(N-1)$, *for* $q \geqslant 2$.

*Proof.* We first show that the map $\Pi_{2q-2} \to B^1(\Gamma, \Pi_{2q-2})$ defined by $p \mapsto p\gamma - p$ is a linear isomorphism. We need only show that it is injective. Suppose that $p\gamma - p = 0$, so that $p(\gamma z)\gamma'(z)^{1-q} = p(z)$. If $p(z_0) = 0$, then $p(\gamma z_0)\gamma'(z_0)^{1-q} = 0$, and either $p(\gamma z_0) = 0$ or $\gamma'(z_0)^{1-q} = (cz_0 + d)^{2q-2} = 0$. In the latter case $\gamma(z_0) = \infty$. If $\gamma(z_0) = \infty = \delta(z_0)$, then $\gamma^{-1}\delta \in \Gamma_{z_0}$ (the stabilizer of $z_0$ in $\Gamma$). Since $\Gamma$ has been assumed non-elementary, $\Gamma_{z_0}$ has infinite index in $\Gamma$. Therefore there are infinitely many $\gamma \in \Gamma$ for which $\gamma(z_0) \neq \infty$ and so $p(\gamma z_0) = 0$. Thus $p$ has infinitely many roots, so $p(z) \equiv 0$. This shows that $\dim B^1(\Gamma, \Pi_{2q-2}) = 2q - 1$.

If $\chi \in Z^1(\Gamma, \Pi_{2q-2})$, then $\chi$ is determined by its values on a set of generators of $\Gamma$. Therefore $\dim Z^1(\Gamma, \Pi_{2q-2}) \leqslant (2q-1)N$, and

$$\begin{aligned} \dim H^1(\Gamma, \Pi_{2q-2}) &= \dim Z^1(\Gamma, \Pi_{2q-2}) - \dim B^1(\Gamma, \Pi_{2q-2}) \\ &\leqslant (2q-1)(N-1). \end{aligned}$$

■

We shall need the following two Lemmas, whose proofs will not be given here.

2.3.2 LEMMA. *Let $v(z)$ be a measurable function on $\mathbb{C}$, which is bounded on compact sets and such that $v(z) = 0(|z|^{2q-4})$, as $z \to \infty$. Let $p(z)$ be a polynomial of degree $2q-2$, with distinct zeros, and let*

$$f(z) = -\frac{p(z)}{\pi}\iint_{\mathbb{C}} \frac{v(\zeta)}{(\zeta - z)\,p(\zeta)}\, d\xi\, d\eta \qquad (\zeta = \xi + i\eta).$$

*Then $f(x)$ is continuous, and*

(i) $f_{\bar{z}} = v$ *(in the distributional sense),*

(ii) $f(z) = O(|z|^{2q-3}\log|z|)$, *as* $z \to \infty$,
(iii) $|f(z) - f(\omega)| \leqslant c|z - \omega|\,|\log|z - \omega||$.

For proof see [**30**].

2.3.3 LEMMA. *If* $\infty \in L(\Gamma)$ *then*

$$\lambda(z) = O\left(\frac{1}{|z|\log|z|}\right)$$

*(as* $z \to \infty$*). If* $\infty \in \Omega(\Gamma)$, *then*

$$\lambda(z) = O\left(\frac{1}{|z|^2}\right).$$

For proof see [**30, 1**].

*Remark.* Let $\psi \in B_q(\Gamma)$. By Lemmas 2.3.2 and 2.3.3, there is a function $f(z)$ such that $f_{\bar{z}} = \bar{\psi}^{2-2q}$ and $f(z) = O(|z|^{2q-2})$. $f(z)$ is called a *potential* of $\psi$. Let $p_\gamma(z) = f(\gamma z)\,\gamma'(z)^{1-q} - f(z)$. Then $\partial/\partial\bar{z}\, p_\gamma = 0$, so $p_\gamma$ is entire. Also $p_\gamma = O(|z|^{2q-2})$, so $p_\gamma \in \Pi_{2q-2}$. It is easy to check that $\{p_\gamma\}$ is a cocycle. Note that $f(z)$ is only unique up to an additive polynomial in $\Pi_{2q-2}$, and $\{p_\gamma\}$ is only defined modulo a coboundary. However the map

$$\beta_q : B_q(\Gamma) \to H^1(\Gamma, \Pi_{2q-2})$$

given by $\beta_q(\psi) = \{p_\gamma\} + B^1(\Gamma, \Pi_{2q-2})$ is well defined. The cocycle $\{p_\gamma\}$ was first written down by Ahlfors in [**1**], and the map $\beta_q$ was used by Bers in [**8**].

2.3.4 LEMMA. *The following are equivalent.*

(i) $\beta_q(\psi) = 0$.
(ii) $\psi$ *has a potential f such that* $f(\gamma z)\,\gamma'(z)^{1-q} = f(z)$.
(iii) $\psi$ *has a potential f such that f restricted to* $L(\Gamma) = 0$.

*Proof.* (i) implies (ii). Let $f_0$ be a potential for $\psi$. Since $\beta_q(\psi) = 0$, there exists $p_0 \in \Pi_{2q-2}$, such that $f_0(\gamma z)\,\gamma'(z)^{1-q} - f_0(z) = p_0(\gamma z)\,\gamma'(z)^{1-q} - p_0(z)$. Put $f(z) = f_0(z) - p_0(z)$.

(ii) implies (iii). Let $f(z)$ satisfy (ii) and let $z_0$ be a fixed point of the loxodromic element $\gamma \in \Gamma$. Then $f(z_0)\,\gamma'(z_0)^{1-q} = f(z_0)$. Since $\gamma'(z_0)^{1-q} \neq 1$

($\gamma'(z_0)$ is the multiplier of $\gamma$) it follows that $f(z_0) = 0$. Since the loxodromic fixed points are dense in $L(\Gamma)$, $f$ restricted to $L(\Gamma) = 0$.

(iii) implies (i). $p_\gamma(z) = f(\gamma(z))\gamma'(z) - f(z)$ vanishes on the limit set. Since $p_\gamma$ is a polynomial and $L(\Gamma)$ is an infinite set, $p_\gamma \equiv 0$. ■

The following Lemma is equivalent to what Ahlfors calls the core of his paper.

2.3.5 LEMMA. *The map $\beta_2 : B_2(\Gamma) \to H^1(\Gamma, \Pi_2)$ is injective.*

*Proof.* By Lemma 2.3.4, we must show that if $\psi$ has a potential $f$ which vanishes on $L(\Gamma)$, then $\psi = 0$.

Let $g(z)$ be a $C^\infty$ function which vanishes in a neighbourhood of $L(\Gamma)$, and such that $g$ and $fg_{\bar{z}}$ are bounded. For any function $F \in A_2(1, \Omega)$ (i.e. an integrable holomorphic function on $\Omega$), Stokes' formula gives

$$\iint_\Omega gF\bar{\psi}\lambda^{-2}\,dx\,dy = \iint_\Omega gFf_{\bar{z}}\,dx\,dy = -\iint_\Omega fg_{\bar{z}}F\,dx\,dy.$$

If we can choose $g$, so that $g$ tends boundedly to $f$ and $fg_{\bar{z}}$ tends boundedly to 0 then $\langle F, \psi\rangle = 0$. By Theorems 2.2.1 and 2.2.2 this implies that $\psi = 0$.

Let $\delta(z)$ denote the (Euclidean) distance from $z$ to $L(\Gamma)$. Part (iii) in Lemma 2.3.2 together with the fact that $f$ is zero on $L(\Gamma)$ imply

(i) $$|f| \leqslant C\delta \log 1/\delta \qquad (\delta \leqslant \delta_0).$$

Ignoring the condition that $g$ be $C^\infty$ for the moment, let $l(t)$ be 0 for $0 \leqslant t \leqslant \varepsilon$, 1 for $t \geqslant 2\varepsilon$, and linear between $\varepsilon$ and $2\varepsilon$. Set

(ii) $$g(z) = l[(\log\log 1/\delta(z))^{-1}] \qquad \text{if} \quad \delta < \delta_0,$$

and $g(z) = 1$ if $\delta \geqslant \delta_0$. For small $\varepsilon$, $g(z)$ is continuous and identically zero near $L(\Gamma)$. If $\delta(z)$ were differentiable, the inequality $|\delta(z) - \delta(\omega)| \leqslant |z - \omega|$ would imply $|\delta_{\bar{z}}| \leqslant 1$, and therefore

$$|g_{\bar{z}}| \leqslant \varepsilon^{-1}\delta^{-1}(\log 1/\delta)^{-1}(\log\log 1/\delta)^{-2}$$

where $g$ is not constant. Since $(\log\log 1/\delta)^{-1} \leqslant 2\varepsilon$ in this region, we obtain

$$|g_{\bar{z}}| \leqslant 4\varepsilon\delta^{-1}(\log 1/\delta)^{-1}$$

and by (i), $|fg_{\bar{z}}| \leqslant 4C\varepsilon$ in $\Omega$.

The argument is concluded by using a smoothing process to replace $\delta$ by a smooth mean value. ■

2.3.6 LEMMA. *If $\Gamma$ is generated by $N$ elements, then* $\dim A_2(\Gamma) \leqslant 3(N-1)$.

*Proof.* Since $\beta_2: B_2(\Gamma) \to H^1(\Gamma, \Pi_2)$ is injective,

$$\dim B_2(\Gamma) \leqslant \dim H^1(\Gamma, \Pi_2) \leqslant 3(N-1).$$

Since $B_2(\Gamma)$ is the dual of $A_2(\Gamma)$, $\dim A_2(\Gamma) = \dim B_2(\Gamma)$. ■

2.3.7 LEMMA. *The map $\beta_q: B_q(\Gamma) \to H^1(\Gamma, \Pi_{2q-2})$ is injective.*

*Proof.* By Lemmas 2.2.4 and 2.3.6 each component $S_i$ of $\Omega/\Gamma$ has finite signature. Let $\bar{S}_i$ denote the compactification of $S_i$. Thus $\bar{S}_i$ is a closed surface.

Suppose $\beta_q(\psi) = 0$, for some $\psi \in B_q(\Gamma)$. By Lemma 2.3.4, $\psi$ has a potential $f$ such that $f(\gamma z)\,\gamma'(z)^{1-q} = f(z)$. Thus $f(z)\,dz^{1-q}$ is invariant.

We shall show that for $\phi \in A_q(\Gamma)$, $\langle \phi, \psi \rangle = 0$. Then Theorem 2.2.1 implies that $\psi = 0$.

$$\langle \phi, \psi \rangle = \iint\limits_{\Omega/\Gamma} \phi\bar{\psi}\lambda^{2-2q}\,dx\,dy = \iint\limits_{\Omega/\Gamma} \phi f_{\bar{z}}\,dx\,dy$$

$$= \sum \iint\limits_{\bar{S}_i} \phi f_{\bar{z}}\,dx\,dy.$$

Thus it suffices to show that $\iint_{\bar{S}_i} \phi f_{\bar{z}}\,dx\,dy = 0$.

Since $\phi(z)\,dz^q$ and $f(z)\,dz^{1-q}$ are invariant, it follows that

$$\phi f\,dz = (\phi\,dz^q)\,(f\,dz^{1-q})$$

is a 1-form on $\bar{S}_i$. We shall replace the 2-form $dx\,dy$ by $dz\,d\bar{z} = 2i\,dx\,dy$. By

Stokes' theorem

$$\iint_{S_i} d(\phi f\, dz) = \int_{\partial S_i} \phi f\, dz = 0.$$

Thus $\iint_{\bar{S}_i} (\phi_{\bar{z}} f + \phi f_{\bar{z}})\, dz\, d\bar{z} = 0$. Since $\phi_{\bar{z}} = 0$, this shows that

$$\iint_{\bar{S}_i} \phi f_{\bar{z}}\, dz\, d\bar{z} = 0. \qquad \blacksquare$$

2.3.8 LEMMA. *If $\Gamma$ is generated by $N$ elements, then*

$$\dim A_q(\Gamma) \leqslant (2q - 1)(N - 1).$$

*Proof.* As for Lemma 2.3.6 $\blacksquare$

Ahlfors' theorem now follows from Lemma 2.2.4 and Lemma 2.3.8.

2.3.9 THEOREM (Bers). *If $\Gamma$ is generated by $N$ elements, then*

$$\text{Area}(\Omega/\Gamma) \leqslant 4\pi(N - 1).$$

*Proof.* Let $\Omega/\Gamma = \bigcup_{i=1}^{n} S_i$. Choose a component $\Omega_i$ in $p^{-1}(S_i)$ and let $\Gamma_i$ be the stabilizer of $\Omega_i$. Then $A_q(\Gamma) = \bigoplus_{i=1}^{n} A_q(\Gamma_i)$. If $S_i$ has signature $(g_i; \nu_{i1}, \ldots, \nu_{ik_i}; t_i)$, then by Lemma 2.2.3,

$$\frac{(q-1)}{2\pi} \text{Area}(S_i) + g_i - 1 \leqslant \dim A_q(\Gamma_i).$$

Thus

$$\sum_{i=1}^{m} \frac{(q-1)}{2\pi} \text{Area}(S_i) + g - 1 \leqslant (2q - 1)(N - 1).$$

Now multiply by $2\pi/(q - 1)$ and let $q \to \infty$. We obtain $\text{Area}(\Omega/\Gamma) \leqslant 4\pi(N - 1)$. $\blacksquare$

2.3.10 THEOREM. *Suppose that $\Gamma$ is generated by $N$ elements and $\Omega/\Gamma$ has $K$*

*components. Then*

(i) $K \leqslant 18(N-1)$.
(ii) $K \leqslant 2(N-1)$ if $\Gamma$ is torsion free.
(iii) $K \leqslant N-1$ if $\Gamma$ is purely loxodromic.

*Proof.* (ii) There are no branching orders. Each $S_i$ has signature $(g_i; - ; t_i)$, and $\text{Area}(S_i) = 2\pi\{2(g_i - 1) + t_i\} \geqslant 2\pi$. Thus

$$2\pi K \leqslant \sum_{i=1}^{K} \text{Area}(S_i) \leqslant 4\pi(N-1),$$

and

$$K \leqslant 2(N-1).$$

(iii) In this case there are no punctures or branching orders.

$$\text{Area}(S_i) = 4\pi(g_i - 1) \geqslant 4\pi, 4\pi K \leqslant 4\pi(N-1),$$

so $K \leqslant N-1$.

(i) An examination of cases shows that for any Fuchsian group $\tilde{\Gamma}$,

$$1 \leqslant \dim A_4(\tilde{\Gamma}) + \dim A_6(\tilde{\Gamma}).$$

Thus

$$K \leqslant \sum_{i=1}^{K} \dim A_4(\tilde{\Gamma}_i) + \sum_{i=1}^{K} \dim A_6(\tilde{\Gamma}_i).$$

But Lemma 2.3.8 implies that

$$\sum_{i=1}^{K} \dim A_4(\tilde{\Gamma}_i) \leqslant 7(N-1)$$

and

$$\sum_{i=1}^{K} \dim A_6(\tilde{\Gamma}_i) \leqslant 11(N-1),$$

so

$$K \leqslant \sum_{i=1}^{K} \dim A_4(\tilde{\Gamma}_i) + \sum_{i=1}^{K} \dim A_6(\tilde{\Gamma}_i) \leqslant 18(N-1). \qquad \blacksquare$$

### 2.4 An algebraic method

There is another way to finish the proof of Ahlfors' theorem after it has been established that $\dim A_2(\Gamma) < \infty$ (Lemma 2.3.6). At that point we need to show that infinitely many triangle groups cannot occur. In order to show this, we may use the following theorems, which are proved by using the Hilbert nullstellensatz and some algebraic number theory.

2.4.1 THEOREM. (Selberg) *Let $\Gamma$ be a finitely generated group of $n \times n$ matrices, with coefficients in a field of characteristic* 0. *Then $\Gamma$ has a subgroup $\Delta$ of finite index with no elements of finite order.*

For proof see [**44, 48**].

*Remark 1.* The theorem is false in characteristic $p > 0$.

*Remark 2.* The theorem shows that a finitely generated, periodic, matrix group is finite (i.e. the Burnside conjecture is true for linear groups. This remains true in arbitrary characteristic.)

2.4.2 THEOREM. *Let $\Gamma$ be a finitely generated group of $n \times n$ matrices, and let $t \neq n$. Then $\Gamma$ has a subgroup $\Delta$ of finite index, such that* trace $(\delta) \neq t$ *for all* $\delta \in \Delta$.

For proof see [**17**].

Note that if $[\Gamma:\Delta] < \infty$, then $\Omega(\Delta) = \Omega(\Gamma) = \Omega$, and $\Omega/\Delta \rightarrow \Omega/\Gamma$ is a finite covering. In fact $\text{Area}(\Omega/\Gamma) = [\Gamma:\Delta]\,\text{Area}(\Omega/\Gamma)$. Thus if the Ahlfors theorem is established for $\Delta$, it will also be true for $\Gamma$.

Theorem 2.4.1 allows us to pass to the case where $\Gamma$ has no elliptic elements. In this case, the only triangle groups which can occur are triply parabolic triangle groups. It is not hard to see that such a triangle group would then be the stabilizer of a component, so that the triangle groups are subgroups of $\Gamma$. Furthermore, any two parabolic triangle groups are conjugate in

PSL(2, $\mathbb{C}$), and therefore they have the same set of traces. Choose some trace $t \neq \pm 2$ which occurs in a parabolic triangle group, and use Theorem 2.4.2 to discard any trace equal to $t$ by dropping down to a subgroup $\Delta$ of finite index. $\Delta$ cannot contain any parabolic triangle group, so Ahlfors' theorem has been established.

## 2.5 Some consequences of Ahlfors' theorem

The first consequence we shall discuss is the existence of a finite-sided fundamental region for $\Gamma$ in the regular set $\Omega$. It is easy to construct such a region if the sides are allowed to be arbitrary (simple) curves. We know that $\Omega/\Gamma = \bigcup_{i=1}^{n} S_i$, where each $S_i$ has finite signature. For each $S_i$, choose a component $\Omega_i$ of $\Omega$ which projects to $S_i$. The projection $\Omega_i \to S_i$ is a (possibly ramified) covering. If we remove branch points, and then cut $S_i$ along finitely many curves so that it becomes simply connected, we may then lift the cut surface up to $\Omega_i$, where it becomes a fundamental region of the stabilizer $\Gamma_i$ of $\Omega_i$. By doing this for all components $S_i$, we obtain a finite-sided fundamental region for $\Gamma$.

The task becomes a little more difficult if we require the fundamental region to have certain properties. For example, we may require its sides to be circular arcs. Beardon and Jørgensen [5] have shown that the Ford polygon is finite-sided. We shall consider the case of the boundary of a Dirichlet region in the upper half-space.

We shall use the notation: $\mathscr{H}^3 = \{(x, y, z) \in \mathbb{R}^3 : z > 0\}$,

$$\overline{\mathscr{H}}^3 = \mathscr{H}^3 \cup \mathbb{C} \cup \{\infty\}, \hat{\mathbb{C}} = \partial\mathscr{H}^3 = \mathbb{C} \cup \{\infty\}.$$

We regard PSL(2, $\mathbb{C}$) as acting in $\overline{\mathscr{H}}^3$, and leaving $\mathscr{H}^3$ and $\hat{\mathbb{C}}$ invariant. $\mathscr{H}^3$ is given the non-Euclidean metric

$$ds^2 = \frac{dx^2 + dy^2 + dz^2}{z^2}.$$

The non-Euclidean distance will be denoted $d(p, q)$. (See Chapter 2 for the geometry in $\mathscr{H}^3$ and the action of PSL(2, $\mathbb{C}$).)

Dirichlet regions are defined as in the 2-dimensional case. Let $z_0$ be a point in $\mathscr{H}$, which is not fixed by any element in $\Gamma$ (different from 1). The

*Dirichlet region* for $\Gamma$, centred at $z_0$, is

$$\mathscr{P} = \{z \in \mathscr{H} : d(z, z_0) \leqslant d(z, \gamma(z_0)), \text{all } \gamma \in \Gamma\}.$$

If $\Gamma$ is discrete, then $\mathscr{P}$ is a fundamental polyhedron for $\Gamma$, which satisfies properties (i)–(iv) in Section 1.2. The faces of $\mathscr{P}$ are polygons contained in some of the non-Euclidean planes (Euclidean hemispheres)

$$\mathscr{S}_\gamma = \{z \in \mathscr{H}^3 : d(z, z_0) = d(z, \gamma(z_0))\}.$$

$S_\gamma$ is the perpendicular bisector of the non-Euclidean segment $[z_0, \gamma(z_0)]$. Since the orbit $\Gamma(z_0)$ accumulates at the limit set $L(\Gamma)$, this is true also of the set $\{S_\gamma : \gamma \in \Gamma\}$.

It is not hard to show that $R = \overline{\mathscr{P}} \cap \Omega$ is a fundamental region (possibly disconnected) for $\Gamma$ in $\Omega$. The fact that the sides $S_\gamma$ of $\mathscr{P}$ accumulate only at $L(\Gamma)$ implies that each connected component of $R$ is a polygon (possibly infinite sided) whose sides are circular arcs.

2.5.1 LEMMA. *Let $\Gamma$ be a Kleinian group, and $K$ a compact subset of $\mathscr{H}^3 \cup \Omega(\Gamma)$. Then only finitely many sides of $\overline{\mathscr{P}}$ intersect* $\Gamma K$.

*Proof.* Only finitely many $\Gamma$-images of $\overline{\mathscr{P}}$ intersect $K$. Furthermore, $\overline{\mathscr{P}} \cap \gamma K \neq \varnothing$ if and only if $\gamma^{-1}\overline{\mathscr{P}} \cap K \neq \varnothing$. Therefore $\overline{\mathscr{P}}$ intersects only finitely many $\Gamma$-images of $K$, say $\gamma_1 K, \ldots, \gamma_n K$. $K' = \bigcup_{i=1}^{n} \gamma_i K$ is a compact subset of $\mathscr{H}^3 \cup \Omega(\Gamma)$, and $\overline{\mathscr{P}} \cap \Gamma K \subset \overline{\mathscr{P}} \cap K'$. Since the sides of $\overline{\mathscr{P}}$ accumulate only at the limit set $L(\Gamma)$, only finitely many sides may intersect $K'$. ■

2.5.2 LEMMA. *Let $\alpha$ be a parabolic transformation, $z_0 \in \mathscr{H}^3$ and $\Sigma$ a horosphere such that $\alpha\Sigma = \Sigma$ and $z_0 \in \Sigma$. Let $\beta = \gamma\alpha\gamma^{-1}$, $d_\alpha = d(z_0, \alpha(z_0))$, $d_\beta = d(z_0, \beta(z_0))$ and $d = \varepsilon d(z_0, \gamma(\Sigma))$, where $\varepsilon$ is $\pm 1$ when $z_0$ is exterior or interior to $\gamma(\Sigma)$. Then*

$$\sinh(d_\beta) = \mathrm{e}^d \sinh(d_\alpha).$$

*Proof.* Let $f(z) = z + 1$, and let $\Sigma_1$ and $\Sigma_2$ be the horospheres $\{z = z_1\}$ and $\{z = z_2\}$. The distances $d_1 = d(p_1, f(p_1))$ and $d_2 = d(p_2, f(p_2))$ are constant for $p_1 \in \Sigma_1, p_2 \in \Sigma_2$. By integrating the metric

$$\mathrm{d}s^2 = \frac{\mathrm{d}x^2 + \mathrm{d}y^2 + \mathrm{d}z^2}{z^2},$$

one finds that $\sinh(d_1) = 1/z_1$ and $\sinh(d_2) = 1/z_2$, while

$$d(\Sigma_1, \Sigma_2) = |\ln z_1/z_2|.$$

Thus $\sinh(d_2) = e^d \sinh(d_1)$, where $d = \varepsilon d(\Sigma_1, \Sigma_2)$, and $\varepsilon$ is $+1$ or $-1$ according as $z_1 > z_2$ or $z_2 > z_1$.

Since any parabolic transformation is conjugate to $f$, the same formula is true if $f$ is an arbitrary parabolic transformation. In this case, $\Sigma_1$ and $\Sigma_2$ are horospheres invariant under $f$, and $\varepsilon$ is $+1$ when $\Sigma_1$ lies inside $\Sigma_2$ and $-1$ when $\Sigma_2$ lies inside $\Sigma_1$.

Now consider $\beta = \gamma\alpha\gamma^{-1}$, as in the statement of the lemma. Note that

$$d_2 = d(z_0, \alpha(z_0)) = d(\gamma(z_0), \gamma\alpha\gamma^{-1}(\gamma z_0)) = d(\gamma(z_0), \beta\gamma(z_0)).$$

Let $\Sigma_1 = \gamma\Sigma$ and let $\Sigma_2$ be the parallel horosphere which passes through $z_0$. Then $d_1 = d(\gamma(z_0), \beta\gamma(z_0)) = d_\alpha$, $d_2 = d(z_0, \beta(z_0)) = d_\beta$, $d = \varepsilon d(z_0, \gamma(\Sigma)) = \varepsilon d(\Sigma_1, \Sigma_2)$ and $\varepsilon$ is $+1$ or $-1$ according as when $\Sigma_1$ lies inside $\Sigma_2$ (which is the same as $z_0$ lying outside $\gamma\Sigma$) or vice versa. ■

We shall use quaternions to describe the action of $\Gamma$ in $\mathscr{H}^3$ (as in Chapter 2 Section 3). Thus we identify the point $(x, y, t) \in \mathscr{H}^3$ with the quaternion $x + yi + tj = z + tj$ (where $z = x + yi$). If $\gamma \in \Gamma$ is represented by the matrix

$$\begin{pmatrix} a & b \\ c & d \end{pmatrix},$$

then

$$\gamma(z + tj) = [a(z + tj) + b][c(z + tj) + d]^{-1} = z^* + t^*j$$

where

$$z^* = \frac{a\bar{c}(|z|^2 + t^2) + a\bar{d}z + b\bar{c}\bar{z} + b\bar{d}}{|cz + d|^2 + |c|^2t^2}$$

$$t^* = \frac{t}{|cz + d|^2 + |c|^2t^2}.$$

2.5.3 LEMMA. *Let $\Gamma$ be a Kleinian group such that*

(i) *the translation $\alpha(z) = z + 1$ belongs to $\Gamma$,*
(ii) *$U = \{x + iy: y > y_0\} \subset \Omega(\Gamma)$, where $y_0 > 0$.*

*Let $\gamma \in \Gamma$ and $\gamma(j) = x + yi + tj$. Then*

(a) $t \leqslant 1$,
(b) $y < y_0 + 1$.

*Proof.* According to the formulae preceding the lemma, $\gamma(j) = z + jt$, where

$$\text{(A)} \qquad z = \frac{a\bar{c} - b\bar{d}}{|c|^2 + |d|^2},$$

and

$$\text{(B)} \qquad t = \frac{1}{|c|^2 + |d|^2}.$$

Equation (A) together with $ad - bc = 1$ imply

$$\text{(C)} \qquad z = \frac{a}{c} - \frac{\bar{d}}{c(|c|^2 + |d|^2)}.$$

Since the translation $\alpha(z) = z + 1$ belongs to $\Gamma$, it follows that either $c = 0$ or $|c| \geqslant 1$ (Chapter 2 Theorem 5.2). If $c = 0$, then $\gamma \in \Gamma_\infty$, the stabilizer of $\infty$. $\Gamma_\infty$ is a Euclidean group which, in the present case, is either the group generated by $\alpha$, or the larger group which also contains a rotation of order 2. In any case, (B) implies that $t \leqslant 1$.

Suppose that $\gamma \in \Gamma \setminus \Gamma_\infty$. $\gamma(\infty) = a/c$ is a limit point of $\Gamma$. Since $U \subset \Omega$, $a/c \notin \Omega$, and equation (C) implies $y < y_0 + 1$. This is valid for $\gamma \in \Gamma_\infty$ also, since $y_0 > 0$. ■

2.5.4 LEMMA. *Same hypothesis as in Lemma 2.5.3.*

(i) *The set $\{t: \gamma(j) = z + tj,\ \gamma \in \Gamma\}$ has no accumulation point in the interval $[0, 1]$ except 0.*
(ii) *The set $\{\gamma(j) = z + tj: t = 1, \gamma \in \Gamma\}$ belongs to finitely many $\Gamma_\infty$-orbits. ($\Gamma_\infty$ is the stabilizer of $\infty$ in $\Gamma$).*

*Proof.* (i) Let $\Sigma$ denote the horosphere $\{z + j : z \in \mathbb{C}\}$. If $\gamma(j) = z + tj$, then $d(j, \gamma^{-1}\Sigma) = d(\gamma(j), \Sigma) = \log(1/t)$. By Lemma 2.5.2, $\beta = \gamma^{-1}\alpha\gamma$ satisfies

$$d(j, \beta(j)) = \frac{1}{t} d(j, \alpha(j)).$$

Since the orbit $\Gamma(j)$ does not accumulate in $\mathscr{H}$, the values $t$ can accumulate only to 0.

(ii) There are only finitely many elements $\beta \in \Gamma$, such that

$$d(j, \alpha(j)) = d(j, \beta(j)).$$

Denote these elements $\alpha_1, \alpha_2, \ldots, \alpha_n$. Some of these may be conjugate to $\alpha$ in $\Gamma$. For each such $i$, choose $\gamma_i \in \Gamma$ so that $\gamma_i^{-1}\alpha\gamma_i = \alpha_i$. Now suppose $\gamma(j) = z + j$, and let $\beta = \gamma^{-1}\alpha\gamma$. Lemma 2.5.2 implies that $\beta = \alpha_i$ (for some $i$). Thus $\alpha_i$ is conjugate to $\alpha$, and $\gamma^{-1}\alpha\gamma = \gamma_i^{-1}\alpha\gamma_i$. Therefore $\gamma\gamma_i^{-1}$ commutes with $\alpha$, which implies that $\gamma\gamma_i^{-1} \in \Gamma_\infty$, $\gamma \in \Gamma_\infty\gamma_i$ and $\gamma(j) \in \Gamma_\infty\gamma_i(j)$. ■

2.5.5 LEMMA. *Let $\gamma(j) = z + tj$, and let $S_\gamma$ be the perpendicular bisector of the non-Euclidean segment $[j, \gamma(j)]$. $S_\gamma$ is a vertical plane if $t = 1$. Otherwise $S_\gamma$ is a (Euclidean)hemisphere with centre $m \in \mathbb{C}$, and radius $r$, where*

(i) $m = \dfrac{z}{1 - t}$

(ii) $r^2 = t(|m|^2 + 1)$.

*Proof.* The first statement is clear. Suppose $S_\gamma$ is a hemisphere with centre $m$ and radius $r$. The reflection in $S_\gamma$ is given by

$$f(\zeta + \tau j) = m + \frac{r^2(\zeta + \tau j - m)}{|\zeta + \tau j - m|^2}.$$

Since $f(j) = \gamma(j) = z + tj$,

$$z + tj = m + \frac{r^2(j - m)}{|j - m|^2} = m\left(1 - \frac{r^2}{|m|^2 + 1}\right) + \frac{r^2 j}{|m|^2 + 1}.$$

Solving for $m$ and $r^2$, we obtain equations (i) and (ii). ■

2.5.6 LEMMA. *Let $\Gamma$ be a Kleinian group such that*

(i) *the translation $\alpha(z) = z + 1$ belongs to $\Gamma$,*

(ii) $U = \{x + iy : y > y_0\} \subset \Omega(\Gamma)$,
(iii) *the point $j$ is not fixed by any element $\gamma \in \Gamma, \gamma \neq 1$.*

*Let $\mathscr{P}$ be the Dirichlet region for $\Gamma$, centred at $j$. Then*

(a) *the set $\{y : \gamma(j) = x + yi + j, \gamma \in \Gamma\}$ has a maximum value $y_{\max}$;*
(b) *there exists $y_1 > 0$, such that if $U_1 = \{x + yi : y > y_1\}$ and $\gamma(j) = x + yi + tj$, then $\gamma\bar{P} \cap U_1 \neq \varnothing$ if and only if $y = y_{\max}$ and $t = 1$;*
(c) *if $\gamma\bar{P} \cap U_1 \neq \varnothing$ then this set is a vertical strip $\{x + yi : r \leqslant x \leqslant s, y > y_1\}$.*

*Proof.* (a) By Lemma 2.5.4 (ii), the set $\{y : \gamma(j) = x + yi + j, \gamma \in \Gamma\}$ is finite, so there is a maximum value. For convenience, we shall assume that $y_{\max} = 0$. (This could be achieved by conjugating $\Gamma$ by a translation.) Thus, if $\gamma(j) = x + yi + j$, then $y \leqslant 0$.

(b) Let $\gamma(j) = x + yi + tj$. By Lemma 2.5.3, $t \leqslant 1$. If $t = 1$, the perpendicular bisector $S_\gamma$ of $[j, \gamma(j)]$ is a vertical plane. If $t < 1$, $S_\gamma$ is a hemisphere whose centre $m$ and radius $r$ are described in Lemma 2.5.5. If $d$ denotes the Euclidean distance from $m$ to $z + tj$, then $d^2 = |z - m|^2 + t^2$. By Lemma 2.5.5, $z = (1 - t)m$, so $d^2 = t^2(|m|^2 + 1)$, while $r^2 = t(|m|^2 + 1)$. Since $t < 1$, $d^2 < r^2$, and $\gamma(j) = z + tj$ lies under the hemisphere $S_\gamma$. Therefore $\gamma P$ is contained under $S_\gamma$ so that $\gamma\bar{P}$ has bounded penetration into $U$.

Suppose $\gamma(j) = x + yi + j$, where $y < 0$. Let $S^+$ and $S^-$ be the perpendicular bisectors of $[\gamma(j), \alpha\gamma(j)]$ and $[\gamma(j), \alpha^{-1}\gamma(j)]$. If $n - \frac{1}{2} \leqslant x < n + \frac{1}{2}$ (where $n$ is an integer) let $T$ be the perpendicular bisector of $[\gamma(j), \alpha^n(j)]$. The vertical planes $S^+$, $S^-$ and $T$ separate $\gamma(j)$ (and $\gamma P$) from $U$. Thus the only regions $\gamma\bar{P}$ which can have infinite penetration into $U$ are those such that $\gamma(j) = x + j$.

Now we want to show that there is a number $y_1$, so that the spherical sides of the regions $\gamma\bar{P}$ intersect $\mathbb{C}$ in the region $y < y_1$. First note that Lemma 2.5.4 (i) implies that there is a value $t_0 < 1$, so that if $\gamma(j) = z + tj$, where $t < 1$, then $t \leqslant t_0$. By Lemma 2.5.3, the points $z$ lie in the region $y < y_0 + 1$. Lemma 2.5.5 (i) says that the centre $m$ of $S_\gamma$ is $m = z/(1 - t)$. Since $t \leqslant t_0$, $m$ lies in the region $y < b$ (where $b = (y_0 + 1)/(1 - t_0)$).

Let us now consider the spherical sides of $\mathscr{P}$. Let $S_\gamma$ be such a side of $\mathscr{P}$, with centre $m = u + iv$ and radius $r$. $\mathscr{P}$ is contained in the region $-\frac{1}{2} \leqslant x \leqslant \frac{1}{2}$. Suppose first that $m$ lies outside the strip $-\frac{1}{2} \leqslant x \leqslant \frac{1}{2}$. Since we are interested in the amount of penetration into $U$, we may suppose that the circle $s_\gamma = \partial S_\gamma$ intersects the set $-\frac{1}{2} \leqslant x \leqslant \frac{1}{2}$, $y > b$. Since $m$ lies under the line $y = b$, one of the corners $(\pm\frac{1}{2}, b)$ must be inside $s_\gamma$. Let $(a, b)$ be this corner. Then

$$(u - a)^2 + (v - b)^2 < r^2 = t(|m|^2 + 1) \leqslant t_0(u^2 + v^2 + 1).$$

From this it follows that

$$\left| m - \frac{a + ib}{1 - t_0} \right|^2 < \frac{t_0}{1 - t_0} + \frac{t_0}{(1 - t_0)^2}(a^2 + b^2).$$

Thus $m$ lies in a bounded region. By Lemma 2.5.5 (ii), $r^2 = t(|m|^2 + 1)$, so $r$ is bounded. Thus, all such circles $s_\gamma$ lie below some line $y = c$.

Now consider a side $S_\gamma$ of $\mathscr{P}$, whose centre $m$ lies in the strip $-\frac{1}{2} \leqslant x \leqslant \frac{1}{2}$. If $m$ lies in the region $y \geqslant 0$, then it lies in the bounded region $-\frac{1}{2} \leqslant x \leqslant \frac{1}{2}$, $0 \leqslant y < b$. Therefore $r$ is bounded. Suppose $m = u + iv$, where $v < 0$. Then

$$v^2 = |m|^2 - u^2 \geqslant |m|^2 - \tfrac{1}{4}.$$

If $|m|^2 - \frac{1}{4} \geqslant r^2$, then $|v| \geqslant r$ so the circle $s_\gamma$ lies below the line $y = 0$. If $|m|^2 - \frac{1}{4} < r^2$, then since $r^2 = t(|m|^2 + 1) \leqslant t_0(|m|^2 + 1)$, it follows that

$$|m|^2 < \frac{t_0 + \frac{1}{4}}{1 - t_0}.$$

Thus $|m|$ is bounded, so $r$ is bounded.

Now consider the sides of $\gamma P$. Let $\gamma(j) = x + yi + tj$. Shifting $\gamma P$ by an element in $\Gamma_\infty$, we may assume that $-\frac{1}{2} \leqslant x \leqslant \frac{1}{2}$. Suppose $t = 1$. If $y < 0$, then $\gamma\mathscr{P}$ is cut off from $U$ by vertical planes. There are only a finite number of $\gamma\mathscr{P}$ with $-\frac{1}{2} \leqslant x \leqslant \frac{1}{2}, y = 0, t = 1$. Each case may be treated by conjugating $\Gamma$ by a translation, so that $\gamma(j)$ is moved to $j$. Then we may use the same argument (and formulas) as in the case of $\mathscr{P}$. Finally suppose $-\frac{1}{2} \leqslant x \leqslant \frac{1}{2}$ and $t < 1$. $\gamma\mathscr{P}$ lies under the sphere $S_\gamma$ (which has already been considered as a possible side of $\mathscr{P}$). Since $m = z/(1 - t)$, $m$ lies in a strip $-d \leqslant x \leqslant d$ (where $d = 1/[2(1 - t_0)]$). The same argument as in the previous paragraph works here to show that the circle $s_\gamma$ lies below a line $y = y_1$.

(c) Let $\gamma(j) = x + j$. Then $\gamma\bar{P}$ intersects $U_1$, since its only barriers are finitely many spherical sides (and no vertical planes). The intersection has no circular edges. Oblique lines are also absent, since the regions $\gamma_1\mathscr{P}$ do not reach $U_1$ if $\gamma_1(j) = x + yi + j$, with $y < 0$.

Therefore the only sides of $\gamma\overline{\mathscr{P}} \cap U_1$ are vertical lines. ■

2.5.7 Theorem. *Let $\Gamma$ be a finitely generated Kleinian group of the second kind, and let $\mathscr{P}$ be a Dirichlet region for $\Gamma$. Then the fundamental region $R = \mathscr{P} \cap \Omega(\Gamma)$ for $\Gamma$ in $\Omega(\Gamma)$ has finitely many sides.*

*Proof.* Let $S = \Omega(\Gamma)/\Gamma$, and let $\phi:\Omega(\Gamma) \to S$ be the canonical projection. According to Ahlfors' theorem, $S$ is a finite union of closed Riemann surfaces with a finite number of punctures. Each puncture has an open neighbourhood $N$ (conformally equivalent to a punctured disc) whose lift $\phi^{-1}(N)$ is a disjoint union of discs, which are each invariant under a parabolic transformation in $\Gamma$. Thus we obtain finitely many such neighbourhoods $N_1, N_2, \ldots, N_t$. $S_0 = S \setminus \bigcup_{i=1}^{t} N_i$ is compact, so Lemma 2.5.1 implies that only finitely many sides of $\bar{\mathscr{P}}$ intersect $\phi^{-1}(S_0)$. We must show that, for each $i$, only finitely many sides of $\bar{\mathscr{P}}$ intersect $\phi^{-1}(N_i)$.

Let $N = N_i$, $U$ a connected component of $\phi^{-1}(N)$, $\alpha$ a primary parabolic element in $\Gamma$ which leaves $U$ invariant, and $p_0$ the centre of the Dirichlet region $\mathscr{P}$. By conjugating $\Gamma$, we may suppose that $U = \{x + iy : y > y_0\}$, $\alpha(z) = z + 1$ and $p_0 = j$. By Lemma 2.5.6 there exists $y_1 > y_0$, so that for each $\gamma \in \Gamma$, the region $U_1 = \{x + iy : y > y_1\}$ either is disjoint from $\gamma\bar{\mathscr{P}}$, or $U_1 \cap \gamma\bar{\mathscr{P}}$ is a strip $\{x + iy : r \leqslant x \leqslant s, y > y_1\}$.

Let $V = \{z = x + iy : z \in U \text{ and } -\frac{1}{2} \leqslant x \leqslant \frac{1}{2}\}$. There can be only finitely many such vertical strips which intersect $V \cap U_1$, because the vertical boundaries accumulate only at the limit set. Only finitely many regions $\gamma\bar{\mathscr{P}}$ intersect $V \setminus U_1$ because this is a relatively compact subset of $\Omega$. Therefore finitely many $\gamma\bar{\mathscr{P}}$ intersect $V$. Since $\gamma\bar{\mathscr{P}} \cap V \neq \varnothing$ if and only if $\bar{\mathscr{P}} \cap \gamma^{-1}(V) \neq \varnothing$, this shows that $\bar{\mathscr{P}}$ intersects finitely many regions $\gamma(V)$, and clearly each such intersection has finitely many sides. Therefore, only finitely many sides of $\bar{\mathscr{P}}$ can meet $\Gamma V = \Gamma U = \phi^{-1}(N)$. ■

2.5.8 THEOREM. *Let $\Gamma$ be a finitely generated, non-elementary group of the second kind, and let $\Delta$ be a finitely generated subgroup which contains a non-trivial subnormal subgroup of $\Gamma$. Then $[\Gamma:\Delta] < \infty$.*

*Proof.* We can show (as in Lemma 1.6.2) that if $\Delta_0$ is a non-trivial normal subgroup of a non-elementary group $\Gamma$, then $L(\Delta_0) = L(\Gamma)$. Since $\Delta$ contains a non-trivial subnormal subgroup, $L(\Delta) = L(\Gamma)$, so that $\Omega(\Delta) = \Omega(\Gamma) = \Omega$. Then $\Omega/\Delta \to \Omega/\Gamma$ is a (possibly ramified) covering, and

$$\text{Area}(\Omega/\Delta) = [\Gamma:\Delta].\text{Area}(\Omega/\Gamma).$$

Since $\Delta$ is finitely generated, $\text{Area}(\Omega/\Delta) < \infty$, and therefore $[\Gamma:\Delta] < \infty$. ■

*Definition.* Let $\Gamma$, $\Delta$ be subgroups of a group $G$. $\Gamma$ and $\Delta$ are *commensurable* ($\Gamma \sim \Delta$) if $\Gamma \cap \Delta$ has finite index in $\Gamma$ and $\Delta$. The *commensurability group* of $\Gamma$

(in $G$) is

$$C(\Gamma) = \{g \in G : g\Gamma g^{-1} \sim \Gamma\}.$$

$C(\Gamma)$ contains all subgroups of $G$ which are commensurable with $\Gamma$. We are interested in the case where $\Gamma$ and $\Delta$ are Kleinian groups and $G = \mathrm{PSL}(2, \mathbb{C})$.

*Remark.* If $\Gamma = \mathrm{PSL}(2, \mathbb{Z})$ (the modular group) and $G = \mathrm{PSL}(2, \mathbb{R})$, it is easy to see that $C(\Gamma) \supset \mathrm{PSL}(2, \mathbb{Q})$, so that $[C(\Gamma):\Gamma] = \infty$, and $C(\Gamma)$ is dense in $\mathrm{PSL}(2, \mathbb{R})$.

2.5.9 THEOREM. *Let $\Gamma$ be a finitely generated non-elementary Kleinian group of the second kind. Suppose that $L(\Gamma)$ is not a circle. Then $[C(\Gamma):\Gamma] < \infty$.*

*Proof.* Let $L = L(\Gamma)$. If $g \in C(\Gamma)$, then $L(g\Gamma g^{-1}) = g(L)$, and since $\Gamma \cap g\Gamma g^{-1}$ has finite index in $\Gamma$ and $g\Gamma g^{-1}$, $L = L(\Gamma) = L(\Gamma \cap g\Gamma g^{-1}) = L(g\Gamma g^{-1}) = g(L)$. Thus $g(L) = L$ for $g \in C(\Gamma)$. Let $\Delta = \{g \in \mathrm{PSL}(2, \mathbb{C}) : g(L) = L\}$. Then $C(\Gamma) \subset \Delta$, so it suffices to show that $[\Delta:\Gamma] < \infty$.

$\Delta$ is a closed subgroup of $\mathrm{PSL}(2, \mathbb{C})$. Therefore the identity component $\Delta_0$ is a connected Lie subgroup of $\mathrm{PSL}(2, \mathbb{C})$. In [**15**] it is shown that a connected Lie subgroup $\Delta_0 \subset \mathrm{PSL}(2, \mathbb{C})$ satisfies one of the following:

(i) $\Delta_0$ has a common fixed point in the upper half-space, or in $\hat{\mathbb{C}} = \mathbb{C} \cup \{\infty\}$.
(ii) $\Delta_0$ is conjugate to $\mathrm{PSL}(2, \mathbb{R})$,
(iii) $\Delta_0 = \mathrm{PSL}(2, \mathbb{C})$.
(iv) $\Delta_0 = \{1\}$.

In the present case, the only possibility is (iv), so that $\Delta$ is discrete. Since $\Delta$ leaves $L$ invariant, it follows that $L(\Delta) = L$, and $\Omega(\Delta) = \Omega(\Gamma) = \Omega$. The projection $\Omega/\Gamma \to \Omega/\Delta$ is a branched covering, and

$$\mathrm{Area}(\Omega/\Gamma) = [\Delta:\Gamma]\ \mathrm{Area}(\Omega/\Delta).$$

Since $\mathrm{Area}(\Omega/\Gamma) < \infty$, $[\Delta:\Gamma] < \infty$. ■

## 2.6 Geometrically finite groups

In this section we shall be interested in Kleinian groups of the first as well as the second kind. As before, $\mathscr{H}^3$ denotes the upper half-space.

*Definition.* A Kleinian group $\Gamma$ is *geometrically finite* if it has a finite sided Dirichlet region in $\mathscr{H}^3$.

We shall soon see that if one Dirichlet region is finite sided, then they all are. If $\Gamma$ is geometrically finite, then it is finitely generated. Unfortunately, the converse is not true.

We shall first prove a theorem about the parabolic fixed points of an arbitrary Kleinian group.

2.6.1 THEOREM. *Let $\Gamma$ be a Kleinian group and $\mathscr{P}$ a Dirichlet region for $\Gamma$. Let $p$ be the fixed point of a parabolic element in $\Gamma$. Then the orbit $\Gamma(p)$ intersects $\partial\mathscr{P}$ in a finite, non-empty set.*

*Proof.* First we show that the intersection is not empty. Suppose that $\mathscr{P}$ is centred at $z_0$. Let $\Sigma$ be the horosphere which passes through $z_0$ and is tangent to $\hat{\mathbb{C}}$ at $p$. Let $B$ denote the open ball bounded by $\Sigma$, and let $E = \mathscr{H}^3 \setminus B$. We claim that $p \in \partial\mathscr{P}$ if and only if $\Gamma(z_0) \subset E$.

If $x, y \in \Sigma$ and $S$ is the non-Euclidean plane which is the perpendicular bisector of the segment $[x, y]$, then $\bar{S}$ passes through $p$. Now suppose $\gamma(z_0) \in B$. The segment $[z_0, \gamma(z_0)]$ may be extended to meet $\Sigma$ in a point $z$. Let $S$ be the perpendicular bisector of $[z_0, z]$, and recall that $S_\gamma$ is the perpendicular bisector of $[z_0, \gamma(z_0)]$. Since $S$ and $S_\gamma$ are perpendicular to the same non-Euclidean line, $\bar{S} \cap \bar{S}_\gamma = \varnothing$. Since $p \in \bar{S}$, we see that $S_\gamma$ separates $p$ from $\mathscr{P}$, so $p \notin \partial\mathscr{P}$. A similar argument shows that if $\gamma(z_0) \in E$, then $S_\gamma$ does not separate $p$ from $z_0$. Therefore, $p \in \partial\mathscr{P}$ if and only if $\Gamma(z_0) \subset E$.

There is a horosphere $\Sigma_0$, contained in $B$, which does not intersect any of its $\Gamma$-images. (This is the corollary of Chapter 2, Theorem 5.2). Let $\Gamma_p$ be the stabilizer of $p$ in $\Gamma$, and $B_0$ the open ball bounded by $\Sigma_0$. If $B_0 \cap \gamma B_0 \neq \varnothing$ (for $\gamma \in \Gamma$) then $\gamma \in \Gamma_p$. Therefore, if $B_0$ contains an element $\gamma(z_0)$, then $\Gamma(z_0) \cap B_0 = \Gamma_p\gamma(z_0)$. The points $\Gamma_p\gamma(z_0)$ all lie on the same horosphere. Thus there is a horosphere $\Sigma_1$ at $p$ (which bounds the open ball $B_1$) such that $\Gamma(z_0) \cap B_1 = \varnothing$, but $\gamma(z_0) \in \Sigma_1$ for some $\gamma \in \Gamma$. Let $\mathscr{P}_\gamma$ be the Dirichlet region centred at $\gamma(z_0)$. By the previous paragraph, $p \in \partial\mathscr{P}_\gamma$. But $\mathscr{P}_\gamma = \gamma\mathscr{P}$, so $\gamma^{-1}(p) \in \partial\mathscr{P}$.

Next we show that $\Gamma(p) \cap \partial\mathscr{P}$ is finite. We may assume that $p \in \partial\mathscr{P}$. Suppose that $\gamma(p) \in \partial\mathscr{P}$, where $\gamma \in \Gamma$. Let $\alpha$ be a parabolic element in $\Gamma$ such that $\alpha(p) = p$, and let $\beta = \gamma\alpha\gamma^{-1}$. For any $\delta \in \Gamma$ $(\delta \neq 1)$, let $S_\delta$ be the perpendicular bisector of $[z_0, \delta(z_0)]$, and let $H_\delta^+$ and $H_\delta^-$ be the connected components of $\mathscr{H}^3 \setminus S_\delta$ such that $z_0 \in H_\delta^+$ and $z_0 \notin H_\delta^-$. Thus $\mathring{\mathscr{P}} = \bigcap_{1 \neq \delta \in \Gamma} H_\delta^+$.
In particular $\mathring{\mathscr{P}} \subset H_{\gamma^{-1}}^+ \cap H_\gamma^+$. Furthermore, since $\gamma[\gamma^{-1}(z_0), z_0] = [z_0, \gamma(z_0)]$,

it follows that $\gamma S_{\gamma-1} = S_\gamma$ and $\gamma H^+_{\gamma-1} = H^-_\gamma$. Now since the fixed point $p \in \partial\mathscr{P}$ and $\gamma(p) \in \partial\mathscr{P}$, it follows that $p \in \partial S_{\gamma-1}$ and $\gamma(p) \in \partial S_\gamma$.

Now let $\Sigma$ be the horosphere at $p$ which passes through $z_0$, and let $r$ be the reflection in $S_\gamma$. Then $r(\gamma(z_0)) = z_0$, since $S_\gamma$ is the perpendicular bisector of $[z_0, \gamma(z_0)]$. Also $r(\gamma\Sigma) = \gamma\Sigma$, since $\gamma(p) \in \partial S_\gamma$. Thus $z_0 = r(\gamma(z_0)) \in r(\gamma\Sigma) = \gamma(\Sigma)$, so $d(z_0, \gamma(\Sigma)) = 0$. Lemma 2.5.2 now implies that $d(z_0, \alpha(z_0)) = d(z_0, \beta(z_0))$. Since $\Gamma$ is discontinuous in $\mathscr{H}^3$, this can happen for only finitely many elements $\beta \in \Gamma$. ■

2.6.2 THEOREM. *Let $\Gamma$ be geometrically finite. Then*

(i) *$\Gamma$ has a finite number of conjugacy classes of elliptic elements;*
(ii) *$\Gamma$ has a finite number of orbits of parabolic fixed points.*

*Proof.* This is evident from consideration of a finite-sided Dirichlet region, using Theorem 2.6.1 for the parabolic case. ■

*Remark.* The analogous theorem is unknown for the case where $\Gamma$ is merely assumed to be finitely generated.

PARABOLIC CUSPS. *Our aim here is to study the behaviour of $\Gamma$ near a parabolic fixed point. This is preliminary to the discussion of the quotient space $M(\Gamma) = (\mathscr{H}^3 \cup \Omega(\Gamma))/\Gamma$.*

Let $p$ be a parabolic fixed point and $\Gamma_p$ the stabilizer of $p$ in $\Gamma$. Let $\Sigma_p$ be a horosphere at $p$, which does not intersect any of its $\Gamma$-images. $\Sigma_p$ is invariant under $\Gamma_p$. The hyperbolic metric in $\mathscr{H}^3$ induces a metric on the horosphere $\Sigma_p$, so that $\Sigma_p$ becomes isometric with the Euclidean plane. Thus $\Gamma_p$ is isomorphic to a discrete Euclidean group. Each such group is determined algebraically and topologically by a signature $\sigma = (g; v_1, \ldots, v_k; t)$ in the same way as for Fuchsian groups. The possible Euclidean signatures are (0;—;2) (1-generator translation group), (0;2,2;1) (infinite dihedral group), (1;—;0) (2-generator translation group), (0;2,2,2,2;0) and some triangle groups (0;3,3,3;0), (0;2,4,4;0), and (0;2,3,6;0). All of these except (0;—;2) and (0;2,2;1) have compact quotient surfaces.

*Definition.* The parabolic fixed point $p$ has *compact type* if $\Sigma_p/\Gamma_p$ is compact. Otherwise $p$ has *non-compact type.*

*Definition.* Let $p$ be a parabolic fixed point of non-compact type. A *pair of horocycles* at $p$ is a pair of open discs $D_1, D_2$ in $\hat{\mathbb{C}}$, such that

(i) $p \in \partial D_1 \cap \partial D_2$,
(ii) $D_1 \cap D_2 = \varnothing$,
(iii) $\bar{D}_1 \cup \bar{D}_2 - \{p\} \subset \Omega(\Gamma)$.
(iv) $D_1 \cup D_2$ is invariant under $\Gamma_p$.

2.6.3 PROPOSITION. *If $\Gamma$ is geometrically finite, then every parabolic fixed point (for $\Gamma$) of non-compact type has a pair of horocycles.*

*Proof.* Let $\mathscr{P}$ be a finite sided Dirichlet region for $\Gamma$, and let $p$ be a parabolic fixed point. By Theorem 2.6.1, we may assume that $p \in \partial\mathscr{P}$, and the orbit $\Gamma(p)$ intersects $\partial\mathscr{P}$ in a finite set. Denote these points: $p = \gamma_1(p), \gamma_2(p), \ldots, \gamma_n(p)$. Choose a horosphere $\Sigma_1$ at $p$, so that for each $i$, $\Sigma_i = \gamma_i\Sigma_1$ intersects only those sides of $P$ which contain $\gamma_i(p)$ on their boundaries. Let $\Pi_i = \Sigma_i \cap \mathscr{P}$. We claim that $\Pi = \bigcup \gamma_i^{-1}\Pi_i$ is a fundamental region for $\Gamma_p$ in $\Sigma_1$.

Consider the polyhedra $\gamma\mathscr{P}$ which have a vertex at $p$. Each of these polyhedra is adjacent to others along the sides $S$ for which $p \in \partial S$. Because of this and the fact that the $\gamma\mathscr{P}$ do not accumulate in $\mathscr{H}^3$, it follows that $\Sigma_1$ is completely covered by these polyhedra. Since $\gamma\mathscr{P} \cap \Sigma_1 = \gamma\Pi_i$, for some $i$, we see that $\Sigma_1$ is covered by the $\Gamma$-images of the $\Pi_i$. In other words, $\bigcup_i \Pi_i$ represents all points of $\Sigma_1$, mod $\Gamma$. It follows that if $\mathscr{P} \cap \gamma\Sigma_1 \neq \varnothing$, then $\gamma\Sigma_1 = \Sigma_i$, for some $i$. Now suppose $\gamma\mathscr{P} \cap \Sigma_1 \neq \varnothing$. Then $\mathscr{P} \cap \gamma^{-1}\Sigma_1 \neq \varnothing$, so that $\gamma^{-1}\Sigma_1 = \Sigma_i$ for some $i$. Since $\Sigma_i = \gamma_i\Sigma_1$, $\gamma\gamma_i\Sigma_1 = \Sigma_1$. Therefore $\gamma\gamma_i \in \Gamma_p$, $\gamma \in \Gamma_p\gamma_i^{-1}$ and $\gamma\Pi_i \subset \Gamma_p\gamma_i^{-1}\Pi_i$. Since $\Pi = \bigcup_i \gamma_i^{-1}\Pi_i$, this shows that $\Pi$ is a fundamental region for $\Gamma_p$ in $\Sigma_1$.

Since $p$ has non-compact type, $\Pi$ (and therefore one of the regions $\Pi_i$) must be non-compact. This shows that $\gamma_i(p)$ is a parabolic vertex of the region $R = \overline{\mathscr{P}} \cap \mathbb{C}$. (Some of the $\Pi_j$ may be compact, so $\gamma_j(p)$ may be an isolated point. We have shown that $\gamma_i(p)$ is actually a vertex of a polygon.)

We may as well suppose (by change of notation) that $\gamma_i(p)$ is $p$. Consider the vertex cycle $p = \delta_1(p), \ldots, \delta_m(p)$ in $R$. Choose a disc $D_1$, invariant under the parabolic transformations in $\Gamma_p$, and such that for each $i$, $D_i = \delta_i D_1$ intersects only those sides of $R$ which are at the vertex $\delta_i(p)$. Let $E_i = D_i \cap R$, $E = \bigcup \delta_i^{-1}E_i$ and let $\Delta_p$ be the parabolic subgroup of $\Gamma_p$. It is easily seen that $D_1 = \Delta_p E$, so that $D_1 \subset \Omega(\Gamma)$. We have found one horocycle.

If $\Gamma_p$ contains an elliptic element $e$ (of order 2) let $D_0 = eD_1$. Otherwise, there must be some $i$ such that $\gamma_i^{-1}R$ has $p$ as a vertex, and the angle of this region near $p$ is exterior to $D_1$. This follows from the fact that $\Pi = \cup \gamma_i^{-1}\Pi_i$ must be an infinite strip, which terminates at both ends in sectors between two circles. Therefore we can find another horocycle $D_0$ by using the same construction as for $D_1$. ■

Let $p$ be a parabolic fixed point, $\Sigma_p$ a horosphere at $p$ which doesn't intersect any of its $\Gamma$-images, and $B_p$ the open ball bounded by $\Sigma_p$.

*Definition.* If $p$ is a parabolic fixed point of compact type, then $B_p$ will be called a *parabolic neighbourhood* of $p$, and $M_p = B_p/\Gamma_p$ will be called a *cusp* for $p$.

If $p$ has non-compact type, suppose that it has a pair of horocycles $D_1, D_2$. Let $S_1, S_2$ be the non-Euclidean planes such that $\partial S_i = \partial D_i$, let $\bar{H}_i$ be the closed set bounded by $S_i \cup D_i$, and let $H_i = \bar{H}_i \setminus \bar{S}_i$. Note that $B_p \cup H_1 \cup H_2$ is invariant under $\Gamma_p$.

*Definition.* Let $p$ be a parabolic fixed point of non-compact type, which has a pair of horocycles. The set $B_p \cup H_1 \cup H_2$ will be called a *parabolic neighbourhood* of $p$, and $M_p = (B_p \cup H_1 \cup H_2)/\Gamma_p$ will be called a *cusp* for $p$.

There exist finitely generated groups $\Gamma$ which have parabolic fixed points of non-compact type without a pair of horocycles. (Jørgensen has given such an example.) We have not defined parabolic neighborhoods or cusps in those cases.

Suppose that $\Gamma_p$ has no elliptic elements. If $p$ has compact type, then $M_p$ is homeomorphic to a solid torus with a central circle removed. If $p$ has non-compact type, then $M_p$ is homeomorphic to a closed cylinder with its central line removed. (Marden calls this a *cusp cylinder* in Chapter 8).

The following result is analogous to Lemma 1.3.1 for Fuchsian groups.

2.6.4 LEMMA. *Let $\Gamma$ be a Kleinian group. Suppose that every parabolic fixed point has a pair of horocycles. Then each parabolic fixed point $p$ has a parabolic neighbourhood $N_p$ such that*

(i) $N_p \cap N_q = \emptyset$, if $p \neq q$,
(ii) $\gamma N_p = N_{\gamma(p)}$, for $\gamma \in \Gamma$.

*Proof.* We proceed inductively, as in the proof of Lemma 1.3.1. Let $p_1, p_2, \ldots$ be a set of representatives for the $\Gamma$-orbits of parabolic fixed points. Choose a horosphere $\Sigma_1$ at $p_1$, which does not intersect its $\Gamma$-images. Let $B_1$ be the open ball, bounded by $\Sigma_1$. If $p_1$ has compact type, put $N_{p_1} = B_1$ and $N_{\gamma(p_1)} = \gamma N_{p_1}$. If $p_1$ has non-compact type, choose a pair of horocycles $D_1, D_2$ at $p$, such that for $\gamma \in \Gamma$, $(D_1 \cup D_2) \cap \gamma(D_1 \cup D_2) \neq \varnothing$ if and only if $\gamma \in \Gamma_{p_1}$. Now define $H_1, H_2$ as before, and put $N_{p_1} = B_{p_1} \cup H_1 \cup H_2$, $N_{\gamma(p_1)} = \gamma N_{p_1}$. In these cases $N_{\gamma(p_1)} \cap N_{\delta(p_1)} = (\gamma N_{p_1}) \cap (\delta N_{p_1}) = \gamma[N_{p_1} \cap (\gamma^{-1}\delta N_{p_1})]$. Therefore, if $N_{\gamma(p_1)} \cap N_{\delta(p_1)} \neq \varnothing$, then $N_{p_1} \cap (\gamma^{-1}\delta N_{p_1}) \neq \varnothing$, so $\gamma^{-1}\delta \in \Gamma_{p_1}$ and $\gamma(p_1) = \delta(p_1)$.

Now suppose we have chosen the $N_{p_i}$, satisfying the required properties, for $i = 1, 2, \ldots, n$. Put $p = p_{n+1}$ and choose a horosphere $\Sigma_p$ at $p$, which does not intersect its $\Gamma$-images. If $p$ has compact type, then the same argument which was used for Lemma 1.3.1 works here. We can find a smaller horosphere $\Sigma'_p$ which is disjoint from $\Gamma(\bigcup_{i=1}^{n} N_{p_i})$. (Note that the sets $\gamma N_{p_i}$ do not accumulate in $\mathscr{H}^3$, because they are disjoint.) Let $B_p$ be the open ball bounded by $\Sigma'_p$, and put $N_p = B_p$, $N_{\gamma(p)} = \gamma N_p$.

Suppose that $p$ has non-compact type, and let $D_1, D_2$ be a pair of horocycles for $p$. By conjugating the group, we may suppose that $p = \infty$, $\Gamma_\infty$ contains a translation $\alpha(z) = z + 2a$, $D_1 = \{x + iy : y > b\}$ and $D_2 = \{x + iy : y < -b\}$, where $b > 0$. Since $D_1 \cup D_2 \subset \Omega(\Gamma)$, $\bigcup_{i=1}^{n} \Gamma(p_i)$ is contained in the strip $-b \leqslant y \leqslant b$. We may suppose that $D_1$ and $D_2$ have been chosen small enough so that they do not meet any of the horocycles chosen at $\gamma(p_i)$, $1 \leqslant i \leqslant$ n. Then all of these horocycles are contained in the strip $-b \leqslant y \leqslant b$.

Let $\Sigma$ be the horosphere $\{(x, y, z) : z = 1\}$, and let

$$R = \{(x, y, z) : -a \leqslant x \leqslant a, \quad -b \leqslant y \leqslant b, \quad z = 1\}.$$

Only finitely many $N_q$ can intersect $R$ (for $q \in \bigcup_{i=1}^{n} \Gamma(p_i)$). Therefore we may find a horosphere $\Sigma' = \{(x, y, z) : z = c\}$ (for some $c > 1$) which does not intersect any $N_q$. Also, the $N_q$ can penetrate only a bounded distance into the regions $\{(x, y, z) : x > b\}$ and $\{(x, y, z) : x < -b\}$. Therefore we can find $d > b$, so that no $N_q$ intersects the vertical planes $y = \pm d$. Now put $B_p = \{(x, y, z) : z > c\}$, $H_1 = \{(x, y, z) : y > d, z \geqslant 0\}$, $H_2 = \{(x, y, z) : u < -d, z \geqslant 0\}$ and $N_p = B_p \cup H_1 \cup H_2$.

*Definition.* $\mathscr{M}(\Gamma) = (\mathscr{H}^3 \cup \Omega(\Gamma))/\Gamma$.
If every parabolic fixed point of non-compact type has a pair of horocycles, we define

$$\mathscr{M}^*(\Gamma) = \mathscr{M}(\Gamma) \setminus \{\bigcup_p M_p\},$$

where the union is taken over all parabolic fixed points $p$, and $M_p = N_p/\Gamma_p$, where the $N_p$ are parabolic neighbourhoods satisfying the conditions of Lemma 2.6.4.

2.6.5 THEOREM. *Let $\Gamma$ be a Kleinian group, such that every parabolic fixed point of non-compact type has a pair of horocycles. Then the following are equivalent.*

(i) $\Gamma$ *is geometrically finite.*
(ii) $\mathcal{M}^*(\Gamma)$ *is compact.*

*Proof.* (i) implies (ii). Let $\mathcal{P}$ be a finite sided Dirichlet region for $\Gamma$. The only limit points in $\partial\mathcal{P}$ are parabolic vertices. The parabolic neighbourhoods $N_p$ cut off these limit points, and $\bar{\mathcal{P}}\setminus\bigcup_p N_p$ is a compact subset of $\mathcal{H}^3 \bigcup \Omega(\Gamma)$. Let $\phi:\mathcal{H}^3 \cup \Omega(\Gamma) \to \mathcal{M}(\Gamma)$ denote the canonical projection. Then $\mathcal{M}^*(\Gamma) = \phi(\bar{\mathcal{P}}\setminus\cup N_p)$ is compact.

(ii) implies (i). Let $\mathcal{P}$ be a Dirichlet region for $\Gamma$. Lemma 2.5.1 implies that only finitely many sides of $\mathcal{P}$ intersect $\phi^{-1}(\mathcal{M}^*(\Gamma)) = \mathcal{H}^3\setminus\bigcup N_p$. Also, since $\mathcal{M}^*(\Gamma)$ is a compact manifold with boundary, it must have a finite number of boundary components. In particular there can be only finitely many orbits of parabolic fixed points. Thus, it suffices to show that for each parabolic fixed point $p$, only finitely many sides of $\mathcal{P}$ intersect $\Gamma N_p$.

Suppose that $p$ has compact type. $N_p$ is a ball bounded by a horosphere $\Sigma_p$, and $\Sigma_p/\Gamma_p$ is compact. Let $R$ be a compact fundamental region for $\Gamma_p$ in $\Sigma_p$. Only finitely many regions $\gamma P$ intersect $R$. Since $\mathcal{P} \cap \gamma R \neq \emptyset$ if and only if $\gamma^{-1}\mathcal{P} \cap R \neq \emptyset$, $\mathcal{P}$ intersects only finitely many $\gamma R$. Therefore $\mathcal{P}$ intersects only finitely many $\gamma N_p$, and it suffices to show that only a finite number of sides of $\mathcal{P}$ intersect any $\gamma N_p$.

We may as well suppose (by change of notation) that $\gamma N_p = N_p$. $\Gamma_p$ contains a 2-generator parabolic subgroup $\langle\alpha, \beta\rangle$. Let $z_0$ be the centre of the Dirichlet region $\mathcal{P}$. For $\gamma \in \Gamma$, let $S_\gamma$ be the perpendicular bisector of $[z_0, \gamma(z_0)]$. Let $Q$ be the closed region bounded by $S_\alpha$, $S_{\alpha^{-1}}$, $S_\beta$ and $S_{\beta^{-1}}$. $\mathcal{P} \subset Q$, and $\mathcal{P}$ can be adjacent to only those $\gamma\mathcal{P}$ such that $\gamma(z_0)$ belongs to one of the neighbouring regions $\alpha^i\beta^j(Q)$, where $-1 \leqslant i,j \leqslant 1$. If $\gamma(z_0)$ belongs to one of these regions then $\gamma(\mathcal{P}) \subset \bigcup_{-2\leqslant i,j\leqslant 2} \alpha^i\beta^j(Q)$. Let $U = \Sigma_p \bigcap \{\bigcup_{-2\leqslant i,j\leqslant 2} \alpha^i\beta^j(Q)\}$. Since $U$ is compact, only finitely many regions $\gamma\mathcal{P}$ intersect $U$. These regions are the only ones which can be adjacent to $\mathcal{P}$ inside $\Sigma_p$. Therefore $\mathcal{P}$ has only finitely many sides which intersect $N_p$.

Now suppose that $p$ has non-compact type. Let $D_1$, $D_2$ be a pair of horocycles at $p$. We may assume that $p = \infty$, $z_0 = j$, $\Gamma_p$ contains the translation

$\alpha(z) = z + 1$, $D_1 = \{x + iy : y > a\}$ and $D_2 = \{x + iy : y < -a\}$, where $a > 0$. By Lemma 2.5.6 only finitely many sides of $\mathscr{P}$ intersect the horosphere $\Sigma_p$ and the regions $y > a$, $y < -a$. Thus there are only finitely many such intersections with $N_p$. Furthermore, only finitely many regions $\gamma\mathscr{P}$ intersect $N_p$ inside the region $-1 \leqslant x \leqslant 1$. Denote these by $\gamma_1\mathscr{P}, \ldots, \gamma_n\mathscr{P}$. If $\mathscr{P} \cap \gamma N_p \neq \varnothing$, then $\gamma^{-1}\mathscr{P} \cap N_p \neq \varnothing$, and $\gamma^{-1} \in \bigcup_{i=1}^{n} \Gamma_p \gamma_i$. Then $\gamma \in \bigcup_{i=1}^{n} \gamma_i^{-1}\Gamma_p$ and $\gamma N_p = \gamma_i^{-1} N_p$. Thus $\mathscr{P}$ intersects only finitely many regions $\gamma N_p$. ■

2.6.6 Corollary. *If one Dirichlet region for $\Gamma$ is finite sided, then they all are.*

2.6.7 Corollary. *If $\Gamma$ is geometrically finite, then the translation lengths of conjugacy classes of loxodromic elements in $\Gamma$ do not accumulate.*

*Proof.* Since $\mathscr{M}^*(\Gamma)$ is compact there is a compact set $K \subset \mathscr{H}^3 \cup \Omega(\Gamma)$, such that $\mathscr{H}^3 \setminus \bigcup_p N_p = \Gamma K$. Since no loxodromic axis $A_\gamma$ may be contained in any $N_p$, there exists $\delta \in \Gamma$, so that $\delta A_\gamma$ intersects $K$. Note that $\delta A_\gamma = A_{\delta\gamma\delta^{-1}}$ carries the same translation length as $A_\gamma$.

Now let $\lambda_0 > 0$, and consider all loxodromic elements $\gamma \in \Gamma$ with translation length $\lambda_0$. For such $\gamma$, there is a conjugate $\delta\gamma\delta^{-1}$ whose axis intersects $K$. These axes cannot accumulate in $\mathscr{H}^3$. Therefore, if there were infinitely many such conjugacy classes, then the axes would accumulate to a point $z \in K \cap \hat{\mathbb{C}}$. Then $z \in L(\Gamma)$, contradicting $K \subset \mathscr{H}^3 \cup \Omega(\Gamma)$. ■

2.6.8 Corollary. *If $\Gamma$ and $\Delta$ are commensurable Kleinian groups, and $\Gamma$ is geometrically finite, then $\Delta$ is also.*

*Proof.* Let $E = \Gamma \cap \Delta$. Then $[\Gamma : E] < \infty$ and $[\Delta : E] < \infty$. $E$ must have the same parabolic fixed points as $\Gamma$ and $\Delta$, and of the same type (compact or non-compact). Therefore the same set of parabolic neighbourhoods $N_p$ may be used for all three groups. Also these groups have the same limit set, therefore the same regular set $\Omega$. Thus $\mathscr{M}^*(G) = [(\mathscr{H}^3 \cup \Omega) \setminus \bigcup N_p]/G$, for $G = \Gamma, \Delta, E$. The inclusions $E \subset \Gamma$ and $E \subset \Delta$ induce finite (possibly ramified) coverings $\phi : \mathscr{M}^*(E) \to \mathscr{M}^*(\Gamma)$, $\psi : \mathscr{M}^*(E) \to \mathscr{M}^*(\Delta)$. Since $\Gamma$ is geometrically finite, $\mathscr{M}^*(\Gamma)$ is compact, so $\mathscr{M}^*(E)$ is compact. Therefore $\mathscr{M}^*(\Delta)$ is compact and $\Delta$ is geometrically finite. ■

2.6.9 Theorem. *If $\mathscr{H}^3/\Gamma$ has finite volume, then $\Gamma$ is geometrically finite.*

For proof see [**13**, **49**].

2.6.10 THEOREM. *Let $\Gamma$ be a non-elementary geometrically finite group, and $\Delta$ a geometrically finite subgroup which contains a non-trivial subnormal subgroup of $\Gamma$. Then $[\Gamma:\Delta] < \infty$.*

*Proof.* If $\Gamma$ is of the second kind, this follows from Theorem 2.5.8. If $\Gamma$ is of the first kind, then since it is geometrically finite, $\mathscr{H}^3/\Gamma$ must have finite volume. Since $\Delta$ contains a subnormal subgroup of $\Gamma$, $L(\Delta) = L(\Gamma) = \hat{\mathbb{C}}$. Thus $\Delta$ is also of the first kind, and $\mathscr{H}^3/\Delta$ has finite volume, since $\Delta$ is geometrically finite. Since $\text{vol}(\mathscr{H}^3/\Delta) = [\Gamma:\Delta]\,\text{vol}(\mathscr{H}^3/\Gamma)$, $[\Gamma:\Delta] < \infty$. ■

## 2.7 Examples of geometrically infinite groups

The examples we shall describe are obtained by a limiting process. For $n = 0, 1, 2, \ldots$ let $\Gamma_n$ be a non-elementary Kleinian group generated by $\{\gamma_{n1}, \gamma_{n2}, \ldots, \gamma_{nk}\}$, such that there is an isomorphism $\phi_n : \Gamma_0 \to \Gamma_n$ with $\phi_n(\gamma_{0j}) = \gamma_{nj}$ $(1 \leqslant j \leqslant k,\ 1 \leqslant n < \infty)$. Suppose that $\lim_{n\to\infty} \gamma_{nj} = \gamma_j$, and let $\Gamma$ be the group generated by $\{\gamma_1, \gamma_2, \ldots, \gamma_k\}$. Then the following conclusions may be drawn.

1. (Chuckrow [**10**],) The map $\phi : \gamma_{0j} \to \gamma_j$ $(1 \leqslant j \leqslant k)$ extends to an isomorphism $\phi : \Gamma_0 \to \Gamma$.
2. (Jørgensen [**24**], Marden [**37**], Yamamoto [**59**]) $\Gamma$ is discrete.
   We shall also need the following facts.
3. (Marden [**36**]). Let $\Gamma$ be a geometrically finite group generated by $\{\gamma_1, \ldots, \gamma_k\}$. Then each $\gamma_i$ has a neighbourhood $N_i$ in PSL(2, $\mathbb{C}$), such that if $\phi : \Gamma \to \Gamma'$ is a homomorphism for which
   (a) $\phi(\gamma_i) \in N_i$ and
   (b) $\phi(\gamma)$ is parabolic when $\gamma$ is parabolic, then
      (i) $\phi$ is an isomorphism,
      (ii) $\Gamma'$ is discrete, and
      (iii) $\Gamma'$ is geometrically finite.
4. (Maskit [**38**]). If $\Gamma$ is a finitely generated, free Kleinian group of the second kind in which every element is loxodromic (or hyperbolic) then $\Gamma$ is a Schottky group.
5. (Mostow [**41**], Marden [**36**]). Let $\Gamma$ be a Kleinian group such that $\mathscr{H}^3/\Gamma$ has finite volume, and let $\Gamma'$ be a Kleinian group, isomorphic to $\Gamma$. Then $\Gamma'$ is conjugate to $\Gamma$.

*Example 1* (Riley). This is an example of a 2-generator group $\Gamma$ which is not geometrically finite. Let $\Gamma_\alpha$ be the group generated by $\gamma_1 = \begin{pmatrix} 1 & 1 \\ 0 & 1 \end{pmatrix}$ and $\gamma_\alpha = \begin{pmatrix} 1 & 0 \\ \alpha & 1 \end{pmatrix}$ (or more precisely, by their images in PSL(2, $\mathbb{C}$)). For large $|\alpha|$, the isometric circles of $\gamma_\alpha$ have small radius and are situated close to $z = 0$. Therefore $\Gamma_\alpha$ is a Schottky-like group (not Schottky only because it contains parabolic elements). In particular, for large $\alpha$, $\Gamma_\alpha$ is a free Kleinian group of the second kind, which is geometrically finite. Also, the only parabolic elements in $\Gamma_\alpha$ are conjugates of powers of $\gamma_1$ and $\gamma_\alpha$. On the other hand, if $|\alpha| < 1$ $(\alpha \neq 0)$, then $\Gamma_\alpha$ is not discrete. This follows from Chapter 2, Theorem 5.2. If $\Gamma$ is discrete and contains $\begin{pmatrix} 1 & 1 \\ 0 & 1 \end{pmatrix}$ and $\begin{pmatrix} a & b \\ c & d \end{pmatrix}$, then either $c = 0$ or $|c| \geqslant 1$.

Now consider the region $R$ consisting of all $\alpha \in \mathbb{C}$ such that

(i) $\Gamma_\alpha$ is discrete;
(ii) $\Gamma_\alpha$ is geometrically finite;
(iii) $\Gamma_\alpha$ is freely generated by $\gamma_1$ and $\gamma_\alpha$;
(iv) the only parabolic elements in $\Gamma_\alpha$ are conjugates of powers of $\gamma_1$ and $\gamma_\alpha$.

$R$ does not intersect the disc $|\alpha| < 1$, and from this it follows that the boundary $\partial R$ is uncountable. By a countability argument, we can show that there exists $\beta \in \partial R$, such that the only parabolic elements in $\Gamma_\beta$ are conjugates of powers of $\gamma_1$ and $\gamma_\beta$. Facts 1 and 2 show that $\Gamma_\beta$ is discrete and free. Fact 3 shows that $\Gamma_\beta$ is not geometrically finite. (Otherwise $R$ would extend to a neighbourhood of $\beta$.)

*Example 2*. An example of a Kleinian group $\Gamma$ of the first kind, which is not geometrically finite. $\Gamma$ will be finitely generated and of the first kind, but $\mathscr{H}^3/\Gamma$ will have infinite volume. The example is similar to Example 1, but we use Schottky groups. We can find a group $\Gamma$ on the boundary of Schottky space, such that $\Gamma$ contains only loxodromic elements. Then $\Gamma$ is free and discrete, as in Example 1. If $\Omega(\Delta) \neq \varnothing$, then Maskit's theorem (fact 4) would imply that $\Gamma$ is Schottky. This would place $\Gamma$ in the interior of Schottky space, rather than on the boundary. Therefore $\Gamma$ is of the first kind. In this case $\Gamma$ is geometrically finite if and only if $\mathscr{H}^3/\Gamma$ has finite volume. But $\mathscr{H}^3/\Gamma$ cannot have finite volume, because it is free, and by using fact 5.

*Example 3* (Bers–Maskit–Greenberg).

*Definition*. A non-elementary group $\Gamma$ is *degenerate* if $\Omega(\Gamma)$ is connected and simply connected.

Bers and Maskit showed that there are degenerate groups on the boundary of Teichmüller space. This is done in a similar fashion to the above examples, but here we use quasi-Fuchsian groups.

2.7.1 THEOREM. *A degenerate group cannot be geometrically finite.*

*Proof.* Suppose $\Gamma$ is a geometrically finite group of the second kind. We want to show that $\Omega(\Gamma)$ cannot be connected and simply connected. We shall give the proof in the simpler case, where $\Gamma$ is assumed to contain no parabolic elements. (This is actually the case in the Bers–Maskit construction.)

By descending to a subgroup of finite index, if necessary, we may assume that $\Gamma$ has no elliptic elements. Then $\Gamma$ is isomorphic to the fundamental group of $\mathring{\mathcal{M}} = \mathcal{H}^3/\Gamma$. Since $\Omega$ is connected and simply connected, $\Gamma \cong \pi_1(\Omega/\Gamma)$. Since $\Gamma$ has no parabolic elements, $\Omega/\Gamma$ is compact. $\mathcal{M} = (\mathcal{H}^3 \cup \Omega)/\Gamma$ is a 3-manifold with boundary $\partial\mathcal{M} = \Omega/\Gamma$. Since the universal covering space $\mathcal{H}^3 \cup \Omega$ of $\mathcal{M}$ is contractible, the higher homotopy groups $\pi_n(\mathcal{M}) = 0$ $(n \geqslant 2)$. Thus the inclusion $\partial\mathcal{M} \hookrightarrow \mathcal{M}$ induces isomorphisms $\pi_n(\partial\mathcal{M}) \to \pi_n(\mathcal{M})$ for all $n$. By a standard theorem in homotopy theory, this implies that $\partial\mathcal{M}$ and $\mathcal{M}$ are homotopically equivalent, and in particular they have the same Euler characteristic: $\chi(\partial\mathcal{M}) = \chi(\mathcal{M})$.

Now let $\hat{\mathcal{M}}$ be the double of $\mathcal{M}$. Then $\chi(\hat{\mathcal{M}}) = 2\chi(\mathcal{M}) - \chi(\partial\mathcal{M})$. But the Euler characteristic of a closed, orientable 3-manifold is 0. Therefore $2\chi(\mathcal{M}) - \chi(\partial\mathcal{M}) = 0$, and since $\chi(\mathcal{M}) = \chi(\partial\mathcal{M})$, we obtain $\chi(\partial\mathcal{M}) = 0$. But this implies that $\partial\mathcal{M}$ is a torus. Thus $\Omega(\Gamma)$ must be the once-punctured sphere, and $\Gamma$ is a 2-generator parabolic group. This contradicts our assumption that $\Gamma$ is non-elementary (as well as the assumption that $\Gamma$ contains no parabolic elements). ■

*Example 4.* (Jørgensen [**25**]). This is an example of a pair of Kleinian groups of the first kind $\Delta \subset \Gamma$ with the following properties.

(i) $\Delta$ is isomorphic to a Fuchsian group of signature $(1; n; 0)$.
(ii) $\Gamma$ is a 2-generator group such that $\mathcal{H}^3/\Gamma$ is compact.
(iii) $\Delta$ is normal in $\Gamma$.
(iv) $\Gamma/\Delta \cong \mathbb{Z}$.

Thus a finitely generated normal subgroup need not be of finite index.

From this example, we may also find two finitely generated groups which intersect in an infinitely generated group. Let $\gamma \in \Gamma$ be a preimage of the generator of $\Gamma/\Delta \cong \mathbb{Z}$. By choosing their fixed points far enough apart, we may find loxodromic elements $\gamma_1, \gamma_2, \ldots, \gamma_n \in \Gamma$, such that $G = \langle \gamma, \gamma_1, \ldots, \gamma_n \rangle$ is a Schottky group. In particular $G$ is a finitely generated group of the second kind. Since $\gamma \in G$, $G/\Delta \cap G \cong \mathbb{Z}$. Thus $\Delta \cap G$ is a normal subgroup of infinite index in $G$. If $\Delta \cap G$ is finitely generated, then Theorem 2.5.8 would imply that $[G:\Delta \cap G] < \infty$. Therefore $\Delta \cap G$ is not finitely generated.

## 2.8 Notes

Ahlfors [**1**] proved the deepest part of Theorem 2.1.1 when he showed that dim $A_2(\Gamma) < \infty$. Unfortunately, this left the possibility that infinitely many triangle groups might occur as component stabilizers. This gap was filled by Greenberg [**17**], using Theorems 2.4.1 and 2.4.2, and by Bers [**7**], using automorphic forms of higher weight. Theorem 2.3.9 was proved by Bers [**8**], and Theorem 2.3.10 by Ahlfors [**2**], who improved an estimate of Bers. We have already mentioned that Beardon and Jørgensen [**5**] proved Theorem 2.5.7 in the case of the Ford polygon. Theorem 2.5.9 is due to Greenberg [**20**]. In this paper, the following result is also proved. Define the signature $\sigma$ of a Kleinian group of the second kind to be the collection of component signatures: $\sigma = (\sigma_1, \ldots, \sigma_n)$. Then, for most pairs of signatures $\sigma, \tau$ (where $\sigma \neq \tau$), no group of signature $\sigma$ can be commensurable with a group of signature $\tau$. Theorem 2.6.5 was proved by Marden [**36**]. Theorem 2.7.1 is due to Greenberg [**16**]. Theorems 2.5.8, 2.6.1, 2.6.10, and Corollary 2.6.7 do not seem to have appeared in print before. The first appearance of geometrically infinite groups occurs in [**16**], where Greenberg showed that degenerate groups are geometrically infinite. Bers and Maskit [**9**], [**39**] had shown the existence of finitely generated degenerate groups. The most concrete example of a geometrically infinite group is due to Jørgensen [**25**], who actually gives the generators of the group.

## REFERENCES

1. L Ahlfors, Finitely generated Kleinian groups. *Amer. J. Math.* **86** (1964), 413–429.
2. L. Ahlfors, Eichler integrals and the area theorem of Bers. *Mich. Math. J.* **15** (1968), 257–263.
3. L. Ahlfors, The structure of a finitely generated Kleinian group. *Acta Math.* **122** (1969), 1–17.
4. A. F. Beardon, Fundamental domains for Kleinian groups, Ann. of Math. Studies, No. 79 (1974), 31–41.
5. A. F. Beardon and T. Jørgensen, Ford polygons for finitely generated Kleinian groups, *Math. Scand.* **36** (1975), 21–26.

6. L. Bers, Automorphic forms and Poincaré series for infinitely generated Fuchsian groups, *Amer. J. Math.* **87** (1965), 196–214.
7. L. Bers, On Ahlfors' finiteness theorem, *Amer. J. Math.* **89** (1967), 1078–1082.
8. L. Bers, Inequalities for finitely generated Kleinian groups, *J. Analyse Math.* **18** (1967), 23–41.
9. L. Bers, On boundaries of Teichmüller spaces and on Kleinian groups I, *Ann. of Math.* **91** (1970), 570–600.
10. V. Chuckrow, On Schottky groups with applications to Kleinian groups, *Ann. of Math.* **88** (1968), 47–61.
11. W. Fenchel and J. Nielsen, Discontinuous groups of non-Euclidean motions (in preparation).
12. L. R. Ford, "Automorphic Functions". Chelsea, New York 1951.
13. H. Garland and M. S. Raghunathan, Fundamental domains for lattices in rank one semisimple Lie groups, *Ann. of Math.* **92** (1970), 279–326.
14. L. Greenberg, Discrete groups of motions, *Canad. J. Math.* **12** (1960), 414–425.
15. L. Greenberg, Discrete subgroups of the Lorentz group, *Math. Scand.* **10** (1962), 85–107.
16. L. Greenberg, Fundamental polyhedra for Kleinian groups, *Ann. of Math.* **84** (1966), 433–441.
17. L. Greenberg, On a theorem of Ahlfors and conjugate subgroups of Kleinian groups, *Amer. J. Math.* **89** (1967), 56–68.
18. L. Greenberg, Fundamental polygons for Fuchsian groups, *J. Analyse Math.* **18** (1967), 99–105.
19. L. Greenberg, Maximal groups and signatures, Ann. of Math. Stud. No. 79 (1974), 207–226.
20. L. Greenberg, Commensurable groups of Moebius transformations, Ann. of Math. Studies No. 79 (1974) 227–237.
21. M. Heins, Fundamental polygons of Fuchsian and Fuchsoid groups. *Ann. Acad. Sci. Fenn.* (1964), 1–30.
22. A. G. Howson. On the intersection of finitely generated free groups, *J. London Math. Soc.* **29** (1954), 428–434.
23. A. Hurwitz, Über algebraische gebilde mit eindeutigen transformationen in sich, *Math. Ann.* **41** (1893), 403–442.
24. T. Jørgensen, On discrete groups of Moebius transformations, *Amer. J. Math.* **98** (1976), 739–749.
25. T. Jørgensen, Compact 3-manifolds of constant negative curvature fibering over the circle (to appear).
26. A. Karrass and D. Solitar, Note on a theorem of Schreier. *Proc. Amer. Math. Soc.* **8** (1957), 696–697.
27. I. Kra, On cohomology of Kleinian groups. *Ann. of Math.* **89** (1959), 533–556.
28. I. Kra, On cohomology of Kleinian groups II, *Ann. of Math.* **90** (1969) 575–589.
29. I. Kra, Eichler cohomology and the structure of finitely generated Kleinian groups, Ann. of Math Stud. No. 66 (1971), 225–263.
30. I. Kra, "Automorphic Forms and Kleinian Groups". Benjamin, 1972.
31. J. Lehner, Discontinuous groups and automorphic functions, *Amer. Math. Soc. Math. Surveys* **18**.
32. A. M. Macbeath, Discontinuous groups and birational transformations, Proceedings of the Summer School in Mathematics, University of St. Andrews, Queen's College, Dundee, 1961.

33. A. M. Macbeath, On a theorem of Hurwitz, *Proc. Glasgow Math. Ass.* **5** (1961), 90–96.
34. A. M. Macbeath, The classification of non-Euclidean plane crystallographic groups, *Canad. J. Math.* **19** (1967), 1192–1205.
35. A. Marden, On finitely generated Fuchsian groups, *Comm. Math. Helv.* **42** (1967), 81–85.
36. A. Marden, The geometry of finitely generated Kleinian groups, *Ann. Math.* **99** (1974), 383–462.
37. A. Marden, Schottky groups and circles *in* "Contributions to analysis", pp. 273–278. Academic Press, New York and London 1974.
38. B. Maskit, A characterization of Schottky groups, *J. Analyse Math.* **19** (1967), 227–230.
39. B. Maskit, On boundaries of Teichmüller spaces and on Kleinian groups II. *Ann. of Math.* **91** (1970), 607–639.
40. W. Massey, "Algebraic Topology: an Introduction". Harcourt, Brace and World, 1967.
41. G. D. Mostow, Strong rigidity of locally symmetric spaces, Ann of Math. Studies No. 78 (1974).
42. J. Nielsen, Über gruppen linearer transformationen. *Mitt. Math. Gesellsch. Hamburg.* **8** (1940), 82–104.
43. C. H. Sah, Groups related to compact Riemann surfaces, *Acta Math.* **123** (1969), 13–42.
44. A. Selberg, On discontinuous groups in higher-dimensional symmetric spaces, Contributions to function theory, Tata Institute of Fundamental Research, Bombay, 1960, pp. 147–164.
45. O. Schreier, Die untergruppen der freien gruppen. *Abh. Math. Sem. Univ. Hamburg* **5** (1928), 161–183.
46. C. L. Siegel, Some remarks on discontinuous groups. *Ann. of Math.* **46** (1945), 708–718.
47. G. Springer, "Introduction to Riemann Surfaces". Addison–Wesley, 1957.
48. B. A. F. Wehrfritz, "Infinite Linear Groups", Springer, Berlin. 1973.
49. N. Wielenberg, On the fundamental polyhedra of discrete Moebius groups (to appear).
50. H. Yamamoto, A theorem on limits of Kleinian groups. *Tôhoku Math. J.* **27** (1975), 273–283.

# 8. Geometrically Finite Kleinian Groups and their Deformation Spaces†

A. MARDEN
*University of Minnesota, Minneapolis, Minnesota U.S.A.*



Poincaré's formulation of the general theory of Kleinian groups in 1883 suggested a geometric approach to the subject via the fundamental polyhedron analogous to his approach to Fuchsian groups. It is this approach interpreted in terms of the 3-manifold $\mathscr{M}(G)$ associated with a given group $G$ that we have adopted here. However the tools required for developing the subject from this point of view are of contemporary design.

After laying the foundation in Section 1 with a rather detailed review of certain properties of discrete groups of Möbius transformations, the two basic surgical techniques from 3-manifold topology are introduced in Section 2. We will illustrate their application in Section 3 by describing the structure of function groups. Here and in the sequel we will restrict our considerations to

† This work was supported in part by the National Science Foundation

torsion-free groups which are geometrically finite. General results from surgery (Waldhausen; see Section 5.2) assert that any such $\mathcal{M}(G)$ can be constructed starting with a ball and working through a succession of intermediate steps $\mathcal{M}(G_i)$. Each $\mathcal{M}(G_i)$ is homeomorphic to the 3-manifold resulting from the identification of two disjoint regions on $\partial\mathcal{M}(G_{i-1})$. This is merely an existence theorem and it is not known how to determine the $G_i$ or the identifications required. Therefore, joining Klein's constructive approach to Poincaré's, given $G$ we should try to determine which of the topologically possible identifications involving $\partial\mathcal{M}(G)$ can be realized in the context of Kleinian groups. Unfortunately, at present only a few possibilities are known to be either excluded or realizable. In Section 4 the main ones known to be realizable are presented. To illustrate their use we will indicate how to construct all possible topological types (as revealed in Section 3) of function groups. In addition we will show that sometimes at least there is a manifold $\mathcal{M}(H)$ which represents the result of doubling $\mathcal{M}(G)$ across some of the components of $\partial\mathcal{M}(G)$. We will return to this matter in Section 6.6 to obtain a necessary and sufficient condition for such doubling to be realized by a group.

The subject of Section 5 is the Isomorphism Theorem which shows how $G$ is uniquely determined by its action on its ordinary set $\Omega(G)$, in fact more generally, merely by its action on those components of $\Omega(G)$ which are not Euclidean discs. This is basic for the construction in Section 6 of the deformation space $\mathbf{T}(G)$. Using the same circle of ideas we present in Section 6.5 a theorem that says roughly that if a 3-manifold $M$ is homeomorphic to some $\mathcal{M}(G)$ then a group $\Gamma$ of homeomorphisms $M \to M$ can be represented as a group of conformal and anti-conformal automorphisms of some $\mathcal{M}(H)$ resulting from deforming $\mathcal{M}(G)$.

So far the discussion has been based on [20] and contains some elaborations of results there. But Section 6.7 and most of Section 7 are based directly on joint work with Earle and describe some of the results we have obtained. This work has also been a guiding factor for our whole presentation. In Section 6.7 a particular holomorphic embedding of Teichmüller space $\mathbf{T}_g$ into the $(3g-3)$-fold product of the space $\mathbf{T}_1$ for the torus is described. The final Section (7) is devoted to the boundary $\partial\mathbf{T}(G)$ of $\mathbf{T}(G)$. First, the notion of cusp and cusp topology is introduced for general $G$. Then, in Section 7.3 we consider the case where $G$ is a Fuchsian surface group and describe how $\mathbf{T}_g$ and the Bers boundary $\partial\mathbf{T}_g$ fit into $\mathbf{T}(G) \cup \partial\mathbf{T}(G)$. We indicate how from the point of view of $\mathbf{T}(G)$ the Teichmüller modular group acts on $\mathbf{T}_g$ and the cusps on $\partial\mathbf{T}_g$ but do not go into any details of this action. The main theorem concerning compactifying moduli space is stated in Section 7.3; using different methods, a similar result has also been announced by Bers [8]. Our report closes with a brief discussion of Schottky space.

Although we will not discuss Ahlfors' finiteness theorem and Bers' inequality (see Chapter 7) we will indicate briefly how these results fit into the structure built here. For torsion-free geometrically finite groups the Bers' inequality and in fact a sharper form of it can be precisely understood in terms of the inclusion of integral homology groups $H_1(\partial\mathcal{M}(G)) \to H_1(\mathcal{M}(G))$. However, when there is torsion this approach is obstructed by unresolved questions in group theory. Application of topological techniques to the case of finitely generated groups which are not geometrically finite is severely restricted too and for the additional reason of lack of knowledge concerning the nature of the "ends" of $\mathcal{M}(G)$. For further details see [20].

## 1. INTRODUCTION

**1.1** We recall from Chapter 2 that the Möbius transformation

$$z \mapsto \frac{az + b}{cz + d}, \quad ad - bc = 1$$

can be "lifted" from the extended complex plane $\mathbb{C} \cup \{\infty\}$ to become an orientation preserving conformal homeomorphism of upper half 3-space $\mathcal{H}^3$ onto itself. The formula is most elegantly expressed in terms of the quaternionic representation

$$\mathbf{h} = z + t\mathbf{j} = x + y\mathbf{i} + t\mathbf{j} + 0\mathbf{k}$$

of the point $(x, y, t) \in \mathbb{R}^3$. The extension is then

$$\mathbf{h} \mapsto \mathbf{h}' = (a\mathbf{h} + b)(c\mathbf{h} + d)^{-1} = \frac{[(az + b)(\bar{c}\bar{z} + \bar{d}) + t^2 a\bar{c}] + t\mathbf{j}}{|cz + d|^2 + t^2|c|^2},$$

and its differential is

$$d\mathbf{h}' = c^{-1}(c\mathbf{h} + d)^{-1}\, cd\mathbf{h}(ch + d)^{-1} \quad (\text{if } c \neq 0).$$

Therefore since $\mathbf{h}' = z' + t'\mathbf{j}$ where $t' = t|c\mathbf{h} + d|^{-2}$,

$$\frac{|d\mathbf{h}'|}{t'} = \frac{|d\mathbf{h}|}{t}$$

showing invariance of the hyperbolic metric.

A Möbius transformation of $\mathscr{H}^3$ preserves the totality of planes and hemispheres in $\mathscr{H}^3$ which are orthogonal to the bounding plane $\partial\mathscr{H}^3 = \{(x, y, t): t = 0\} = \mathbb{C}$. This collection is precisely the totally geodesic, 2-dimensional submanifolds of $\mathscr{H}^3$ in the hyperbolic metric. The geodesic lines consist of those straight lines and circles in $\mathscr{H}^3$ which are orthogonal to $\partial\mathscr{H}^3$ and this collection too is preserved. Each Möbius transformation with two fixed points in $\mathbb{C} \cup \{\infty\}$ preserves the uniquely determined geodesic line in $\mathscr{H}^3$ which connects them. This line is called the *axis* of the transformation. A point of it is fixed by the transformation if and only if it is elliptic, and in this case, every point is fixed.

Given, $P$, $P'$,$Q \in \mathscr{H}^3$ and a geodesic ray $\gamma'$ emanating from $P'$, there is a uniquely determined Möbius transformation which maps $\mathscr{H}^3$ to itself, $P$ to $P'$, and $Q$ to some point on $\gamma'$.

More generally, in $\mathbb{R}^n \cup \{\infty\}$ a (orientation preserving) *Möbius transformation* is defined to be the conformal homeomorphism resulting from an even number of reflections in $(n - 1)$ spheres and planes. In particular the transformation

$$\mathbf{h}' = (\mathbf{h} - \mathbf{j})(\mathbf{h} + \mathbf{j})^{-1}\mathbf{j}$$

maps $\mathscr{H}^3$ onto the open unit ball $\mathbb{B}$ in $\mathbb{R}^3$ with $\mathbb{C} \cup \{\infty\}$. sent to the 2-sphere $\partial\mathbb{B}$ by stereographic projection. In terms of this, the group of all Möbius transformations which preserve $\mathscr{H}^3$, which coincides with the group of all liftings of Möbius transformations from $\mathbb{C} \cup \{\infty\}$, is conjugate to the group preserving $\mathbb{B}$.

**1.2** The following useful fact is generally known. This formulation is similar to that of Srebro [**32**]; it was prepared in conjunction with [**10**].

*Basic convergence property*. Suppose $\{T_n\}$ is a sequence of distinct Möbius transformations such that the corresponding fixed points $(p_n, q_n)$ converge to $(p, q) \in \mathbb{C} \cup \{\infty\}$. (Possibly $p_n = q_n$ and/or $p = q$). Then there is a subsequence $\{T_k\}$ with one of the following properties.

(i) There exists a Möbius transformation $T$ such that $\lim T_k = T$ uniformly on $\mathscr{H}^- = \mathscr{H}^3 \cup \mathbb{C} \cup \{\infty\}$.

(ii) $\lim T_k(z) = q$ for all $z \neq p$, the convergence being uniform on compact subsets of $\mathscr{H}^- - \{p\}$.

(iii) The same as (ii) but with $p$ and $q$ interchanged.

Typical examples are the sequences $\{z + n\}$, $\{k^n z\}$, $\{z \exp(i/n)\}$.

**1.3** Let $G$ be a group of Möbius transformations.

$G$ is said to be *discrete* if possibility (i) above cannot occur in $G$. That is, if there is no sequence $\{T_k\}$ in $G$ such that $\lim T_k(z) = z$ for all $z \in \mathbb{C} \cup \{\infty\}$, or even for three distinct points $z$. Alternatively, if no sequence of matrices in $SL(2, \mathbb{C})$ corresponding to elements of $G$ converges to the identity matrix.

The point $\xi \in \mathscr{H}^- = \mathscr{H}^3 \cup \mathbb{C} \cup \{\infty\}$ is called a *limit point* of $G$ if for some $\xi' \in \mathscr{H}^-$ there exists a sequence $\{T_k\}$ of distinct elements of $G$ such that $\lim T_k(\xi') = \xi$.

The basic convergence property implies that $G$ is discrete if and only if it has no limit points in $\mathscr{H}^3$, (also, if and only if each compact subset of $\mathscr{H}^3$ meets at most a finite number of its $G$-images). Thus for a discrete group the set $\Lambda(G)$ of limit points lies in $\mathbb{C} \cup \{\infty\}$. It turns out that $\Lambda(G)$ is a closed set and is the closure of the set of fixed points of elements of $G$ (see Chapter 2.3). The complement of $\Lambda(G)$ in $\mathbb{C} \cup \{\infty\}$ is called the *ordinary set* $\Omega(G)$. If $\Omega(G) \neq \varnothing$, for historical reasons $G$ is called a *discontinuous group* or *Kleinian group*. In analogy with Fuchsian groups the terms Kleinian group of the first or second kind are also used to distinguish between the cases $\Omega(G) = \varnothing$, $\Omega(G) \neq \varnothing$.

The groups for which card $\Lambda(G) \leqslant 2$ are called *elementary*. They are completely classified (see Chapter 2.3). If $G$ is Kleinian and card $\Lambda(G) \geqslant 3$, $\Lambda(G)$ is a nowhere dense, perfect set.

**1.4** A fundamental property of discrete groups was recently discovered by Jørgensen.

JØRGENSEN'S INEQUALITY [13]. *If $X$ and $Y$ generate a non-elementary discrete group then*

$$|\operatorname{tr}^2 X - 4| + |\operatorname{tr}(XYX^{-1}Y^{-1}) - 2| \geqslant 1.$$

Here $\operatorname{tr} X$ is the trace of either matrix representation of the coefficients of $X$ (they differ only in sign); the inequality is the same whichever is used.

Jørgensen's inequality is a precise expression of the statement that $X$ cannot get too close to the identity. Among a number of important applications the following two are particularly beautiful. The second completes some earlier work of Siegel.

COROLLARY 1 (Jørgensen [15]). *A non-elementary group is discrete if and only if every two-generator subgroup is discrete.*

*Proof.* Take $Y_1$, $Y_2$, $Y_3$ in the group $G$ to be loxodromic with no common fixed point. Assume there is a sequence $\{X_n\}$ in $G$ approaching the identity. Then for each $n$ at least one of the three groups $\langle X_n, Y_i \rangle$ is non-elementary, i.e. $X_n$ and $Y_i$ have no common fixed point. When $n$ is large this violates the inequality.

COROLLARY 2 (Jørgensen [15]). *A non-elementary group preserving the upper half plane is discrete if and only if every one-generator subgroup is discrete, i.e. if and only if every elliptic element has finite order.*

**1.5** Because a discrete group $G$ acts properly discontinuously on $\mathscr{H}^3 \cup \Omega(G)$ we can form the quotient

$$\mathscr{M}(G) = \Omega(G) \cup \mathscr{H}/G$$

which is an orientable 3-manifold with boundary

$$\partial\mathscr{M}(G) = \Omega(G)/G.$$

Only in the case $G$ is Kleinian is $\partial\mathscr{M}(G) \neq \varnothing$. In addition $\mathscr{M}(G)$ has a conformal structure induced from the natural structure on $\Omega(G) \cup \mathscr{H}^3$.

*Example* 1. $G = \langle z \mapsto kz, k > 1 \rangle$. This notation means that $G$ is (the elementary group) generated by $z \mapsto kz$. In $\mathscr{H}$, $G$ is generated by $\mathbf{h} \mapsto k\mathbf{h}$. Erect two hemispheres orthogonal to the plane $\{(x, y, t): t = 0\}$ along the circles $\{|z| = 1\}$, $\{|z| = k\}$ respectively and let $\mathscr{P}$ denote the compact set in $\mathscr{H}^-$ bounded by these two hemispheres and the annulus $\{1 \leqslant |z| \leqslant k\}$. $\mathscr{M}(G)$ is obtained from $\mathscr{P}$ by identifying the two hemispheres. In this way we see that $\mathscr{M}(G)$ is a *solid torus*. That is, $\mathscr{M}(G)$ is homeomorphic to the compact set bounded by a torus in Euclidean 3-space.

This example is easily elaborated upon as follows. Fix $g$ pairs of circles $\{\gamma_1, \gamma'_1, \ldots, \gamma_g, \gamma'_g\}$ exterior to one another in $\mathbb{C}$. Choose any $g$ Möbius transformations $\{T_i\}$ with the property that $T_1$ maps the exterior of $\gamma_i$ onto the interior of $\gamma'_i$. Then $G = \langle T_1, \ldots, T_g \rangle$ is a free group on $g$ generators and a special kind of Kleinian group called a *classical Schottky group*. For $g \neq 1$ it is non-elementary and $\Omega(G)$ is connected and infinitely connected. Using the above construction we see that $\mathscr{M}(G)$ is a *handlebody of genus g*. That is, $\mathscr{M}(G)$ is homeomorphic to the compact set bounded by a sphere with $g$ handles in Euclidean 3-space.

*Example* 2. $G = \langle z \mapsto z + 1 \rangle$. In $\mathscr{H}^3$, $G$ is generated by the unit translation $\mathbf{h} \mapsto \mathbf{h} + 1$. Let $\mathscr{P}$ denote the closed set $\{(x, y, t)\colon 0 \leqslant x \leqslant 1, t \geqslant 0\}$ in $\mathscr{H}^-$. $\mathscr{M}(G)$ is obtained from $\mathscr{P}$ by identifying its two opposite faces. We see that $\mathscr{M}(G) \cong \{0 < |z| \leqslant 1\} \times (0, 1)$. For the factor $\{0 < |z| \leqslant 1\}$ is (conformally in fact) the quotient of each half-plane $\{(x, t, y)\colon t \geqslant 0,\ y = \tan \pi(c - 1/2), 0 < c < 1\}$ with respect to $G$. The boundary $\partial\mathscr{M}(G)$ is conformally the twice punctured sphere.

*Example* 3. $G = \langle z \mapsto z + 1, z \mapsto z + i \rangle$. Let $\mathscr{P}$ denote the half column $\{(x, y, t)\colon t \geqslant 0,\ 0 \leqslant x \leqslant 1,\ 0 \leqslant y \leqslant 1\}$. Then $\mathscr{M}(G)$ is obtained from $\mathscr{P}$ by identifying each pair of opposite faces. Using the method of Example 2 we see that $\mathscr{M}(G) \cong \{0 < |z| \leqslant 1\} \times \{|z| = 1\}$. $\partial\mathscr{M}(G)$ is a torus.

*Example* 4. $G$ is a finitely generated Fuchsian group of the first kind preserving the upper half plane $\mathscr{H}$. Let $\mathscr{P}_\theta$ denote the half plane in $\mathscr{H}^-$ that intersects $\{(x, y, t)\colon t = 0\}$ along the $x$-axis making the inclination $\theta$ to the positive $y$-axis. Then $G$ acting in $\mathscr{H}^-$ preserves each $\mathscr{P}_\theta$ and $\mathscr{P}_\theta/G$ is conformally equivalent to $\mathscr{P}_0/G = \mathscr{H}/G$. Consequently $\mathscr{M}(G) \cong (\mathscr{H}/G) \times [0, 1]$. The boundary $\partial\mathscr{M}(G)$ is the union of two surfaces. The map $z \mapsto \bar{z}$ induces an anticonformal involution of $\mathscr{M}(G)$ leaving the leaf $\mathscr{P}_{\pi/2}/G$ pointwise fixed and interchanging the two boundary components.

*Example* 5. Riley's knots and links. Riley ([**30**] and personal communication) has shown that the complement with respect to the 3-sphere, $S^3$, of each of several knots and links is homeomorphic to some $\mathscr{M}(G)$. The groups appearing are torsion free (i.e. have no elliptic elements) and of course are discrete, but they are not Kleinian. A few of these knots and links are illustrated below. For these examples, the groups are subgroups of $\mathrm{PSL}(2, \mathcal{O}_d)$, where $\mathcal{O}_d$ denotes the ring of integers in $Q(\sqrt{-d}\,)$, for values of $d$ as indicated.

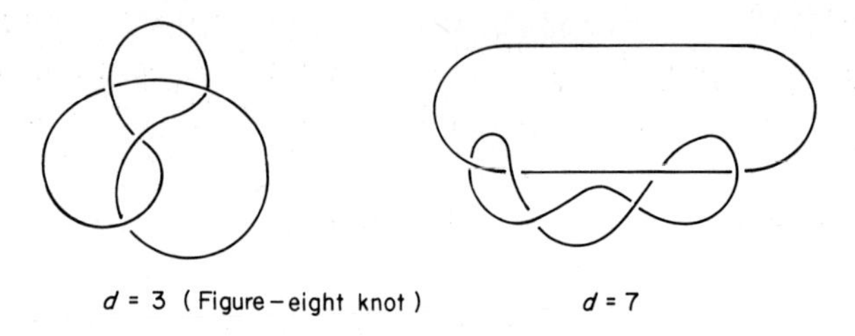


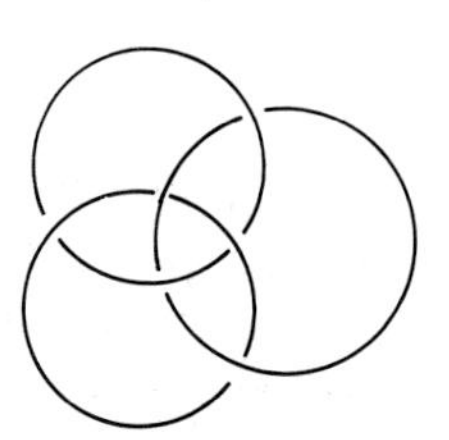


*Figure 1.*

**1.6** Suppose $G$ is a discrete group without torsion. Let $p \in \mathbb{C} \cup \{\infty\}$ be a parabolic fixed point. The *maximal parabolic subgroup*

$$M_p = \{T \in G\colon T(p) = p\}$$

consists entirely of parabolic tranformations and is either infinite cyclic or free abelian of rank two.

We recall from Chapter 2 that a *horosphere* at $p$ is an open Euclidean ball in $\mathscr{H}^3$ internally tangent to $\mathbb{C}$ at $p$, or if $p = \infty$, a half space $\{(x, y, t)\colon t > s\}$. As shown on p. 65, there exists $r > 0$ such that all horospheres $\mathscr{H}_p$ at $p$ of (Euclidean) radius $< r$ or, if $p = \infty$, all half spaces with height $s > r$ have the following two properties:

(i) $T(\mathscr{H}_p^-) = \mathscr{H}_p^-$ for all $T \in M_p$ ($\mathscr{H}_p^-$ denotes closure).

(ii) if $T(\mathscr{H}_p^-) \cap \mathscr{H}_p^- \neq \varnothing$ for some $T \in G$ then $T \in M_p$.

Referring back to Example 3 of Section 1.5 we see that if $M_p$ is of rank 2 then $\mathscr{T} = \mathscr{H}_p/M_p$ is homeomorphic to $\{0 < |z| < 1\} \times S^1$. $\mathscr{T}$ is called a *solid cusp torus*. Its boundary $\partial\mathscr{T}$ is called a *cusp torus*. $\mathscr{T}$ is naturally embedded in $\mathscr{M}(G)$. Note that $\mathscr{T}$ is determined not simply by $M_p$ but by the conjugacy class of $M_p$ in $G$. Once the parameter $r$ is fixed for one group $M_p$, it is determined for each conjugate group. Given any compact set $K$ in $\mathscr{M}(G)$, for all sufficiently small $r$ (or sufficiently large if $p = \infty$) $\mathscr{T} \cap K = \varnothing$. Thus the solid cusp tori corresponding to different conjugacy classes in $G$ of rank two maximal parabolic subgroups may be assumed to be mutually disjoint.

**1.7** The situation when $M_p$ is infinite cyclic is much more complicated. Certainly it is true as in Example 2 of Section 1.5 that $\mathscr{H}_p/M_p$ is homeomorphic to $\{0 < |z| < 1\} \times (0, 1)$. But as it stands this is not terribly useful. Rather we must have further information available. This will be explained below.

Now $\partial\mathscr{M}(G)$ is a union of Riemann surfaces. A *puncture* on a Riemann surface is an isolated ideal boundary component which has a neighbourhood conformally equivalent to the once-punctured disc. Suppose $\xi$ is a puncture on the component $R$ of $\partial\mathscr{M}(G)$ and $\gamma$ is a simple loop from $O \in R$ retractable to $\xi$. Then if $O^* \in \Omega(G)$ is any point over O, the lift of $\gamma$ from $O^*$ terminates at $T(O^*)$ for some parabolic $T \in G$. A different lift will determine a conjugate transformation.

The parabolicity of $T$ can be proved as follows. Suppose $O^*$ lies in the component $\omega$ of $\Omega(G)$. If $\omega$ is simply connected, $\omega/\langle T\rangle$ is an annulus or once-punctured disc (if $G$ is non-elementary). The natural projection $\omega/\langle T\rangle \to R$ has a single valued inverse in a neighbourhood $N \supset \gamma$ of $\xi$ taking it to a neighborhood of one of the boundary components of $\omega/\langle T\rangle$. This boundary component must therefore be a puncture. And this can occur only if $T$ is parabolic (for if $T$ is loxodromic, $\omega/\langle T\rangle$ is embedded in the torus $\mathbb{C} \cup \{\infty\} - \{p,q\}/\langle T\rangle$ where $p$, $q$ are the fixed points of $T$). The same proof works in the general case with $\omega$ replaced by the lift of $N$ in $\omega$ which contains $O^*$.

Because $T$ is parabolic *and* comes from a puncture of $\partial\mathscr{M}(G)$ it has a *horocycle* in $\omega$. That is if $p$ is its fixed point, for all sufficiently small $r$ (or large $r$ if $p = \infty$) there is an open Euclidean disc (or half plane) $D$ of radius $r$ such that

(i) $p \in \partial D$ and $D \subset \omega$,

(ii) $T(D) = D$,

(iii) if $S(D) \cap D \neq \varnothing$ for $S \in G$ then $S$ is a power of $T$.
By taking $D$ a little smaller if necessary $\partial D - \{p\}$ projects to a *circle* in $R$, retractable to $\xi$, and $D$ itself projects to a neighbourhood of $\xi$.

The important notion we have to introduce is that of two punctures $\xi$, $\xi'$ on $\partial\mathscr{M}(G)$ being *paired*. This means that there is in $\mathscr{M}(G)$ a *solid cusp cylinder*

$\mathscr{C}$, homeomorphic to $\{0 < |z| < 1\} \times [0, 1]$, bounded by a *cusp cylinder* $\partial\mathscr{C}$, homeomorphic to $\{|z| = 1\} \times [0, 1]$, with the following property. $\mathscr{C} \cap \partial\mathscr{M}(G)$ is the union of two once-punctured discs, one about each of $\xi, \xi'$.

In particular the two loops $\gamma, \delta$ of $\partial\mathscr{C} \cap \partial\mathscr{M}(G)$, each of which is retractable to one of the punctures, are freely homotopic in $\mathscr{M}(G)$. Consequently, if in the manner described above $\gamma$ determines the parabolic transformation $T \in G$, then there is a lift of $\delta$ which determines $T$ as well.

This in turn implies that there is a double horocycle at the fixed point $p$ of $T$: there are two disjoint discs tangent at $p$, one is a horocycle corresponding to $\xi$, the other to $\xi'$ (see Fig. 2). It is now clear that a third puncture cannot also be paired with $\xi$. The element $T$ is a generator of $M_p$, and the conjugacy class of $M_p$ in $G$ is determined by the pair $(\xi, \xi')$ alone.

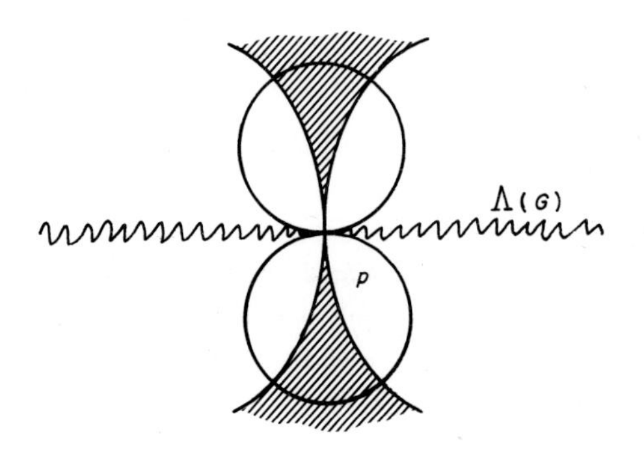


*Figure 2.*

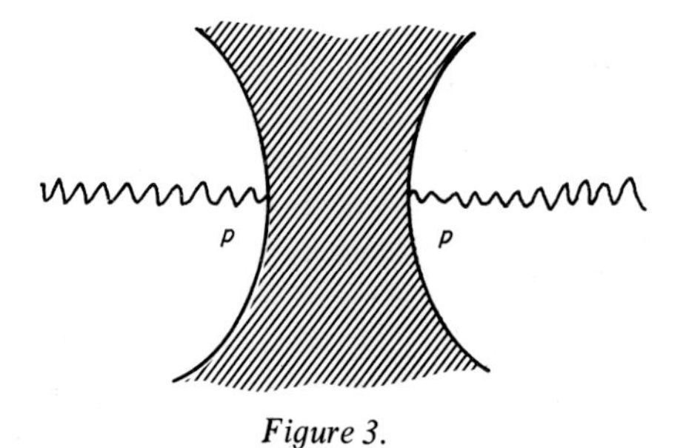


*Figure 3.*

Excising from $\mathscr{M}(G)$ a solid cusp cylinder has the same topological effect on $\partial\mathscr{M}(G)$ as pulling open $p$ in $\Omega(G)$ (Fig. 3).

The solid cusp cylinder $\mathscr{C}$ can be constructed, for example, as follows. We may assume $p = \infty$ and $M_\infty$ is generated by $T: z \mapsto z + 1$. The two horocycles at $\infty$ can be taken as the half planes $\{z: y = \operatorname{Im} z > N\}$, $\{z: y < -N\}$ for sufficiently large $N$. Set $\mathscr{C}^* = \{(x, y, t): y^2 + t^2 > N^2,\ t > 0\}$. Then $\mathscr{C} = \mathscr{C}^*/M_\infty$ is a solid cusp cylinder. Given a compact set $K$ in $\mathscr{M}(G)$, for sufficiently large $N$, $\mathscr{C} \cap K = \varnothing$. Therefore solid cusp cylinders corresponding to two pairs of punctures can be taken mutually disjoint.

**1.8** Given a discrete group $G$, fix a point $O \in \mathscr{H}$, not the fixed point of any (elliptic) element of $G$. For $T \in G$ define

$$E_T = \{x \in \mathscr{H}^3: d(x, O) = d(x, T(O))\} = \text{perpendicular bisector of } [O, T(O)]$$

$$\mathscr{E}_T = \{x: d(x, O) < d(x, T(O))\} = \text{component of } \mathscr{H}^3 - E_T \text{ containing } O.$$

Then

$$\mathscr{P} = \{x \in \mathscr{H}^3: d(O, x) < d(x, T(O)), \text{ for all } T \neq \text{id} \in G\} = \cap \mathscr{E}_T$$

is the (*Poincaré or Dirichlet*) *fundamental polyhedron* for $G$ with centre at O. We refer to Chapter 7, pp. 237–243 for its basic properties. In particular, the closure of $\mathscr{P}$ in $\mathscr{H}^3 \cup \Omega(G)$ with its opposite faces identified provides a model of $\mathscr{M}(G)$.

**1.9** Unless explicitly indicated otherwise, in the remainder of this chapter we will assume that

(1) $G$ is torsion free,
(2) $G$ is non-elementary,
(3) $G$ has a finite sided fundamental polyhedron, i.e. $G$ is *geometrically finite.*

As a consequence of (1), $\pi_1(\mathscr{M}(G))$ is isomorphic to $G$ and as a consequence of (2), no component of $\partial\mathscr{M}(G)$ is a sphere, once- or twice-punctured sphere, or a torus. In addition $\Omega(G)$ has one, two or infinitely many components each of which is either simply or infinitely connected. The most important assumption is usually (3). The reason is as follows.

PROPOSITION [20]. *$G$ is geometrically finite if and only if $\mathscr{M}(G)$ has the following structure. There are a finite number of mutually disjoint solid cusp cylinders $\{\mathscr{C}_i\}$ and solid cusp tori $\{\mathscr{T}_j\}$ such that $\mathscr{M}_0(G) = \mathscr{M}(G) - \cup\mathscr{C}_i - \cup\mathscr{T}_j$ is* *compact. The $\{\mathscr{C}_i\}$ are in one-one correspondence with the cyclic maximal parabolic subgroups, and the $\{\mathscr{T}_i\}$ with the rank two subgroups.*

The proposition implies that if $\mathscr{P}$ has a finite number of sides any reasonably defined fundamental polyhedron does, in particular a Poincaré polyhedron with a different centre. Selberg proved that if $\mathscr{P}$ has finite hyperbolic volume, then $\mathscr{P}$ has a finite number of sides. For a direct proof we refer to Wielenberg [**35**] who also proved a stronger local form of the result. Chapter 7, theorem 2.6.5 contains more details.

All of the examples of Section 1.5 satisfy assumptions (1) and (3). In the case of Riley's groups (Example 5) each curve (component of a link) in $S^3$ corresponds to a solid cusp torus $\mathscr{T}$ in $\mathscr{M}(G)$ (think of it as corresponding to the missing central curve $\{z = 0\} \times S^1$ in $\mathscr{T}$). The interior of $\mathscr{M}(G) - \cup \mathscr{T}_i$ is homeomorphic to the required knot or link complement in $S^3$.

## 2. SURGICAL TECHNIQUES

**2.1** We will frequently shift between $\mathscr{M}(G)$ or $\partial\mathscr{M}(G)$ on the one hand and $\mathscr{H}^3 \cup \Omega(G)$ or $\Omega(G)$ on the other. The natural projection $\mathscr{H}^3 \cup \Omega(G) \to \mathscr{M}(G)$ will be denoted by $\pi$. Usually the correspondence will involve lifting a loop $\gamma \subset \mathscr{M}(G)$ to $\mathscr{H}^3 \cup \Omega(G)$. If the lift begins at O* over the origin of $\gamma$, it terminates at $T(\text{O*})$ for some $T \in G$. We will say that $\gamma$ *determines* $T$. Different lifts of $\gamma$ determine conjugate elements of $G$ so we may also speak of the *conjugacy class determined by* $\gamma$.

The loop $\gamma$ will be called *trivial* in $\mathscr{M}(G)$ (respectively $\partial\mathscr{M}(G)$) if it is contractible in $\mathscr{M}(G)$ (respectively $\partial\mathscr{M}(G)$). Since any map of the closed disc into $\mathscr{M}(G)$ can be lifted, to say that $\gamma$ is trivial is equivalent to saying that one and hence all lifts of $\gamma$ are closed curves in $\mathscr{H} \cup \Omega(G)$, or that each $T$ determined by $\gamma$ is the identity. A simple loop $\gamma \subset \partial\mathscr{M}(G)$ is trivial in $\partial\mathscr{M}(G)$ if and only if it bounds a topological disc there (for a simple proof see [**22**]). Therefore $\gamma$ is non-trivial in $\partial\mathscr{M}(G)$ but trivial in $\mathscr{M}(G)$ if and only if each lift of $\gamma$ is a simple loop dividing $\Omega(G)$ into two parts, neither of which is simply connected. In fact, since the limit set $\Lambda(G)$ is a perfect set, each part is infinitely connected. For examples, consider Schottky groups.

Two loops are freely homotopic in $\mathscr{M}(G)$ if and only if they determine the same conjugacy class in $G$. For $\gamma$ freely homotopic to $\delta$ implies $\gamma$ is homotopic to $\rho\delta\rho^{-1}$ for some arc $\rho$ joining the origin of $\gamma$ to that of $\delta$. Then the lifts of $\gamma$ and $\rho\delta\rho^{-1}$ from the same point O* determine the same element $T \in G$. But if the lift of $\rho$ from O* terminates at $O_1^*$, the lift of $\delta$ from $O_1^*$ also determines $T$. We remark that it is possible to have two loops in $\partial\mathcal{M}(G)$ which are freely homotopic in $\mathcal{M}(G)$ but not in $\partial\mathcal{M}(G)$. For example, consider Fuchsian or Schottky groups. Two disjoint loops in $\partial\mathcal{M}(G)$, non-trivial and freely homotopic there, bound an annular region in $\partial\mathcal{M}(G)$ (see [**22**]).

**2.2** We will state the following two facts in a form directly applicable to Kleinian groups. For details and references see [**20**] (we are, of course, in the piecewise linear situation).

DEHN'S LEMMA AND THE LOOP THEOREM (PAPAKYRIAKOPOULOS)

*Suppose $\gamma \subset \partial\mathcal{M}(G)$ is a loop which is non-trivial in $\partial\mathcal{M}(G)$ but contractible in $\mathcal{M}(G)$. Let $N \supset \gamma$ be an arbitrarily small neighbourhood of $\gamma$. Then there is a simple loop $\gamma_0 \subset N$ which is also non-trivial in $\partial\mathcal{M}(G)$ but bounds a disc $D \subset \mathcal{M}(G)$ with $D \cap \partial\mathcal{M}(G) = \gamma_0$. If $\gamma$ is a simple loop then $\gamma_0$ can be taken to be $\gamma$.*

For Kleinian groups a typical source for $\gamma$ is the projection of a simple loop lying in a non-simply connected component $\omega$ of $\Omega(G)$, separating it into two infinitely connected parts. The Loop Theorem says there is a simple loop $\gamma_0$ in a neighbourhood of $\gamma$ which bounds a disc $D$ in $\mathcal{M}(G)$. Each lift $\gamma_0^*$ of $\gamma_0$, for instance in $\omega$, is simple and bounds a lift $D^*$ of $D$ in $\mathcal{H}$. The natural projection $D^* \cup \gamma_0^* \to D \cup \gamma_0$ is a homeomorphism. $\gamma_0^*$ divides $\omega$ into two infinitely connected parts while $D^*$ divides $\mathcal{H}^3$.

A disc $D$ arising as above also splits $G$ into a (non-trivial) free product. This is seen as follows; there are two cases. Assume first that $\mathcal{N} = \mathcal{M}(G) - D$ is connected. Fix a component $\mathcal{N}^*$ of $\pi^{-1}(\mathcal{N})$ and let $G_1 \cong \pi_1(\mathcal{N})$ denote the subgroup stabilizing it. Given a lift $D^*$ of $D$ which lies in $\partial\mathcal{N}^*$ there is a unique $T \in G$, $T \notin G_1$, such that $\mathcal{N}^*$ and $T(\mathcal{N}^*)$ are adjacent along $D^*$ (they are necessarily disjoint). Examining the action of $G_1$ and $T$ we see that $G = \langle G_1, T\rangle \cong G_1 * \langle T\rangle$. Note too that $\mathcal{N} \subset \mathcal{M}(G_1)$ can be retracted onto $\mathcal{M}(G_1)$; every complementary component of $\mathcal{N}$ in $\mathcal{M}(G_1)^\circ$ is a ball.

If there are two components $\mathcal{N}_1, \mathcal{N}_2$ of $\mathcal{M}(G) - D$, fix a lift $\mathcal{N}_1^*$ of $\mathcal{N}_1$ and a lift $D^* \subset \partial\mathcal{N}_1^*$ of $D$. Let $\mathcal{N}_2^*$ denote the lift of $\mathcal{N}_2$ which is adjacent to $\mathcal{N}_1^*$ along $D^*$. Then $G = \langle G_1, G_2\rangle \cong G_1 * G_2$ where $G_i$ is the subgroup stabilizing $\mathcal{N}_i^*$. As before, $\mathcal{N}_i \subset \mathcal{M}(G_i)$ is retractable onto $\mathcal{M}(G_i)$.

We refer back to the classical Schottky groups of Example 1 in Section 1.5 for an explicit picture of the splitting of a group by disks $D^*$, there realized as hemispheres in $\mathcal{H}^3$. Their projections $D$ not divide $\mathcal{M}(G)$.

**2.3** CYLINDER THEOREM (WALDHAUSEN). *Suppose* $\gamma, \delta \subset \partial\mathcal{M}(G)$ *are disjoint loops which are non-trivial but freely homotopic in* $\mathcal{M}(G)$. *Let* $N_1 \supset \gamma, N_2 \supset \delta$ *be arbitrarily small disjoint neighbourhoods. Then there exist simple loops* $\gamma_0 \subset N_1$, $\delta_0 \subset N_2$, *which are non-trivial in* $\mathcal{M}(G)$, *and a cylinder* $\mathscr{C} \subset \mathcal{M}(G)$ *with* $\mathscr{C} \cap \partial\mathcal{M}(G) = \gamma_0 \cup \delta_0$. *If* $\gamma$ *and/or* $\delta$ *are simple loops then we can take* $\gamma_0 = \gamma$ *and/or* $\delta_0 = \delta$.

A typical source of $\gamma$ and $\delta$ is as follows. Suppose $T \in G$ preserves the dis-distinct components $\Omega_1$, $\Omega_2$ of $\Omega(G)$. Fix $x_i \in \Omega_i$ and take $\gamma$ and $\delta$ to be the projections of an arc in $\Omega_i$ joining $x_i$ to $T(x_i)$, $i = 1, 2$.

If for example $G$ is a Fuchsian group of the first kind then any $T$ will work and $\gamma$, $\delta$ can be taken as geodesics in the two-dimensional hyperbolic metric. If, say, $\gamma$ is a simple loop so must $\delta$ be as well. In this case $\mathscr{C}$ can be taken as the projection of a hemisphere or plane in $\mathscr{H}^3$ (orthogonal to $\mathbb{C}$) depending on whether or not $T$ has a fixed point at $\infty$.

## 3. THE STRUCTURE OF FUNCTION GROUPS

**3.1** The purpose of this chapter is to show the techniques of Section 2 in action. A *function group* is a Kleinian group $G$ for which there is a component $\Omega_0(G)$ of $\Omega(G)$ invariant under $G$. Set $S_0 = \Omega_0(G)/G$ and denote the remaining components of $\partial\mathcal{M}(G)$ by $S_1, \ldots, S_n$. Function groups are characterized by the property that the inclusion $\pi_1(S_0) \to \pi_1(\mathcal{M}(G))$ is surjective, i.e. every loop in $\mathcal{M}(G)$ is freely homotopic to a loop on $S_0$. Because of this property the three-dimensional structure of $\mathcal{M}(G)$ can not only be completely determined, but it is determined merely from $\partial\mathcal{M}(G)$, actually from $S_0$ alone. This state of affairs is in sharp contrast to the general case. In its two-dimensional setting, the geometric description of $G$ is given in Maskit [**26**]. Our description appears in [**20**] to which the reader is referred for more details.

If $G$ is Fuchsian we have seen directly (Example 4, Section 1.5) that $\mathcal{M}(G) \cong S_0 \times [0, 1]$.

**3.2** *Case* 1. If two components $\Omega_0(G)$ and $\Omega_1$ of $\Omega(G)$ are invariant under $G$ then $\mathcal{M}(G) \cong S_0 \times [0, 1]$ and (Maskit [**26**]) $G$ is quasi-Fuchsian. Here we don't need to assume that $G$ is geometrically finite, it is a consequence.

First, each of $\Omega_0, \Omega_1$ is simply connected (Accola [**3**]). For if $\Omega_0$, for example, is not then by Dehn's Lemma and the Loop Theorem there is a disc $D \subset \mathcal{M}(G)$ such that $D \cap S_0$ is a non-trivial simple loop in $S_0$. $D$ must divide $\mathcal{M}(G)$.

Otherwise there would be a loop $\gamma$ with intersection number $\gamma \times D \neq 0$. But then $\gamma$ could not be homologous much less freely homotopic to a loop on $S_1 = \Omega_1/G$ because such would necessarily be disjoint from $D$. So let $\mathcal{N}$ denote the component of $\mathcal{M}(G) - D$ which is not adjacent to $S_1$. Then every loop in $\mathcal{N}$ is freely homotopic to a loop on $S_1$ and consequently to a loop on $D$; that is, every loop in $\mathcal{N}$ in contractible to a point in $\mathcal{N}$. Thus each component of $\pi^{-1}(\mathcal{N})$ is homeomorphic to $\mathcal{N}$. This implies $\mathcal{N} \cap S_0$ is a disc and therefore $\partial D = \partial(\mathcal{N} \cap S_0)$ is trivial in $S_0$, a contradiction.

For simplicity we will outline the remainder of the proof only for the case $S_0$ is a compact surface of genus $g$. Fix a system of loops $\gamma_1, \ldots, \gamma_{2g}$ on $S_0$ so that $\gamma_i$ intersects $\gamma_{i-1}, \gamma_{i+1}$ exactly once, transversely, and is disjoint from the other $\gamma_j$. Then $S_0 - \cup \gamma_i$ is a disc. By the Cylinder Theorem, each $\gamma_i$ together with a simple loop $\gamma_i'$ on $S_1$ bounds a cylinder $\mathscr{C}_i$ in $\mathcal{M}(G)$. Matters can be arranged so that $\mathscr{C}_i$ intersects $\mathscr{C}_{i-1}'$ and $\mathscr{C}_{i+1}$ each transversely in an arc and is disjoint from the other $\mathscr{C}_{i'}$. Let $\mathscr{C}_i = \mathscr{C}_i \times (-\varepsilon, \varepsilon)$ be a thin neighbourhood about $\mathscr{C}_i$. Then $\partial(\cup \mathscr{C}')$ is a cylinder and $S_0 - S_0 \cap (\cup \mathscr{C}_i')$ a disc. As above we find that $S_1 - S_1 \cap (\cup \mathscr{C}_i')$ is also a disc. Therefore, the interior of $\mathcal{M}(G) - \cup \mathscr{C}_i'$, having a sphere contained in its boundary, is a ball. This shows $\mathcal{M}(G) \cong S_0 \times [0, 1]$.

Now if $G_0$ is a Fuchsian surface group of genus $g$ we can take a piecewise linear, quasi-conformal homeomorphism $\mathcal{M}(G_0) \to \mathcal{M}(G)$. The lift of this to $\mathscr{H}^3 \cup \Omega(G_0)$ is also quasi-conformal and therefore by Gehring's extension theorem (see [**11**]) extends to be quasi-conformal on all $\mathbb{C}$. This is the simplest case of the Isomorphism Theorem which will be presented in Section 5.

**3.3** *Case* 2. $G$ is a *B-group*, i.e. $\Omega(G)$ has only one invariant component $\Omega_0(G)$ and this is simply connected. Let $\mathcal{M}_0(G)$ denote the manifold resulting from the removal of all solid cusp cylinders which do not meet $S_0$. A study of the inclusion $H_1(\partial \mathcal{M}_0) \to H_1(\mathcal{M}_0)$ shows $\partial \mathcal{M}_0(G) \neq S_0$. Exactly the same proof as in Case 1 shows that $\mathcal{M}_0(G) \cong S_0 \times [0, 1]$. By the Cylinder Theorem, if there are $n$ pairing cylinders there are $n$ mutually disjoint simple loops $\{\gamma_i\}$ in $S_0$ each of which together with a simple loop around a pairing cylinder bounds a cylinder $\mathscr{C}_i$ in $\mathcal{M}_0(G)$. $\mathscr{C}_i$ divides $\mathcal{M}_0(G)$ if and only if $\gamma_i$ divides $S_0$. Let $M$ be a component of $\mathcal{M}_0(G) - \cup \mathscr{C}_i$ and $M^*$ a component of $\pi^{-1}(M)$. The stabilizer $G_0$ of $M^*$ is quasi-Fuchsian and $\mathcal{M}(G_0)$ can be retracted onto $M$. The transformations in $G$ determined by the loops $\gamma_i$ are called *accidental parabolic* because they are parabolic in $G$ yet act in $\Omega_0(G)$ like hyperbolic transformations in the unit disc, i.e. they are not determined by punctures of $S_0$.

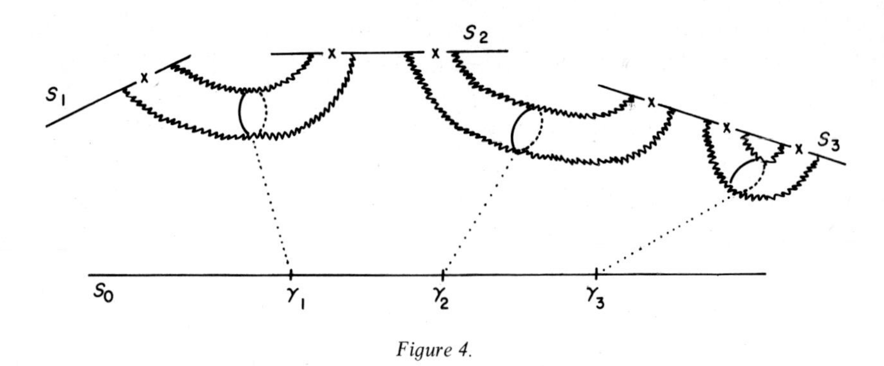


*Figure 4.*

### 3.4 Free groups: Schottky and Schottky-like groups

Before discussing the most general function groups we will pause to give an alternate group-theoretic characterization of two classes of groups which include those already discussed. We will first describe all free groups $G$ (as always, geometrically finite).

Let $\mathcal{M}_0(G)$ denote the compact manifold resulting from the removal of all solid cusp cylinders from $\mathcal{M}(G)$. Suppose first that $\partial\mathcal{M}_0$ is a single surface of genus $g$. Since $G$ is free the inclusion $\pi_1(\partial\mathcal{M}_0) \to \pi_1(\mathcal{M}_0) \cong G$ cannot be injective. Consequently by Dehn's Lemma and the Loop Theorem there exists a simple loop $\gamma_1$, non-trivial in $\partial\mathcal{M}_0$, which bounds a disc $D_1$ in $\mathcal{M}_0$. Assume first that $\gamma_1$ can be chosen so that it does not divide $\partial\mathcal{M}_0$ and therefore $D_1$ does not divide $\mathcal{M}_0$ either. Take a thin neighbourhood $D_1^* = D_1 \times (-\varepsilon, \varepsilon)$ about $D_1$ and set $\mathcal{M}_1 = \mathcal{M}_0 - D_1^*$. $\mathcal{M}_1$ is compact and $\partial\mathcal{M}_1$ is connected of genus $g - 1$. Repeat the procedure with $\mathcal{M}_1$ instead of $\mathcal{M}_0$, etc. always assuming a non-dividing $\gamma_i$ can be found. After exactly $g$ steps we end up with a closed ball $\mathcal{M}_g$.

The proof that $\gamma_1$, $\gamma_2$, etc. can be taken as non-dividing loops follows the same lines. If we cannot find such a $\gamma_1$, for example, then $D_1^*$ either divides $\mathcal{M}_0$ into two parts $M_1, M_2$ neither of which is a ball or $\mathcal{M}_0 - D_1^*$ has two boundary components. The latter case will be disposed of below. In the former, the total genus of $\partial M_1 \cup \partial M_2$ is $g$ and a simple loop in $\partial M_i \cap \partial\mathcal{M}_0$ divides $\partial M_i$ if and only if it divides $\partial\mathcal{M}_0$, $1 = 1, 2$. Furthermore $\pi_1(\mathcal{M}_0) \cong \pi_1(M_1) * \pi_1(M_2)$ so that each of $\pi_1(M_i)$ is also free. After a finite number of steps we obtain $M_n$ with $\pi_1(M_n)$ free and $\partial M_n$ a torus. Since no non-trivial simple loop on a torus is dividing this gives a contradiction.

The proof that $\partial(\mathcal{M}_0 - D_1^*)$ is indeed connected is also along these lines. If it is not, then take one of its components $R$ and apply Dehn's Lemma and the Loop Theorem, but only to $R$ and pieces thereof, successively cutting it down (the fundamental group of every manifold obtained is free). We end up with a submanifold $M$ of $\mathcal{M}_0$ where one component of $\partial M$ is a sphere and another is an additional component of $\partial\mathcal{M}_0$. This is impossible.

Thus, we have proved that *if $G$ is a free group then $\mathcal{M}_0(G)$ is a handlebody of genus $g$*, i.e. homeomorphic to the solid obtained by attaching $g$ solid handles to a closed ball. In particular those groups for which $\Omega(G)$ is connected fall into this category.

A Schottky group $H$ of genus $g$ is one constructed by taking $g$ pairs of mutually exterior, mutually disjoint Jordan curves $\{(\alpha_i, \beta_i)\}$ in $\mathbb{C}$ and postulating the existence of $g$ Möbius transformations $\{T_i\}$ where $T_i$ maps the exterior of $\alpha_i$ onto the interior of $\beta_i$. The group $H$ is defined as $\langle T_1, \ldots, T_g\rangle$. Dehn's Lemma and the Loop Theorem shows that $\mathcal{M}(H)$ is a handlebody of genus $g$. It is easily seen that all Schottky groups arise as quasi-confirmal deformations of classical Schottky groups (see Example 1, Section 1.5).

If $G$ is free and purely loxodromic then there are no pairing cylinders, $\mathcal{M}_0(G) = \mathcal{M}(G)$, and we see very directly that $G$ is a Schottky group (Maskit [**24**]). In fact, under these circumstances, as Maskit has pointed out, it is not even necessary to assume $G$ is geometrically finite—that is a consequence Ahlfors' Finiteness Theorem implies that $\partial\mathcal{M}(G)$ has at most a finite number of punctures and under the assumption $G$ is purely loxodromic, there can be none after all.

In general it can be shown that $G$ can be determined like a Schottky group except that the Jordan curves involved will be tangent at parabolic fixed points and their common exterior may not be connected. A special case is that of any finitely generated Fuchsian group containing parabolic elements.

### 3.5. Surface groups

If $G$ is isomorphic to $\pi_1(S)$ for a compact surface $S$ and $\mathcal{M}_0(G)$ is the compact manifold resulting from removing all solid cusp cylinders from $\mathcal{M}(G)$, then $\mathcal{M}_0(G) \cong S \times [0, 1]$.

The proof of this is a repetition of the arguments of Section 3.2, using the additional facts, (a) if a subgroup $H$ is a surface group of no larger genus than $S$, then $H = G$, and (b) $G$ is not a non-trivial free product of subgroups.

Maskit [**25**] has given an example of a group $G$ of this type for which $\partial\mathcal{M}(G)$ is the union of four triply-punctured spheres.

**3.6** The general case of a function group is understood by putting together the previous results. We start by applying Dehn's Lemma and the Loop Theorem to $S_0 = \Omega_0(G)/G$. We can find mutually disjoint, simple loops $\{\gamma_i\}$, no two of which bound an annulus in $S_0$, non-trivial in $S_0$, which bound discs $\{D_1\}$ in $\mathscr{M}(G)$. Choose a maximal set in the sense that the inclusion into $\pi_1(\mathscr{M}(G))$ of the fundamental group of each component of $S_0 - \cup\gamma_i$ is injective. In general, we cannot require the $\gamma_i$ to be non-dividing.

Let $M$ be a component of $\mathscr{M}(G) - \cup D_i$ and fix a component $M^*$ of $\pi^{-1}(M)$. Then the stabilizer $G_0$ of $M^*$ is a $B$-group, a free abelian group of rank two, or a cyclic parabolic or loxodromic group. Furthermore, $\mathscr{M}(G_0)$ can be retracted onto $M$. In fact, each simple loop $\gamma_i \subset \partial M$ bounds a disc in $\partial\mathscr{M}(G_0)$.

The geometry of this decomposition will be seen in a different light in Section 4 where methods for constructing all these possibilities will be exhibited.

## 4. BUILDING GROUPS

**4.1** In Section 3 we saw how function groups could be described by breaking them into simpler parts: quasi-Fuchsian groups and balls. In this section we will present some typical examples of the reverse process: building more complicated groups from simpler parts. Such processes, called combination theorems, are well known especially through Maskit's work. Our point of view is to start with a topological building operation and then try to carry it out in the context of Kleinian groups. Later on when we take a look at deformation theory, we will see that once one (geometrically finite) group representing a particular topological model is constructed, all groups representing it can be obtained. However it is far from known just what topological operations can be so realized. Here, making use of our knowledge from Section 3, we will indicate how to construct all topological types of function groups. Then we will discuss how the corresponding manifolds can be doubled.

Combination techniques, since they give rise to new compact manifolds from old (after solid cusp tori and cylinders are removed), preserve the class of geometrically finite groups.

**4.2 Free products** (This construction is due to Klein.)

Given $\mathscr{M}(G_1)$, $\mathscr{M}(G_2)$, and closed Euclidean discs $D_i \subset \partial\mathscr{M}(G_i)$ (if $\mathscr{M}(G_1) = \mathscr{M}(G_2)$, then take $D_1 \cap D_2 = \varnothing$) there is a group $G_3$ such that $\mathscr{M}(G_3)$ is homeomorphic to the manifold resulting by identifying $D_1$ and $D_2$. By a Euclidean disc $D$ we mean a region with the property that each component of $\pi^{-1}(D)$ can be mapped onto the unit disc by a Möbius transformation.

To construct $G_3$ assume first $\mathcal{M}(G_1) \neq \mathcal{M}(G_2)$; we allow the possibility $G_1 = G_2$. Start with $G_1$ and a fixed component $D_1^*$ of $\pi^{-1}(D_1)$. Find a conjugate $G_2'$ of $G_2$ so that the complement of $D_1^*$ is a component of $\pi^{-1}(D_2)$, $\mathcal{M}(G_2) = \mathcal{M}(G_2')$. Then $G_3 = \langle G_1, G_2' \rangle \cong G_1 * G_2$ has the desired properties.

If $\mathcal{M}(G_1) = \mathcal{M}(G_2)$, also fix a component $D_2^*$ of $\pi^{-1}(D_2)$. Choose any Möbius transformation $T$ that maps the exterior of $D_1^*$ onto $D_2^*$. Then set $G_3 = \langle G, T \rangle \cong G * \langle T \rangle$. If $D_1$ and $D_2$ lie on the same component of $\partial\mathcal{M}(G)$, this process adds a handle to that component; if $D_1, D_2$ are in different components this operation has the effect of connecting them.

If instead of being disjoint, $D_1$ and $D_2$ in $\partial\mathcal{M}(G)$ are tangent at $p$, then $T$ can be taken to be parabolic with fixed point $p$. The effect of adjoining $T$ is to add a pair of punctures to the corresponding component of $\partial\mathcal{M}(G)$.

A Schottky group is obtained by adding a succession of $g$ handles to a sphere. If $G_1$, $G_2$ are Fuchsian groups representing surfaces of genus $g_i$ with $n_i$ punctures respectively, then $G_3$ is a function group and $\Omega_0(G_3)/G_3$ has genus $g_1 + g_2$ with $n_1 + n_2$ punctures.

**4.3 Amalgamation over a cyclic subgroup.** (This construction is due to Maskit.)

By a *horocyclic disc D* about a puncture $p$ *on* $\partial\mathcal{M}(G)$ we mean something (c.f. Section 1.7) with the properties, (a) $D$ is conformally equivalent to the once punctured disc and, (b) each component of $\pi^{-1}(D)$ is a Euclidean disc or half plane invariant under a member of the conjugacy class in $G$ determined by $p$. Given two manifolds $\mathcal{M}(G_1)$, $\mathcal{M}(G_2)$, punctures $p_1$ on $\partial\mathcal{M}(G_1)$ and $p_2$ on $\partial\mathcal{M}(G_2)$, and horocyclic discs $D_1$ about $p_1$ and $D_2$ about $p_2$ there is a group $G_3$ such that $\mathcal{M}(G_3)$ is homeomorphic to the result of joining $\mathcal{M}(G_1)$ to $\mathcal{M}(G_2)$ by identifying $D_1$ and $D_2$.

To accomplish this, fix a component $D_1^*$ of $\pi^{-1}(D_1)$ and let $M_p$ denote the maximal cyclic parabolic subgroup stabilizing $D_1^*$. We can find a conjugate $G_2'$ of $G_2$ such that, (a) the exterior of $D_1^*$ is a component $D_2^*$ of $\pi^{-1}(D_2)$, and (b) $M_p$ is also the maximal parabolic subgroups of $G_2'$ stabilizing $D_2^*$. Now set $G_3 = \langle G_1, G_2' \rangle$. Group theoretically, $G_3$ is the free product of $G_1$ and $G_2'$ amalgamated over the common subgroup $M_p$.

This process can also be carried out for two punctures $p_1, p_2$ on the same manifold $\mathcal{M}(G)$. As above fix components $D_i^*$ of $\pi^{-1}(D_i)$ where $D_i$ is a neighbourhood of $p_i$, $i = 1, 2$. Let $M_i$ denote the corresponding maximal parabolic subgroups. There exists a Möbius transformation $T$ which (a) sends the exterior of $D_1^*$ to $D_2^*$, and (b) conjugates $M_1$ and $M_2$: $M_2 = TM_1T^{-1}$. Set $G_3 = \langle G, T \rangle$. $\mathscr{M}(G_3)$ is homeomorphic to the manifold obtained by identifying $D_1$ and $D_2$.

Suppose $G_1$, $G_2$ are Fuchsian groups representing $n_i$-punctured surfaces of genus $g_i$, $i = 1, 2$. The group $G_3$ resulting from identifying one puncture on $\partial\mathscr{M}(G_1)$ with one on $\partial\mathscr{M}(G_2)$ is a $B$-group with $\Omega_0(G_3)/G_3$ an $(n_1 + n_2 - 2)$-punctured surface of genus $g_1 + g_2$. There are two other components of $\partial\mathscr{M}(G_3)$ and these are $n_1$-punctured surfaces of genus $g_1$. On the other hand the result of identifying two punctures in the same component of $\partial\mathscr{M}(G_1)$ gives a $B$-group $G_3$ with $\Omega_0(G_3)/G_3$ an $(n_1 - 2)$-punctured surface of genus $g_1 + 1$. The other component of $\partial\mathscr{M}(G_3)$ remains an $n_1$-punctured surface of genus $g_1$.

If we identify two punctures on $\partial\mathscr{M}(G)$ which are paired to get a new group $G_3$ then $\partial\mathscr{M}(G_3)$ has exactly two fewer punctures than $\partial\mathscr{M}(G)$ but there is one additional solid cusp torus instead. Doing this to a Fuchsian group representing a once-punctured surface of genus $g$ gives a $2g + 1$ generator group $G_3$ with $\partial\mathscr{M}(G_3)$ a single compact surface of genus $2g$. Applying this technique in succession we can eliminate all punctures on $\partial\mathscr{M}(G)$ obtaining solid cusp tori instead.

Referring to our description of function groups in Section 3 we see that a model group for each of the possibilities can be obtained by applying in succession the methods of Section 4.2 and above. For more details see Maskit [**21**]. These models have the additional property that every component of the ordinary set other than the invariant one is a Euclidean disc in $\partial\mathscr{H}$.

### 4.4 Doubling $\mathscr{M}(G)$ across part of $\partial\mathscr{M}(G)$.

Suppose $\partial\mathscr{M}(G) = \cup S_i$ where the components $S_i$, $1 \leqslant i \quad m$, have the property that each component of $\pi^{-1}(S_i)$ is a Euclidean disc (i.e. equivalent to one under a Möbius transformation). Then with the sole exception that $G$ is a Fuchsian group, there is a group $H$ such that $\mathscr{M}(H)$ is homeomorphic to the manifold resulting from doubling $\mathscr{M}(G)$ across $S_1, \ldots, S_m$.

To construct $H$, fix a component $D_1^*$ of $\pi^{-1}(S_1)$ and let $J_1$ denote the reflection in the circle $\partial D_1^*$. The first step is to set $G_1 = \langle G, J_1GJ_1 \rangle$. $\mathscr{M}(G_1)$ is homeomorphic to the manifold obtained by doubling $\mathscr{M}(G)$ across $S_1$. The surfaces $S_2, \ldots, S_m$ each appear twice in $\partial\mathscr{M}(G_1)$; denote them by $S_{2j}, \ldots, S_{mj}$, $j = 1, 2$. Of course, each component of $\pi^{-1}(S_{ij})$, $1 \leqslant i \leqslant m$, is a disc.

The next step is to identify $S_{21}$ and $S_{22}$. Fix a component $D_2^*$ of $\pi^{-1}(S_{21})$. Then $J_1(D_2^*)$ is a component of $\pi^{-1}(S_{22})$. Let $J_2$ denote the reflection in the circle $\partial D_2^*$. $J_1 J_2$ maps $D_2^*$ onto the exterior of $J_1(D_2^*)$ and conjugates the subgroups of $G_1$ that stabilize $D_2^*$ and $J_1(D_2^*)$. Set $G_2 = \langle G_2, J_1 J_2 G_2 J_2 J_1 \rangle$. $\mathscr{M}(G_2)$ is homeomorphic to the manifold obtained from $\mathscr{M}(G_1)$ by identifying $S_{21}$ and $S_{22}$. The third step proceeds exactly as the second, etc. After $m$ steps we obtain the desired group $H$. Note that $J_1 H J_1 = H$, and $J_1$ determines an anti-conformal involution of $\mathscr{M}(H)$ that interchanges the two halves.

We can now see exactly why this process fails if and only if $G$ is Fuchsian. For it is only in this case that $G_1 = G$ and in the second step, $J_2 = J_1$ so that $J_2 J_1 = \mathrm{id}$. For group theoretic reasons too, the double of $\mathscr{M}(G)$ when $G$ is Fuchsian cannot be represented by a Kleinian group. If $G$ represents the surface $S$ then the double of $\mathscr{M}(G)$ is homeomorhic to $S \times S^1$. The centre of $\pi_1(S \times S^1)$ is non-trivial which cannot be the case for a Kleinian group.

As we saw in Section 4.3, corresponding to every function group there is a model $G$ for which all the components of $\Omega(G)$ except the invariant one, $\Omega_0(G)$, are discs. So if $G$ is not Fuchsian, the manifold resulting by doubling $\mathscr{M}(G)$ across $\partial\Omega(G) - (\Omega_0(G)/G)$ is homeomorphic to some $\mathscr{M}(H)$. To a pair of punctures on $\partial\mathscr{M}(G) - (\Omega_0(G)/G)$ corresponds a solid cusp torus in $\mathscr{M}(H)$. $\partial\mathscr{M}(H)$ consists of two components which are anticonformally equivalent by an involution $J$ of $\mathscr{M}(H)$. The fixed set of $J$ is a union of surfaces in the interior $\mathscr{M}(H)^0$ conformally equivalent to $\partial\mathscr{M}(G) - (\Omega_0(G)/G)$, and each is totally geodesic.

## 5. THE ISOMORPHISM THEOREM

**5.1** The following result tells how each group fits geometrically into its isomorphism class. It is an elaboration of Theorem 8.1 in [**20**]. Terminology: An isomorphism $\varphi\colon G \to H$ is *induced* by a homeomorphism $f$ if $f \circ T(z) = \varphi(T) \circ f(z)$ for all $T \in G$ and $z$ in the domain of $f$. An isomorphism $\varphi\colon G \to H$ is a *conjugation* if $\varphi(T) = ATA^{-1}$ for some Möbius transformation $A$ and all $T \in G$. The prefix "anti-"signifies orientation reversing. The term *Euclidean disc* includes all regions equivalent to one under a Möbius transformation.

Isomorphism Theorem. *Suppose $G$ is a discrete, torsion free, geometrically finite group and $\varphi\colon G \to H$ is an isomorphism onto a discrete group $H$. If $\Omega(G) = \varnothing$ then $\varphi$ is a conjugation or anti-conjugation.*

*If $\Omega(G) \neq \varnothing$ assume in addition that $\varphi$ is induced by a quasi-conformal or anti-quasi-conformal homeomorphism $f$ of $\Omega(G)$ into $\Omega(H)$. Then $H$ is also geometrically finite and* a) *the map $f$ has a quasi-conformal or anti-quasi-conformal extension to all $\mathbb{C} \cup \{\infty\}$, and* b) *the projection of $f$ to $\partial\mathscr{M}(G) \to$*

*$\partial\mathcal{M}(H)$ extends to a homeomorphism $\mathcal{M}(G) \to \mathcal{M}(H)$ which induces the isomorphism $\varphi_*: \pi_1(\mathcal{M}(G)) \to \pi_1(\mathcal{M}(H))$. Moreover,*

(i) *If $G$ is not Fuchsian and $f$ is conformal or anti-conformal on all components of $\Omega(G)$ with the possible exception of those $\{\Omega_d\}$ for which both $\Omega_d$ and $f(\Omega_d)$ are Euclidean discs then $\varphi$ is a conjugation or anti-conjugation.*

(ii) *If $f$ is conformal or anti-conformal on all $\Omega(G)$ then $f$ is the restriction of some Möbius or anti-Möbius transformation.*

This Theorem strongly affirms the rigidity of discrete groups of finite volume (thus incorporating Mostow's Rigidity Theorem [29]) and also those Kleinian groups which are not Fuchsian but for which all components of $\Omega(G)$ are Euclidean discs under deformations which preserve this property.

**5.2** We will outline the main ingredients of the proof; for details refer to [**20**]. To simplify the discussion assume all components of $\Omega(G)$ are simply connected. By Dehn's Lemma and the Loop Theorem, any finitely generated Kleinian group is the free product of a finite number of such groups (see Sections 2.2 and 3.6).

First, we will explain why $H$ is geometrically finite. The solid cusp tori in $\mathcal{M}(H)$ are completely determined by $\varphi$ and $f$ determines all solid cusp cylinders involving $f(\Omega(G))$. Hence $f$ determines a map $f_*$ of $\partial\mathcal{M}_0(G)$ into $\partial\mathcal{M}_0(H)$. The topological double $M$ of $\mathcal{M}_0(G)$ across $\partial\mathcal{M}_0(G)$ is compact without boundary. $f$ determines an isomorphism of $\pi_1(M)$ onto the fundamental group of the result of doubling $\mathcal{M}_0(H)$ across $f_*(\partial\mathcal{M}_0(G))$. Consequently there is an isomorphism between the third homology groups. But it is these which tell about compactness.

Next, assuming $f$ is sufficiently smooth (otherwise it is homotopic to a suitable map) it projects and extends to a smooth, quasi-conformal (or anti-, if it is orientation reversing) homeomorphism $f_*: \mathcal{M}(G) \to \mathcal{M}(H)$. This is the heart of the Theorem and it is a result of Waldhausen [**34**] (also see the survey [**31**]). His Theorem, stated for our situation, says that an isomorphism $\varphi_*: \pi_1(\mathcal{M}_0(G)) \to \pi_1(\mathcal{M}_0(H))$ which acts properly with respect to the injections from $\partial\mathcal{M}_0(G)$ and $\partial\mathcal{M}_0(H)$ (if these are $\neq \emptyset$) is induced by a homeomorphism (we ignore here an exceptional case). It is worthwhile to recall Waldhausen's method of proving this. He shows that there exists a finite sequence of surfaces $\{S_k\}$ as follows. $S_1 \subset \mathcal{M}_0(G)$ with the properties, (a) $S_1 \cap \partial\mathcal{M}(G) = \partial S_1 \neq \emptyset$ (unless $\partial\mathcal{M}_0 = \emptyset$), (b) $S_1$ does not divide $\mathcal{M}_0(G)$, (c) the inclusion $\pi_1(S_1) \to \pi_1(\mathcal{M}_0)$ is injective. Take a thin neighbourhood $S_1^* = S_1 \times (-\varepsilon, \varepsilon)$ and set $M_1 = \mathcal{M}_0(G) - S_1^*$. The next surface $S_2$ lies in $M_1$ as $S_1$ lies in $\mathcal{M}_0$ (but now in all cases $\partial M_1 \neq \emptyset$), etc. At the last step $M_n$ is a ball. Using $\varphi_*$ one can construct a corresponding decomposition of $\mathcal{M}_0(H)$. The map $f_*$ is found by painstakingly reconstructing $\mathcal{M}_0(G)$ and $\mathcal{M}_0(H)$ with pieces corresponding under $f_*$.

Once the quasi-conformal (or anti-) homeomorphism $f_*: \mathcal{M}(G) \to \mathcal{M}(H)$ is found it can be lifted to a homeomorphism $f$ of $\mathcal{H}^3 \cup \Omega(G)$. By Gehring's Extension Theorem there is a uniquely determined quasi-conformal (or anti-) extension from $\mathcal{H}^3$ to all $\mathbb{C} \cup \{\infty\}$ and in particular from $\Omega(G)$ to $\mathbb{C} \cup \{\infty\}$ (this is the only way we have of getting at the limit set).

In the case $\Omega(G) = \varnothing$, Mostow has shown that the extension $f$ is conformal or anti-conformal on $\mathbb{C} \cup \{\infty\}$. The statement of the Theorem in this case is known as Mostow's Rigidity Theorem [**29**]; actually, as it is presented here, it is an elaboration of that. The case when $G$ is non-Fuchsian yet all components of $\Omega(G)$ and $f(\Omega(G))$ are discs is an application of this result. For then as in Section 4.4. $\mathcal{M}(G)$ and $\mathcal{M}(H)$ can be doubled across their entire boundary and then represented by new groups, necessarily of finite volume. Using $f$, $\varphi$ extends to an isomorphism between the new groups.

When $G$ is Kleinian, it is a result of Ahlfors [**5**] that its limit set has zero area on $\mathbb{C}$. Therefore if $f$ is known to be conformal (or anti-) on $\Omega(G)$ it is the restriction of a Möbius (or anti-Möbius) transformation. If $f$ is conformal (or anti-) except on those components $\{\Omega_d\}$ or $\Omega(G)$ for which both $\Omega_d$ and $f(\Omega_d)$ are discs, then $\mathcal{M}(G)$ and $\mathcal{M}(H)$ must first be reflected in the components of $\partial\mathcal{M}(G)$ and $\partial\mathcal{M}(H)$ arising from these corresponding discs. The original map $f: \Omega(G) \to \Omega(H)$ determines a new map $f_1: \Omega(G_1) \to \Omega(H_1)$ which induces an isomorphism between the new groups $G_1$, $H_1$ equal to the original $\varphi$ on the subgroups $G$, $H$. But now $f_1$ is conformal (or anti-) on all $\Omega(G_1)$.

## 6. DEFORMATION SPACES

**6.1** As an abstract group a (torsion free, geometrically finite) Kleinian group has a certain finite presentation by generators $A_1, \ldots, A_N$ and relations

$$R_i(A_1, \ldots, A_N) = 1 \tag{1}$$

with each $R_i$ a word in the letters $A_1, \ldots, A_N$. In addition, there are a finite number of parabolic transformations which with their conjugates and powers give all parabolic transformations of $G$. Each of these finite number of elements (they can be chosen in many ways) can be expressed as a word $P_j$ in

$A_1, \ldots, A_N$ for which (if the determinants are one)

$$\text{(2)} \qquad \operatorname{tr}^2 P_j(A_1, \ldots, A_N) = 4.$$

Of course, those relations of (2) which come from rank two parabolic subgroups are a consequence of (1).

We can associate a Möbius transformation $B(z) = (az + b)/(cz + d)$ or the corresponding $2 \times 2$ matrix in PSL $(2, \mathbb{C})$ with the point $(a, b, c, d)$ in complex three-dimensional projective space $\mathbb{P}_3$. Conversely, points $(a, b, c, d) \in \mathbb{P}_3$ with $ad - bc \neq 0$ correspond to Möbius transformations.

Using unnormalized matrices for the transformations $A_i$, writing each $A_i$ as $(\text{mat } A_i)\,((\det A_i)^{1/2}\, I_2)^{-1}$ where mat $A_i$ is the $2 \times 2$ matrix of coefficients and $I_2$ is the identity matrix, and then clearing determinants from the denominator, (1) and (2) can be rewritten as homogeneous polynomials in the product space $\mathbb{P}_3^N$.

Let $\mathbf{V}(G)^*$ denote the algebraic variety in $\mathbb{P}_3^N$ consisting of those points which satisfy (1) and (2) viewed as homogeneous polynomials. Let $X$ denote the subvariety consisting of those points at least one of whose coordinates $(a, b, c, d)$ in $\mathbb{P}_3^N$ satisfies $ad - bc = 0$. Our basic space is the noncompact, Zariski open subset

$$\mathbf{V}(G) = \mathbf{V}(G)^* - X.$$

It is known $\mathbf{V}(G)$ can be represented as an affine algebraic variety but we do not need this fact.

**6.2** Each point of $\mathbf{V}(G)$ corresponds to an $N$-tuple of Möbius transformations $(B_1, \ldots, B_H)$ which satisfy (1) and (2). Therefore, the correspondence $\theta: A_i \to B_i$ determines a homomorphism $\theta$ of $G$ onto the group $H$ generated by $B_1, \ldots, B_N$. If $T \in G$ is parabolic then $\theta(T)$ is either parabolic in $H$ or is the identity. Conversely, every homomorphism $\theta: G \to \text{PSL}\,(2, \mathbb{C})$ with this property on parabolic transformations corresponds to a point of $\mathbf{V}(G)$. For this reason points of $\mathbf{V}(G)$ can be written as $\theta$ or as pairs $(H, \theta)$ where $\theta$ is a homomorphism of $G$ onto $\theta(G) = H$.

The natural topology of $\mathbf{V}(G)$ determines the topology of "pointwise" convergence on the groups $H$. That is, $\theta_n \to \theta$ in $\mathbf{V}(G)$ if and only if $\lim \theta_n(A_i) = \theta(A_i)$, $1 \leqslant i \leqslant N$, convergence as Möbius transformations. This in turn is equivalent to the condition that for each $T \in G$, $\lim \theta_n(T) = \theta(T)$. Such convergence is not necessarily uniform throughout $G$.

**6.3** Define the *deformation space* $\mathbf{T}(G)$ as

$$\mathbf{T}(G) = \{(H, \theta) \in \mathbf{V}(G) \colon H \text{ is Kleinian and } \theta \text{ is induced by an orientation preserving homeomorphism } f_* \colon \mathcal{M}(G) \to \mathcal{M}(H) \text{ or equivalently by } f \colon \Omega(G) \to \Omega(H)\}.$$

Now each $T \in \mathrm{PSL}\,(2, \mathbb{C})$ acts freely on $\mathbf{V}(G)$ as the birational transformation $T \colon \theta \to T\theta T^{-1}$ and to normalize the groups with respect to conjugation we must consider the coset space $\mathbf{V}_0(G) = \mathbf{V}(G)/\mathrm{PSL}\,(2, \mathbb{C})$ which is in general not a Hausdorff space. The normalized space $\mathbf{T}_0(G) = \mathbf{T}(G)/\mathrm{PSL}\,(2, \mathbb{C})$ can alternatively be defined by requiring, for example, that $A_1$ be loxodromic, the attractive and repulsive fixed points of $\theta(A_1)$ to be $\infty$ and 0 respectively, and that the attractive fixed point of $\theta(A_2)$ be 1.

An equivalent definition of $\mathbf{T}_0(G)$ in the manner of Teichmüller space is as follows. The pair $(\mathcal{M}(H), h)$ means that $h \colon \mathcal{M}(G) \to \mathcal{M}(H)$ is an orientation preserving homeomorphism. Two pairs are said to be equivalent $(\mathcal{M}(H_1), h_1) \equiv (\mathcal{M}(H_2), h_2)$ if and only if $h_1 h_2^{-1} \colon \mathcal{M}(H_2) \to \mathcal{M}(H_1)$ is homotopic to a conformal map. With this notation, in view of the Isomorphism Theorem,

$$\mathbf{T}_0(G) = \{\text{equivalence classes } (\mathcal{M}(H), h)\}.$$

**6.4** Let $g_k$ denote the genus of the $k$th component of $\partial\mathcal{M}(G)$ and $b_k$ the number of its punctures. The central theorem here is this (see [**20**]):

Deformation Theorem. *$\mathbf{T}(G)$ is a connected open subset of $\mathbf{V}(G)$ and a complex analytic manifold of dimension $\Sigma(3g_k + b_k - 3) + 3$. Similarly for $\mathbf{T}_0(G)$ in $\mathbf{V}_0(G)$, and $\mathbf{T}(G)$ is biholomorphically equivalent to $\mathbf{T}_0(G) \times \mathrm{PSL}\,(2, \mathbb{C})$.*

Note that if $\partial\mathcal{M}(G) = \varnothing$, or if all its components are triply punctured spheres, then $\mathbf{T}_0(G)$ reduces to a single point which, of course, is what the Isomorphism Theorem gives. It is a remarkable fact that although $\mathcal{M}(G)$ may have much internal structure not reflected on its boundary, the dimension of $\mathbf{T}(G)$ depends only on its boundary. For instance let $G_1$ be a two generator Schottky group and construct a four generator group $G_2$ as follows. Take two Fuchsian groups $F_1$. $F_2$ each representing two triply-punctured spheres. Using Section 4.3, pair the three punctures on one component of $\partial\mathcal{M}(F_1)$ with the three on a component of $\partial\mathcal{M}(F_2)$. This results in a $B$-group $G_2$ such that $\partial\mathcal{M}(G_2)$ is the union of a surface of genus two and two triply-punctured spheres. Both $\mathbf{T}_0(G_1)$ and $\mathbf{T}_0(G_2)$ have dimension three, yet the internal structure of $\mathcal{M}(G_1)$ and $\mathcal{M}(G_2)$ is quite different.

**6.5** We will briefly outline the proof which is divided into two parts. The first part begins by fixing a fundamental polyhedron $\mathscr{P}$ with centre at $O \in \mathscr{H}^3$ for $G$; $\mathscr{P}$ is taken to be closed in $\mathscr{H}^-$. Assume for simplicity that $G$ has no parabolic elements. Locate the finite set of transformations $\{T_i^{\pm 1}\}$ in $G$ for which $T_i(\mathscr{P}) \cap \mathscr{P} \neq \varnothing$. This means the hyperbolic planes which form the perpendicular bisector of the segments $[0, T_i^{\pm 1}(0)]$ contain either a face of $\mathscr{P}$ or simply a vertex of $\mathscr{P}$. Now vary the generators of $G$ slightly in $\mathbf{V}(G)$ in any manner thereby obtaining a new group $G_\varepsilon$ and point $(G_\varepsilon, \theta_\varepsilon) \in \mathbf{V}(G)$. The transformations $T_i$ move slightly to $\theta(T_i)$ and consequently the corresponding hyperbolic planes determine a polyhedron $\mathscr{P}_\varepsilon$ close to $\mathscr{P}$. One proves that $\mathscr{P}_\varepsilon$ is a fundamental polyhedron for $G_\varepsilon$ which is therefore discrete as well. Furthermore $\mathscr{M}(G)$ and $\mathscr{M}(G_\varepsilon)$ are homeomorphic. By investigating how $\mathscr{P}_\varepsilon \cap \partial\mathscr{H}^3$ differs from $\mathscr{P} \cap \partial\mathscr{H}^3$ it can be shown that for sufficiently small $\varepsilon$, $\theta_\varepsilon\colon G \to G_\varepsilon$ is induced by a quasi-conformal map $\Omega(G) \to \Omega(G_\varepsilon)$ of arbitrarily small dilatation. The proof works equally well in a neighbourhood of each point in $\mathbf{T}(G)$. Bers has called this property "strong stability".

The second part of the proof involves applying a result of Bers [7]. Let $S_1, \ldots, S_n$ denote the components of $\partial\mathscr{M}(G)$. According to [7] there is a natural local injection of maximal rank of $\mathbf{T}_*(S_1) \times \ldots \times \mathbf{T}_*(S_n) \times \mathrm{PSL}(2, \mathbb{C})$ into $\mathbf{T}(G)$ where $\mathbf{T}_*(S_j)$ denotes the $(3g_j + b_j - 3)$-dimensional Teichmüller space of $S_j$. Bers' map is constructed by taking quasi-conformal deformations of $G$; these are determined on $\Omega(G)$ and hence depend only on $\partial\mathscr{M}(G)$. The coefficients of the generators of the groups corresponding to the image points in $\mathbf{T}(G)$ are holomorphic functions on the product space. That the points $(H, \theta)$ of $\mathbf{T}(G)$ are strongly stable says in effect that all points of a small neighbourhood of $(H, \theta)$ in $\mathbf{V}(G)$ are determined by quasi-conformal deformations of $H$. Consequently, the Bers map is in fact a local homeomorphism and gives a natural complex structure to $\mathbf{T}(G)$. However, it is a global homeomorphism only if all the components of $\Omega(G)$ are simply connected. In general, it is only a covering map.

**6.6** We will next present a useful variant of the Deformation Theorem.

THEOREM *Let $(R_1, \ldots, R_n)$ be a collection of abstract Riemann surfaces of finite type, $G$ a Kleinian group, and $f\colon (R_1, \ldots, R_n) \to \partial\mathscr{M}(G)$ a quasi-conformal homeomorphism. Then there exists $(H, \theta) \in \mathbf{T}_0(G)$ uniquely determined by the existence of a conformal map $h\colon \partial\mathscr{M}(H) \to (R_1, \ldots, R_n)$ such that $f \circ h\colon \partial\mathscr{M}(H) \to \partial\mathscr{M}(G)$ extends to a homeomorphism $W\colon \mathscr{M}(H) \to \mathscr{M}(G)$.*

*Now assume in addition that $\Gamma$ is a group of conformal and anti-conformal* *automorphisms of the collection* $(R_1, \ldots, R_n)$, $\Gamma'$ *is a group of orientation preserving and reversing homeomorphisms of* $\mathcal{M}(G)$, *and* $\varphi\colon \Gamma \to \Gamma'$ *is an isomorphism such that* $f \circ \gamma$ *is homotopic to* $\varphi(\gamma) \circ f$ *for each* $R_i$ *and each* $\gamma \in \Gamma$. *Then there exists a group* $\Delta$ *of conformal and anti-conformal automorphisms of* $\mathcal{M}(H)$ *and an isomorphism* $\psi\colon \Delta \to \Gamma$ *such that*

(i) $(h \circ \delta)(x) = \psi(\delta) \circ h(x)$, all $\delta \in \Delta$ *and* $x \in \partial\mathcal{M}(H)$,
(ii) $W \delta W^{-1}$ *is homotopic to* $\varphi\psi(\delta)$, *all* $\delta \in \Delta$.

In short, this Theorem says that if there exists a Kleinian group $G$ which provides a topological model for the configuration $(R_1, \ldots, R_n; \Gamma)$, there is a unique group in $\mathbf{T}_0(G)$ which provides a conformal model.

For the proof, start with the necessarily finite group $\Gamma'' = \{f\gamma f^{-1} : \gamma \in \Gamma\}$ which consists of quasi-conformal and anti-quasi-conformal automorphisms of $\partial\mathcal{M}(G)$. Each element $f\gamma f^{-1}$ is homotopic on $\partial\mathcal{M}(G)$ to the restriction of $\varphi(\gamma)$. Because of this, each element of $\Gamma''$ has a well determined $G$-conjugacy class of lifts to $\Omega(G)$. By the Isomorphism Theorem, each lift extends to all $\partial\mathcal{H}^-$.

Take the Beltrami differential $\mu = (f^{-1})_{\bar{z}}/(f^{-1})_z$ on $\partial\mathcal{M}(G)$, lift it to $\Omega(G)$, and solve the corresponding Beltrami equation to get a global quasi-conformal homeomorphism $F$ on $\partial\mathcal{H}^-$. $FTF^{-1}$ is a Möbius transformation for all $T \in G$ but even more is true. Namely $FTF^{-1}$ is a Möbius or anti-Möbius transformation for all lifts $T$ of elements of $\Gamma''$ because of the behaviour of $f$ and hence $\mu$ with respect to $\Gamma''$.

Consequently, $F$ uniquely determines the point $(H, \theta) \in \mathbf{T}_0(G)$ where $\theta(T) = FTF^{-1}$, $T \in G$. If $S$ is a lift of an element of $\Gamma''$, for $S_1 = FSF^{-1}$ it is true that $S_1 \theta(T) S_1^{-1} = \theta(STS^{-1})$.

Under the projected map $F_* : \partial\mathcal{M}(G) \to \partial\mathcal{M}(H)$, the group $\Gamma''$ is sent to the group $\Delta = F_* \Gamma'' F^{-1}$ of conformal and anti-conformal automorphisms which acts not only on $\partial\mathcal{M}(H)$ but on all $\mathcal{M}(H)$. Because $\mu$ was derived from $f^{-1}$, the map $h = f^{-1}F_*^{-1} : \partial\mathcal{M}(H) \to (R_1, \ldots, R_n)$ is conformal. For $\delta \in \Delta$, set $\psi(\delta) = h\delta h^{-1}$. $\psi$ is the isomorphism $\Delta \to \Gamma$ that sends $\delta \in \Delta$ to $f^{-1}F_*^{-1}\delta F_* f \in \Gamma$. Hence for each $\delta \in \Delta$, the restriction of $\varphi\psi(\delta)$ to each component of $\partial\mathcal{M}(G)$ is homotopic to $F_*^{-1}\delta F_*$.

The map $F_* : \partial\mathcal{M}(G) \to \partial\mathcal{M}(H)$ can also be extended to a homeomorphism $F_* : \mathcal{M}(G) \to \mathcal{M}(H)$ inducing the isomorphism $\theta_* : \pi_1(\mathcal{M}(G)) \to \pi_1(\mathcal{M}(H))$ by the Isomorphism Theorem. Since they induce the same automorphism of $\pi_1(\mathcal{M}(G))$ (as well as the same isomorphisms among the groups $\pi_1(S_i)$ for the components $S_i$ of $\partial\mathcal{M}(G)$), the maps $F_*^{-1}\delta F_*$ and $\varphi\psi(\delta)$ are homotopic $\mathcal{M}(G)$.

**6.7** We will apply Theorem 6.6 to display a curious relation between an aspect of the classification problem and a geometric problem involving the ordinary set.

THEOREM. *Suppose $S_1, \ldots, S_n$ is a subset of the components of $\partial\mathscr{M}(G)$ and G is not quasi-Fuchsian. There exists a Kleinian group $H^*$ with the property that $\mathscr{M}(H^*)$ is homeomorphic to the manifold M resulting from doubling $\mathscr{M}(G)$ across $S_1, \ldots, S_n$ if and only if the following condition is satisfied: there exists a point $(\mathscr{M}(H), f) \in \mathbf{T}_0(G)$ such that all components of $\pi^{-1}(f(S_i))$ are (equivalent to) Euclidean discs. If one such point exists in $\mathbf{T}_0(G)$ then any complex structure can be prescribed on $\partial\mathscr{M}(G) - \cup S_j$ and there is a unique point $(\mathscr{M}(H), f) \in \mathbf{T}_0(G)$ such that*

(i) *all components of $\pi^{-1}(f(S_j))$ are discs, and*
(ii) *the prescribed complex structure is realized on $\partial\mathscr{M}(H) - \cup f(S_j)$.*

We have seen in Section 4.4 that the above condition is sufficient. There are at least three obvious situations where Euclidean discs cannot be obtained.

(a) A component of $\pi^{-1}(S_j)$ is not simply connected. Of course this case is obvious but note too that $M$ would then contain non-contractible spheres.

(b) $G$ is not quasi-Fuchsian but contains a loxodromic element $T$ which preserves two distinct components $\{\pi^{-1}(S_j), 1 \leqslant j \leqslant n\}$.

(c) One of these components is preserved by an accidental parabolic transformation, that is, a parabolic transformation that is determined by a simple loop in some $S_j$ not retractable to a puncture there.

Again the impossibility for (b) and (c) are geometrically obvious but note too that if doubling were possible in (b) there would exist in $H$ a non-parabolic abelian subgroup of rank two and in (c) four punctures of $\partial\mathscr{M}(H)$ would be paired.

Although a $B$-group cannot be doubled across its invariant component, it can be across the others.

COROLLARY (Maskit [**28**]). *Suppose G is a function group with invariant component $\Omega_0(G)$ and $g: \Omega_0(G)/G \to R$ is a quasi-conformal map onto a Riemann surface R. There exists a unique point $(\mathscr{M}(H), f) \in T_0(G)$ such that*

(i) *the map $g \circ f^{-1}: \Omega_0(H)/H \to R$ is conformal, and*
(ii) *each component of $\Omega(H) - \Omega_0(H)$ is a Euclidean disc.*

Maskit has called function groups with this property "Koebe groups".

**6.8** The proof of Theorem 6.7 is based on the following fact for an anti-Möbius transformation $T$. The fixed point set for $T$ consists of either one or two points or it is a circle or line in $\mathbb{C}$. This will be applied in the following way. If $H_0$ is a non-elementary torsion free Kleinian group such that $TST^{-1} = S$ for all $s \in H_0$, then $T$ preserves pointwise the limit set of $H_0$ and therefore $H_0$ must in fact be a Fuchsian group.

One way of obtaining the topological double $M$ is to allow $G$ to act on the lower half space $\mathscr{H}'$. Let $\mathscr{U} = \mathscr{H}^3 \cup \overline{\mathscr{H}}^3 \cup \{\pi^{-1}(S_j): 1 \leqslant j \leqslant n\}$. Then $\mathscr{U}/G$ is the interior $M^\circ$ of $M$. Attaching two copies of $\Omega(G) - \{\pi^{-1}(S_j), 1 \leqslant j \leqslant n\}$ to $\mathscr{U}$ (actually the two "sides" of this set) and calling the result $\mathscr{U}'$ we obtain $\mathscr{U}'/G = M$. Note however that unless $G$ is a quasi-Fuchsian group with $\{S_j\}$ consisting of exactly one component of $\partial\mathscr{M}(G)$, $\mathscr{U}$ is not the universal cover of $M^\circ$. This representation of $M$ has the property that there is an anti-conformal involution $J$ interchanging the two sets of components of $\partial M$ corresponding to $\partial\mathscr{M}(G) - \cup S_j$. Possibly $\partial M = \varnothing$.

Suppose there exists a group $H^*$ with $\mathscr{M}(H^*)$ homeomorphic to $M$. Assume first that $\partial M \neq \varnothing$. Then according to Theorem 6.6, using the configuration $(\partial M; J)$ as the model, we may assume that there is an anti-conformal involution $\delta$ of $\mathscr{M}(H^*)$ and a conformal map $h: \partial\mathscr{M}(H^*) \to \partial M$ which extends to a homeomorphism $h: \mathscr{M}(H^*) \to M$ such that $h \circ \delta$ is homotopic to $J \circ h$.

Now $J$ fixes pointwise the surfaces $S_j$ which form the relative boundary of the natural embedding of $\mathscr{M}(G)$ in $M$, while $h^{-1} \circ J \circ h$ fixes the surfaces $h^{-1}(S_j)$ in $\mathscr{M}(H^*)$ which form the relative boundary of $h^{-1}(\mathscr{M}(G))$. Consider for example $h^{-1}(S_1)$ and the subgroup $H_0$ of $H^*$ that stabilizes a given component $S_1^*$ of $\pi^{-1}(h^{-1}(S_1))$. Because $h^{-1}Jh$ is homotopic to $\delta$, the lift $J^*$ of $h^{-1}Jh$ that fixes $S_1^*$ induces the same automorphism of $H^*$ as the corresponding lift $\delta^*$ of $\delta$ which is an anti-Möbius transformation. So from the fact that $J^*SJ^* = S$ for all $S \in H_0$ follows that also $\delta^*S\delta^* = S$ for all $S \in H_0$. This implies that $H_0$ is a Fuchsian group.

We conclude that the subgroup $H$ of $H^*$ that stabilizes a given lift of $h^{-1}(\mathscr{M}(G)) \subset \mathscr{M}(H^*)$ determines a point in $\mathbf{T}(G_0)$ with the properties required to prove the first part of the Theorem.

Next, if $\partial M = \varnothing$, then if $\mathscr{M}(H^*)$ is homeomorphic to $M$ there is an involution $J'$ of $\mathscr{M}(H^*)$ which we may assume is anti-quasi-conformal. But now $\mathscr{M}(H^*)$ has finite volume. Hence by the Isomorphism Theorem $J'$ is homotopic to an anti-conformal homeomorphism which induces the same automorphism of $\pi_1(\mathscr{M}(H^*))$. The proof is then completed as above.

The second statement of Theorem 6.7 is proved similarly. Uniqueness is implied by the Isomorphism Theorem.

**6.9** The following example is of importance in [**10**]. Suppose $G$ is a $B$-group such that $S_0 = \Omega_0(G)/G$ is a compact surface of genus $g$ and all other components are triply-punctured spheres—there are $2g - 2$ of them. By the Isomorphism Theorem $G$ is uniquely determined up to conjugation by the conformal structure on $S_0$. Now form a new group $H$ and manifold $\mathscr{M}(H)$ either by doubling $\mathscr{M}(G)$ across $\partial\mathscr{M}(G) - S_0$ or by identifying in the manner of Section 4.3 horocyclic discs about each pair of punctures (the two constructions are topologically identical). Then $\partial\mathscr{M}(H)$ has two components each a compact surface of genus $g$; $S_0$ is naturally identified with one of them, denote the other by $S_1$. The structure of $\mathscr{M}(H)$ is as indicated. Each cusp torus $\mathscr{T}_i$, $1 \leqslant i \leqslant 3g$, can be obtained by projecting a horocyclic sphere in $\mathscr{H}^3$. Thus each $\mathscr{T}_i$ has a natural conformal structure.

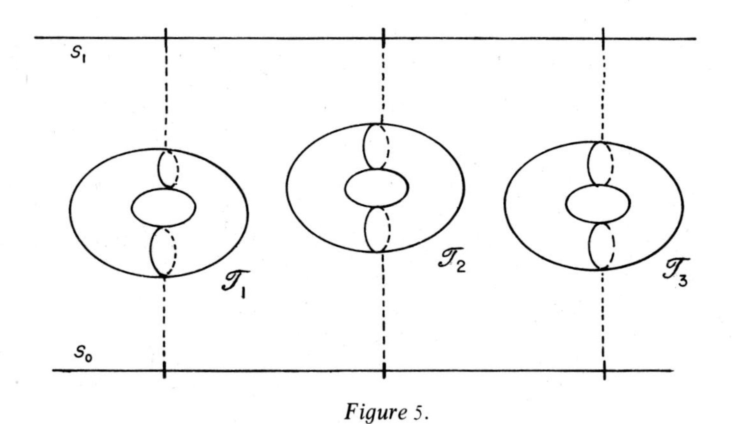


*Figure* 5.

The deformation space $\mathbf{T}_0(H)$ is biholomorphically equivalent to $\mathbf{T}_g \times \mathbf{T}_g$ where $\mathbf{T}_g$ is the Teichmüller space of genus $g$. Let $\mathbf{T}' \subset \mathbf{T}_0(H)$ be the $3g - 3$ dimensional submanifold biholomorphically equivalent to $T_g$ resulting from fixing the conformal structure on $S_0$.

Each group corresponding to a point of $\mathbf{T}'$ is obtained from the fixed $G$ by adjoining $3g - 3$ parabolic transformations. We can thereby canonically designate a set of generators for each $\pi_1(\mathscr{T}_i)$ that arises. Thus each corresponding $\mathscr{T}_i$ is uniquely specified by a point in the upper half plane $\mathscr{U}$, the Teichmüller space of a torus. In this manner we obtain an injection $F$ of $\mathbf{T}'$ into $\mathscr{H}^{3g-3}$. Because of Jørgensen's inequality implying that elements in a discrete group can't get too close to the identity, we see that $F(\mathbf{T}')$ is a proper subset of $\mathscr{H}^{3g-3}$.

THEOREM. [**10**]. *$F$ is a biholomorphic map of $\mathbf{T}_g$ onto a proper open subset of $\mathscr{H}^{3g-3}$.*

This representation of $\mathbf{T}_g$ is extremely useful in studying the cusps on $\partial\mathbf{T}_g$ (see Section 7) and forms the basis for the analytic structure given to the compactification of moduli space in [**10**]. In particular, the embedding has the property that certain Dehn twists act as translations on it. A different embedding, but one also with this property, was given by Maskit [**27**].

## 7. BOUNDARIES OF DEFORMATION SPACES

**7.1** Let $\partial\mathbf{T}(G)$ denote the relative boundary of $\mathbf{T}(G)$ in $\mathbf{V}(G)$. The basic fact is this:

THEOREM. *Suppose for the point $(H, \theta) \in \mathbf{V}(G)$ and a sequence $\{(G_n, \theta_n)\} \in \mathbf{T}(G)$, $\lim (G_n, \theta_n) = (H, \theta)$ in the topology of $\mathbf{V}(G)$. Then*

(i) *$\theta: G \to H$ is an isomorphism and*
(ii) *$H$ is discrete.*

Chuckrow proved (i) in the fundamental paper [**9**] and several years ago we noticed that her methods also imply (ii). Recently, Yamamoto [**36**] published a proof also based on [**9**]. The simplest proof of the Theorem is due to Jørgensen [**13**] as a consequence of his inequality.

Although for $(H, \theta) \in \partial\mathbf{T}(G)$, $H$ is discrete, it need not be Kleinian.

**7.2** 7.2.1 *Definition.* A *cusp* in $\partial\mathbf{T}(G)$ is a point $(H, \theta) \in \partial\mathbf{T}(G)$ where $H$ is a Kleinian group which is geometrically finite.

Cusps play a special role on the boundary, at least in those few cases one understands. A major open problem, initiated by Bers with a less restrictive definition of cusp, is whether cusps are dense on $\partial\mathbf{T}(G)$ (in in fact they even exist there in general!).

We will combine a definition with a theorem as follows. The proof will appear in [**10**]. Denote the set of cusps on $\partial\mathbf{T}$(G) by $\mathbf{C}$(G).

7.2.2 *Definition.* The sequence $\{(G_n, \theta_n)\}$ in $\mathbf{T}(G) \cup \mathbf{C}(G)$ is said to converge to $(H, \theta) \in \mathbf{C}(G)$ in the *cusp topology* if one of the following equivalent conditions is satisfied.

(i) There exists a point $\mathrm{O} \in \mathscr{H}^3$ such that the fundamental polyhedron $\mathscr{P}_n$

for $G_n$ with centre at O converges to the polyhedron $\mathscr{P}$ for $H$ with centre at O, convergence in the Euclidean topology in $\mathbb{R}^3$. (If this condition is fulfilled for one point $O \in \mathscr{H}^3$, it is satisfied for any point in $\mathscr{H}$).

(ii) If $\{T_n\} \in G$ is a sequence for which $\{\theta_n(T_n)\}$ converges to a Möbius transformation $T_\infty$ then $T_\infty \in H$ and $T_n = \theta^{-1}(T_\infty)$ for all large $n$.

(iii) Given a neighbourhood $V$ of the punctures on $\partial\mathscr{M}(H)$ then for all large $n$ there is a conformal map $f_n$ of $\Omega(H) - \{\pi^{-1}(V)\}$ into $\Omega(G_n)$ which induces the isomorphism $\theta_n\theta^{-1}: H \to G_n$.

The only groups $G$ for which $\partial\mathbf{T}(G)$ is even partly understood are Fuchsian and Schottky groups (see Sections 7.3, 7.4). Jørgensen has made a deep study of $\partial\mathbf{T}(G)$ for $G$ a Fuchsian group representing a once-punctured torus and has discovered some remarkable and beautiful groups in $\mathbf{V}(G) - \mathbf{T}(G)$ if not on $\partial\mathbf{T}(G)$. He too recognized the equivalence of (i) and (ii) and has shown in [**14**] how the question of cusp convergence can be interpreted analytically in terms of how the coefficients of $\theta_n\theta^{-1}(T)$ converge to those of $T$ for each parabolic transformation $T \in H$. A discussion of boundary convergence is also contained in Abikoff [**1** and **2**].

**7.3** Suppose $G$ is a Fuchsian group representing a compact surface $S$ of genus $g$. The deformation space $\mathbf{T}(G)$ of dimension $6g - 3$ is biholomorphically equivalent to $T_g \times T_g \times \mathrm{PSL}(2, \mathbb{C})$. By a result of Kra [**16**], $\mathbf{V}(G) - (\mathbf{T}(G) \cup \partial\mathbf{T}(G))$ has a non-empty interior.

What lies on $\partial\mathbf{T}(G)$? Several years ago we noticed (see [**10**]) that if $H$ is a geometrically finite $B$-group and $\theta: G \to H$ an isomorphism then $(H, \theta)$ is the limit in the cusp topology of a sequence in $\mathbf{T}(G)$. From the point of view of $G$, one way of obtaining a sequence in $\mathbf{T}(G)$ which approaches in the cusp topology a cusp on $\partial\mathbf{T}(G)$ is as follows. Take $n \leqslant 3g - 3$ mutually disjoint simple loops $\{\gamma_i\}$ which

(i) are non-trivial and determine distinct free homotopy classes in $\mathscr{M}(G)$,

(ii) lie on *one* component of $\partial\mathscr{M}(G)$.

Remove a thin annulus about $\gamma_i$ and sew in ever thicker ones (larger moduli), $1 \leqslant i \leqslant n$. The procedure determines a sequence in $\mathbf{T}(G)$ a subsequence of which has the asserted property. The operation itself is called *pinching*. It is not known if condition (ii) can be dropped; it is not known whether all geometrically finite groups isomorphic to $G$ lie on $\partial\mathbf{T}(G)$.

The group of automorphisms Aut $G$ of $G$ acts as a birational transformation on $\mathbf{V}(G)$ preserving $\mathbf{T}(G)$. The orbit of a given $x \in \mathbf{T}(G)$ under Aut $G$ has no accumulation point in $\mathbf{T}(G) \cup \partial\mathbf{T}(G)$.

In the normalized space $\mathbf{V}_0(G)$, $\mathbf{T}_0(G) \cong \mathbf{T}_g \times \mathbf{T}_g$ also has a relative boundary $\partial\mathbf{T}_0$. Fix the conformal structure of one of the components of $\partial\mathcal{M}(G)$. This gives a slice $\mathbf{T}' \cong \mathbf{T}_g$ of $\mathbf{T}_0(G)$ whose closure $\mathbf{T}' \cup \partial\mathbf{T}'$ in $\mathbf{T}_0(G) \cup \partial\mathbf{T}_0(G)$ is compact. In fact, $\partial\mathbf{T}'$ is just the Bers boundary of Teichmüller space $\mathbf{T}_g$.

Fixing a Bers slice $\mathbf{T}' \equiv \mathbf{T}_g$, it is possible to define a projection of $\mathbf{T}_0(G) \cup \mathbf{C}(G)$ onto $\mathbf{T}_g \cup \mathbf{C}$ where $\mathbf{C}$ denotes the cusps on $\partial\mathbf{T}_g$. The projection of the normalized group Aut $G$/Inner Aut $G$ gives an action of the *Teichmüller modular group* $\Gamma$ on $\mathbf{T}_g \cup \mathbf{C}$. Suppose for instance $\tau \in \Gamma$ is a Dehn twist. Then given $x \in \mathbf{T}_g$, $\lim \tau^n(x)$ is a cusp $c$ but the convergence is not in the cusp topology, it is a "tangential" convergence. Approaching $c$ in the cusp topology, which it is possible to do, has to do with "pinching" rather than "twisting". Either way $c$ is being approached but the "tangential" approach is distinguished by its destruction of the fundamental groups (e.g. in $\mathbf{T}_0(G)$ there is no convergence of $\tau^n(x)$).

In the case $g = 1$, $\mathbf{T}_1$ is the upper half plane $\mathcal{H}$ and $\Gamma$ can be interpreted to be the classical modular group PSL(2, $\mathbb{Z}$) so that $\mathbf{T}_1/\Gamma$ is the once-punctured sphere. The set of cusps $\mathbf{C}$ comprises the rational points on $\mathbb{R}$ plus the point at $\infty$. Each cusp corresponds to the doubly-punctured sphere. The cusp toplogy reduces to the ordinary horocyclic topology, i.e. precisely that induced by horocyclic discs about the punctures. The transformation $\tau: z \to z + 1$, for example, is a Dehn twist.

Full details will appear in [**10**]. The main result is as follows.

THEOREM. $\mathbf{T}_g \cup \mathbf{C}/\Gamma$ *is a simply connected, compact analytic space.*

The compactness is due to Harvey (these proceedings and [**12**]) and the simply connectivity is an extension of Maclachlan's result [**18**]. By matching our work with some forthcoming work of Knudson-Mumford it is in fact true that our compactification is a projective algebraic variety. Bers [**8**] using different methods has also obtained a compactification of moduli space which is an analytic space and, via Knudson-Mumford, is a projective variety as well.

COROLLARY [**10**]. *If* $g \geqslant 4$ *the Teichmüller modular group is finitely presented.*

McCool [**19**] has given a purely algebraic proof of this which includes the case $g = 3$ (the case $g = 2$ is already known).

**7.4** Suppose $G$ is a Schottky group of genus $g$. It follows from Chuckrow [9] that most points of $\partial\mathbf{T}(G) \subset \mathbb{P}_3^g$, namely those that correspond to groups without parabolic elements, are discrete but not Kleinian. The Isomorphism Theorem implies such groups cannot be geometrically finite. All free Fuchsian groups of rank $g$ correspond to points of $\mathbf{T}(G) \cup \partial\mathbf{T}(G)$. However, it is not known whether all free groups of rank $g$ which are geometrically finite also correspond to points there.

A particularly interesting open subset $\mathscr{S}$ of $\mathbf{T}(G)$, called the *classical Schottky space*, consists of those points $(H, \theta) \in \mathbf{T}(G)$ with the following property. There exist $g$ pairs of circles $(C_j, C_j')$ in $\mathbb{C}$, the total collection forming the boundary of a $2g$-connected region, and a set of generators $T_1, \ldots, T_g$ of $H$ such that $T_j$ maps the exterior of $C_j$ onto the interior of $C_j'$.

We showed [21] that every point of the closure $\mathscr{S}^-$ of $\mathscr{S}$ in $\mathbf{T}(G) \cup \partial\mathbf{T}(G)$ is Kleinian and therefore $\mathscr{S} \neq \mathbf{T}(G)$. Zarrow [37] proved that $\mathscr{S}$ is connected and he has found specific points $(\mathrm{H}, \theta) \in \mathbf{T}(\mathrm{G})$ which do not lie in $\mathscr{S}$.

## REFERENCES


1. W. Abikoff, On boundaries of Teichmüller spaces and on Kleinian groups III, *Acta Math.* **134** (1975) 211–237.
2. W. Abikoff, Degenerating families of Riemann surfaces, *Ann. of Math.* **105** (1977) 29–44.
3. R. D. M. Accola, Invariant domains for Kleinian groups. *Amer. J. Math.* **88** (1966) 329–336.
4. L. V. Ahlfors, Finitely generated Kleinian groups, *Amer. J. Math.* **86** (1964) 413–429 and **87** (1965), 759.
5. L. V. Ahlfors, Fundamental polyhedrons and limit sets of Kleinian groups. *Proc. Nat. Acad. Sci. U.S.A.* **55** (1966) 251–254.
6. L. Bers, On boundaries of Teichmüller spaces and on Kleinian groups. *I. Ann. Math.* **91** (1970) 570–600.
7. L. Bers, Spaces of Kleinian Groups. *in* "Maryland Conference in Several Complex Variables I." Lecture Notes in Math, No. 155, Springer-Verlag.
8. L. Bers, Deformations and moduli of Riemann surfaces with nodes and signatures. *Math. Scand.* **36** (1975) 12–16.
9. V. Chuckrow, On Schottky groups with applications to Kleinian groups, *Ann. of Math.* **88** (1968) 47–61.
10. C. Earle and A. Marden, To appear.
11. F. W. Gehring, Rings and quasiconformal mappings in space, *Trans. Amer. Math. Soc.* **103** (1962) 353–393.
12. W. Harvey, Chabauty spaces on discrete groups, *in* "Discontinuous Groups and Riemann Surfaces" (ed. L. Greenberg) Ann. of Math. Stud. No. 79, 1974.
13. T. Jørgensen, On discrete groups of Möbius transformations, *Amer. J. Math.* **98** (1976) 839–749.
14. T. Jørgensen, On reopening of cusps, to appear.

15. T. Jørgensen, A note on subgroups of SL(2, ℂ), to appear in *Quart. J. Math.*
16. T. Jørgensen, to appear.
17. I. Kra, Deformations of Fushsian groups II. *Duke math. J.* **38** (1971), 449–508.
18. C. Maclachlan, Modulus space is simply connected. *Proc. Amer. Math. Soc.* **29** (1971), 85–86.
19. McCool, Some finitely presented subgroups of the automorphism group of a free group, to appear.
20. A. Marden, The geometry of finitely generated Kleinian groups. *Ann. of Math.* **99** (1974) 383–462.
21. A. Marden, Schottky groups and circles, *in* "Contributions to Analysis," (L. V. Ahlfors *et al.* eds.) Academic Press, New York and London 1974.
22. A. Marden, I. Richards, B. Robin, On the regions bounded by homotopic curves. *Pacific J. Math.* **16** (1966) 337–339.
23. B. Maskit, Construction of Kleinian groups, *in* 'Proc. Conference on Complex Analysis at Minneapolis." Springer–Verlag, New York, 1965.
24. B. Maskit, A characterization of Schottky groups. *J. Analyse Math.* **19** (1967) 227–230.
25. B. Maskit, On a class of Kleinian groups. *Ann. Acad. Sci. Fenn.* **442** (1969).
26. B. Maskit, On boundaries of Teichmüller spaces II. *Ann. of Math.* **91** (1970) 607–639.
27. B. Maskit, Moduli of Riemann surfaces, *Bull. Amer. Math. Soc.* **80** (1974) 773–777.
28. B. Maskit, On the classification of Kleinian groups: I-Koebe groups. *Acta math.* **135** (1975) 249–270.
29. G. D. Mostow, "Strong Rigidity of Locally Symmetric Spaces," Ann. of Math. Studies, No. 78, 1973.
30. R. Riley, A quadratic parabolic group. *Math. Proc. Camb. Philos. Soc.* **77** (1975) 281–288.
31. G. P. Scott, "An introduction to 3-manifolds," Univ. of Maryland Lecture notes, 1974.
32. U. Srebro, Groups of linear fractional transformations. *Duke Math. J.* **34** (1967) 49–52.
33. T. W. Tucker, A correction to a paper of A. Marden. *Ann. of Math.* **102** (1975) 565–566.
34. F. Waldhausen, On irreducible 3-manifolds which are sufficiently large. *Ann. of Math.* **87** (1968) 56–88.
35. N. J. Wielenberg, Discrete Möbius Groups: fundamental polyhedra and convergence. *Amer. J. Math.* to appear.
36. H. Yamamoto, A theorem on limits of Kleinian groups, *Tôhoku Math. J.* **27** (1975) 273–283.
37. R. Zarrow, Classical and non-classical Schottky groups. *Duke Math. J.* **42** (1975) 717–724.

# 9. Spaces of Discrete Groups

W. J. Harvey

*King's College, London, England*



## INTRODUCTION

The theme of these notes is the study of moduli of surfaces and Fuchsian groups, approached rather in the spirit of Weil's fundamental work on deformations of discrete groups. It is assumed that the reader has already acquired an understanding of surface topology and the fundamental facts about discontinuous groups of Möbius transformations as presented in Chapters 1 and 2 of this volume. Some knowledge of topological and Lie groups would be useful but not essential.

In subject matter I have kept fairly strictly to the content of my Cambridge lectures, but I have attempted to provide, at least in outline, proofs of all stated results. In particular in Section 1 there is a description of the global parametrization of Teichmüller space using decompositions in the manner of Nielsen and Fenchel. The local structure of the space of unmarked Fuchsian groups is approached in Section 2 via a generalization of a theorem of Macbeath relating variation of group to that of an associated finite polygon. Together with the compactness theorem of Chabauty, this makes possible the main result of Section 3, which is an embedding of the moduli space for a fixed group of finite area as a dense open submanifold of a compact (real) analytic space.

There is of course a close family relationship between this course and those of Birman, Earle and Marden in this volume. The connection between the material in Section 3 and the study of moduli described in Marden's chapter is especially marked—I have not attempted to make the link explicit here, as that would have involved too many preliminaries on the complex analytic aspects of Teichmüller theory which are not otherwise involved in the discussion.

I must acknowledge here my indebtedness to Lipman Bers whose lectures on degeneration of Riemann surfaces introduced me to this subject, and to D. Mumford, whose paper [**36**] suggested that I might be able to contribute to it.

## 1. SPACES OF HOMOMORPHISMS

### 1.1 Moduli of elliptic curves

The analytic study of spaces of Riemann surfaces is to some extent modelled on the situation for surfaces of genus 1, or in other words elliptic curves defined over the complex numbers. This theory is largely classical although many aspects of the deeper arithmetic structure are of modern vintage (see for example the collection [**1**]). It will be beneficial for us to examine as a preliminary the rough outline of this case, which presents many of the basic features appearing in higher genus while remaining elementary in those aspects which concern us here.

Let $\Gamma$ denote a lattice in $\mathbb{C}$, which means that $\Gamma$ is a free Abelian group of rank two, generated by complex numbers $\lambda_1, \lambda_2$ which are linearly independent over $\mathbb{R}$. Then $\Gamma$ is a discrete subgroup of the topological group $\mathbb{C}$, and the quotient space $\Gamma\backslash\mathbb{C}$ with its inherited complex structure is a compact Riemann surface of genus 1. Conversely a Riemann surface $X$ of genus 1 can always be represented in this way by choosing a basis for the holomorphic differentials on $X$ consisting of a single differential $\omega$, and integrating it over a basis $\{a_1, b_1\}$ for the first homology group $H_1(X, \mathbb{Z})$, which is a free Abelian group of rank 2. This yields a pair $\int_{a_1}\omega$, $\int_{b_1}\omega$ of numbers which generate a lattice $L$ such that $X$ is isomorphic to $L\backslash\mathbb{C}$ (for more details the reader is referred to Siegel's book [**45** I]). We denote by $\mathscr{R}$ the space of generators $\boldsymbol{\lambda} = (\lambda_1, \lambda_2)$ (so that $\lambda_1/\lambda_2$ is non-real). Two elements $\boldsymbol{\lambda}, \boldsymbol{\lambda}' \in \mathscr{R}$ generate the same lattice up to a change of basis if and only if $\boldsymbol{\lambda}' = A\boldsymbol{\lambda}$ for some matrix $A \in \mathrm{GL}_2(\mathbb{Z})$. Moreover if two generating sets $\boldsymbol{\lambda}$, $\boldsymbol{\lambda}'$ are such that $\boldsymbol{\lambda}' = c\boldsymbol{\lambda}$, with $c \in \mathbb{C}^*$ a non-zero constant, then the corresponding Riemann surfaces are conformally equivalent and it is natural therefore to identify them. One builds a space of equivalence classes by passing from $\boldsymbol{\lambda} = (\lambda_1, \lambda_2)$ to $\tau = \lambda_1/\lambda_2 \in \mathbb{C} - \mathbb{R}$. Since the points $\tau$ and $-\tau$ represent the same surface (with opposite orientation) it suffices to take as parameter space for the classes of generating sets (often thought of as tori with a "marking", that is a distinguished set of generators for the fundamental group) the upper half plane $\mathscr{H} = \{\tau : \mathrm{Im}\ \tau > 0\}$. The final step in constructing the so-called *moduli space* of tori, which is to be a space whose points correspond to conformal isomorphism classes of tori, is to consider the result on $\mathscr{H}$ of the action of $\mathrm{GL}_2(\mathbb{Z})$ on $\mathscr{R}$ mentioned above. In fact one sees that

$$\begin{pmatrix} \lambda_1' \\ \lambda_2' \end{pmatrix} = \begin{pmatrix} a & b \\ c & d \end{pmatrix} \begin{pmatrix} \lambda_1 \\ \lambda_2 \end{pmatrix}$$

with $\tau = \lambda_1/\lambda_2$, $\tau' = \lambda_1'/\lambda_2'$ both in $\mathscr{H}$ if and only if $\tau' = (\alpha\tau + b)/(c\tau + d)$, with $ad - bc = 1$, and so the action on $\mathscr{H}$ is that of the discontinuous *elliptic modular group* $G = \mathrm{PSL}_2(\mathbb{Z})$ of Möbius transformations. It follows that the quotient $G\backslash\mathscr{H} = \mathscr{X}$ carries a complex structure by projection from $\mathscr{H}$. Using the theory of modular forms [**1, 45**] one can show that adjoining a single point to $\mathscr{X}$ corresponding to the *cusp* at $\infty$ of $G$ results in a compact space $\overline{\mathscr{X}}$ identifiable with the Riemann sphere $\mathbb{P}_1(\mathbb{C})$ by the mapping induced by the celebrated $G$-automorphic function $j(\tau)$.

One can also proceed from $\mathscr{R}$ to $\mathscr{X}$ by first identifying pairs $\lambda \in \mathscr{R}$ modulo the action of $GL_2(\mathbb{Z})$ and then taking the quotient of that by the action of $\mathbb{C}^*$. The intermediate space $\mathscr{S}$ is the set of (unmarked) lattices in $\mathbb{C}$ since $GL_2(\mathbb{Z})$ has the effect of making all choices of marking equivalent, and by using the fact that $\mathscr{R}$ is itself isomorphic to $GL_2(\mathbb{R})$ one can view $\mathscr{S}$ as a homogeneous space

$$\mathscr{S} \cong GL_2(\mathbb{Z})\backslash GL_2(\mathbb{R}),$$

with the natural topology. It can be shown, using the association to each lattice $L$ of the pair $(g_2, (L), g_3(L))$ where $g_2(L) = 60\sum \lambda^{-4}, g_3(L) = 140\sum \lambda^{-6}$ (over $\lambda \in L - \{0\}$) that $(x, y) \in \mathbb{C}^2$ corresponds to a lattice $L$ if and only if

$$g_2^3 - 27\, g_3^2 \neq 0;$$

which implies that the space $\mathscr{S}$ is the complement in $\mathbb{C}^2$ of the algebraic curve $Y = \{x^3 - 27y^2 = 0\}$. The fundamental group of $\mathscr{S}$ is a certain central $\mathbb{Z}$-extension of $SL_2(\mathbb{Z})$ known to topologists as the group of the *trefoil knot* (see [**43**], [**35**]); it is also isomorphic to Artin's *braid group* B(3) on three strings (see Chapter 6), which is the fundamental group of the configuration space $X_3$ of triples of distinct points in $\mathbb{C}$.

*Exercise.* Construct a homeomorphism of the configuration space $X_3$ onto $\mathbb{C} \times \{\mathbb{C}^2 \backslash Y\}$ which exhibits the isomorphism of B(3) with the trefoil group. (c.f. [**43**] Section 1.5).

These results will each have a counterpart in the theory of Fuchsian groups and surfaces of higher genus.

## 1.2 The deformation space of a discrete group

In analogy with the preceding section, we fix an isomorphism class of object—in this case a discrete subgroup $\Gamma$ of some Lie group $\mathscr{L}$ (usually the group of Möbius transformations)—and construct a space consisting of all objects isomorphic to it in an appropriate geometrical sense. Where possible we shall use a general setting, only invoking the special nature of Fuchsian and Kleinian groups where it seems appropriate (or unavoidable).

*Definition.* The *deformation space* of $\Gamma$ in $\mathscr{L}$, denoted by $\mathscr{R}(\Gamma, \mathscr{L})$ (or simply $\mathscr{R}(\Gamma)$), is the space of all injective homomorphisms $\theta : \Gamma \to \mathscr{L}$ such that $\theta(\Gamma)$ is closed (and hence discrete).

It is natural to view $\mathscr{R}(\Gamma, \mathscr{L})$ as a subset of the space $\mathrm{Hom}(\Gamma, \mathscr{L})$ of all homomorphisms of $\Gamma$ into $\mathscr{L}$ and to use the topology induced on them as subsets of $\mathscr{L}^{\Gamma}$ with the product topology. If we assume for simplicity, as we shall from now on, that $\Gamma$ is *finitely presented*, then a choice of generators $\gamma_1, \ldots, \gamma_n$ with basic relators $\{w_j(\gamma_1, \ldots, \gamma_n)\}$ for $\Gamma$ leads to a bijective mapping between $\theta \in \mathrm{Hom}(\Gamma, \mathscr{L})$ and points $g = (g_1, \ldots, g_n) \in \mathscr{L}^n$ with $g_i = \theta(\gamma_i)$ which satisfy the equations $w_j(g) = \mathrm{e}$ in $\mathscr{L}$. This is readily shown to be a homeomorphism. A different choice of presentation for $\Gamma$ results in an analytically equivalent model of $\mathrm{Hom}(\Gamma, \mathscr{L})$, as can be seen by writing the new set of generators as words in the old ones, and we shall identify $\mathrm{Hom}(\Gamma, \mathscr{L})$ and the subspace $\mathscr{R}(\Gamma, \mathscr{L})$ with the subspaces of $\mathscr{L}^n$ corresponding to a convenient choice of presentation.

If $\Gamma$ is *cocompact* in $\mathscr{L}$ (that is, if the quotient space $\Gamma\backslash\mathscr{L}$ is compact) then one can show that some neighbourhood of the identity homomorphism $\mathrm{id}_{\Gamma}$ in $\mathrm{Hom}(\Gamma, \mathscr{L})$, consists entirely of points $\theta$ of $\mathscr{R}(\Gamma, \mathscr{L})$ with $\theta(\Gamma)$ also cocompact [**48**]. However in the case where $\mathscr{L}$ is $\mathrm{PSL}(2, \mathbb{R})$, the group of real Möbius transformations, one can give a more precise form to the local (and even global) structure of $\mathscr{R}(\Gamma, \mathscr{L})$ for $\Gamma$ *any* finitely generated Fuchsian group, and this we shall discuss in the next sections.

*Exercise.* Verify the result of Weil above for $\Gamma$ a lattice in $\mathbb{R}^n$.

*Remark.* If the quotient space $\Gamma\backslash\mathscr{L}$ is not compact then $\mathscr{R}(\Gamma, \mathscr{L})$ is no longer open in $\mathrm{Hom}(\Gamma, \mathscr{L})$, as we shall see later for Fuchsian groups.

### 1.3 The normalized deformation space

If $\theta \in \mathscr{R}(\Gamma, \mathscr{L})$ then one obtains for each element $\alpha$ in the group $\mathrm{Aut}\,\mathscr{L}$ of *continuous automorphisms* of $\mathscr{L}$ a new isomorphism $\theta' = \alpha^*(\theta) \in \mathscr{R}(\Gamma, \mathscr{L})$ defined by

$$\alpha^*(\theta)(\gamma) = \alpha(\theta(\gamma)) \quad \text{for} \quad \gamma \in \Gamma.$$

Moreover, if $X$ denotes the symmetric space associated to $\mathscr{L}$, the quotient manifolds $\theta(\Gamma)\backslash X$ and $\theta'(\Gamma)\backslash X$ are isometric. One terms such a deformation of $\theta$ *trivial*, and it is a reasonable step to remove them from consideration by taking the quotient of $\mathscr{R}(\Gamma, \mathscr{L})$ with respect to this action of $\mathrm{Aut}\,\mathscr{L}$. This led us in the case of lattices in $\mathbb{C}$ to the space $\mathscr{R}/\mathbb{C}^* = \mathscr{H}$. If for example we consider for $\mathscr{L} = \mathrm{PSL}_2(\mathbb{R})$ a Fuchsian group $\Gamma$ and an isomorphism $\tilde{\imath} : \Gamma \to \mathscr{L}$

given for $t \in \mathscr{L}$ by conjugation

$$\gamma \mapsto \tilde{t}(\gamma) = t\gamma t^{-1},$$

then the image group represents a Riemann surface $\tilde{t}(\Gamma)\backslash\mathscr{H}$ conformally isomorphic to $\Gamma\backslash\mathscr{H}$. Conversely any directly conformal map between Riemann surfaces $\Gamma\backslash\mathscr{H}$ and $\Gamma'\backslash\mathscr{H}$ can be lifted to a biholomorphic self-mapping of $\mathscr{H}$; this map must be a Möbius transformation by a standard application of the Schwarz Lemma, and it determines an inner automorphism of $\mathscr{L}$ which takes $\Gamma$ to $\Gamma'$ as above. Allowing the possibility of *orientation reversing* maps leads us to consider conjugation by elements of $PGL_2(\mathbb{R})$, the "*Hyperbolic*" *group* of fractional linear and anti-linear self mappings of $\mathscr{H}$. This is the group $\mathrm{Aut}(\mathscr{L})$ in the present example.

The action of $\mathrm{Aut}\ \mathscr{L}$ on $\mathrm{Hom}(\Gamma, \mathscr{L})$ is smooth. We denote by $\mathbf{T}(\Gamma)$ the quotient of $\mathscr{R}(\Gamma, \mathscr{L})$ with the induced topology. It is immediate from the definitions that if $\mathscr{L}$ is $PSL_2(\mathbb{R})$, the action is proper and free of fixed points. Furthermore the differential of the quotient map is everywhere surjective [**11**], so that local sections exist by the implicit function theorem. This shows that:

1.3.1 PROPOSITION. *$\mathscr{R}(\Gamma, \mathscr{L})$ is a principal fibre bundle over $\mathbf{T}(\Gamma)$ with fibre* $\mathrm{Aut}\ \mathscr{L}$ *if $\Gamma$ is Fuchsian.*

An important result of Weil (op. cit.) states that for a large class of Lie groups $\mathscr{L}$, which specifically *excludes* the group of Möbius transformations, all deformations $\theta \in \mathscr{R}(\Gamma, \mathscr{L})$ near the identity element are obtained by conjugation with an element of $\mathscr{L}$, if $\Gamma$ has quotient $\Gamma\backslash\mathscr{L}$ compact. This answered a conjecture of Selberg in his fundamental paper [**41**], which marked the beginning of the general study of deformations of discrete groups. Further developments in this direction may be found in the survey article [**42**]. The result of Weil means that $\mathbf{T}(\Gamma)$ is then a discrete set. It is in fact now known by results of Mostow (see Chapter 12) that $\mathbf{T}(\Gamma)$ is a single point for a wide class of Lie groups $\mathscr{L}$ and discrete subgroups $\Gamma$ with the property that the quotient space $\Gamma\backslash\mathscr{L}$ has *finite volume* in the measure induced by projecting the Haar measure from $\mathscr{L}$.

### 1.4 Deformation spaces for Fuchsian groups

We shall devote the rest of this chapter to a detailed study of the case where $\mathscr{L}$ is the real Möbius group, $PSL_2(\mathbb{R})$. A good deal more can be said than in the general situation because our knowledge of the geometry of $PSL_2(\mathbb{R})$ and the symmetric space $\mathscr{H}$ is more complete.

Assume first that $\Gamma$ is a Fuchsian *surface group*, that is $\Gamma$ is isomorphic to the fundamental group of a compact surface of genus $g \geqslant 2$. Such groups exist for each value of $g$, either by the uniformization theorem or by Poincaré's theorem on fundamental polygons. Now the isomorphisms $\theta : \Gamma \to \mathscr{L}$ all define compact Riemann surfaces $\theta(\Gamma)\backslash\mathscr{H} = S_\theta$ homeomorphic to $\Gamma\backslash\mathscr{H} = S$. Moreover a theorem of Nielsen (discussed in Chapters 1 and 6) implies that there is a homeomorphism $f: \mathscr{H} \to \mathscr{H}$ such that $f$ *induces* $\theta$, in the sense that for $\gamma \in \Gamma$ we have

$$(1.1) \qquad \theta(\gamma) = f \circ \gamma \circ f^{-1}.$$

Such an $f$ also projects to a homeomorphism between $S$ and $S_\theta$. Conversely any homeomorphism of surfaces $\phi : S \to S'$ can be lifted to a homeomorphism $\Phi$ of the universal covering $\mathscr{H}$ onto itself which makes the diagram below commute.

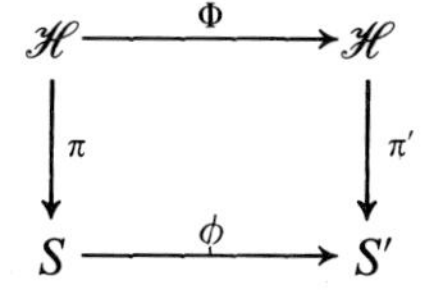


One obtains an isomorphism $r$ between the groups $\Gamma, \Gamma'$ of covering transformations of $\pi, \pi'$ respectively, given by the rule

$$\gamma \mapsto \Phi \circ \gamma \circ \Phi^{-1} = r(\gamma).$$

Since the lifting process is unique only up to conjugation by a Möbius transformation, it is more meaningful to regard the homeomorphism $\phi$ as determining an element $[r] \in \mathbf{T}(\Gamma)$. This discussion means that one may regard $\mathbf{T}(\Gamma)$ as the *Teichmüller space* of the Riemann surface $S = \Gamma\backslash\mathscr{H}$ (see Chapter 5, Sections 8 and 9, and [**5**]).

In order to expand on this relationship between deformations of the Fuchsian groups and those of their quotient surfaces, we need to adjust slightly the definitions of deformation spaces given above. Let $\Gamma$ now be a Fuchsian group with signature $(g; n; \nu_1, \ldots, \nu_k; m)$, which we recall (see Chapter 7,

Section 1.4) means that there is a presentation for $\Gamma$ of the following form:

$$(1.2)\quad \begin{cases} \text{generators: } A_1, B_1, \ldots, A_g, B_g, P_1, \ldots, P_n, X_1, \ldots, X_k, F_1, \ldots, F_m; \\ \text{relations: } \prod_{j=1}^{g} [A_j, B_j] P_1 \ldots P_n X_1 \ldots X_k F_1 \ldots F_m = \text{id}, \end{cases}$$

$$X_i^{\nu_i} = \text{id}, \qquad i = 1, \ldots, k.$$

The quotient surface $\Gamma \backslash \mathscr{H}$ has genus $g$, $n$ punctures, $k$ prescribed branch points and $m$ boundary curves; and the elements $\{P_1, \ldots, P_n\}$, (resp. $\{F_1, \ldots, F_m\}$) form a complete set of conjugacy classes of *boundary* parabolic (resp. hyperbolic) elements, which means that they correspond to simple (primary) loops around the various boundary components of $\Gamma \backslash \mathscr{H}$.

If $\Gamma$ has no torsion then the elements $\{X_i\}$ are absent. The signature in the torsion-free case will be abbreviated to read $(g; n; m)$.

We shall also use the term (signature of) *type* $(g; d)$ to describe any signature $(g; n; m)$ such that $n + m = d$.

*Definition.* A homomorphism $\theta : \Gamma \to G$ between Fuchsian groups is called *allowable* if the following condition holds:
(*): $\theta(\gamma)$ is a parabolic (resp. boundary hyperbolic) element if and only if $\gamma$ is of that type.

The term *type-preserving* is also used for this.

*Definition.* The set of allowable homomorphisms in $\mathscr{R}(\Gamma, \mathscr{L})$ is denoted $\mathscr{R}_0(\Gamma)$. A group $G$, together with an isomorphism $\theta \in \mathscr{R}_0(\Gamma)$ with $\theta(\Gamma) = G$, is called a *marked group*.

*Note 1.* If $\Gamma$ is a surface group, then the conditions are void, so $\mathscr{R}_0 = \mathscr{R}$.

*Note 2.* It is important to notice that this condition is a natural one to impose from the geometric point of view, because it is precisely the hypothesis necessary in order to extend the Nielsen theorem and interpret the isomorphisms $\theta$ as induced by homeomorphisms between the underlying quotient surfaces which preserve the type (and number) of boundary components (see Theorem 1.4.2. below). As an example of what can go wrong if one omits the restriction, we observe that the free group on two generators occurs as the fundamental group of both the one-holed torus and the three-holed sphere, and so non-allowable isomorphisms need not respect the signature of a group.

One can interpret the condition (*) in another way. We associate to each hyperbolic element $\gamma$ of a group $\Gamma$ its *axis* $A_\gamma$ (which we recall is the geodesic in $\mathscr{H}$ joining the two fixed points of $\gamma$); then the *boundary* hyperbolic elements $T$ of $\Gamma$ are distinguishable from the others by the property that the axis $A_T$ does not intersect that of any other $\gamma \in \Gamma$ which lies outside the maximal cyclic subgroup of $\Gamma$ containing $T$. This follows easily from examination of the Nielsen region $K(\Gamma)$ (Chapter 7, Section 1). In other words, if $T$ is a *primary* hyperbolic boundary element than $A_S \cap A_T = \varnothing$ unless $S$ is a power of $T$. Now consider the following condition on a homomorphism $\theta: \Gamma \to \Gamma'$, known as the *intersecting axes property*:

(**): the axes of $\gamma, \delta \in \Gamma$ cross if and only if those of $\theta(\gamma), \theta(\delta) \in \Gamma'$ cross.

This condition has the advantage of being applicable to infinitely generated groups.

1.4.1. Proposition. *An isomorphism $\theta$ between Fuchsian groups (finitely generated) is allowable if and only if it has the intersecting axes property* (**).

We refer to Marden [**31**] (see also Tukia [**47**]) for a proof, noting that the sufficiency of property (**) is clear from the characterization given of boundary elements. It is an instructive exercise to try to prove the result even for the simplest case, which is that of a group of signature type (0, 3).

The quotient of the space $\mathscr{R}_0(\Gamma)$ by the action of Aut $\mathscr{L}$ is denoted by $\mathbf{T}_0(\Gamma)$. One can identify $\mathbf{T}_0(\Gamma)$ with the *Teichmüller space* of the group $\Gamma$ as defined in Chapter 5, (Sections 5–8) by the result below, extending the Nielsen theorem mentioned in note 2.

1.4.2 Theorem. *Let $\theta : \Gamma \to \Gamma'$ be an allowable isomorphism between Fuchsian groups. Then there is a homeomorphism $f: \mathscr{H} \to \mathscr{H}$ which induces $\theta$ in the sense of* (1.1).†

A short proof of the theorem and an interesting discussion of the history behind it may be found in Marden (op. cit.). It can be shown that there is even a *quasiconformal* $f_\theta$ which induces $\theta$; setting $\mu_\theta$ to be the Beltrami derivative of $f_\theta$, we therefore have an association $\theta \mapsto \mu_\theta$ between $\mathscr{R}_0(\Gamma)$ and the space $\mathbf{M}(\Gamma)$ of Beltrami differentials with norm $<1$ (Chapter 5, Section 6). Conversely, given $\mu \in \mathbf{M}(\Gamma)$, by the existence of solutions to the Beltrami equation (Chapter 4, Theorem 6.1) there is a homeomorphism $\omega_\mu: \mathscr{H} \to \mathscr{H}$ which is a

† Such an isomorphism is termed *geometric*.

solution to $\omega_{\bar{z}} = \mu\omega_z$ and this leads to a deformation $\theta_\mu$ of $\Gamma$ given by $\gamma \mapsto \omega_\mu \circ \gamma \circ \omega_\mu^{-1}$, for $\gamma \in \Gamma$ (see for example Chapter 5, Proposition 10.1). In fact one can normalize the $\omega_\mu$, since it is unique only up to a Möbius transformation, and thereby define a real analytic homeomorphism onto $\mathbf{T}_0(\Gamma)$ from the Teichmüller space of $\Gamma$ as defined in Chapter 5, Section 8.

We now state that basic theorem on the global structure of $\mathbf{T}_0(\Gamma)$.

1.4.3 THEOREM. *If $\Gamma$ has presentation as at* (1.2), *then $\mathbf{T}_0(\Gamma)$ is homeomorphic to a cell in $\mathbb{R}^N$ with $N = 6g - 6 + 2n + 2k + 3m$.*

This result goes back to Klein and Fricke, who used it as the basis for a proof (in certain cases only) via the "continuity method" that the space of Riemann surfaces of genus $g$ is a continuum of dimension $6g - 6$ (see [**13**] vol. II). Other proofs have been given by Fenchel and Nielsen (manuscript unpublished but alive) and more recently by Keen and others [**19**, **50**]. If one exploits the equivalence with Teichmüller space as defined in Chapter 5. Teichmüller's theorem [**3**] provides another proof.

Certain aspects of the geometrical approach to the proof of theorem 1.4.3 merit our attention. The techniques employed by the authors cited involve the same basic idea: that of a dissection of the surface $S = \Gamma\backslash\mathscr{H}$ into a union of subsurfaces by cutting along simple closed curves in $S$, the pieces having some fixed (elementary) topological type. One then combines parametrizations for the various subgroups representing the subsurfaces, with adjustments to ensure that where two pieces are neighbours (i.e. they have at least one common boundary curve) the appropriate parameters marry accordingly.

In the next four sections we shall in essence outline one way of carrying out this procedure, using as basic building block the sphere with three holes. With slight modifications the method would work for arbitrary groups but we restrict ourselves to the torsion-free case in the interests of brevity.

### 1.5 Teichmüller space for a three-holed sphere

Let $G$ be a hyperbolic triangle group which represents a three-holed sphere. Thus $G$ has a presentation $\langle V_1, V_2, V_3; V_1V_2V_3 = 1\rangle$, with each $V_i$ a primary hyperbolic boundary element whose axis $b_i$ in $\mathscr{H}$ projects to a simple loop on $S = G\backslash\mathscr{H}$ passing around a hole.

1.5.1 PROPOSITION. *The rule $[\theta] \mapsto (t_1, t_2, t_3)$, where $t_i = |\mathrm{tr}\ \theta(V_i)|$†, $i = 1, 2, 3$,*

† The function tr on $\mathscr{L}$ is defined only modulo $\pm 1$.

*defines a homeomorphism (real analytic) of* $\mathbf{T}_0(G)$ *onto the 3-cell* $\{2 < t_i: i = 1, 2, 3\}$.

*Proof.* (c.f. **[13]** II pp. 289–291.) One can verify directly that the group $G(t_1, t_2, t_3)$ generated by transformations $\{V_i\}$ whose matrices are

$$\bar{V}_3 = \begin{pmatrix} \frac{1}{2}(-t_1 - \sqrt{t_1^2 - 4}) & 0 \\ 0 & \frac{1}{2}(-t_1 + \sqrt{t_1^2 - 4}) \end{pmatrix}$$

$$\bar{V}_2 = \begin{pmatrix} \dfrac{1}{2}\left(t_2 + \dfrac{t_1 t_2 + 2t_3}{\sqrt{t_1^2 - 4}}\right) & \dfrac{J}{\sqrt{t_1^2 - 4}} \\ -\dfrac{J}{\sqrt{t_1^2 - 4}} & \dfrac{1}{2}\left(t_2 - \dfrac{t_1 t_2 + 2t_3}{\sqrt{t_1^2 - 4}}\right) \end{pmatrix}$$

with $J = (t_1^2 + t_2^2 + t_3^2 + t_1 t_2 t_3 - 4)^{\frac{1}{2}}$, has the stated traces for $V_1$, $V_2$ and $V_3$ (with $V_1^{-1} = V_2 V_3$) and represents a surface of type (0, 3) (see exercise below). Conversely it is easy to transform any given (marked) group in $\mathscr{R}_0(G)$ into this form by using conjugation within $\mathscr{L}$ to move the fixed points of the generators to the requisite positions. No two distinct triples can represent conjugate marked groups. ■

*Note.* These parameters may be given an attractive geometric flavour. Recall (Chapter 7, Section 1.4) that the trace of a hyperbolic element and its translation length $\lambda$ are related by

$$|\mathrm{tr}(V)| = 2 \cosh \tfrac{1}{2} \lambda(V).$$

It therefore follows that the variables $t_i$ are analytic functions of the *minimal lengths* $\{\lambda(\theta(V_i))\}$ *of geodesics* drawn around the holes of the surface corresponding to the deformation class $[\theta]$, and the lengths could equally well be used as parameters.

*Exercise 1.* Prove that $\mathrm{tr}(V_1 V_2^{-1} V_1^{-1} V_2) = J^2 + 2 > 2$ for $V_i$ as above.

*Exercise 2.* Prove that the axes $l_i$ of $V_i$, $i = 1, 2, 3$, satisfy $l_i \cap l_j = \varnothing$ if $i \neq j$ and deduce that the surface $G\backslash\mathscr{H}$ really does have three boundary components.

One can derive similarly the structure of the spaces $\mathbf{T}_0(G)$ for $G$ any group of signature type (0, 3) without torsion. Details are left to the reader. We summarize the facts as follows.

1.5.2 THEOREM. *Let G be any group of signature type* (0, 3). *Then* $\mathbf{T}_0(G)$ *is real-analytically a cell of dimension* $m \leqslant 3$, *where m is the number of hyperbolic boundary generators.*

This yields immediately a description of $\mathcal{R}_0(G)$ in these cases; since $\mathbf{T}_0(G)$ is contractible, Proposition 1.3.1 implies the following

1.5.3 COROLLARY. *If G is of type* (0, 3) *then* $\mathcal{R}_0(G)$ *is real-analytically homeomorphic to* $\mathbf{T}_0(G) \times \text{Aut}\,\mathcal{L}$.

In other words $\mathcal{R}_0(G)$ is a disjoint union of two spaces homeomorphic to the product of a cell and a circle, since $\text{Aut}\,\mathcal{L} \cong \mathcal{L} \cup \mathcal{L}$ and

$$\mathcal{L} = \mathrm{PSL}_2(\mathbb{R}) \cong S^1 \times \mathbb{R}^2. \qquad \blacksquare$$

We return now to the case where $G$ is a purely hyperbolic group of type (0, 3). It is a question of some interest to determine what the boundary of the space $\mathbf{T}_0(G)$ represents: if one (or more) of the parameters $t_i \to 2$, what happens to the group $G = G(\mathbf{t})$? For this purpose the normalization by fixed points used in Proposition 1.5.1 is not always suitable as the limiting points on the boundary of $\mathbf{T}_0(G)$ may represent groups having one or more parabolic generators. A detailed discussion of the problem occurs in [**13** Vol. II, pp. 425–429]. We shall only need for our purposes the following result, which states that topologically the set of all marked groups of type (0, 3) is homeomorphic to $[2, \infty)^3$.

1.5.4 PROPOSITION. *Let* $G_1$ *be any Fuchsian group with signature type* (0, 3) (*without torsion*), *and* $r: G \to G_1$ *an isomorphism, not necessarily type-preserving. Then* $[r]$ *belongs to the closure of* $\mathbf{T}_0(G)$. *In other words, any marked group representing a sphere with three boundary components is a limit of marked groups isomorphic* (*geometrically*) *to G.*

*Proof.* Given $G_1$ and generators $U_1$, $U_2$, $U_3$ (hyperbolic or parabolic) construct a fundamental polygon $\mathcal{P}_1$ in $\mathcal{H}$ with sides paired by the $\{U_i\}$ (e.g. a Ford region—see Chapter 2, Section 7). Given any $\varepsilon > 0$ it is possible† to construct a polygon which is close to $\mathcal{P}_1$ and which has sides paired by trans-

† See Section 1.9 for more details.

formations $V_1(\varepsilon)$, $V_2(\varepsilon)$, $V_3(\varepsilon)$ such that:

(i) $V_1(\varepsilon)$, $V_2(\varepsilon)$, $V_3^{-1}(\varepsilon) = V_1(\varepsilon)V_2(\varepsilon)$ are all hyperbolic;
(ii) $|\mathrm{tr}\, V_i(\varepsilon) - \mathrm{tr}\, U_i| < \varepsilon, \quad i = 1, 2;$
(iii) the axes of $V_1(\varepsilon)$ and $V_j(\varepsilon)$ are non-intersecting for $i \neq j$.

(This follows from Lemma 1.5.2 if all $U_i$ are hyperbolic). But for each group $G_\varepsilon = \langle V_1(\varepsilon), V_2(\varepsilon)\rangle$ the map $r: G \to G_\varepsilon$ taking $V_i$ to $V_i(\varepsilon)$ is in $\mathscr{R}_0(G)$. ■

*Remark.* Fundamentally, this result holds because such marked Fuchsian groups $(r, G_1)$ correspond to points $\mathbf{t} = t_1, t_2, t_3$ with

$$\phi(t) = t_1^2 + t_2^2 + t_3^2 + t_1t_2t_3 \leqslant 4,$$

and this set is naturally closed and connected.

If $(r, G_1) \in \mathscr{R}_0(G)$ then $\phi\ (G_1) < 4$.

## 1.6 Combination theorems

In this section we shall introduce the two basic combination methods needed in order to build other Fuchsian groups of finite type, used for this purpose by Klein and Fricke [**13**] and by Nielsen and Fenchel. We describe initially two basic examples; each illustrates a well-known construction in combinatorial group theory, which is given here an additional geometric structure mirroring its algebraic one.

1.6.1 LEMMA. *Let G and G′ be groups with signature* (0; 3), *generated by hyperbolic boundary elements* $\{V_i\}$, $\{V_i'\}$, *and assume that* $V_3 = V_1V_2$ *coincides with* $V_3' = V_1'V_2'$. *Assume furthermore that the Nielsen convex regions* $K(G)$, $K(G')$ *are separated by the axis* $l_3$ *of* $V_3$ *(which forms part of the boundary of each region in* $\mathscr{H}$*). If* $\mathscr{P}$, $\mathscr{P}'$ *denote fundamental polygons for G, G′ respectively, chosen so that the edges* $\mathscr{P} \cap l_3$ *and* $\mathscr{P}' \cap l_3$ *coincide, then the group* $G_1 = \langle G, G'\rangle$ *is discrete and has* $\mathscr{P} \cap \mathscr{P}' = \mathscr{P}_1$ *as fundamental region.*

*Remark.* This implies that $G_1$ has presentation given by the join of those for $G$ and $G'$ together with the linking relation $V_3 = V_3'$; in other words, denoting by $H$ the subgroup $\langle V_3\rangle$, we have that $G_1 = G \ast_H G'$ *is the geometric free product of G and G′ with the subgroup H amalgamated.*

*Proof.* To prove the discreteness, we show first that if $\gamma \in G_1$ then $\gamma \mathring{\mathscr{P}}_1 \cap \mathring{\mathscr{P}}_1 = \varnothing$ except when $\gamma = \text{id}$. Clearly we may assume $\gamma \notin H$ and write $\gamma = g_1 g_2 \cdots g_n$ in such a way that $g_i \in G - H$ or $G' - H$, and $g_i \in G$ if and only if $g_{i+1} \in G'$ for $i = 1, \ldots, n-1$; this is the "normal form" for an element $\gamma$. Now using the mapping properties of $G$, $G'$ it is easily seen that if $z \in \mathring{\mathscr{P}}_1$ then $g_1(z)$ lies in the semi-disc bounded by the axis $l_3$ which lies *outside* the Nielsen region for the subgroup to which $g_1$ belong; also $g_1(z) \notin \mathscr{P}_1$. Inductively we find that $\gamma(z) \notin \mathring{\mathscr{P}}_1$, so that $\gamma \neq \text{id}$ which means that no word in normal form is the identity. Since $h\mathring{\mathscr{P}}_1 \cap \mathring{\mathscr{P}}_1 = \varnothing$ for $h \in H - \{\text{id}\}$, we know that $\mathring{\mathscr{P}}_1 \cap \gamma\mathring{\mathscr{P}}_1 = \varnothing$ for all $\gamma$ in $G_1 - \{\text{id}\}$. It is elementary to show that $\bigcup_{g \in G_1} g\mathscr{P}_1$ fills up $\mathscr{H}$. ■

The construction is depicted in Fig. 1. $G_1$ has signature (0; 4), and represents a 4-holed sphere.

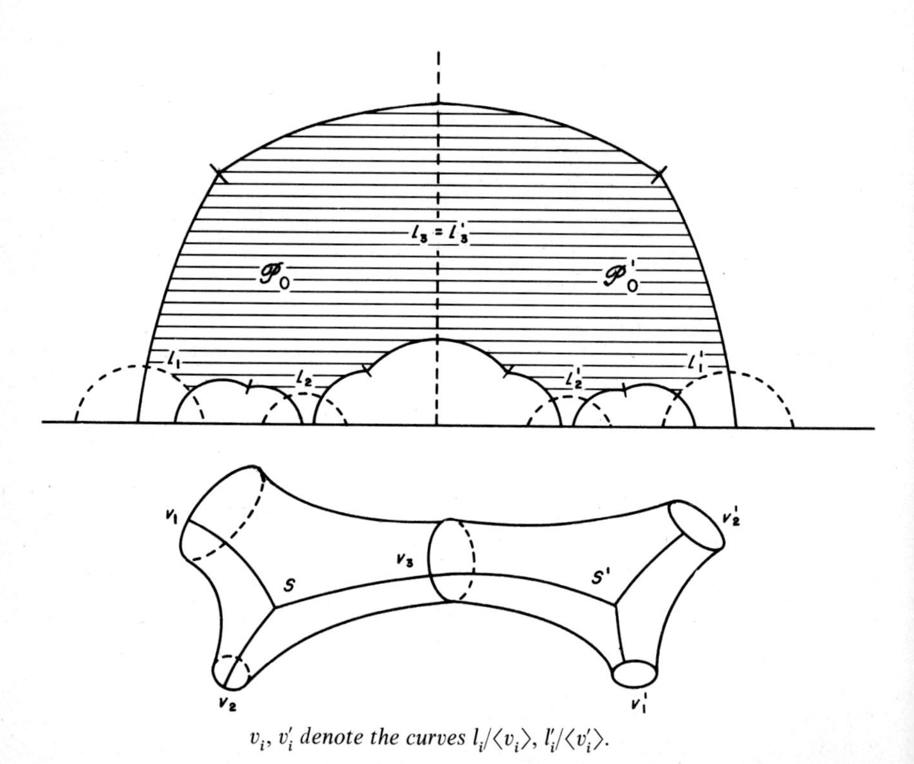


*$v_i$, $v_i'$ denote the curves $l_i/\langle v_i \rangle$, $l_i'/\langle v_i' \rangle$.*

*Fig. 1. $\mathscr{P}_0$, $\mathscr{P}_0'$ denote the intersections of $\mathscr{P}$, $\mathscr{P}'$ with their Nielsen regions $K(G)$, $K(G')$, and the axes of $V_i$, $V_i'$ are $l_i$, $l_i'$ respectively.*

1.6.2 LEMMA. *Let* $G = \langle V_1; V_2, V_3 \colon V_1 V_2 V_3 = 1 \rangle$ *be a group with signature* (0, 3) *and* $V_1$, $V_2$ *hyperbolic, and let* $T$ *be a Möbius transformation such that* $T V_1 T^{-1} = V_2^{-1}$. *Let* $\mathscr{P}$ *be a fundamental polygon for* $G$ *which intersects the axes* $l_1, l_2$ *of* $V_1, V_2$ *in arcs* $\alpha_1, \alpha_2$ *respectively with* $T(\alpha_1) = \alpha_2$. *Then the group* $G_1 = \langle G, T \rangle$ *is discrete and has as fundamental polygon the set* $\mathscr{P} \cap B_1 \cap B_2$, *where* $B_1$ *and* $B_2$ *are the hyperbolic half planes bounded by* $l_1$ *and* $l_2$ *which do not contain a fixed point of* $V_3$.

*Remark*. This construction is known as a *geometric H.N.N. extension of G*. The proof is analogous to that of Lemma 1.6.1, and we omit it. ■

We note that the axis of $T$ intersects those of $V_1$ and $V_2$, and the group $G_1 \cong \langle V_1, T, V_3 : V_1 T V_1^{-1} V_3 = 1 \rangle$ has signature (1; 1) (see Fig. 2).

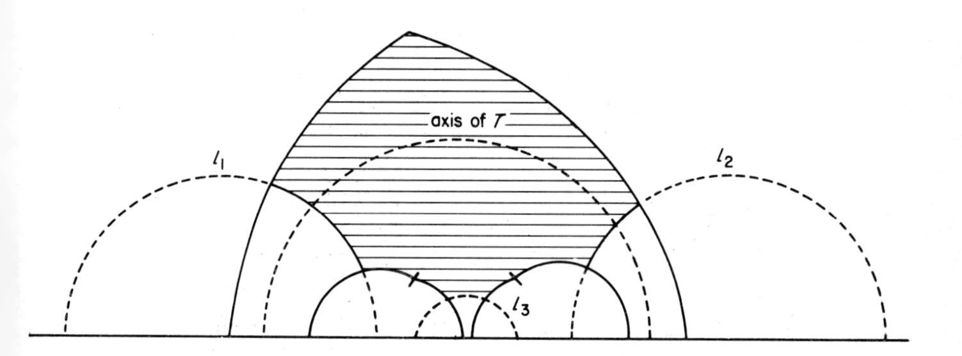


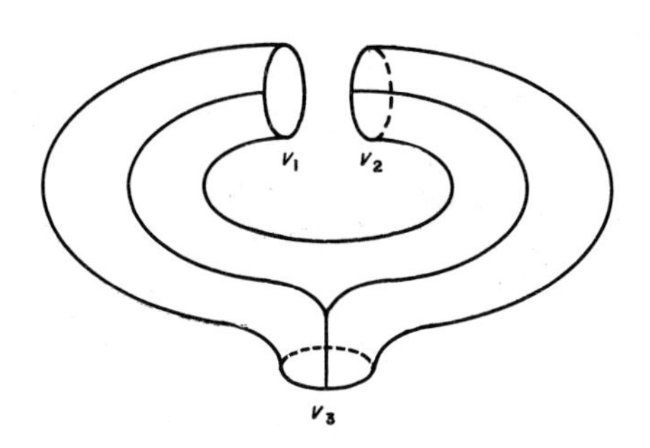


*Fig. 2. The intersection of* $\mathscr{P}$ *with* $K(G_1)$ *is shaded.*

One can profitably view these processes in terms of the topology of the quotient surfaces. In 1.6.1, the two spheres $S = G\backslash\mathscr{H}$ and $S' = G'\backslash\mathscr{H}$ are joined by identifying the holes $v_3$ on $S$ and $v'_3$ on $S'$. This is the situation of van Kampen's theorem (Chapter 1, Section 3.8). In 1.6.2 one has a single sphere $S = G\backslash\mathscr{H}$ with three holes, two of which are being identified by the transformation $T$. This case can also be regarded as derived from the other in a certain sense by first forming a group $A$ by an infinite chain of amalgamated free products of copies of $G$, the amalgamation being taken with respect to the injective homomorphisms $H \cong H_1 \xrightarrow{i_1} G$, $H \cong H_2 \xrightarrow{i_2} G$, with subsequent adjunction of an element $t$ which makes the diagram

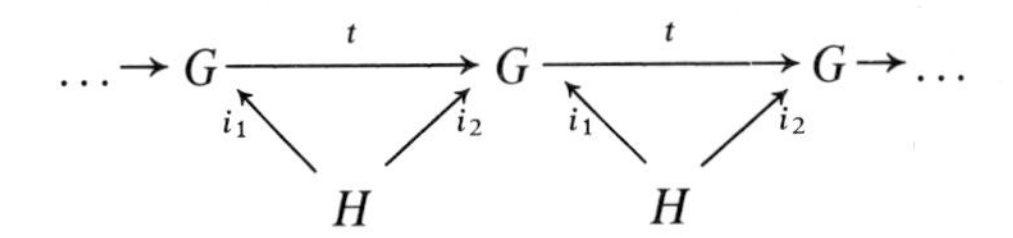


commute. Then $G_1 \cong A * \langle t \rangle$.

As we shall see below, these combination methods work in much greater generality—they are the prototypes of Maskit's combination theorems I and II, [**32**], which extend the original method of Klein. In particular the groups to be combined can be of arbitrary signature provided they are of the second kind† (i.e. have free boundary sides on their fundamental regions which can then be connected to others of linked).

## 1.7 Decomposition of finitely presented Fuchsian groups

The converse operation to combination is that of decomposition and our aim in this section is to describe how to decompose any finitely generated Fuchsian group (assumed to be torsion-free for simplicity) into spheres with three boundary components.

We start from a fixed group $\Gamma$ and a presentation for $G$ which takes the form (c.f. 1.2)

$$\text{Generators: } A_1, B_1, \ldots, A_g, B_g, P_1, \ldots, P_n, F_1, \ldots, F_m;$$

$$\text{Relations: } \prod_{i=1}^{g} [A_i, B_i] P_1 \ldots P_n F_1 \ldots F_m = 1,$$

where $[A_i, B_i] = A_i B_i A_i^{-1} B_i^{-1}$.

† This refers to Fuchsian groups only. More general combination techniques apply if one requires only the looser class of Kleinian groups (see [**33**] and Chapter 8).

We assume that this presentation corresponds to a choice of normal fundamental polygon $\mathscr{P}$ for $\Gamma$ with edges paired by the generators. Such a situation can be attained for any finitely presented Fuchsian group by commencing with a Dirichlet region—necessarily finite-sided by Chapter 7, Theorem 1.12—and reducing it to canonical form by transforming the surface symbol in standard fashion (see for example Siegel [**45**] or a text on surface topology). Figure 3 below illustrates the case $g = 2, n = m = 1$.

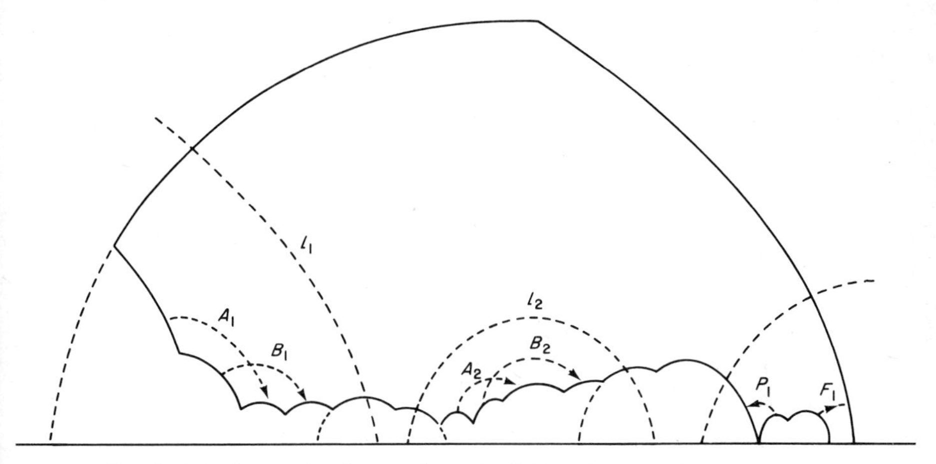


*Fig. 3. A fundamental polygon of type* (2; 1, 1), *and a partial decomposition.*

Let $C_j = [A_j, B_j]$ for $j = 1, \ldots, g$. Then if $(g, n) \neq (1; 1)$, each $C_j$ is hyperbolic and the axis $l_j$ of $C_j$ divides $\mathscr{H}$ into two parts, $\mathscr{H}_j$ and $\mathscr{H}'_j$, and $\mathscr{P}$ into $\mathscr{P}_j$, $\mathscr{P}'_j$ in such a way that $\mathscr{P}_j$ is a fundamental set for the action in $B_j$ of the subgroup $\Gamma_j$ generated by $\langle A_j, B_j \rangle$, and $\mathscr{P}'_j$ is a fundamental set in $B'_j$ for the complementary subgroup (^ denotes omission) $\Gamma'_j = \langle A_1, B_1, \ldots, \hat{A}_j, \hat{B}_j, \ldots A_g, B_g, P_1, \ldots, P_n, F_1, \ldots, F_n \rangle$. The group $\Gamma$ may be written as an amalgamated product $\Gamma_j \mathop{*}_{H_j} \Gamma'_j$, where $H_j = \langle C_j \rangle$, along the lines of Lemma 1.6.1.

The axes of $A_j$ and $B_j$ intersect because the group $\Gamma_j$ has type (1; 1) (if they did not intersect the group $\Gamma_j$ would have type (0; 3) as we have seen in 1.5). Now we remove the element $B_j$ from the set of generators for $\Gamma$, thereby introducing two new boundary components (the sides $b_j$, $b'_j$ of $\mathscr{P}$ which were paired by $B_j$). The resulting subgroup $G_j$ of $\Gamma$ has signature type $(g - 1; n + 2)$ therefore, and we see that $\Gamma = G_j * \langle B_j \rangle$ is an example of the H.N.N. extension.

*Note 1*. We could equally well approach this question by making dissections of the underlying surface $S = \Gamma\backslash\mathscr{H}$ which involve cutting off handles or cutting along a non-dividing cycle which loops once around one handle.

*Note 2*. Instead of splitting up $\Gamma$ with respect to the element $C_j$ one could use any (primitive) element such that its axis divides the domain $\mathscr{P}$. Such elements represent null cycles in the homology of $S$. Similarly one can replace $B_j$ in the second process by any primitive element which does not divide $\mathscr{P}$.

*Definition.* A group $\Gamma$ is called *simply decomposable* if it can be expressed either as a free product with amalgamation or as a H.N.N. extension. A list of subgroups $\{G_i\}$ together with data for combining them (including the adjunction of further elements $\{t_j\}$ where necessary to effect an H.N.N. extension) is referred to as a *decomposition* of $\Gamma$ if it arises from a finite set of such splittings of $\Gamma$. The individual $G_i$ are called *constituent* subgroups.

We remark that the most effective way to group together all the information involved in such a decomposition is to define an associated *decomposition graph* $\mathscr{K}$ for $\Gamma$ which is constructed as follows: the *vertex set* of $\mathscr{K}$ consists of the constituent subgroups $\{G_i\}$, and the *edges* are the glueing operations between vertices. Thus an edge exists between $G_i$ and $G_j$ $(i \neq j)$ when $G_i \cap G_j \neq \varnothing$ and it carries with it as label the common subgroup $H_{ij} = G_i \cap G_j$, while an edge between $G_i$ and $G_i$ arises from the existence in $G$ of an element $T$ which conjugates some pair of cyclic subgroups of $G_i$ into each other. The graph $\mathscr{K}$, together with the labelling data $\mathscr{G}$ is now a *graph of groups* $(\mathscr{G}, \mathscr{K})$ in the sense of Bass–Serre **[43]** and $\Gamma$ is isomorphic to the "*fundamental group of* $(\mathscr{G}, \mathscr{K})$" if $(\mathscr{G}, \mathscr{K})$ is a decomposition graph for $\Gamma$.

1.7 PROPOSITION. *Every finitely generated (torsion free) Fuchsian group admits a decomposition with all the constituent (vertex) subgroups of type* (0; 3).

*Proof* (Sketch). Using the operations described at the start of this section we can decompose $\Gamma$ first into the subgroups $\{\Gamma_j, j = 1, \ldots g\}$ of type (1; 1) together with a subgroup $\Gamma_0 = \langle c_1, \ldots, c_g, P_1, \ldots P_n, F_1, \ldots F_m \rangle$ of type $(0; g + n + m)$ (except when $g = 2, n = m = 0$). Next we apply the operation of cutting out $\langle B_j \rangle$ from $\Gamma_j, j = 1, \ldots g$, giving $g$ three-holed sphere groups $\Gamma_j^*$. It remains to dissect $\Gamma_0$ and this can be done in the following way. First split off the parabolic elements in pairs—any pair $P_i, P_j$ generate a subgroup of signature (0; 2; 1) with one generator $Q_{ij} = P_i P_j$ hyperbolic—until at most one remains. This leaves a group $\Gamma_0^* = \langle c_1, \ldots, c_g, Q_{12}, \ldots, Q_{2l-1,2l}, P_n,$

$F_1, \ldots F_m\rangle$ (or without $P_n$ if $n$ is even). Now $P_n$, $F_1$ generate a group of signature (0; 1; 2) with $F_1$ and $P_n \circ F_1$ hyperbolic, and so we are left, after splitting this off, with an $N$-holed sphere where $N = g + m + [n/2] \geqslant 4$. This is easily decomposed into $N - 2$ three-holed spheres. ■

*Note 3.* The process involves repeatedly the fact that we have a canonical geometric presentation of $\Gamma$; we know precisely which elements of $\Gamma$ can be parabolic. One needs to check at each stage of the decomposition process that the axes of the generators (hyperbolic) of groups to be split off do have the right intersection properties. This follows from Proposition 1.4.1, but can be proved directly without it (see Fricke–Klein II).

*Note 4.* The decomposition could equally be effected by choosing a suitable collection of non-intersecting closed curves on the quotient surface having the property that no two curves cobound a cylinder and no curve bounds a disc, and cutting the surface along them (c.f. Note 2 of this Section). This is called a *partition* of $S$.

*Note 5.* If $\Gamma = \Gamma_1 *_H \Gamma_2$ with $H$ infinite cyclic, then the Euler characteristics of the surfaces $S$, $S_1$ and $S_2$ are related by $\chi(S) = \chi(S_1) + \chi(S_2)$, while if $\Gamma = \Gamma_1 * \langle t\rangle$ is an H.N.N. extension then $\chi(S) = \chi(S_1)$. (Using the relationship between the Poincaré area of a surface and the Euler characteristic, one can recover these facts directly from the statements about fundamental regions for $G$ in terms of those for constitutent subgroups.) Since a three-holed sphere has $\chi = -1$, it is natural to call a decomposition into groups of type (0; 3) *maximal.*

*Exercise.* Determine all non-isomorphic maximal decompositions for the signatures (2; 0), (3; 0).

### 1.8 Parametrization of $\mathscr{R}_0(\Gamma)$ and $T_0(\Gamma)$

We wish to characterize the deformations of $\Gamma$ in terms of those of the constituent subgroups of some decomposition. To do this we treat each simple combination first of all.

1.8.1 PROPOSITION. *Let $G_1 = G *_H G'$, with $G$, $G'$ satisfying the hypotheses of Lemma* 1.6.1. *Then there is a real-analytic homeomorphism between* $\mathscr{R}(G_1)$ *and* *the fibre product* $\mathscr{R}(G) \underset{\mathscr{R}(H)}{\times} \mathscr{R}(G')$, *so that* $\mathscr{R}(G_1)$ *is an analytic manifold*† *of dimension 9.*

*Proof.* The relation $G_1 \cong G *_H G'$ means that there is a commutative diagram

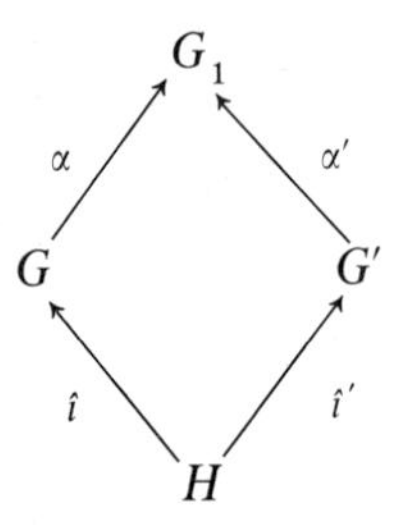


with the property that if $\Gamma$ is another group such that $G \xrightarrow{\beta} \simeq \Gamma \xleftarrow{\beta'} \simeq G'$ and $\beta \circ \imath' = \beta' \circ \imath'$, then there is a homomorphism $\phi : G_1 \to \Gamma$ which satisfies $\beta = \phi \circ \hat{\alpha}$, $\beta' = \phi \circ \alpha'$. Now $\imath, \imath', \alpha, \alpha'$ are injective so there are automatically induced mappings $\hat{\imath}, \hat{\imath}', \hat{\alpha}, \hat{\alpha}'$ between the various $\mathscr{R}_0$ spaces which are real analytic, and the diagram

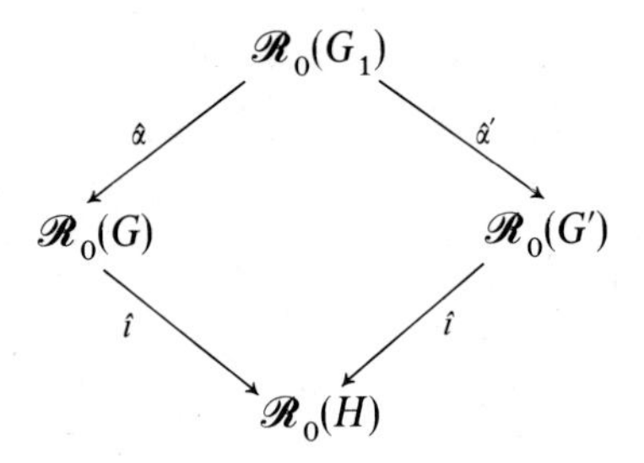


commutes.

To show that $\mathscr{R}_0(G_1)$ is the fibre product one verifies trivially that $\hat{\imath}, \hat{\imath}'$ are surjective, and that given two elements $r \in \mathscr{R}(G)$, $r' \in \mathscr{R}(G')$ with $\hat{\imath}(r) = \hat{\imath}'(r')$ (i.e. $r \circ \hat{\imath} = r' \circ \hat{\imath}'$) there is a unique $r_1 \in \mathscr{R}_0(G_1)$, with $r_1 = \hat{\alpha}(r) = \hat{\alpha}'(r')$ by Lemma 1.6.1. For the analyticity one needs to examine the equations defining the various $\mathscr{R}_0$-spaces and verify that the set $\{(r, r') \in \mathscr{R}_0(G) \times \mathscr{R}_0(G')$: $\hat{\imath}(r) = \hat{\imath}'(r)\}$ is an analytic submanifold near any given point. This is elementary

† Note a manifold here need not necessarily be connected.

in our case: by application of the implicit function theorem to the functions $\phi$: $\mathscr{L}^3 \to \mathscr{L}^3$, $\psi$: $\mathscr{L}^3 \times \mathscr{L}^3 \to \mathscr{L}$ given by $\phi(\mathbf{X}) = X_1 . X_2 . X_3$, $\psi(\mathbf{X}, \mathbf{X}') = \phi(\mathbf{X}) . \phi(\mathbf{X}')^{-1}$, $(\mathbf{X} = X_1, X_2, X_3)$ we see that, since $\mathscr{R}_0(G) \cong \mathbf{X}: \phi(\mathbf{X}) = E\}$, $\mathscr{R}_0(G') = \{\mathbf{X}': \phi(\mathbf{X}) = E\}$, and $\mathscr{R}_0(G_1) = \{(\mathbf{X}, \mathbf{X}'): \psi(\mathbf{X}, \mathbf{X}') = E), X_3 = X'_3\}$, it follows that $\mathscr{R}_0(G_1)$ is an analytic submanifold of $\mathscr{L}^6$ of dimension 3 dim $\mathscr{L} = 9$, which equals $\dim \mathscr{R}_0(G) + \dim \mathscr{R}_0(G') - \dim \mathscr{R}_0(H)$. ■

1.8.2 PROPOSITION. *Let $G_1 = \langle G, T \rangle$ with $G$, $T$ satisfying the hypotheses of Lemma* 1.6.2. *Then there is a real analytic homeomorphism between $\mathscr{R}_0(G_1)$ and a submanifold $\mathscr{Y}$ of $\mathscr{R}(G) \times \mathscr{L}$ of dimension* 6.

*Proof.* In this case we have a commutative diagram as below,

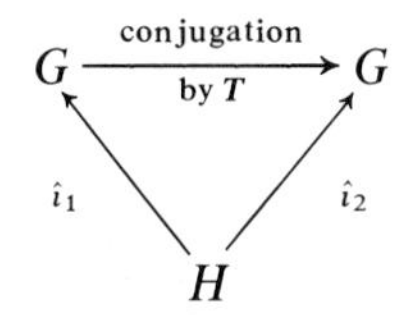


with $H = \langle V \rangle$, and $T$ conjugating the generators $V_1 = \hat{\imath}_1(V)$, $V_2 = \hat{\imath}_2(V)$ of $G$. Therefore if $\mathscr{Y}$ denotes the submanifold of $\mathscr{R}_0(G) \times \mathscr{L}$ defined by $\{(r, t): \tilde{t} \circ \hat{\imath}_1(r) = \hat{\imath}_2(r)\}$ (with $\tilde{t}$ denoting conjugation by $t$) then $G_1$ corresponds to the point (id, $T$) $\in \mathscr{Y}$ and $\mathscr{R}_0(G_1)$ is mapped bijectively onto $\mathscr{Y}$. The result follows in a manner analogous to the previous proposition. We note that $\dim \mathscr{Y} = \dim \mathscr{R}(G) = 6$.

The local structure of $\mathscr{R}_0(\Gamma)$ can now be derived from these facts.

1.8.3 THEOREM. *Let $\Gamma$ be a finitely presented (torsion free) Fuchsian group, with signature $(g; n; m)$. Then $\mathscr{R}_0(\Gamma)$ is a real analytic manifold of dimension $d = 6g - 3 + 2n + 3m$.*

*Proof.* We need only observe that any $\Gamma$ admits a decomposition into subgroups of type (0; 3) given by Proposition 1.7 and the local analytic structure of $\mathscr{R}_0(\Gamma)$ follows from Propositions 1.8.1 and 1.8.2. The dimension $d$ is readily computed from the signature using the specific choice of decomposition given in 1.7; it is of course independent of the choice. ■

*Note*. This proves the theorem of Weil (1.2) that $\mathcal{R}_0(\Gamma)$ is open in $\mathrm{Hom}(\Gamma, \mathcal{L})$.

The same methods enable us to complete the proof of Theorem 1.4.8. We must use the normalization by the action of Aut $\mathcal{L}$ to reduce the parameters involved in a decomposition to a suitable form, and it is again convenient to do this for each simple combination first.

Let $G_1 \cong G *_H G'$ be a decomposition of a group $G_1$ of signature (0; 0, 4). The groups $G$ and $G'$ can each be transformed by conjugation within $\mathcal{L}$ into the form given in Proposition 1.5.1. Let $\mathbf{t}$, $\mathbf{t}$ denote the associated trace parameters for $G$, $G'$ respectively, so we have $G \sim G(\mathbf{t})$, $G' \sim G(\mathbf{t}')$, and the elements $V_3 \in G$ and $V'_3 \in G'$ (which are identified in the amalgamation with the generator $V$ of $H$) are conjugate to the generators of $G(\mathbf{t})$ and $G(\mathbf{t}')$ respectively which have as axis the imaginary axis $\mathcal{I} = \{\mathrm{Re}\, z = 0\}$ in $\mathcal{H}$. Conjugation by the reflection in $\mathcal{I}$ turns $G(\mathbf{t})$ into a group $\overline{G}(\mathbf{t})$ which has the same trace parameters but with Nielsen convex region in the *right* half plane. Since the groups $G$ and $G'$ have common generator $V$ we must have $t_3 = t'_3$, and the only matter remaining is to specify the canonical form of $G *_H G'$ among all possible combinations of $\overline{G}(\mathbf{t})$ with a conjugate of $G(\mathbf{t}')$. It is easily seen geometrically that if $U$ is any Möbius (or "anti-Möbius") transformation, then $U \,.\, G(\mathbf{t}')U^{-1}$ and $\overline{G}(\mathbf{t})$ satisfy the hypotheses of the combination Lemma 1.6.1 if and only if $U$ fixes 0 and $\infty$ and preserves orientation. Consequently for some transformation $z \mapsto U_\tau(z) = \tau z$, where $\tau > 0$, we have that $G_1$ is conjugate under Aut $\mathcal{L}$ to a product specified uniquely by the parameters $(\mathbf{t}, \mathbf{t}', \tau) \in \mathbb{R}^7$ with $t_3 = t_3{}'$, $2 < t_i$, $t'_i$ for $i = 1, 2, 3$ and $0 < \tau$. Conversely, one can construct for any such point a group of signature (0; 0; 4) by the above procedure. This proves the result below.

1.8.4 PROPOSITION. *The Teichmüller space for a 4-holed sphere is an open cell of dimension 6.*

Next we let $G_1 = \langle G, T\rangle$ be an H.N.N. decomposition for a group $G_1$ of signature (1; 0; 1). Then $G$ is a three-holed sphere group, and so it is conjugate to some group $G(\mathbf{t})$ as before. The Möbius transformation $T$ is conjugate within $\mathcal{L}$ to one of the form $z \mapsto \tau z$ with $\tau > 1$, and we may specify that it permutes the axes of $V_2$ and $V_3$ as in Lemma 1.6.2. A simple geometrical argument and computation with Möbius transformations then shows that the space $\{(\mathbf{t}, \tau) \in \mathbb{R}^4 : t_2 = t_3, 2 < t_i, 1 < \tau\}$ parametrizes the normalized marked groups in $\mathbf{T}_0(G_1)$ real analytically, and we obtain

1.8.5 PROPOSITION. *The Teichmuller space for a one-holed torus is an open cell of dimension 3.*

Using the method of dissection for an arbitrary finitely generated Fuchsian group $\Gamma$ described in Section 1.7 we see using an induction on the number of vertex groups in a decomposition graph of $\Gamma$ that the above results yield a real analytic homeomorphism from $\mathbf{T}_0(\Gamma)$ onto a cell in $\mathbb{R}^N$ with $N = 6g - 6 + 3m$ if $\Gamma$ has signature $(g; 0; m)$. To extend this to the case where $\Gamma$ has parabolic elements one needs to verify using Proposition 1.5.4 that it is valid to set the corresponding trace parameters to equal 2 for constituent subgroups having parabolic elements. The proof of Theorem 1.4.2 is then finished in the torsion-free case; if $\Gamma$ has torsion, one must consider separately the class of groups of type (0; 3) having elliptic elements. ■

*Note.* It is apparent that the parameters $\tau$ introduced above provide a measurement of how boundary curves on constituent subsurfaces of $\Gamma\backslash\mathscr{H}$ are joined. They are often referred to as the *twist* parameters.

*Exercise 1.* If $G^{(\tau)}$ denotes the group obtained from a fixed group $G = G(\mathbf{t}_0)$ by adjunction of a transformation $T$ with multiplier $\tau > 1$ as in 1.8.4, determine all values of $\tau$ for which $G^{(\tau)} = G^{(2)}$.

One obtains in the same way as Corollary 1.5.3 a global description of $\mathscr{R}_0(\Gamma, \mathscr{L})$:

1.8.6 COROLLARY. *If $\Gamma$ is a finitely generated Fuchsian group, then $\mathscr{R}_0(\Gamma, \mathscr{L})$ is homeomorphic to the direct product $\mathbf{T}_0(\Gamma) \times$ Aut $\mathscr{L}$. Thus $\mathscr{R}_0(\Gamma)$ is a disjoint union of two spaces isomorphic to $\mathbb{R}^{N+2} \times S^1$.*

*Exercise 2.* Using the results of this section, prove that if $\Gamma = \langle \gamma_1, \ldots \gamma_n \rangle$ and $\Gamma\backslash\mathscr{H}$ is compact then there is a neighbourhood $U$ of the identity $e \in \mathscr{L}$, and a compact set $K \subseteq \mathscr{H}$ such that every element $\theta \in \mathrm{Hom}\,(\Gamma, \mathscr{L})$ with $\theta(\gamma_i) \in U . \gamma_i$ for $i = 1, 2, \ldots, n$ satisfies $\theta(\Gamma) . K = \mathscr{H}$ and $\theta^{-1}(U) = \{e\}$. (c.f. Weil's theorem, Section 1.3.).

## 1.9 Fundamental regions and the Nielsen theorem

The local structure of $\mathscr{R}_0(\Gamma, \mathscr{L})$ is, as we have seen, closely related to the shape of a particular type of fundamental domain for the group $\Gamma$. One aspect of the situation which has not yet been explicitly mentioned, although it is touched on briefly in the discussion of Section 1.5 for groups of signature (0; 3), is the fact that the normalized parametrization given in proposition 1.5.1 has the property that all the groups possess a fundamental polygon $\mathscr{P}$ of *canonical fixed shape* from which the parameters can be read off. Given a group $\Gamma$ of signature (0; 0; 3) with normalized generating set as in (1.5.1) one may construct $\mathscr{P}$ as follows. From a point $p$ inside the Nielsen convex region of $\Gamma$ one draws non-Euclidean rays $\alpha_2$, $\alpha_3$ from $p$ which intersect the axes of the generators $V_2$ and $V_3$ respectively and which are such that the angle $\alpha_2\alpha_3$ is less than $\pi$. The images $\alpha'_3 = V_3^{-1}(\alpha_3)$ and $\alpha'_2 = V_2(\alpha_2)$ are rays emanating from $V_3^{-1}(p)$ and $V_2(p)$ and one now draws from $V_3^{-1}(p)$ a ray $\alpha_1$ intersecting the axis of $V_1$, whose image under $V_1^{-1}$ is then a ray $\alpha'_1$ from $V_1^{-1}V_3^{-1}(p) = V_2(p)$. The polygon $\mathscr{P}$ bounded by these rays (see Fig. 4) is a fundamental region for $\Gamma$, and it can be made canonical by giving a rule for determining the initial point $p$ and the three rays $\alpha_1$, $\alpha_2$, $\alpha_3$ (one such choice is to let $p$ be the intersection of the axes of $V_2^{-1}V_3$ and $V_2V_3^{-1}$ and to determine the rays as the perpendiculars to the axes in question).

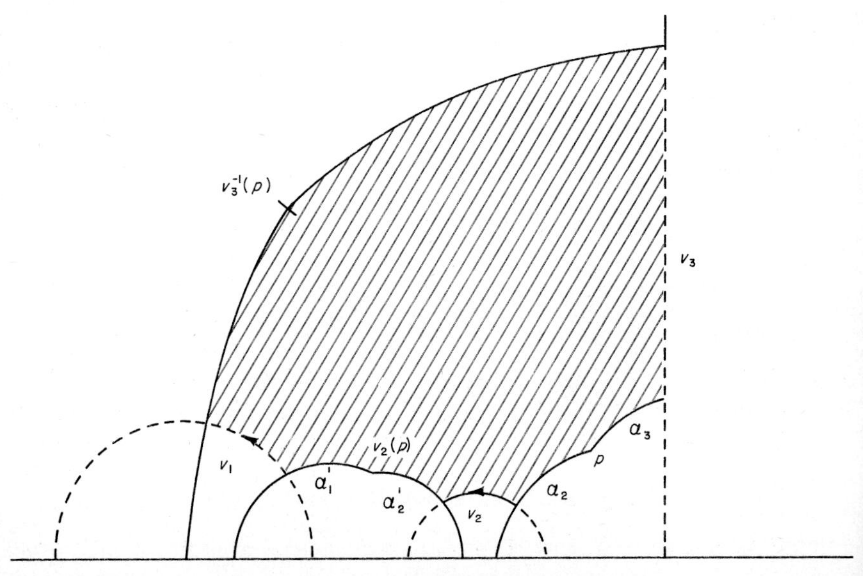


*Fig. 4. A canonical polygon of type (0; 3).*

We note that $\mathscr{P}(\Gamma)$ has six sides and the truncated polygon $\mathscr{P}_0(\Gamma)$ obtained by deleting the parts outside the axes has nine. A triangulation of $\mathscr{P}$ (or $\mathscr{P}_0$) is obtained by joining any interior point to each of the vertices.

It is possible from this to sketch a proof of the Nielsen theorem 1.4.2. Let $\theta : \Gamma \to \Gamma'$ be an allowable isomorphism. Then $\Gamma'$ admits an analogous polygon $\mathscr{P}_0(\Gamma')$; one sees immediately that $\theta$ is induced geometrically by the homeomorphism obtained by choosing a mapping $f$ of a triangulation for $\mathscr{P}(\Gamma)$ onto the one for $\mathscr{P}(\Gamma')$ which pairs the sides appropriately and then extending it to be compatible with the actions of $\Gamma$ and $\Gamma'$: thus for $\gamma \in \Gamma$ we define $f$ on the set $\gamma(\mathscr{P}(\Gamma))$ by setting

$$f(\gamma(z)) = \gamma'(f(z)), \qquad \text{with } \gamma' = \theta(\gamma), \qquad z \in \mathscr{P}(\Gamma).$$

This proves Theorem 1.4.2 for groups with signature (0; 0; 3).

To complete the proof for an allowable isomorphism $\theta$ between arbitrary torsion-free finitely generated Fuchsian groups, one may use the decomposition given in Section 1.7 of $\Gamma$, $\Gamma'$ into isomorphic graphs $\mathscr{G}$, $\mathscr{G}'$ with vertex subgroups of type (0; 3), together with the extension of the above argument to all groups of type (0; 3) to reduce the questions to one of piecing together homeomorphisms on a finite union of subpolygons for constituent (vertex) subgroups $\{\Gamma_i\}$, $\{\Gamma_i'\}$ of $\Gamma$, $\Gamma'$ respectively, which must then be extended so as to be compatible with the actions of any pairs of elements $T \in \Gamma$, $\theta(T) = T' \in \Gamma'$ which correspond to self-linkings of vertex groups besides being compatible with respect to paired subgroups $\Gamma_i$, $\Gamma_i' = \theta(\Gamma_i)$. Details are left to the reader as an exercise. ■

A full proof along similar lines which also treats the torsion case may be found in [**33**].

## 2. THE SPACE OF DISCRETE SUBGROUPS

### 2.1 Chabauty's topology

In Section 1.1 we saw that the space of lattices in $\mathbb{C}$ forms a real manifold with a natural topology of homogeneous space under $\mathrm{GL}_2(\mathbb{R})$. We shall be concerned in this chapter with the space of all discrete subgroups of the real Möbius group. A strict analogy with the elementary case would require the groups concerned to have compact quotient space, but we shall soon see that this restricted subspace would not suffice for one purpose we have in mind, which is to study limits of sequences $\{\Gamma_n\}$ of Fuchsian groups corresponding to a process of degeneration which results in the appearance of parabolic elements not in the groups $\Gamma_n$. Since Fuchsian groups with parabolic elements automatically have non-compact quotient space we must at least allow groups with quotient space of finite volume.

There is a fundamental theorem due to Mahler on convergence of sequences of lattices which states that given $\varepsilon > 0$ and $K > 0$ constants, then the lattices $L \subseteq \mathbb{C}$ such that

$$\text{(2.1)} \qquad \begin{array}{l} \text{(i) if } \omega \in L, \text{ then } |\omega| < \varepsilon \text{ implies that } \omega = 0, \\ \text{(ii) } \mathrm{vol}(L\backslash\mathbb{C}) \leqslant K, \end{array}$$

form a compact set. In seeking to generalize this result, Chabauty [9] defined a topology $\mathscr{T}$ on the space $\mathbf{S}(\mathscr{L})$ of all discrete subgroups of a locally compact topological group $\mathscr{L}$ as follows.

*Definition.* Let $\Gamma \subset \mathscr{L}$ be a discrete subgroup. *A neighbourhood basis* at the point $\Gamma \in \mathbf{S}(\mathscr{L})$ is composed of sets $N(\Gamma, K, U)$ with $U$ any open neighbourhood of the identity $e$ in $\mathscr{L}$ and $K$ any compact subset of $\mathscr{L}$, where

$$\text{(2.2)} \qquad N(\Gamma, K, U) = \{G \in \mathbf{S}(\mathscr{L}) : G \cap K \subseteq \Gamma U, \Gamma \cap K \subseteq GU\}$$

We refer to the topology $\mathscr{T}$ as the *Chabauty topology.*

This rather mysterious way of defining closeness becomes more familiar if one redrafts it in the language of sequences.

2.1 Lemma. *A sequence* $\{\Gamma_n, n = 1, 2, \ldots\}$ *converges to a group* $\Gamma$ *in the topology* $\mathscr{T}$ *if and only if the following conditions hold:*

(i) *if a sequence* $\{\gamma_{n_k} \in \Gamma_{n_k}\}$ *of elements converges to* $\gamma \in \mathscr{L}$ *then* $\gamma \in \Gamma$;
(ii) *for any element* $\gamma \in \Gamma$ *there is a sequence* $\{\gamma_n \in \Gamma_n\}$ *which converges to* $\gamma$.

*In other words,* $\overline{\lim}\, \Gamma_n = \underline{\lim}\, \Gamma_n = \Gamma$.

The proof is straightforward and we leave it to the reader. ∎

*Exercise 1.* Show that the Chabauty topology is the same as that generated by taking as basis for the open sets the families

(a) for $U$ open in $\mathscr{L}$, $A_U = \{\Gamma \in \mathbf{S}(\mathscr{L}) : \Gamma \cap U \neq \varnothing\}$,
(b) for $K$ compact in $\mathscr{L}$, $B_k = \{\Gamma \in \mathbf{S}(\mathscr{L}) : \Gamma \cap K = \varnothing\}$.

*Exercise 2.* Show that there is a uniform structure on $\mathscr{L}$ which determines the Chabauty topology (see [8] Chap. 8).

*Exercise 3*. Show that the topology $\mathscr{T}$ on $\mathbf{S}(\mathscr{L})$ is Hausdorff.

*Note*. One can in fact [**12**] use this topology on the larger space $\mathscr{F}(\mathscr{L})$ of closed subgroups of $\mathscr{L}$. It turns out that $\mathscr{F}(\mathscr{L})$ is compact, which is a useful indication of the feasibility of constructing a *compact* space from the collection of all groups of given signature. I am grateful to Heiner Zieschang for pointing this out to me.

### 2.2 Euler–Poincaré measure and Chabauty's theorem

We recall [**8**] that there is an invariant Haar measure on a Lie group $\mathscr{L}$ which induces by projection a measure on any quotient space $\Gamma\backslash\mathscr{L}$ with $\Gamma$ discrete. Equivalently, for our special situation $\mathscr{L} = \mathrm{PSL}_2(\mathbb{R})$ we may consider the *Poincaré area* $\mathrm{d}\mu = y^{-2}\,\mathrm{d}x\,\mathrm{d}y$ on the homogeneous space $\mathscr{H}$, as defined in Chapter 2, Section 1, which projects to an area measure on the surface $\Gamma\backslash\mathscr{H}$. If $\mathscr{P}$ denotes any (measurable) fundamental set for $\Gamma$ in $\mathscr{H}$, it is known (Chapter 7, Section 1.5) that $\iint_{\mathscr{P}} \mathrm{d}\mu = \mu(\mathscr{P})$ is independent of the choice of $\mathscr{P}$ and we denote it by $\mu(\Gamma)$. If $\Gamma$ is finitely generated and of the first kind then $\mu(\Gamma)$ is finite and has value computable in terms of the signature; for groups with hyperbolic boundary elements we obtain the same result if we use the notion of *truncated fundamental set* $\mathscr{P}_0$ for $\Gamma$ acting on its Nielsen convex region $K(\Gamma)$. In fact, (see Chapter 7, Theorem 1.5.4), if $\Gamma$ has signature $(g; n; v_1 \dots v_k; m)$ then

$$\mu_0(\Gamma) = \mu(\mathscr{P}_0) = 2\pi\left[2g - 2 + n + m + \sum_{i=1}^{k}\left(1 - \frac{1}{v_i}\right)\right].$$

We remark that up to a constant multiple $\mu_0(\Gamma)$ is the *Euler characteristic* of the quotient surface $\Gamma\backslash\mathscr{H}$ in the torsion-free case. We note also that $\mu$ is an *Euler–Poincaré measure* in the sense defined by Serre [**44**].

The fundamental compactness result of Chabauty, [op. cit.] can now be formulated for $\mathscr{L}$.†

2.2.1 Theorem. *Let $U$ be a neighbourhood of $e$ in $\mathscr{L}$ and $M$ a real constant. Then the set $\Omega = \Omega(U, M)$ of all discrete groups $\Gamma \in \mathbf{S}(\mathscr{L})$ such that*

(i) $\Gamma \cap U = \{e\}$ *for all* $\Gamma \in \Omega$,
(ii) $\mu(\Gamma) \leqslant M$,

† The result is valid for any locally compact group $\mathscr{L}$ which is a countable union of compact sets.

*is sequentially compact. Moreover if* $\Gamma_n \in \Omega$ *converge to* $\Gamma$ *then*

$$\mu(\Gamma) \leqslant \liminf_{n\to\infty} \mu(\Gamma_n). \tag{2.3}$$

*Proof.* Let $\Gamma_n$, $n = 1, 2, \ldots$, be a sequence in $\Omega$. Using an exhaustion of $\mathscr{L}$ by compact subsets $K_\nu$, it follows from the Bolzano–Weierstrass theorem that some subsequence of the sequence of finite sets $G_{n,\nu} = \Gamma_n \cap K_\nu$ converges inside $K_\nu$ and consequently by a diagonal argument there is a subsequence which converges on $\mathscr{L}$ to a set $\Gamma^*$. One shows easily that $\Gamma^*$ is a group, and hypothesis (i) implies that $\Gamma^* \cap U = \{e\}$ so $\Gamma^*$ is discrete.

To show that $\mu(\Gamma^*) \leqslant M$, take $W$ an open set in $\mathscr{H}$ which is $\Gamma^*$-*irreducible* (that is, $W \cap \gamma(W) = \varnothing$ for $\gamma \in \Gamma^* - \{e\}$) and a compact set $K$ with $K \subseteq U$. We shall prove that for sufficiently large $n$, $K$ is also $\Gamma_n$-irreducible. If we take $V$ a symmetric neighbourhood of $e$ small enough to ensure that $V \subseteq U$ and $VK \subseteq W$, then for $n$ sufficiently large $\Gamma_n \in N(\Gamma^*, C, V)$ with $C = \{t \in \mathscr{L}: t(K) \cap K \neq \varnothing\}$. Now if there are $z, w \in K$ and $\gamma \in \Gamma_n$ with $\gamma z = w$, then there is $\gamma^* \in \Gamma^*$ with $\gamma^* \in V.\gamma$ and it follows that $\gamma^*(W) \cap W \neq \varnothing$, which implies $\gamma^* = e$. Therefore $\gamma = e$ since $V \subseteq U$, which proves that $K$ is $\Gamma_n$-irreducible. Finally we observe that $W$ can be chosen equal in measure to $\mu(\Gamma^*)$ and $K$ may be taken with $\mu(K)$ arbitrarily close to $\mu(\Gamma^*)$; hence

$$\mu(\Gamma^*) \leqslant \liminf_{n'\to\infty} \mu(\Gamma_{n'}) \leqslant M. \qquad \blacksquare$$

*Note.* In the case $\mathscr{L} = \mathbb{R}^n$ with $\{\Gamma_n\}$ lattices, one sees directly that $\mu(\Gamma^*) = \lim_{n'\to\infty} \mu(\Gamma_{n'})$. In general this is not the case, as we shall see in the next section.

### 2.3 The local structure of $\mathbf{S}(\mathscr{L})$ and the subspaces $\mathbf{S}_0(\Gamma)$

Although the topology $\mathscr{T}$ on $\mathbf{S}(\mathscr{L})$ is apparently rather weak, it turns out that there is a geometric relationship between neighbouring groups, and one can show that the groups close to a discrete group with *compact* quotient space are all geometrically isomorphic to it. This is embodied in the result below, which is valid for any Lie group $\mathscr{L}$ although the proof given here uses the geometry of $\mathscr{H}$.

2.3.1 THEOREM. [**21**] *Let* $\Gamma \in \mathbf{S}(\mathscr{L})$ *have compact quotient space and assume that* $\Gamma_n \to \Gamma$ *in* $\mathbf{S}(\mathscr{L})$. *Then there are (allowable) isomorphisms* $\phi_n : \Gamma \xrightarrow{\cong} \Gamma_n$ *for all large n.*

*Proof.* Take a set $A = \{a_1, \ldots a_k\}$ of generators for $\Gamma$ which correspond to a

*Dirichlet region* $\mathscr{P}(\Gamma, z_0)$.† Then for each $a_v \in A$, there is a sequence $\gamma_{n,v} \to \alpha_v$ as $n \to \infty$, with $\gamma_{n,v} \in \Gamma_n$. If we write

$$\gamma_{n,v} = \phi_n(a_v) \quad \text{for each } v = 1, \ldots k,$$

then $\phi_n$ is unambiguously defined‡ for all large enough values of $n$ and to show that it determines a homomorphism it suffices to verify that each of the finitely many basic relations in $\{a_1, \ldots, a_k\}$ goes over to the same one in $\{\gamma_{n,1}, \ldots, \gamma_{n,k}\}$. But if $R(a_1, \ldots, a_k) = e$ then $R(\gamma_{n1}, \ldots \gamma_{nk}) \to e$ as $n \to \infty$, and for large enough $n$ we therefore have $R(\gamma_{n1} \ldots \gamma_{nk}) = e$ since $\Gamma_n \to \Gamma$ and $\Gamma$ is discrete.

To show that $\phi_n$ is injective we construct a homeomorphism of $\mathscr{H}$ which induces $\phi_n$. Let $\mathscr{P}_n(z_0)$ denote the polygon formed by taking intersections of the (Hyperbolic) half-planes as below;

$$(2.4) \qquad \mathscr{P}_n(z_0) = \{z \in \mathscr{H} : d(z, z_0) < d(z, \gamma_{n,v}(z_0)), \qquad v = 1, \ldots, k\}.$$

LEMMA 1. *With the exception of points $z_0$ lying on a certain real analytic set $F = F(\Gamma)$ in $\mathscr{H}$, the Dirichlet region $\mathscr{P}(\Gamma, z_0)$ has the property that if $\gamma\overline{\mathscr{P}} \cap \overline{\mathscr{P}} \neq \varnothing$ with $\gamma \in \Gamma$, $\gamma \neq e$, then either $\gamma$ pairs two edges of $\mathscr{P}$ or $\gamma$ is elliptic and fixes a vertex of $\mathscr{P}$.*

*Proof.* This will follow if we can show that three translates of $\mathscr{P}$ meet at any non-elliptic vertex. Let $v$ be a vertex of $\mathscr{P}$ at which $N$ faces meet. Then $v$ is equidistant from $N$ points of the $\Gamma$-orbit of $z_0$. If $N \geqslant 4$ this means that four points (at least) $\gamma_1(z_0), \ldots, \gamma_4(z_0)$ lie on a circle in $\mathscr{H}$, which occurs if and only if the cross-ratio of the points is real. It follows that such points $z_0$ lie in the set $F = F(\Gamma)$ given by the condition that for some 4-tuple in $\Gamma$ the expression $\operatorname{Im}(\gamma_2(z_0), .;., \gamma_4(z_0))$ vanishes. $F$ is a countable union of analytic arcs§—locally it is a finite union. ∎

It follows from this that the polygon $\mathscr{P}_n(z_0)$ has the same number of sides as $\mathscr{P}$, provided that $z_0 \notin F(\Gamma)$ and $n$ is sufficiently large.

LEMMA 2. *$\mathscr{P}_n(z_0)$ is a fundamental polygon for $\phi_n(\Gamma)$ if $n$ is large.*

† See Chapter 7, Section 1.2 for the definition of $\mathscr{P}$.
‡ See exercise 1 below.
§ This follows from Exercise 2 below.

*Proof.* It is easily seen that as $n \to \infty$, $\mathscr{P}_n \to \mathscr{P}$ in the sense that given any neighbourhood $V$ of $e$, $\mathscr{P}_n \subseteq V\mathscr{P}$ for all $n$ sufficiently large. From the very definition, $\mathscr{P}_n \subseteq \mathscr{P}(\phi_n(\Gamma), z_0)$, so the images of $\mathscr{P}_n$ under $\phi_n(\Gamma)$ cover $\mathscr{H}$. To see that $\mathscr{P}_n$ is eventually $\phi_n(\Gamma)$-irreducible, let $z_n$, $w_n$ be points of $\mathscr{P}_n$ with $\gamma_n(z_n) = w_n$. Then using the compactness of the closure $\overline{\mathscr{P}}_n$ and the fact that $\mathscr{P}_n \to \mathscr{P}$ as $n \to \infty$ we obtain $z, w \in \overline{\mathscr{P}}$ and $\gamma \in \Gamma$ with $\gamma(z) = w$, which implies that either $\gamma \in \{a_1, \ldots a_k\}$ and so $\gamma_n \in \{\gamma_{n,v}\}$ for large $n$, or $\gamma = e$ and $z = w$ in which case $\gamma_n \to e$ and so $\gamma_n = e$ for $n$ sufficiently large. ∎

It is now elementary to contruct homeomorphisms $f_n : \mathscr{H} \to \mathscr{H}$ which induce $\phi_n$: triangulate $\mathscr{P}$ and $\mathscr{P}_n$ by joining $z_0$ to the vertices, and define $f_n$ to be (hyperbolic) "affine" between the paired triangles and extended so as to be $(\Gamma, \Gamma_n)$ compatible. This gives the injectivity of $\phi_n$.

Assume that $\phi_n(\Gamma) \neq \Gamma_n$ for infinitely many $n$. Then there exist $\delta_n \in \Gamma_n - \phi_n(\Gamma)$, and points $z_n$, $w_n \in \mathscr{P}_n$ with $\delta_n(z_n) = w_n$. Using the same argument as above we conclude that $\delta_n \to e$ and hence that $\delta_n = e$ for large enough $n$. ∎

*Exercise 1.* Verify that the rule $\phi_n$ in the proof of 2.3.1 is well defined for sufficiently large values of $n$. [*Hint.* Choose $W$ a neighbourhood of $e$ small enough to ensure $W^4 \subset V$, $W^{-1} \subset V$, and $a_v^{-1} W a_v \subset V$ for $v = 1, \ldots, k$ and such that $W$ contains no subgroup of $\mathscr{L}$. If $\Gamma_n \in \mathrm{N}(\Gamma, C, W)$ with suitably chosen compact $C$ then $\phi_n : \Gamma \to \Gamma_n$ is unambiguous.]

*Exercise 2* (Singerman). Prove that the cross-ratio $\{\gamma_1(z), \gamma_2(z); \gamma_3(z), \gamma_4(z)\}$ is a non-constant meromorphic function of $z$ unless two of the Möbius transformations $\{\gamma_j\}$ have a common fixed point. Deduce that the imaginary part cannot vanish identically if the $\{\gamma_j\}$ lie in a discrete group, except when each $\gamma_j$ is a power of a fixed elliptic transformation. (This last result also appears in Fricke–Klein [**13**] vol. I pp. 251–254).

There are two important consequences of the above proof.

2.3.2 Corollary. *The polygon $\mathscr{P}_n(z_0)$ defined by* (2.4) *is the Dirichlet polygon for $\phi_n(\Gamma)$ at $z_0$ if $\Gamma_n$ is sufficiently close to $\Gamma$ and $z_0 \notin F(\Gamma)$.*

*Proof.* This follows at once from Lemma 2. ∎

The second provides an extension of Macbeath's theorem and of Chabauty's result (2.3) for groups of finite type.

2.3.3 THEOREM. *Let $\Gamma$ be any finitely generated Fuchsian group and assume that $\Gamma_n \to \Gamma$ in $\mathbf{S}(\mathscr{L})$. Then there are (geometric) injective homomorphisms $\phi_n: \Gamma \to \Gamma_n$ for all large values of n, and*

$$\mu_0(\Gamma) \leqslant \liminf_{n\to\infty} \mu_0(\Gamma_n). \tag{2.5}$$

*Proof.* We proceed as in the proof of Theorem 2.3.1. In this case however the polygons $\mathscr{P}$ and $\mathscr{P}_n(z_0)$ defined as in (2.4) may have free sides and vertices on the real axis, and we use instead the intersections of these with the corresponding Nielsen convex regions $K^*(\Gamma)$, $K^*(\phi_n(\Gamma))$, which results in convex relatively compact sets $\mathscr{P}^*$, $\mathscr{P}_n^*$ which are polygons except for horocyclic edges corresponding to parabolic generators. It follows as before that for large $n$ the sets $\mathscr{P}_n^*$ are $\phi_n(\Gamma)$-irreducible and have the same number of edges as $\mathscr{P}^*$ and the patterns of identification of edges are isomorphic. The argument for injectivity of $\phi_n: \Gamma \to \Gamma_n$ goes over verbatim, using

LEMMA 2′. *$\mathscr{P}_n^*$ is a fundamental polygon for $\phi_n(\Gamma)$ on $K^*(\phi_n(\Gamma))$.*

The proof is the same as for Lemma 2.

The final statement (2.5) will follow from the next Lemma.

LEMMA 3. *If $\phi_{n'}(\Gamma) \neq \Gamma_{n'}$ for infinitely many $n'$, then as $n' \to \infty$ $\phi_{n'}(\Gamma)$ is a constituent subgroup of a decomposition of $\Gamma_{n'}$.*

*Proof.* Writing $n$ for $n'$, we shall show that for all large $n$ there is a fundamental region for $\Gamma_n$ which contains $\mathscr{P}_n^*$. If $\mathscr{D}_n$ denotes the Dirichlet region for $\Gamma_n$ at $z_0$, then either $\mathscr{P}_n^* \subseteq \mathscr{D}_n$ and we are finished, or there are (for an infinite sequence of values of $n$) pairs of points $z_n \in \mathscr{P}_n^*$, $w_n \in \mathscr{D}_n$ with $\gamma_n z_n = w_n$ for some $\gamma_n \in \Gamma_n - \phi_n(\Gamma)$. After passage to a subsequence we have $z_n \to z_0 \in \overline{\mathscr{P}^*}$ and $w_n \to w_0 \in \overline{\mathscr{H}}$. More precisely $w_0 \in \tilde{\mathscr{P}}$ (the Euclidean closure of $\mathscr{P}$) and *either* (i) $w_0 \in \tilde{\mathscr{P}} \cap \mathscr{H}$ which implies that $\gamma_n \to \gamma \in \Gamma$ with $\gamma(z_0) = w_0$, *or* (ii) $w_0 \in \tilde{\mathscr{P}} \cap \partial\mathscr{H}$ and $\gamma_n$ tends to the constant map $t(z) = w_0$. In the first case we have $\gamma_n = \phi_n(\gamma)$ for all large $n$, which is a contradiction. In the second case there is for every $n$ a $\delta_n \in \Gamma_n$ and an associated line $l_n = \{z: d(z, z_0) = d(z, \delta_n z_0)\}$ which intersects $\mathscr{P}_n^*$ with $d(z_n, z_0) > d(z_n, \delta_n(z_0))$. Since $z_n \to z_0 \in \overline{\mathscr{P}^*}$, $\delta_n$ must converge to some element of $\Gamma$ and again we obtain a contradiction.

There is therefore a dissection of the truncated Dirichlet region $\mathscr{D}_n$ into the union of $\mathscr{P}_n^*$ with finitely many convex subpolygons of $\mathscr{D}_n$, separated by the axes of boundary hyperbolic elements of $\phi_n(\Gamma)$. This gives the claimed decomposition of $\Gamma_n$ by application of Section 1.7. ■

*Definition.* For each Fuchsian group $\Gamma$ we denote by $\mathbf{S}_0(\Gamma)$ the subspace of $\mathbf{S}(\mathscr{L})$ consisting of groups allowably isomorphic to $\Gamma$, with the topology induced by $\mathscr{T}$.

If $\Gamma$ has compact quotient space, Theorem 2.3.1 states that $\mathbf{S}_0(\Gamma)$ is open in $\mathbf{S}(\mathscr{L})$. The example below shows that this is not so for the larger class of groups with finite quotient measure.

*Example* [**23**]. Let $\Gamma$ denote the group with signature $(0; 3; 0)$—often referred to as a *parabolic triangle* group—having as fundamental region in $\mathscr{H}$ the union of the triangle $T$ bounded by arcs of $l_1 = \{|z - \frac{1}{2}| = \frac{1}{2}\}$, $l_2 = \{\operatorname{Re} z = 0\}$, and $l_3 = \{\operatorname{Re} z = 1\}$ with its reflection in one of the sides. $\Gamma$ is a subgroup of index 2 in the group $\tilde{\Gamma} \subseteq \mathrm{PGL}_2(\mathbb{R})$ generated by reflections in the three lines $l_i$. Now let $\tilde{\Gamma}_n$, for $n \geqslant 1$, be the group generated by reflections in the circles $l_1$, $l_2$, $l_n = \{z : |z - n - 1| = n\}$ and $l'_n = \{\operatorname{Re} z = 2n + 1\}$, with $\Gamma_n \subseteq \mathscr{L}$ the subgroup of orientation-preserving elements. Then $\Gamma_n$ is discrete with signature $(0; 4; 0)$ for each $n$ and it is a simple exercise to show that $\Gamma_n$ tends to $\Gamma$ as $n \to \infty$. See Fig. 6. (p. 343).

*Note.* It is also easy to verify that in this example $\lim_{n \to \infty} \mu(\Gamma_n) = 2\mu(\Gamma)$, which shows that Chabauty's result (2.3) cannot be sharpened as it stands. However the discussion above does not bring out the full picture of what happens as $n \to \infty$; we shall prove later using Theorem 2.3.3 that in some sense there *is* a conservation of measure in the limit.

### 2.4 The relation between $\mathbf{S}_0(\Gamma)$ and $\mathscr{R}(\Gamma)$

*Definition.* Let $\Gamma$ be a finitely generated Fuchsian group. We denote by $\operatorname{Aut}(\Gamma)$ the group of *allowable* automorphisms of $\Gamma$, and by $\operatorname{Aut}^+(\Gamma)$ the subgroup of those which are sense-preserving (that is they map the basic relator (c.f. (1.2)) of $\Gamma$ into a conjugate of itself rather than of its inverse).

The group $\operatorname{Aut}(\Gamma)$ acts on the space $\mathscr{R}_0(\Gamma, \mathscr{L})$ of allowable homomorphisms by right translation: if $\alpha \in \operatorname{Aut}(\Gamma)$, the action $\rho_\alpha$ is defined by the rule that if $r \in \mathscr{R}_0(\Gamma)$, then

$$r \xrightarrow{\rho_\alpha} r \circ \alpha.$$

Notice the analogy with the action of $GL_2(\mathbb{Z}) \cong \operatorname{Aut} L$ on the space of marked lattices. One sees immediately that the $\rho_\alpha$ are analytic homeomorphisms of $\mathscr{R}_0(\Gamma)$.

2.4.1 PROPOSITION. *The action of* $\operatorname{Aut}(\Gamma)$ *on* $\mathscr{R}_0(\Gamma)$ *is discontinuous and without fixed points.*

*Proof.* Let $\Gamma = \langle \gamma_1, \ldots \gamma_n \rangle$. If $r \in \mathscr{R}_0(\Gamma)$, there is a neighbourhood $V$ of $e$ such that $tV \cap V = \varnothing$ for all $t \in r(\Gamma) - \{e\}$. Let $N(r, V) = N$ denote the set

$$\{\theta \in \mathscr{R}_0(\Gamma) : \theta(\gamma_i) \in r(\gamma_i) . V \quad \text{for} \quad i = 1, \ldots, n\}.$$

Then $N$ is open and contains $r$. If $\alpha \in \operatorname{Aut} \Gamma$ and $\alpha \neq id$, then for some $v$ we have $\alpha(\gamma_v) \neq \gamma_v$ and so $\rho_\alpha(r) \notin N$. It follows that $\operatorname{Aut}(\Gamma)$ is discontinuous at $r$ in the classical sense. There are obviously no fixed points. ∎

*Note.* One can with more effort show using the results of the previous section that for every $r \in \mathscr{R}_0(\Gamma)$ there is an open set $N_r$ with $r \in N_r$ and $N_r \cap \rho_\alpha N_r = \varnothing$ for all $\alpha \neq id$. This is done in [**24**] for the case where $\Gamma$ has compact quotient, and we sketch a proof in Theorem 2.4.2 below.

There is a mapping of $\mathscr{R}_0(\Gamma)$ into $\mathbf{S}_0(\Gamma)$ defined by

$$r \mapsto r(\Gamma),$$

and we denote it by $\pi$. We observe that $\pi$ is surjective and the fibre $\pi^{-1}(r(\Gamma))$ consists of the $\operatorname{Aut}(\Gamma)$-orbit of $r$.

2.4.2 THEOREM. *The mapping* $\pi$ *is a local homeomorphism, and induces a homeomorphism of* $\mathscr{R}_0(\Gamma)/\operatorname{Aut}\Gamma$ *onto* $\mathbf{S}_0(\Gamma)$.

*Proof.* We employ the method and notation of Theorem 2.3.3. If $\Gamma_n$, $n = 1, 2, \ldots$ is a sequence in $\mathbf{S}_0(\Gamma)$ converging to $\Gamma_0$ then, using $\Gamma_0$ as base point and choosing as generating set $E = \{\gamma_1, \ldots, \gamma_n\} = \{\gamma \in \Gamma - \{e\} : \gamma\bar{\mathscr{P}} \cap \bar{\mathscr{P}} \neq \varnothing\}$ with $\mathscr{P} = \mathscr{P}(\Gamma, z_0)$ as defined there, we obtain a sequence of (allowable) injective homomorphisms $\phi_n : \Gamma_0 \to \Gamma_n$. In this case the $\phi_n$ must also be surjective for all large $n$, because $\Gamma_n \cong \Gamma_0 \cong \phi_n(\Gamma_0)$ and $\phi_n(\Gamma_0)$ is a constituent subgroup of $\Gamma_n$, making it impossible for $\phi_n(\Gamma_0)$ to be geometrically isomorphic to $\Gamma_n$ unless they coincide. Therefore we have constructed a sequence $\phi_n \in \mathscr{R}_0(\Gamma_0)$ with the property that $\pi(\phi_n) = \Gamma_n$ and $\phi_n \to \text{id}$ as $n \to \infty$.

We also observe that if $\alpha \in \mathrm{Aut}(\Gamma_0)$ then the isomorphisms $\phi_n \circ \alpha$ satisfy $\pi(\phi_n \circ \alpha) = \Gamma_n$ and converge to the element $\alpha \in \pi^{-1}(\Gamma_0)$. One sees easily that the neighbourhood $N_0 = N(\Gamma_0, C, W)$ is bijective with each open set $\mathcal{N}_\alpha = \mathcal{N}(\alpha, E, W) = \{r \in \mathcal{R}_0 : r(\gamma) \in \alpha(\gamma)W \text{ for } \gamma \in E\}$, if $C$ is compact and contains the set $\{t \in \mathcal{L} : t\overline{\mathcal{P}}^* \cap \overline{\mathcal{P}}^* \neq \emptyset\}$ and $W$ is a neighbourhood of $e$ small enough to ensure that the set $W\overline{\mathcal{P}}^* = W^*$ satisfies $\{t \in \Gamma_0 - \{e\} : tW^* \cap W^* \neq \emptyset\} = E$. It is also true that the sets $\mathcal{N}_\alpha$ are then disjoint, and the argument above applied to sequences converging to any point $\Gamma$ of $\mathcal{N}_0$ proves bicontinuity of each $\pi|_{\mathcal{N}_\alpha}$. ■

This result has immediate implications for the structure of $\mathbf{S}_0(\Gamma)$.

2.4.3 COROLLARY. *The space* $\mathbf{S}_0(\Gamma)$ *is a connected real manifold of dimension equal to* $\dim \mathcal{R}_0(\Gamma, \mathcal{L})$.

*Proof.* The only points to be made are that $\mathcal{R}_0 = \mathcal{R}_0^+ \cup \mathcal{R}_0^-$ is a disjoint union of connected manifolds and that $\mathrm{Aut}^+ \Gamma$ is the subgroup which preserves $\mathcal{R}_0^+$. Consequently $\mathcal{R}_0/\mathrm{Aut}\,\Gamma \simeq \mathcal{R}_0^+/\mathrm{Aut}^+\,\Gamma \simeq \mathbf{S}_0(\Gamma)$. ■

2.4.4 COROLLARY. $\mathbf{S}_0(\Gamma)$ *is a* $K(G, 1)$ *space and has as fundamental group an extension*

$$1 \to \mathbb{Z} \to G \to \mathrm{Aut}^+\,\Gamma \to 1. \tag{2.6}$$

*Proof.* This follows from the fact that $\mathcal{R}_0^+(\Gamma) \simeq \mathbb{R}^{d+2} \times S^1$ with $d$ as in Theorem 1.4.3, so that $\mathbf{S}_0(\Gamma)$ has $\mathbb{R}^{d+3}$ as universal covering and $\pi_1(\mathcal{R}_0^+) \cong \mathbb{Z}$. ■

*Notes.* One can show **[22]**, **[26]** that the extension (2.6) is central and does not split. If $\Gamma$ has signature $(1;1)$, then $\mathrm{Aut}^+\,\Gamma \cong \mathrm{GL}_2^+(\mathbb{Z})$ and $G$ is Artin's braid group $B_3$ (see Section 1.1).

*Exercise 1.* Show that $\mathbf{S}_0(\Gamma)$ is not closed in $\mathbf{S}(\mathcal{L})$ for any Fuchsian group $\Gamma$.

*Exercise 2.* Use 2.4.2 and 2.3.1 to give a proof of Weil's result on the local structure of $\mathcal{R}_0(\Gamma)$ (Section 1.2).

## 2.5 Compact subsets of $\mathbf{S}_0(\Gamma)$

We should like to know what is the *boundary* of the manifold $\mathbf{S}_0(\Gamma)$. However, in seeking to visualize the structure at infinity of $\mathbf{S}_0(\Gamma)$ it is important to have a precise description of the compact subsets which are in some sense large. We begin by deducing a result of Mumford [**36**], which has been a stimulus for much recent work on degenerating Fuchsian groups.

2.5.1 THEOREM. *Let* $\Gamma$ *be a Fuchsian group with signature* $(g;0)$. *For any* $\varepsilon > 0$ *the set* $\mathscr{K}_\varepsilon(\Gamma)$ *of groups* $\Gamma' \in \mathbf{S}_0(\Gamma)$ *satisfying*

$$|\mathrm{tr}\,\gamma| \geqslant 2 + \varepsilon, \qquad \text{for all } \gamma \in \Gamma' \text{ with } \gamma \neq e, \tag{2.7}$$

*is compact.*

*Proof.* It follows from (2.7) that there is an open set $V_\varepsilon$ in $\mathscr{L}$ which contains all parabolic elements (and $e$) such that $\Gamma' \cap V_\varepsilon = \{e\}$ for any $\Gamma' \in \mathscr{K}_\varepsilon(\Gamma)$. Therefore by Chabauty's theorem any sequence in $\mathscr{K}_\varepsilon(\Gamma)$ must have a subsequence converging to a discrete group $\Gamma_0$ which contains no parabolic elements, and since $\mu(\Gamma') = 4\pi(g-1)$ for all $\Gamma' \in \mathbf{S}_0(\Gamma)$ it also follows that $\mu(\Gamma_0) = 4\pi(g_0 - 1)$ with $2 \leqslant g_0 \leqslant g$. Hence $\Gamma_0$ has compact quotient space, and $\Gamma_0 \in \mathbf{S}_0(\Gamma)$ by Macbeath's Theorem 2.3.1. ■

*Exercise.* Formulate and prove an analogous result for groups of compact type with torsion.

*Remark 1.* The interest of this theorem stems from the interpretation of $|\mathrm{tr}\,\gamma|$ for a hyperbolic element $\gamma \in \Gamma'$ as the length of a minimal closed geodesic in the surface $\Gamma' \backslash \mathscr{H}$ obtained, we recall, by projecting the hyperbolic line segment on the axis of $\gamma$ from a point $z$ to $\gamma(z)$. In this view the theorem states that the set of groups representing surfaces with a lower bound on the length of any minimal geodesic is compact.

*Remark 2.* With the help of Theorem 2.3.3 one can prove an extension of a sort to *all* groups of finite signature: given $\varepsilon > 0$ and $M > 0$ constants, the set of groups $\mathscr{K}_{\varepsilon,M} \subseteq \mathbf{S}(\mathscr{L})$ defined by

$$\mathscr{K}_{\varepsilon,M} = \{\Gamma : \mu_0(\Gamma) \leqslant M, \qquad |\mathrm{tr}^2\,\gamma - 4| \geqslant \varepsilon \text{ for all } \gamma \in \Gamma \text{ with } \mathrm{tr}\,\gamma \neq 2\}$$

is closed.

*Remark 3.* It is of interest to note that the result (2.5.1) does *not* extend to all groups with quotient of finite area since for example the space of all parabolic triangle groups (signature $(0;3;0)$), which forms a single $\mathscr{L}$-conjugacy class, is not compact although all the groups have the same discrete set of traces and therefore satisfy a condition of the type $|\mathrm{tr}\,\gamma| \geqslant 2 + \varepsilon$ for all hyperbolic elements $\gamma$. There is however a result of this kind for the space of *conjugacy classes* of groups of finite area, as we shall see in Chapter 3.

## 2.6 The relative boundary of $\mathbf{S}_0(\Gamma)$ in $\mathbf{S}(\mathscr{L})$

By putting together some facts on decompositions from Section 1 with the results of Section 2.3, we can characterize most of the boundary groups of $\mathbf{S}_0(\Gamma)$ which are Fuchsian purely in terms of the topological degeneration of $S = \Gamma\backslash\mathscr{H}$.

Let $\Gamma_1 \in \partial\,\mathbf{S}_0(\Gamma)$ be a non-elementary Fuchsian group. Then Lemma 3 of Section 2.3 implies that $\Gamma_1$ is geometrically isomorphic to a constituent subgroup $\Gamma'$ of some decomposition for $\Gamma$ (see Section 1.7); the injective mapping $\phi : S_1' \to S$ which induces the isomorphism takes $S_1$ onto a sub-surface $S' \subseteq S$ bounded by $k$ loops, some of which correspond to boundary curves of $S_1$ and some to punctures. It is even true that every loop in $\partial S'$ that is not in $\partial S$ corresponds to a puncture of $S_1$ but we shall not need this fact. For our purposes we need only to describe some of the signatures possible for $\Gamma_1$ in terms of that of $\Gamma$. We introduce a partial ordering on signatures as follows:

*Definition.* Let $\Gamma, \Gamma'$ be non-elementary. We write $\Gamma' \lhd \Gamma$ if $\Gamma$ has a decomposition graph with a vertex group geometrically isomorphic to $\Gamma'$.

This means that the signatures $(g'; m', n')$ and $(g; m, n)$ satisfy $g' \leqslant g$, $m' \leqslant m$, $2g' - 2 + m' + n' \leqslant 2g - 2 + m + n$, and if the edge number of the vertex $\Gamma'$ in the decomposition graph is $k'$, then $k' \leqslant n' \leqslant n + k'$. We extend the ordering $\lhd$ as follows: Under a geometric isomorphism $\Gamma'$ can be changed to a group with signature $(g'; m' + k; n' - k)$ for any $k \leqslant n'$. We say that $\Gamma_0 \prec \Gamma$ if the signature of $\Gamma_0$ is of this form with $k' \leqslant k \leqslant n'$, and we write $\mathscr{E}(\Gamma)$ for the finite set of signatures (groups) obtainable in this way from $\Gamma$.

We can now formulate an approximation theorem for $\mathbf{S}_0(\Gamma)$.

2.6.1 THEOREM. *The boundary of* $\mathbf{S}_0(\Gamma)$ *contains the union of the spaces* $\mathbf{S}_0(\tilde{\Gamma})$ *for* $\tilde{\Gamma} \in \mathscr{E}(\Gamma)$.

*Proof.* Let $\Gamma_0 \in \mathbf{S}_0(\tilde{\Gamma})$ and denote by $\Gamma'$ a corresponding constituent subgroup of $\Gamma$. Under the isomorphism $\theta:\Gamma' \to \Gamma_0$ some (maximal) primary set $\{T_1, \ldots, T_k\}$ of hyperbolic generators become parabolic. We employ the technique of decomposing $\Gamma'$ and $\Gamma_0$ into triangle groups, together with the method of Proposition 1.5.4, to construct for any given $\varepsilon > 0$ a marked group $r_\varepsilon(\Gamma') \in \mathscr{R}_0(\Gamma')$ satisfying

(i) $2 < |\mathrm{tr}\, r_\varepsilon(T_\alpha)| < 2 + \varepsilon$ for $\alpha = 1, \ldots, k$;
(ii) $|\mathrm{tr}\,(r_\varepsilon(T)) - \mathrm{tr}\,\theta(T)| < \varepsilon$, for all $T \notin \{T_\alpha\}$ in a suitable finite generating set for $\Gamma'$;
(iii) if $\langle T_\alpha \rangle$, $\langle T_\beta \rangle$ are joined by an edge in the decomposition graph of $\Gamma'$ then $\mathrm{tr}\, r_\varepsilon(T_\alpha) = \mathrm{tr}\, r_\varepsilon(T_\beta)$;
(iv) there is a fundamental polygon for $r_\varepsilon(\Gamma')$ lying in an $\varepsilon - nbd$ of $D(\Gamma_0, i)$, the Dirichlet region for $\Gamma_0$ based at $i$.

One can now construct a combination group $\Gamma_\varepsilon$ containing $r_\varepsilon(\Gamma')$ which lies in $\mathbf{S}_0(\Gamma)$, and using Lemma 2.1 it follows that as $\varepsilon \to 0$ $\Gamma_\varepsilon \to \Gamma_0$ in $\mathbf{S}(\mathscr{L})$. It is in fact true that the fundamental polygons $D(\Gamma_\varepsilon, i)$ tend to $D(\Gamma_0, i)$. ■

*Note 1.* The last step of the construction allows considerable freedom of choice. We shall later sharpen this to prove a stronger approximation property (Theorem 3.1.1) for $\mathbf{S}_0(\Gamma)$.
*Note 2.* It can be shown that this yields the whole boundary of $\mathbf{S}_0(\Gamma)$ except for the elementary groups. For $\Gamma$ of finite volume this will in fact follow from later considerations.
*Note 3.* There is a structure of *stratified analytic space* (in the sense of Whitney) on the space $\mathbf{S}_0(\mathscr{L})$, induced by the partial ordering $\prec$ on the allowable isomorphism classes of groups. The simplest aspect of it is the subset of groups with signature (0; 3) treated in Proposition 1.5.4.

## 3. THE MODULI SPACE

### 3.1 Preliminaries

In this part we study the spaces $\mathscr{X}^+(\Gamma)$ of *conjugacy classes* of Fuchsian groups with finite area. Our results concern primarily the topological properties of $\mathscr{X}^+$—the important question of complex structure is neglected as this would lead too far afield.

To motivate the rather intricate construction of a compactification for $\mathscr{X}^+$, we return to the example of elliptic curves (Section 1.1), where one embeds $\mathscr{X}_1 \cong \mathbb{C}$ inside $\mathbb{P}_1(\mathbb{C})$ by adjoining a point $\infty$ corresponding to the cusp of the modular group. Identifying a lattice $\Lambda = \langle 1, \tau \rangle$, where $\tau \in \mathscr{H}$, with the elliptic curve $C_\tau$ given by

$$y^2 = 4x^3 - g_2(\Lambda)x - g_3(\Lambda),$$

with $g_2$, $g_3$ as defined in Section 1.1, it is evident from the $q$-expansions of $g_2, g_3$ ($q = \exp 2\pi i \tau$) that as $\tau \to i\infty$ inside $0 \leqslant \operatorname{Re} \tau < 1$ the curve acquires a node; $C_\infty$ is (c.f. [**45**, I])

$$\left\{y^2 = 4\left(x + \frac{\pi^2}{3}\right)^2 \cdot \left(x + \frac{2\pi^2}{3}\right)\right\}.$$

In a similar way, it transpires that the adjunction of certain singular curves to the moduli space $\mathscr{X}_g$ with $g \geqslant 2$ results in a compact space. Here we describe this process in terms of the underlying Fuchsian groups.

### 3.2 The Teichmüller modular group

We begin with the space $\mathbf{T}_0(\Gamma)$ for a finitely generated group $\Gamma$. Parallel to the action of $\operatorname{Aut}(\Gamma)$ on $\mathscr{R}_0(\Gamma)$ given in Section 2.4, the allowable automorphisms of $\Gamma$ also act on $\mathbf{T}_0(\Gamma)$ but not effectively since the subgroup $\operatorname{Inn}(\Gamma)$ consisting of *inner automorphisms* of $\Gamma$ fixes every element of the space. The quotient group $\operatorname{Aut}(\Gamma)/\operatorname{Inn}(\Gamma)$, denoted $\operatorname{Mod}(\Gamma)$, is the algebraic *mapping class group* of $\Gamma$ (c.f. Chapter 6). It acts as a group of self-homeomorphisms of $\mathbf{T}_0(\Gamma)$ by the rule ($r \in \mathscr{R}_0(\Gamma)$, $\alpha \in \operatorname{Aut} \Gamma$ define points $[r] \in \mathbf{T}_0$ and mapping classes $\hat{\alpha}$)

$$(3.1) \qquad [r] \xrightarrow{\hat{\alpha}} [r \circ \alpha].$$

In fact the action is *real analytic* (or algebraic), because one can lift an allowable automorphism of $\Gamma$ to the free group of the presentation (1.2) for $\Gamma$ and it is then sufficient to consider "Nielsen automorphisms" (see for example [**28**]) which do act biregularly on the set of traces parametrizing $\mathbf{T}_0(\Gamma)$.

It is important to note that by virtue of the Fenchel–Nielsen Theorem, 1.4.2 one can replace the action of $\operatorname{Aut} \Gamma$ by a suitable group of $\Gamma$-compatible† homeomorphisms of $\mathscr{H}$, and a theorem of Marden [**29**] asserts that such a homeomorphism is isotopic to the identity if and only if it induces an inner automorphism of $\Gamma$. We can interpret this as follows (c.f. Chapter 5, Section 4; Chapter 1, Section 3; Chapter 6).

† By this we mean that each $f$ projects to a self-homeomorphism of $\Gamma \backslash \mathscr{H}$.

3.2.1 PROPOSITION [**27**]. *The algebraic and topological mapping class groups are isomorphic, and* Mod(Γ) *with the action* (3.1) *is the Teichmuller modular group.*

With some low-dimensional exceptions [**24**], the action is effective. As we wish to form the quotient, the next result, which goes back to Fricke, is crucial.

3.2.2 THEOREM. Mod Γ *acts properly discontinuously on* $\mathbf{T}_0(\Gamma)$.

*Proof.* Let $K, L \subseteq \mathbf{T}_0(\Gamma)$ be compact and assume that the set $A$ of $\hat{\alpha} \in \mathrm{Mod}(\Gamma)$ with $\hat{\alpha}K \cap L \neq \varnothing$ is infinite. Then there is a sequence of points $z_n \in K$ with distinct $\hat{\alpha}_n \in \mathrm{Mod}(\Gamma)$ and $\hat{\alpha}_n(z_n) \in L$. After passage to a subsequence, $z_n \to z$, $\hat{\alpha}_n(z_n) \to w$ and the elements $\hat{\alpha}_n^{-1} \circ \hat{\alpha}_{n-1} = \hat{\beta}_n$ have the property that $\{\hat{\beta}_n(z)\}$ accumulates at $z \in K$. But each $\beta_n$ permutes the traces of the groups $\{\Gamma_z\}$ represented by the point $z$, and it is known (Chapter 7, Theorem 1.4.2) that the set of traces in $\Gamma_z$ is discrete. From this it follows that for all large $n$, $\beta_n$ fixes the trace of every element of $\Gamma_z$.

Now a fundamental lemma on isomorphisms of non-elementary subgroups of Lie groups shows that $\beta_n$ must then extend to an automophism of $\mathscr{L}$ [**39, 41**] and so only finitely many classes $\{\hat{\alpha}_n\}$ are distinct. ∎

*Definition.* The quotient space $\mathbf{T}_0(\Gamma)/\mathrm{Mod}\,\Gamma$ is the *reduced moduli space* for Γ, denoted $\mathscr{X}(\Gamma)$. $\mathbf{T}_0(\Gamma)/\mathrm{Mod}^+(\Gamma)$ is called the *moduli space for* Γ or the *Riemann space of* Γ; we denote it by $\mathscr{X}^+(\Gamma)$.

The points of $\mathscr{X}^+(\Gamma)$ are $\mathscr{L}$-conjugacy classes of groups allowably isomorphic to Γ, and Theorem 3.2.2 implies that they form a connected real manifold. For the case of a surface group this represents the completion of the "continuity method" due to Klein and Poincaré, proving an assertion of Riemann that the number of parameters on which the conformal class of a Riemann surface with genus $g \geqslant 2$ depends is $6g - 6$. $\mathscr{X}(\Gamma)$ is obtained from $\mathscr{X}^+(\Gamma)$ by identifying surface classes which are mirror images of each other.

### 3.3 Topological properties of $\mathscr{X}^+(\Gamma)$†

There is for fixed Γ a commutative diagram of spaces

† Most of the results here apply *mutatis mutandis* to $\mathscr{X}(\Gamma)$.

$$
\begin{array}{ccc}
\mathscr{R}_0^+(\Gamma) & \longrightarrow & \mathbf{S}_0(\Gamma) \\
\downarrow & & \downarrow \\
\mathbf{T}_0(\Gamma) & \longrightarrow & \mathscr{X}^+(\Gamma),
\end{array}
\tag{3.2}
$$

where the vertical (resp. horizontal) arrows represent quotient maps modulo $\text{Aut}^+ \mathscr{L}$ (resp. $\text{Aut}^+ \Gamma$). We insert here a brief discussion of the properties of $\mathscr{X}^+(\Gamma)$ which are directly related to this picture. Further details may be found [**24**] and [**27**].

Recall that $\mathscr{R}_0^+ \to \mathbf{T}_0$ is a principal bundle with structure group $\mathscr{L}$.

3.3.1 PROPOSITION. *The projection* $\mathbf{T}_0(\Gamma) \to \mathscr{X}(\Gamma)$ *is branched over an analytic subset* $\mathscr{I}(\Gamma)$, *corresponding to fixed points of* $\text{Mod}^+ \Gamma$.

*Proof.* If [r] is fixed by $\hat{\alpha} \in \text{Mod}^+ \Gamma$, then $[r \circ \alpha] = [r]$, so for some $t \in \mathscr{L}$ we have $\tilde{t} \circ r = r \circ \alpha$ which implies that $t$ normalizes the group $r(\Gamma)$. Conversely any element of the *normalizer* $N = \mathscr{N}(r(\Gamma))$ in $\mathscr{L}$ determines a mapping class which fixes $[r]$. Now $N$ is Fuchsian and the index $[N : r(\Gamma)]$ is finite, being the number of conformal self-homeomorphisms of the Riemann surface $\mathscr{H}/r(\Gamma)$. One can show that the finite group in $\text{Mod}^+(\Gamma)$ must fix an analytic submanifold of $\mathbf{T}_0(\Gamma)$ isomorphic to $\mathbf{T}_0(N)$ [**14**]. ■

This leads to a stratification of the *branch locus* $\mathscr{I}(\Gamma)$ into submanifolds fixed by the various finite subgroups [**27**].

An important unsolved problem deserves mention here. Since any subgroup of $\text{Mod}^+ \Gamma$ (or $\text{Mod}\,\Gamma$) fixing some point of $\mathbf{T}_0(\Gamma)$ must be finite, it is natural to ask whether every finite group $G$ in $\text{Mod}\,\Gamma$ has non-void fixed set; this is the *Hurwitz–Nielsen Realization Problem.* A positive answer for cyclic $G$ was given by Nielsen, and for solvable groups by Fenchel. The latter proof works for arbitrary $\Gamma$ [**27**]. Recently Maclachlan has shown that the answer is yes if the type of $\Gamma$ is either (2, 0) or $(g, n)$ with $g$ arbitrary and $1 \leqslant n \leqslant 4$.

As regards the mapping $S_0(\Gamma) \to \mathscr{X}^+(\Gamma)$, we note that outside the branch set $\mathscr{I}(\Gamma)$ it is a fibration. Over a point $x \in \mathscr{X}^+(\Gamma)$ representing a conjugacy class $\{\Gamma_x\}$ the fibre is $\mathscr{L}/\mathscr{N}(\Gamma_x)$.

An interesting topological property of $\mathscr{X}^+(\Gamma)$ follows from Theorem 3.2.2 and a known result on the structure of the modular group.

3.3.2 PROPOSITION [**25**]. $\mathscr{X}^+(\Gamma)$ *is simply connected if* $\Gamma$ *is a surface group.*

*Proof*. By Armstrong's result (see Chapter 1, [**26**]) on fundamental groups of branched coverings, $\pi_1(\mathscr{X}^+)$ is isomorphic to $\text{Mod}^+(\Gamma)$ modulo the normal closure of the set of elements with fixed points. But $\text{Mod}^+(\Gamma)$ is generated by elements of finite order if $\Gamma$ has signature $(g, 0)$ (see Chapter 6, Corollary 4.6). ■

*Remark*. The result holds for all torsion-free $\Gamma$, with some exceptions when the genus is 2 [**26**], and also for the spaces $\mathscr{X}(\Gamma)$.

*Exercise*. Show that if $\Gamma$ has sufficiently many distinct periods then $\mathscr{X}^+(\Gamma)$ is not simply connected.

### 3.4 The global algebraic structure of $\mathscr{X}^+(\Gamma)$; stable curves and groups

It is known that $\mathbf{T}_0(\Gamma)$ carries a complex structure (see Chapter 5), and the holomorphic nature of the action of $\text{Mod}^+ \Gamma$ implies that the moduli space $\mathscr{X}^+(\Gamma)$ is a complex analytic space. If $\Gamma$ is torsion-free and of finite area, one can show that $\mathscr{X}^+(\Gamma)$ is a quasi-projective variety (see for example [**7**]). Now at least in the fundamental case of surface groups all this is superceded by a result of Mumford (following some unpublished work with A. Mayer) that there is a "coarse moduli scheme" of non-singular complete curves of genus $g$, which admits a compactification $\overline{\mathscr{M}_g}$ consisting of curves with certain simple singularities allowed. The scheme $\overline{\mathscr{M}_g}$ is known to be projective and irreducible, which implies that $\mathscr{X}^+(\Gamma)$ is a dense open subset of a compact algebraic manifold. It is noteworthy however that the proof† [**10**] of the irreducibility of the moduli scheme uses the fact that $\mathbf{T}_0(\Gamma)$ is connected, and further evidence that our approach is vauable comes from the fact that there is a natural description of the Mayer–Mumford compactification in the framework developed in Section 2.6. Marden's article in this volume and various recent research announcements ([**2**], [**7**]) demonstrate that one can hope to recover the full strength of Mumford's results for curves over $\mathbb{C}$ using Teichmüller theory and Kleinian groups. We shall describe next this process of embedding $\mathscr{X}^+(\Gamma)$ in a compact space, beginning from the motivating concept of stable curve.

*Definition*. A *stable curve of genus g* is a complete algebraic curve $C$ (in some projective space) having at worst ordinary double points and satisfying

† For arbitrary characteristic—in characteristic O it is classical.

(i) each rational component of $C$ has at least three double points,
(ii) the arithmetic genus of $C$ is $g$.

A topological model of a stable curve consists of a collection of compact surfaces joined to form a connected space by identifying pairwise a finite set of points. By (i), each component of $C$ is a hyperbolic Riemann surface and (ii) implies that the total Poincaré area of the component surfaces is $4\pi(g-1)$.

The appropriate concepts for Fuchsian groups may be formulated in similar fashion:

*Definition.* Let $(\mathscr{G}, K)$ be a connected graph of Fuchsian groups with vertex groups $G_i$ of finite area and edge groups parabolic cyclic, and with total signature $(g; n)$.† A *marked stable group* with signature $(g, n)$ of type $(\mathscr{G}, K)$ is a collection of allowable isomorphisms $r_i \in \mathbf{R}_0(G_i, \mathscr{L})$ for all $G_i$ in $\mathscr{G}$.

We note the resemblance to decomposition graphs. Here however the vertex groups are all of the first kind and the identifications given by the edges are between parabolic subgroups, so the group $\pi_1(\mathscr{G}, K)$ is never Fuchsian unless $K$ is simply a single point, although it does have a Fuchsian presentation.

We can also form the space $\mathbf{T}_0(\mathscr{G}, K)$ of *normalized marked stable groups* and the *moduli spaces* $\mathscr{X}^+(G, K)$ and $\mathscr{X}(G, K)$ in the same way.

There is a partial ordering on the family of isomorphism types of stable group with fixed total signature, induced by the ordering $\prec$ on signatures defined in Section 2.6.

*Definition.* We write $(\mathscr{G}', K') \prec (\mathscr{G}, K)$ if there is a family of decompositions of vertex groups in $(\mathscr{G}, K)$ such that the expanded graph of groups is isomorphic to $(\mathscr{G}', K')$.

*Remark.* This condition implies that there is a continuous mapping from the topological model of $(\mathscr{G}, K)$ onto that of $(\mathscr{G}', K')$.

*Definition.* $\overline{\mathscr{X}} = \overline{\mathscr{X}}^+(g; n)$ denotes the union of all moduli spaces $\mathscr{X}^+(\mathscr{G}, K)$ for $(\mathscr{G}, K)$ of total signature $(g; n)$, with the following topology.

A sequence $\{\mathbf{x}_\nu\}$ in $\overline{\mathscr{X}}$ converges to $\mathbf{x} \in \mathscr{X}^+(\mathscr{G}, K)$ if and only if

(i) for each vertex class $[G_i] \in \mathbf{x}$, there is a sequence of vertex classes $[G_{i,\nu}] \in \mathbf{x}_\nu$ with $G_{i,\nu} \to G_i$ in $\mathbf{S}_0(\mathscr{L})$ as $\nu \to \infty$;

† which means that $\pi_1(\mathscr{G}, K)$ is of signature type $(g; n)$.

(ii) if for some sequence of vertex classes $[G_\nu] \in \mathbf{x}_\nu$ the groups $G_\nu$ converge to $G \in \mathbf{S}_0(\mathscr{L})$, then $[G]$ is a vertex class of $\mathbf{x}$.

*Note.* If $\mathbf{x}_\nu \to \mathbf{x}$ and $\mathbf{x}_\nu \in \mathscr{X}(\mathscr{G}_\nu, K_\nu)$, then for all large $n$, $(\mathscr{G}, K) \prec (\mathscr{G}_\nu, K_\nu)$. Moreover, $\mathbf{x}$ is uniquely determined.

We refer to the space $\overline{\mathscr{X}}(g; n)$ as the *full moduli space* of groups with signature $(g; n)$. For convenience we also use the notation $\overline{\mathscr{X}}(\Gamma)$, when $\Gamma$ has signature $(g; n; 0)$. It is immediate that $\overline{\mathscr{X}}(\Gamma)$ contains $\mathscr{X}(\Gamma)$ as $\mathscr{X}(\mathscr{G}_0, K_0)$, where $(\mathscr{G}_0, K_0)$ is the graph with solitary vertex group $[\Gamma]$ and no edges.

3.4.1 THEOREM. *$\overline{\mathscr{X}} = \overline{\mathscr{X}}(g; n)$ is a Hausdorff space and $\mathscr{X} = \mathscr{X}(\Gamma)$ is a dense open subset.*

*Proof.* We first note that the restriction of the topology of $\overline{\mathscr{X}}$ to $\mathscr{X}$ is the same as that induced by the projection from the Chabauty space $\mathbf{S}_0(\Gamma)$; in other words, the inclusion $\mathscr{X} \to \overline{\mathscr{X}}$ is continuous and open. This follows using Lemma 2.1. The Hausdorff property now follows easily after observing that similarly the topology restricts on each subspace $\mathscr{X}(\mathscr{G}, K)$ to the product of the topologies on the vertex spaces $\mathscr{X}(G_i)$.

To show that $\mathscr{X}$ is dense in $\overline{\mathscr{X}}$ we must construct a sequence in $\mathscr{X}$ which approximates any given stable group $(\mathscr{G}, K)$. Let the vertex conjugacy classes be $[G_1], \ldots, [G_l]$ with connecting edges $\{e_\delta\}$. By the argument of Theorem 2.6.1 it is possible for $\varepsilon > 0$ to approximate each group $G_1, \ldots, G_l$ by a group $\Gamma_{i\varepsilon} = r_{i\varepsilon}(G_i)$,† $i = 1, \ldots, l$, of the second kind in such a way that the subgroups $H \subseteq G_i$, $H' \subseteq G_j$ linked by an edge $e_\alpha$ in $K$ have approximating subgroups $r_i(H)$, $r_j(H')$ with equal traces. One can now choose a set of conjugates $\{\tilde{\Gamma}_i = t_i \Gamma_i t_i^{-1}\}$ in $\mathscr{L}$ and a set of glueing elements $t_\alpha \in \mathscr{L}$ such that the data $\{\tilde{\Gamma}_i, \langle t_\alpha \rangle\}$ and graph $K$ determine a Fuchsian group $\tilde{\Gamma}_\varepsilon \in \mathbf{S}_0(\Gamma)$. It follows that since the conjugates $t_i^{-1} \tilde{\Gamma}_\varepsilon t_i$ approach $\Gamma_i$ for $i = 1, \ldots, l$, condition (i) for $[\tilde{\Gamma}_\varepsilon] \in \overline{\mathscr{X}}(\Gamma)$ to approach the stable group $(\mathscr{G}, K)$ is satisfied. To verify condition (ii), we let $G$ be any non-elementary limit of conjugates of $\{\tilde{\Gamma}_\varepsilon\}$ as $\varepsilon \to 0$. Then by Theorem 2.3.3, $G$ has Dirichlet polygon $\mathscr{D}$ which is a limit of polygons $\mathscr{D}_\varepsilon$ for some $\epsilon$-family of constituent subgroups of $\tilde{\Gamma}_\varepsilon$. But $\mathscr{D}_\varepsilon$ is a union of subpolygons, conjugates of which tend to Dirichlet polygons for $G_i$, $i = 1 \ldots l$, and no other limits within $\mathbf{S}(\mathscr{L})$ can occur. ■

*Note.* The direction of approach to $(\mathscr{G}, K)$ depends on $k$ parameters where $k$ is precisely the number of edges $\{e_\alpha\}$—there is one twist parameter for each

† We suppress the approximating factor $\varepsilon$ usually.

$t_\alpha$ and these can be taken as angles between 0 and $2\pi$ after taking into account the action of the *Dehn twist* $\tau_\alpha$ (in Mod $\Gamma$) about the loop in $S$ corresponding to the edge $e_\alpha$.

### 3.5 Uniform discreteness; a fundamental geometric lemma

A valuable indication that the space $\mathscr{X}(\Gamma)$ has better convergence properties than $\mathbf{S}_0(\Gamma)$ comes from a lemma due to Kazdan and Margulis [**18**], valid for all semisimple Lie groups $\mathscr{L}$ without compact factors; there is a neighbourhood $V$ of the identity $e$ in $\mathscr{L}$ such that for every discrete group $\Gamma$ some conjugate $t^{-1}\Gamma t$ of it satisfies the condition

$$t^{-1}\Gamma t \cap V = \{e\}. \tag{3.3}$$

This property (known as *uniform discreteness*) is what is needed to apply Chaubauty's compactness Theorem 2.2.1; given any sequence of groups in $\mathbf{S}_0(\Gamma)$, some sequence of conjugates has a subsequence converging to a group $\tilde{\Gamma}$ of measure at most $\mu(\Gamma)$. There is a snag however: if $\mu(\tilde{\Gamma}) < \mu(\Gamma)$, what has happened to the missing piece of surface? To answer this we must first reinterpret the uniform discreteness in appropriate form: if we consider the Dirichlet polygon $\mathscr{D}$ for a group $\Gamma$ at $z_0 \in \mathscr{H}$, and project from $\mathscr{L}$ to $\mathscr{H}$ by the mapping $T \overset{p}{\mapsto} T(z_0)$, then the condition (3.3) for $\Gamma$ torsion free implies that for $U$ satisfying $UU^{-1} \subset V$ the open set $p(U)$ is contained in $\mathscr{D}$. This implies that for variable $\Gamma$ the surface $S_\Gamma$ always includes a piece $p(U)$.

The result that is necessary for a full convergence picture is the following simple lemma from non-Euclidean geometry which went unsuspected until the above connection was noticed. An independent proof is due to Shinnar and Sturm [**46**].

3.5.1 Lemma [**15**]. *For $A > 0$, $\mathscr{T}_A$ denotes the set of non-Euclidean triangles with area $A$. There exists a $\delta = \delta_A > 0$ such that every triangle $\Delta \in \mathscr{T}_A$ contains a disc of radius $\delta$.*

*Proof.* We show first that any triangle which contains as *largest* subdisc a disc of radius $\delta$ has area bounded above by some function $M(\delta)$ of $\delta$.

For given $\delta > 0$ the "angle of parallelism" $\theta_\delta$ (between a perpendicular $\mathbf{b}$ to a geodesic $\mathbf{a}$ and the line $\mathbf{c}$ intersecting $\mathbf{b}$ at distance $\delta$ from $\mathbf{a}$ which is tangent to $\mathbf{a}$ at $\infty$) satisfies the formula $\cot \frac{1}{2}\theta_\delta = e^\delta$.

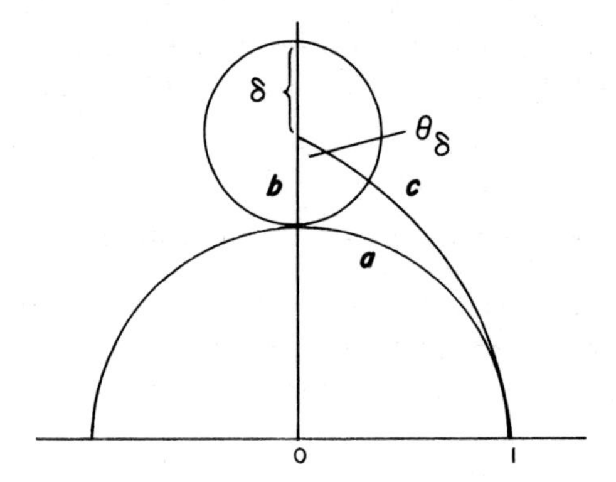


Now a triangle with in-radius $\delta$ may be subdivided into six triangles by adding the three lines joining the in-centre to the vertices and the three perpendiculars to the sides, and each subtriangle has area at most $(\pi/2 - \theta_\delta)$. It follows that the area of the triangle is at most $M(\delta) = 6(\pi/2 - 2 \operatorname{arc\,cot} e^\delta) = 6 \sinh \delta$.

This implies the desired result since $M(\delta)$ is $O(\delta)$ as $\delta \to 0$. ■

*Remark 1.* The same result holds for convex hyperbolic $n$-gons with $n$ fixed.

*Remark 2.* For a hyperbolic triangle with angles $\alpha$, $\beta$, $\gamma$ the in-radius $r$ is given by

$$\tanh^2 r = \frac{\cos^2 \alpha + \cos^2 \beta + \cos^2 \gamma + 2 \cos \alpha \cos \beta \cos \gamma - 1}{2(1 + \cos \alpha)(1 + \cos \beta)(1 + \cos \gamma)}$$

as a consequence of formulae well-known in hyperbolic geometry. I am indebted to my colleague J. A. Tyrrell for this identity and for the observation that the right hand side is $\leqslant \frac{1}{4}$, with equality occurring only when $\alpha = \beta = \gamma = 0$.

*Exercise 1.* Determine a sharper bound for the area in terms of the in-radius.

*Exercise 2.* Use the Lemma to deduce the result of Kazdan and Margulies for Fuchsian groups.

The key result on convergence of Fuchsian groups is now an easy deduction.

3.5.2 THEOREM. *Let* $(\Gamma_n, \mathscr{P}_n)$ *be a sequence of groups of finite area and polygons* $\mathscr{P}_n$ *with at most N sides which are* $\Gamma_n$*-irreducible and have area* $\geqslant A > 0$ *for all n. Then for some sequence of elements* $T_n \in \mathscr{L}$, *a subsequence of the sequence of conjugates* $\{T_n \Gamma_n T_n^{-1}\}$ *converges to a group* $\tilde{\Gamma}$ *with area* $\mu(\tilde{\Gamma}) \leqslant \liminf \mu(\Gamma_n)$.

*Proof.* The Lemma implies that there is a sequence of $T_n \in \mathscr{L}$ such that a fixed ball $\Delta$ of radius $\delta = \delta_A$ lies inside $T_n(\mathscr{P}_n)$ for every $n$, and the sequence of conjugates $\{T_n \Gamma_n T_n^{-1} = \Gamma_n'\}$ is therefore uniformly discrete with respect to the open set $U = \{T \in \mathscr{L} : T\Delta \cap \Delta \neq \varnothing\}$. The convergence now follows by Chabauty's Theorem 2.2.1. ∎

A typical situation in which the theorem is useful occurs as follows. Assume that a sequence of simple loops $l_n$ is given on the surfaces $S_n = \Gamma_n \backslash \mathscr{H}$ with lengths tending to 0, corresponding to a sequence of elements $\{\gamma_n \in \Gamma_n\}$. After passing to a subsequence the loops may be assumed to be either (i) all dividing cycles or (ii) all non-dividing. In case (i) there is for each $n$ by the decomposition procedure of Section 1 a pair of constituent subgroups of $\Gamma_n$ with type (0, 3)—perhaps with torsion—having as intersection the group $\langle \gamma_n \rangle$. Assigning to each $\Gamma_n$ the convex fundamental polygons constructed for these two subgroups as in Section 1.9, one to the left of the axis of $\gamma_n$ and the other to the right, we may apply the theorem twice to obtain a sequence of $\{\Gamma_n\text{'s}\}$ for which *two* conjugate sequences converge. The conjugates of $\gamma_n$ must tend to a parabolic element in each case, and there is an implicit pairing of these two cyclic subgroups in the two classes of limit group, and of the corresponding punctures on the two limit surfaces. In case (ii) there is a subgroup $G_n = \langle \gamma_n, U_n \rangle$ of signature type (1; 1) in $\Gamma_n$ for each $n$, and if the limit of the conjugate subsequence is non-elementary there are *two* parabolic limit subgroups which correspond to limits of $\gamma_n$ and $U_n \gamma_n U_n^{-1}$. Figures 5(i), 5(ii) below illustrate these two situations (c.f. [**15**]).

We can now deduce a result similar to Theorem 2.5.1 for groups with parabolic elements, proved originally by Bers [**6**] using other techniques.

3.5.3 THEOREM. *Let* $\Gamma$ *be of finite area. Then for any* $\varepsilon > 0$ *the set* $\mathscr{K}_\varepsilon$ *of groups* $[\Gamma']$ *in* $\mathscr{X}(\Gamma)$ *satisfying the condition*

$$|\operatorname{tr} \gamma| \notin (2, 2 + \varepsilon) \qquad \textit{for all } \gamma \in \Gamma', \tag{3.4}$$

*is compact.*

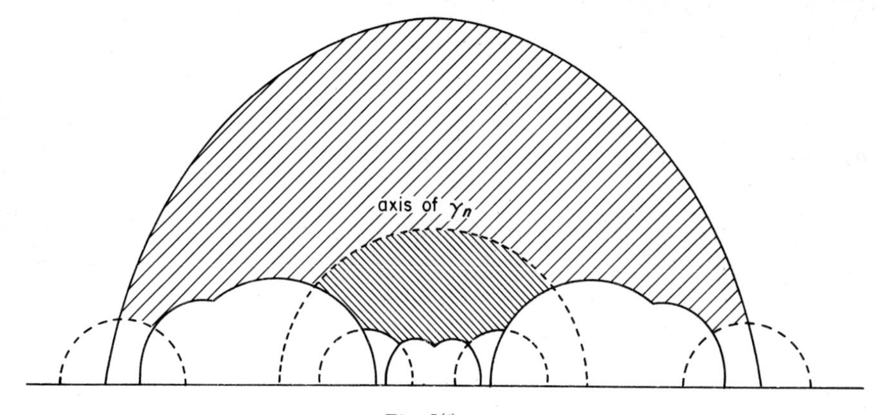


*Fig.* 5(*i*)

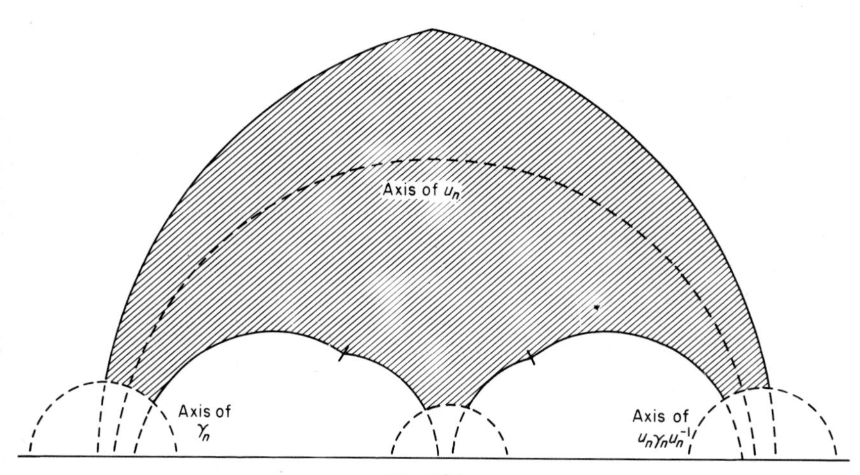


*Fig.* 5(*ii*)

*Proof.* By Theorem 3.5.2 some conjugate subsequence of a given sequence must converge. The condition (3.4) precludes the possibility of parabolic elements being limits of hyperbolic elements so any limit group of finite area has the same signature as $\Gamma$ by Lemma 3 of Section 2.3. ∎

### 3.6 The compactness of $\mathscr{X}(\Gamma)$

After the preliminaries, we are in a position to prove

3.6.1 THEOREM. $\overline{\mathscr{X}}(\Gamma)$ *is sequentially compact if* $\Gamma$ *is a group with finite area.*

*Proof.* By the definition of convergence in $\overline{\mathscr{X}}$, and the finiteness of the type and number of classes of decomposition graphs for $\Gamma$ it suffices to prove that any sequence in a fixed subspace $\mathscr{X}(\mathscr{G}, K)$ has a subsequence converging in $\overline{\mathscr{X}}$ and for this we need only consider sequences of vertex groups inside a fixed $\mathscr{X}(\Gamma_0)$. So let $[\Gamma_n] \in \mathscr{X}(\Gamma_0)$, $n = 1, 2, \ldots,$ be a sequence of conjugacy classes. By application of Theorem 3.5.2 with a chosen sequence of Dirichlet polygons we obtain a subsequence $\Gamma'_n$ of some conjugate sequence which converges to $\tilde{\Gamma}_1$ with $\mu(\tilde{\Gamma}_1) \leqslant \mu(\Gamma_0)$. If equality holds we are finished, so assume that $\mu(\Gamma_0) - \mu(\Gamma_1) = A > 0$. Then by Theorem 2.3.3, there are injections $\phi_n: \tilde{\Gamma}_1 \to \Gamma'_n$ for large $n$ and p.l. homeomorphisms between the Dirichlet regions $\mathscr{P}$ for $\tilde{\Gamma}_1$ and $\mathscr{P}_n$ for $\phi_n(\tilde{\Gamma}_1)$ which induce $\phi_n$. It follows that there is for large $n$ a decomposition of $\mathscr{P}_n$ into a convex subset $Q_n$, of measure $\mu(\tilde{\Gamma}_1)$ (which is p.l. homeomorphic to the Dirichlet region $\mathscr{P}$ for $\tilde{\Gamma}_1$) and a complementary set of measure $A$ with $\leqslant k$ components where $k$ is the edge number of $\tilde{\Gamma}_1$ in $\Gamma_0$. An inductive argument will show that by dealing with these convex subpolygons one by one and taking successive conjugates and convergent subsequences, one can build up piece by piece a graph with vertex groups $\{\tilde{\Gamma}_\nu, \nu = 1, \ldots, m\}$ which represent—up to conjugacy—all limits of some subsequence $\{[\Gamma_{n'}]\}$.

We describe the inductive step. It is assumed that a partial decomposition graph $(\mathscr{G}^{(\nu)}, K^{(\nu)})$ has been obtained, consisting of vertex classes $[G_1], \ldots,$ $[G_{\nu'}]$ together with a set of edges linking them, and that an associated subsequence of $\{\Gamma_n\}$ has the property that each $\Gamma_n$ possesses a constituent subgroup $\Gamma_n^{(\nu)}$ such that the sequence of classes $\{[\Gamma_n^{(\nu)}], n \in \mathbb{N}\}$ approximates the stable group $(\mathscr{G}^{(\nu)}, K^{(\nu)})$ as in Theorem 3.4.1. Now $(\mathscr{G}^{(\nu)}, K^{(\nu)})$ has total area less than $\mu(\Gamma_0)$, and so there are free edges in the decompositions of $\{\Gamma_n\}$ which may be assumed, after passing to a subsequence, to have a pattern (number and location) independent of $n$. Choosing a free edge corresponds to taking a cycle in $S_n(= \Gamma_n\backslash\mathscr{H})$ which can be realized geometrically as a geodesic in $\mathscr{H}$ separating a fundamental polygon $\mathscr{P}_n$ for $\Gamma_n$ into two parts, one of them fundamental for $\Gamma_n^{(v)}$ in its Nielsen region. The remainder has area at least $2\pi$ (unless there is torsion†), and by the discussion following Theorem 3.5.2 we obtain a conjugate subsequence converging to a group $\tilde{\Gamma}_{v'+1}$ of measure $\mu \geqslant \pi/3$ which has a parabolic generator belonging to the limit of the chosen edges (and therefore linked with a parabolic subgroup of the vertex class $[\tilde{\Gamma}_i]$ where the free edge lies). This implies the inductive assumption with a further limit class, $[\tilde{\Gamma}_{v'+1}]$, and the result follows since $\mu(\Gamma) < \infty$. ■

*Exercise 1.* Let $\Gamma_0$ have finite area. Show that there is a constant $k > 0$ such that for any $\Gamma \in \mathbf{S}_0(\Gamma_0)$ there exists a partition of $\Gamma$ (maximal) with all traces (lengths) in the decomposition less than $k$.

*Exercise 2.* Formulate and prove analogues of Theorems 3.5.3 and 3.6.1 for groups of the second kind.‡

As an illustration we return to the example of Section 2.3. The sequence of groups $\Gamma_n$ with signature (0; 4; 0) illustrated in Fig. 6, converges as $n \to \infty$ to a triangle group $\Gamma$. In terms of quotient surfaces one might imagine that two punctures are coalescing, but this is not really correct; in fact there is a dividing cycle, which corresponds to the hyperbolic element $R_n . R_1 = T_n \in \Gamma_n$, shrinking to zero as $n \to \infty$. The axis of $T_n$ divides the fundamental polygon into two parts, a bounded portion which converges to the fundamental polygon for $\Gamma$, and an unbounded one which tends to infinity with $n$.

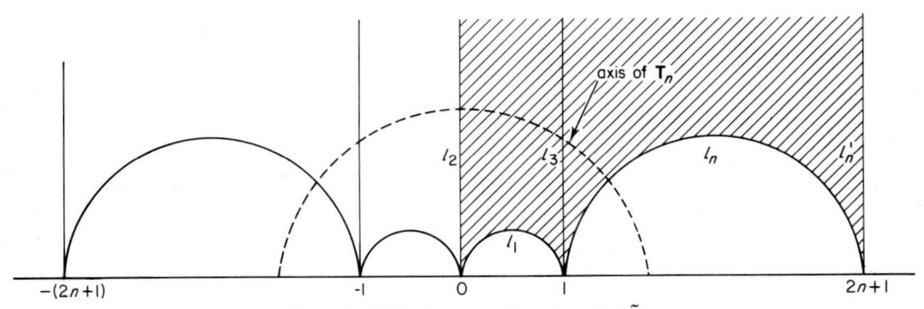


*Fig. 6. A fundamental region for $\tilde{\Gamma}_n$.*

† If there is 2-torsion in $\Gamma$, the decomposition may involve infinite dihedral subgroups which have zero area. This is easily dealt with and for the rest there is a lower bound $2(1 - \frac{1}{2} - \frac{1}{3}) = \pi/3$ for the area.

‡ See [**34**] for an independent approach to this result.

*Exercise 3.* Compute the matrix of $T_n$ and show that $\mathrm{tr}(T_n) \to 0$ as $n \to \infty$.

*Exercise 4.* Find explicitly a sequence of conjugate groups $\Gamma'_n$ which converges so as to reproduce the missing part of the limit of the surfaces $\mathscr{H}/\Gamma_n$ in the above example.

As a final illustration we consider a sequence of groups $\Gamma_n \cong \Gamma_{1,1}$, the group $\langle \gamma, u \rangle$ of a punctured torus, as pictured in Fig. 7 below. Here $\Gamma_n = \langle \gamma_n, u_n \rangle$; and $\{\delta_n = u_n \gamma_n u_n^{-1}\}$, $\{\gamma_n\}$ converge to parabolic elements stabilizing 1, 0 respectively, so that $\Gamma_n$ tends to the triangle group $\Gamma$ of the previous example.

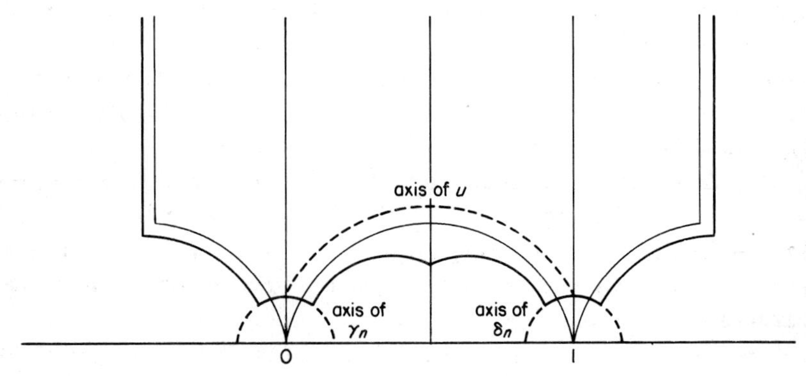


*Fig. 7. A convergent sequence of punctured tori.*

*Exercise 5.* Describe the structure of $\overline{\mathscr{T}}(1;1)$ in detail (i) using the isomorphism with $\overline{\mathscr{T}}(1;0)$ (ii) directly using the parametrization of $\mathbf{T}_0(\Gamma_{1;1})$ given by the traces of $\gamma$ and $u$ above.

### 3.7 Final comments

The space $\overline{\mathscr{T}}(\Gamma)$ has a natural structure of real analytic space, obtainable either by a careful examination of the local moduli at points on the boundary as discussed at the end of Section 3.5 or by going back to the Teichmüller space $\mathbf{T}_0(\Gamma)$, adjoining to it the spaces $\mathbf{T}_0(\mathscr{G}, K)$ of signature equal to $\mathrm{sig}(\Gamma)$ and taking the quotient by the action of $\mathrm{Mod}^+\Gamma$, which acts in a natural way on the union by virtue of the fact that it permutes the set of decompositions while acting on the vertex groups of a decomposition as the corresponding modular group. The process is similar to the adjunction of boundary components to the symmetric space on which an arithmetic group acts (see e.g. [**40**, **37**]).

This approach is topologically identical with that described in Chapter 8, and it is closely related to the work of Bers and Abikoff [**7**, **2**]. A challenging problem in the theory of moduli is the conjecture of Bers ([**4**], [**5**]) that the union of the spaces $\mathbf{T}_0(\mathscr{G}, K)$ is *dense* as a subset of the boundary $\partial\mathbf{T}_0(\Gamma)$, viewed via the embedding of $\mathbf{T}_0(\Gamma)$ inside the space $B_2(\Gamma)$ of quadratic differentials for $\Gamma$ (see Chapter 5, Section 11). We note without proof that the image of the stable boundary components $\mathbf{T}_0(\mathscr{G}, K)$ is precisely the set of (marked) *non-degenerate B-groups in* $\partial\mathbf{T}_0(\Gamma)$ (often termed *regular*). These are the "*cusps*" of $\mathbf{T}_0(\Gamma)$.

As a final analogy with the situation for the genus 1 case we observe that each cusp (and its boundary subspace $\mathbf{T}_0(\mathscr{G}, K)$) has as stabilizer in $\mathrm{Mod}^+ \Gamma$ the group generated by the mapping classes which correspond to *Dehn twists* about the loops in $S_\Gamma$ determining $(\mathscr{G}, K)$. If the decomposition is maximal this group is free (abelian) on $(3g - 3 + n)$ generators for $\Gamma$ of type $(g; n)$ (c.f. Chapter 6 and Section 3.4).

It might be possible to study the structure of $\mathrm{Mod}\, \Gamma$ by determining its action on some 1-complex embedded in $\overline{\mathbf{T}}_0(\Gamma)$, following the approach in [**43**] to the structure of discrete groups. To do this one would need to know more about the subgroups of $\mathrm{Mod}\, \Gamma$ stabilizing the various classes of maximal partition of $\Gamma$.

The role of maximal partitions in surface theory is becoming ever more pervasive, as can be seen by the recent work of Thurston on foliations of surfaces and the related work of Strebel, Bers and others on quadratic differentials and the geometry of Teichmüller space (see for example the research announcement by Hubbard and Masur [**16**]).

We append here some brief remarks on how the methods used in these notes apply to moduli of Kleinian groups. The definitions of $\mathrm{Hom}(\Gamma, \mathscr{L})$ and $\mathscr{R}(\Gamma, \mathscr{L})$ for $\Gamma$ a finitely presented Kleinian group and $\mathscr{L} = \mathrm{PGL}_2(\mathbb{C})$ are of course those of Section 1.2. However a fundamental difficulty appears in consideration of the appropriate definition for $\mathscr{R}_0(\Gamma, \mathscr{L})$: one needs to specify some al ebraic framework which determines the geometric type of $\Gamma$. This will involve at least a list of primary parabolic generators which correspond to punctures on the Riemann surfaces defined by the action of $\Gamma$ on $\mathbb{P}_1(\mathbb{C})$, but more structure is necessary unless $\Gamma$ is of special type. The definition of Bers (see [**4**], [**5**]) requires that a homomorphism $\theta\colon\Gamma \to \Gamma'$ should be induced by a (quasiconformal) homeomorphism of $\mathbb{P}_1(\mathbb{C})$; using it one can give a satisfactory local and global structure to the deformation spaces of $\Gamma$ if $\Gamma$ is *geometrically finite* (see Chapter 8). Maskit has given a more general definition of the deformation space, which unfortunately does not relate intrinsically to the framework of $\mathscr{R}(\Gamma, \mathscr{L})$. A suitable algebraic definition might well involve some class of representation of $\Gamma$ as the fundamental group of some 1-complex of groups (c.f. Chapter 10 for such a structure if $\Gamma$ is a function group). We close this discussion by citing a result, due to Chuckrow, Marden, Jørgensen and Yamamoto, which states (c.f. Chapter 8, theorem 7.1) that the subspace $\mathscr{R}(\Gamma, \mathscr{L})$ is closed in $\text{Hom}(\Gamma, \mathscr{L})$.

## REFERENCES


1. "Modular Functions of One Variable" (Proceedings International Summer School, University of Antwerp). Lecture Notes in Mathematics Nos. 320, 349, 350. Springer Verlag, Heidelberg, 1973.
2. W. Abikoff, Augmented Teichmüller spaces. *Bull. Amer. Math. Soc.* **82** (2) (1976), 333–334.
3. L. Bers, Quasi-conformal mappings and Teichmüller's theorem, *in* "Analytic Functions". Princeton University Press, 1960, pp. 89–119.
4. L. Bers, On boundaries of Teichmüller spaces and on Kleinian groups I, *Ann. of Math.* **91** (2) (1970), 570–600.
5. L. Bers, Uniformisation, moduli and Kleinian groups. *Bull. London Math. Soc.* **4** (1972), 257–300.
6. L. Bers, On Mumford's compactness theorem. *Israel J. Math.* **12** (4) (1972), 400–407.
7. L. Bers, Deformations and moduli of Riemann surfaces with nodes and signature, *Math. Scand.* **36** (1975), 12—16.
8. N. Bourbaki, "Intégration", Chapitres 7 and 8. Hermann, Paris, 1963.
9. C. Chabauty, Limites d'ensembles et geometrie des nombres. *Bull. Soc. Math. France*, **78** (1950), 143–151.
10. P. Deligne and D. Mumford, The irreducibility of the space of curves of given genus. *Publ. Math. I.H.E.S.* **36** (1969), 75–110.
11. C. J. Earle and J. Eells, A fibre bundle description of Teichmüller theory. *J. Differential Geometry* **3** (1), 1969, 19–43.
12. J. Flachsmeyer, Verschiedene Topologisierungen im Raum der abgeschlossenen Mengen, *Math. Nachr.* **26** (1964), 321–337.
13. R. Fricke and F. Klein, "Automorphen Functionen". Teubner, Leipzig, 1897, 1912.
14. W. J. Harvey, On branch loci in Teichmüller space. *Trans. Amer. Math. Soc.* **153** (1971), 387–399.
15. W. J. Harvey, Chabauty spaces of discrete groups, *in* "Discontinuous Groups and Riemann Surfaces." Ann. of Math. Studies, No. 79, pp. 239–247.
16. J. H. Hubbard and H. Masur, On the existence and uniqueness of Strebel differentials. *Bull. Amer. Math. Soc.* **82** (1976), 77–79.
17. T. Jorgensen, On discrete groups of Mobius transformations. *Amer. J. Math.* **98** (1976), 739–749.
18. D. A. Kazdan and G. A. Margulis, A proof of Selberg's conjecture. *Math. U.S.S.R.-Sb.*, **4** (1968), 147–152.
19. L. Keen, Intrinsic moduli on Riemann surfaces. *Ann. of Math.* **84** (1966), 405–420.

20. L. Keen, On Fricke moduli,† *in* Advances in the theory of Riemann surfaces, Ann. Math. Studies, No. 66, 1971, pp. 205–224.
21. A. M. Macbeath, Groups of homeomorphisms of a simply connected space, *Ann. of Math.* **79** (1964), 473–488.
22. A. M. Macbeath, The fundamental groups of certain subgroup spaces, Ann. of Math. Studies, No. 79, 1974, pp. 289–296.
23. A. M. Macbeath and S. Swierczkowski, Limits of lattices in a compactly generated group. *Canad. J. Math.* **12** (1960), 426–436.
24. A. M. Macbeath and D. Singerman, Spaces of subgroups and Teichmüller space. *Proc. Lond. on Math. Soc.* **31** (3) (1975), 211–256.
25. C. Maclachlan, Modulus space is simply connected. *Proc. Amer. Math. Soc.*, **29** (1971), 185–186.
26. C. Maclachlan, Modular groups and fiber spaces over Teichmüller spaces, Ann. of Math. Studies, No. 79, 1974, pp. 297–314.
27. C. Maclachlan and W. J. Harvey, On mapping class groups and Teichmüller spaces. *Proc. London Math. Soc.* **30** (3) (1975), 495–512.
28. W. Magnus, A. Karrass and D. Solitar, "Combinatorial Group Theory". Wiley Interscience, New York, 1966.
29. A. Marden, On homotopic mappings of Riemann surfaces. *Ann. of Math.*, **90** (1969), 1–8.
30. A. Marden, The geometry of finitely generated Kleinian groups. *Ann. of Math.*, **99** (1974), 383–462.
31. A. Marden, "Isomorphisms between Fuchsian Groups", Lecture Notes in Mathematics No. 505, pp. 56–78. Springer, New York, 1975.
32. B. Maskit, On Klein's combination theorem, *Trans. Amer. Math. Soc.* **120** (1965), 499–509.
33. B. Maskit, On boundaries of Teichmüller spaces and on Kleinian groups II, *Ann. of Math.* **91** (1970), 607–639.
34. J. P. Matelski, A compactness theorem for Fuchsian groups of the second kind, *Duke Math. J.* **43** (1976), 829–840.
35. J. Milnor, "Introduction to algebraic *K*-theory", Ann. Math. Studies, No. 72, 1971.
36. D. Mumford, A remark on Mahler's compactness theorem, *Proc. Amer. Math. Soc.* **28** (1971), 289–294.
37. D. Mumford, A new approach to compactifying locally symmetric varieties, *in* "Discrete Subgroups of Lie Groups", pp. 211–224. Tata Inst. Bombay, Oxford Univ. Press, Bombay, 1975.
38. J. Nielsen, Untersuchungen zur Topologie der geschlossenen zweiseitigen Flachen, *Acta Math.* **50** (1927), 189–358.
39. J. H. Sampson, Sous-groupes conjugués d'un groupe linéaire, *Ann. Inst. Fourier (Grenoble)* **26** (1976), 1–6.
40. I. Satake, On compactifications of quotient spaces for discontinuous groups, *Ann. of Math.* **72** (2) (1960), 555–580.
41. A. Selberg, On discontinuous groups in higher dimensional symmetric spaces, Contributions to function theory Bombay, 1960, pp. 147–164.
42. A. Selberg, Recent developments in the theory of discontinuous groups, XVth Scand. Congress, Lecture Notes in Mathematics, No. 118, Springer, 1970, pp. 99–120.

† See also correction in *Proc. Am. Math. Soc.* **40** (1973), 60–62.

43. J-P. Serre, Arbres, amalgames et $SL_2$, cours au Collège de France 1968–69, *Asterix, Soc. Math. France* (to appear).
44. J-P. Serre, Cohomologie des groupes discretes, *in* "Prospects in Mathematics", Ann. of Math. Studies, No. 70, 1971, pp. 77–169.
45. C. L. Siegel, "Topics in complex function theory I, II". Wiley Interscience, 1971.
46. J. Sturm and M. Shinnar, "The maximal inscribed ball of a Fuchsian group", Ann. of Math. Studies, No. 79, 1974, pp. 439–443.
47. P. Tukia, Discrete subgroups of the unit disc and their isomorphisms, *Ann. Acad. Sci. Fenn. Ser A1* **504** (1972), 1–45.
48. A. Weil, On discrete subgroups of Lie groups I, *Ann. of Math.* **72** (2) (1960), 369–384, II *ibid.* **75** (1962), 578–602.
49. A. Weil, Remarks on the cohomology of groups, *Ann. of Math.* **80** (1964), 149–157.
50. H. Zieschang, E. Vogt. and H. D. Coldewey, "Flächen und ebene discontinuerliche Gruppen". Lecture Notes in Mathematics No. 122, Springer, Berlin, 1970.

# 10. On the Classification of Function Groups

BERNARD MASKIT

*State University of New York, Stony Brook, New York 11794*



## 1. INTRODUCTION

**1.1** One would like to be able to classify all Kleinian groups, where we regard two Kleinian groups as being the same if they are conjugate in PSL(2; ℂ). There is not much hope of finding such a classification; one needs fairly strong finiteness conditions, and one has to weaken the equivalence relation.

We first weaken the equivalence relation. If $G$ and $G^*$ are conjugate in PSL(2; ℂ), then there is a conformal homeomorphism of the 2-sphere which conjugates $G$ into $G^*$ .We will say that $G$ and $G^*$ are *qc-conjugate* if there is a quasiconformal homeomorphism of the 2-sphere which conjugates $G$ into $G^*$. If we can classify Kleinian groups up to *qc*-conjugacy, then the groups which appear in each class are just the quasiconformal deformations of any one group in the class. These spaces of deformations have been studied by Bers [**4**], Kra [**14**], and myself [**17**]; see also the expository article by Bers [**5**].

In addition to weakening the equivalence relation, we also need to impose some finiteness condition on the Kleinian groups to be classified. An obvious condition along these lines is to require our Kleinian groups to be finitely generated. This seems especially fitting in view of Ahlfors' finiteness theorem [**2**] (see Chapter 7, Section 2), which asserts that a finitely generated Kleinian group represents a finite number of finite Riemann surfaces. On the other hand, there is Bers' proof of the existence of finitely generated degenerate groups [**6**], and Greenberg's proof that such groups do not have finite sided fundamental polyhedra [**10**]. This leads one to expect that the proper condition to impose on our Kleinian groups is that they have a finite sided *fundamental polyhedron* i.e. they are *geometrically finite* (c.f. chapters 7 and 8).

One would expect the classification of groups with a finite sided fundamental polyhedron to proceed roughly as follows. The stabilizer of each component is again such a group, and it has an invariant component. First one classifies those Kleinian groups which have both a finite sided fundamental polyhedron and an invariant component, and then one sees how such groups fit together to get the general case.

The first step of this programme has actually been carried out. The details appear elsewhere [**17**]; we outline the procedure in these notes. The second step in the programme remains murky, although there is a uniqueness theorem due to Marden [**16**], a decomposiiton theorem due to Abikoff and myself [**1**], and a structure theorem concerning intersections of component subgroups [**19**].

**1.2** Finitely generated Kleinian groups with an invariant component arise naturally as uniformizations of closed Riemann surfaces, or equivalently of algebraic curves.

An *algebraic curve* is the set of pairs of complex numbers $(z, w)$ satisfying $P(z, w) = 0$, where $P$ is an irreducible polynomial. A *uniformization* of the curve is a pair of functions $(z(t), w(t))$, where $P(z(t), w(t)) = 0$, and every root of $P$ corresponds to at least one value of $t$. The classical uniformizing variable is the multivalued inverse function from the curve into $\mathbb{C}$, where the different branches of the multi-valued function are related by fractional linear transformations.

In modern terminology, a uniformization of a closed Riemann surface $\bar{S}$ is a plane domain $\Delta$, together with a group $G$ of fractional linear transformations acting on $\Delta$, where $\Delta/G$ is conformally equivalent to $\bar{S}$ with a finite number of points deleted, and where the covering map $p:\Delta \to \Delta/G$ is branched over finitely many points.

Using Ahlfors' finiteness theorem, one sees that the finitely generated Kleinian groups with an invariant component are precisely the groups that arise in the uniformizations of closed Riemann surfaces; we call this class $\mathscr{C}_1$.

As part of the classification mentioned above, we also obtain a classification of $\mathscr{C}_1$, but under a weaker equivalence relation.

**1.3** An isomorphism $\psi: G \to G^*$ between Kleinian groups is called *type preserving* if

(i) $\psi$ preserves the square of the trace of every elliptic element, and
(ii) both $\psi$ and $\psi^{-1}$ preserve parabolic elements.

The first condition is related to the fact that an elliptic cyclic group can have many algebraic generators, but it has only two geometric generators; these are the rotations through smallest angles.

Let $G$ and $G^*$ be groups in $\mathscr{C}_1$, with invariant components $\Delta$ and $\Delta^*$, respectively. A *weak similarity* is an orientation-preserving homeomorphism $\phi: \Delta \to \Delta^*$, where $g \mapsto \phi \circ g \circ \phi^{-1}$ defines an isomorphism, called the *induced isomorphism*, between $G$ and $G^*$.

The existence of the induced isomorphism is equivalent to the statement that $\phi$ is the lifting of a homeomorphism $\phi_0: \Delta/G \to \Delta^*/G^*$. This means that the classification of groups in $\mathscr{C}_1$ up to weak similarity is the classification of uniformizations of closed Riemann surfaces, where we do not distinguish between topologically equivalent coverings of topologically equivalent surfaces.

Our classification of groups in $\mathscr{C}_1$ is actually somewhat stronger. A *similarity* is a weak similarity in which the induced isomorphism is type-preserving. We will classify groups in $\mathscr{C}_1$ up to similarity.

**1.4** We denote the class of groups which have both a finite sided fundamental polyhedron and an invariant component by $\mathscr{C}_0$. The following theorem reduces the classification of groups in $\mathscr{C}_0$ up to *qc*-conjugacy to classification of groups in $\mathscr{C}_1$ up to similarity.

THEOREM [**18**]. *If $G$ and $G^*$ are similar groups in $\mathscr{C}_0$, then $G$ and $G^*$ are qc-conjugate.*

## 2. FUCHSIAN GROUPS

**2.1** Before proceeding, we recall the classification of finitely generated Fuchsian groups of the first kind.

**2.2** To every finitely generated Fuchsian group $G$ of the first kind, we assign a *signature* $(g, n; v_1, \ldots, v_n)$ as follows (c.f. Chapter 7, Section 1.5).

We assume that $G$ operates on the unit disc $\mathscr{U}$. We delete from $\mathscr{U}$ the fixed points of the elliptic elements of $G$; call the resulting domain $\mathscr{U}'$. Then $G$ operates freely on $\mathscr{U}'$, and $\mathscr{U}'/G$ is a closed Riemann surface $\bar{S}$, of genus $g$, from which a finite number of points $x_1, \ldots, x_n$ have been removed. For each point $x_i$, let $N_i$ be a simply connected neighbourhood of $x_i$, where $N_i$ contains no other $x_j$. Let $\tilde{N}_i$ be a connected component in $\mathscr{U}'$ of the preimage of $N_i' = N_i - \{x_i\}$. Then $p: \tilde{N}_i \to N_i'$ is a covering of the topological punctured disc $N_i'$. We define $v_i$ to be the index of this covering.

Note that the subgroup of $G$ which keeps $\tilde{N}_i$ invariant is cyclic; it is elliptic if $v_i < \infty$, and it is parabolic if $v_i = \infty$.

**2.3** The signature $(g, n; v_1, \ldots, v_n)$ satisfies the following:

(i) $g \geqslant 0$

(ii) $n \geqslant 0$

(iii) $2 \leqslant v_i \leqslant \infty$.

Standard considerations of curvature and area lead to the additional inequality

(iv)
$$2g - 2 + \sum_{i=1}^{n} (1 - 1/v_i) > 0,$$

where $1/\infty = 0$.

If $\sigma = (g, n; v_1, \ldots, v_n)$ is any set of "integers" satisfying (i)–(iv), then there is a Fuchsian group having signature $\sigma$.

**2.4** We remark that the signature $(g, n; v_1, \ldots, v_n)$ of a Fuchsian group is of course well defined only up to permutation of the $n$ symbols $v_1, \ldots, v_n$. We regard two signatures which differ by such a permutation to be the same.

**2.5** If $G$ and $G^*$ are Fuchsian groups having the same signature, then $G$ and $G^*$ are *qc*-conjugate; more precisely, there is a quasiconformal homeomorphism which conjugates $G$ into $G^*$ and which preserves $\mathscr{U}$.

It is essentially immediate from the definition that if $G$ and $G^*$ are *qc*-conjugate, then $G$ and $G^*$ have the same signature.

**2.6** If we start with a finitely generated Fuchsian group of the first kind $G$, then $T(G)$ the Teichmüller space of $G$ is the set of all type-preserving isomorphisms of $G$ onto other Fuchsian groups. Each such isomorphism can be realized by a *qc*-conjugation, and one uses quasiconformal mappings to define the topology and analytic structure on $T$. (See Chapter 5.)

## 3. BASIC GROUPS

**3.1** The Fuchsian groups act discontinuously on the entire unit disc. There are also groups of fractional linear transformations which operate discontinuously everywhere on the other classical simply-connected domains: the plane and the sphere. These, along with the groups operating on the punctured plane comprise the *elementary* groups.

**3.2** If $G$ operates on either the sphere or the plane, then, exactly as in the Fuchsian case, we can assign a signature $(g, n; v_1, \ldots, v_n)$ to $G$. Using the classification of elementary groups (see Ford [**9**]), one sees that the signatures of the finite groups are precisely: $(0, 0)$; $(0, 2; v, v)$, $v < \infty$; and $(0, 3; v_1, v_2, v_3)$, $1/v_1 + 1/v_2 + 1/v_3 > 1$. The groups which operate discontinuously on the entire plane have signatures: $(0, 2; \infty, \infty)$, $(0, 3; v_1, v_2, v_3)$, $1/v_1 + 1/v_2 + 1/v_3 = 1$; $(0, 4; 2, 2, 2, 2)$; and $(1, 0)$.

As in the Fuchsian case, the signatures are defined only up to permutation of $v_1, \ldots, v_n$, and we regard two signatures which differ by such a permutation as being the same.

Also as in the Fuchsian case, two of these elementary groups (i.e. groups with 0 or 1 limit point) have the same signature if and only if they are *qc*-conjugate.

**3.3** Combining the lists of signatures in 3.2 and 2.3 we see that except for $(0, 1; v)$ and $(0, 2; v_1, v_2)$, $v_1 \neq v_2$, all possible signatures with $g \geqslant 0$, $n \geqslant 0$, $2 \leqslant v_i \leqslant \infty$ do occur. For topological reasons, the two exceptional cases cannot occur.

**3.4** We need to look not only at Fuchsian and elementary groups, but at all groups which are similar to these groups. This prompts the following definitions.

A parabolic element $g$ of a group $G$ in $\mathscr{C}_1$ is called *accidental* if there is a weak similarity $\phi$ between $G$ and some other group $G^*$ in $\mathscr{C}_1$, so that $\phi \circ g \circ \phi^{-1}$ is not parabolic.

One easily sees that elementary groups have no accidental parabolic elements. It was shown by Ahlfors [2] that Fuchsian groups have no accidental parabolic elements. Of course *qc*-conjugations and similarities both preserve accidental parabolic elements.

**3.5** A Kleinian group $G$ in $\mathscr{C}_1$ is called a *basic group* if

(i) the invariant component $\Delta$ is simply connected, and
(ii) $G$ contains no accidental parabolic elements.

The basic groups are in some sense all known [**20**] (see also Bers [**5**] and Kra–Maskit [**15**]).

If $G$ is a basic group, then $G$ is either

(i) elementary, or
(ii) quasi-Fuchsian (i.e. $G$ is *qc*-conjugate to a finitely generated Fuchsian group of the first kind), or
(iii) degenerate (i.e. $G$ has only the one component $\Delta$, and $\Delta$ is hyperbolic).

One easily sees that elementary and Fuchsian groups have finite sided fundamental polyhedra. It was shown by Marden [**16**] that quasi-Fuchsian groups have finite sided fundamental polyhedra, and it was shown by Greenberg [**10**] that degenerate groups do not.

## 4. PRECISELY INVARIANT SETS AND FACTOR SUBGROUPS

**4.1** Let $G$ be a group in $\mathscr{C}_1$ with invariant component $\Delta$. If $H$ is any subgroup of $G$, then $H$ has a *distinguished invariant component* $\Delta(H) \supset \Delta$. In what follows, the invariant component of $H$ will be understood to be $\Delta(H)$, unless specifically stated otherwise.

**4.2** A *factor subgroup* $H$ of a group $G$ in $\mathscr{C}_1$ is a subgroup which satisfies the following:

(i) $H$ is a basic group.
(ii) If the fixed point of a parabolic element $g \in G$ lies in $\Lambda(H)$, the limit set of $H$, then $g \in H$.
(iii) $H$ is a maximal subgroup of $G$ satisfying (i) and (ii) above.

These are also known as structure subgroups.

We remark that it is not obvious that $G$ contains any factor subgroups.

THEOREM [**21**]. *Every group $G$ in $\mathscr{C}_1$ contains at least one factor subgroup;*

*G contains only finitely many conjugacy classes of factor subgroups, and every factor subgroup is finitely generated.*

**4.3** In general, if $G$ is a Kleinian group, $H$ a subgroup of $G$, and $A$ a subset of $\hat{\mathbb{C}} = \mathbb{C} \cup \{\infty\}$, we say that $A$ is *precisely invariant under* $H$ *in* $G$ if

(i) $g(A) = A$, for all $g \in H$;
(ii) $g(A) \cap A = \phi$, for all $g \in G - H$.

**4.4** For groups in $\mathscr{C}_1$, precisely invariant sets are most easily obtained as follows.

Let $S = \Delta/G$, let $Y$ be a connected subset of $S$, and let $A$ be a connected component of the preimage of $Y$. Let $H$ be the stabilizer of $A$; i.e.

$$H = \{g \in G : g(A) = A\}.$$

Then every element of $G$ either belongs to $H$, or else it maps $A$ onto some other connected component of $p^{-1}(Y)$. Hence $A$ is precisely invariant under $H$ in $G$. In this case, we say that $A$ *covers* $Y$, and we also say that $H$ is a *covering subgroup* of $Y$, or that $H$ *lies over* $Y$.

**4.5** The relation between factor subgroups and precisely invariant sets is given by the following:

THEOREM [**21**]. *Let $G$ be a group in $\mathscr{C}_1$ with invariant component $\Delta$. Then there are simple disjoint loops $w_1, \ldots, w_k$, which divide $S$ into subsurfaces $Y_1, \ldots, Y_s$, so that the following hold.*

(i) *Every covering subgroup of each $Y_i$ is a factor subgroup of $G$.*
(ii) *Every factor subgroup of $G$ is a covering subgroup of some $Y_i$.*
(iii) *Two factor subgroups are conjugate in $G$ if and only if they cover the same $Y_i$.*
(iv) *Each covering subgroup of each $w_j$ is either trivial, or elliptic cyclic, or parabolic cyclic.*
(v) *Every elliptic or parabolic element of $G$ lies in at least one factor subgroup.*

Let $W$ cover some $w_j$. If $W$ is stabilized by a finite subgroup of $G$, then $W$ is a loop. If $W$ is stabilized by a parabolic subgroup of $G$, then $W$ becomes a loop, which we again call $W$, after we adjoin the parabolic fixed point. The loop $W$ is called a *structure loop*.

Similarly, if $A$ covers some $Y_i$, then $A$ is called a *structure region*.

The following was also shown in [**21**].

(vi) *Let $W$ be a structure loop on the boundary of the structure region $A$. Let $H$ be the stabilizer of $A$, and let $J$ be the stabilizer of $W$. Let $B$ be the topological disc bounded by $W$ which does not intersect $A$. Then $B$ is precisely invariant under $J$ in $H$.*

**4.6** The above theorem gives us the following picture of the action of $G$ on $\Delta$. An elaboration of this in certain special cases can be found in [**23**].

There is a collection of simple disjoint loops, the structure loops, lying inside $\Delta$, except that some of these loops may touch the boundary of $\Delta$ at one point, a parabolic fixed point. In general, this is an infinite collection of loops, but the spherical diameter of any sequence of these loops tends to zero.

The structure loops cut up $\Delta$ into structure regions. If we focus on any one structure region $A$, then $A$ is stabilized by a factor subgroup $H$, and if $g$ is any element of $G$ which is not in $H$, then $g(A) \cap A = \emptyset$.

The boundary of $A$ consists of a collection of structure loops, together with $\Lambda(H)$, the limit set of $H$. Since $H$ is a basic group, $\Lambda(H)$ is connected. The structure loops on the boundary of $A$ fall into finitely many classes under the action of $H$. If we project the structure loops onto $S_0 = \Delta(H)/H$, then we get a finite collection of simple disjoint loops on $S_0$. If $W$ is stabilized by a parabolic cyclic group, then $W$ projects on $S_0$ to a simple loop which bounds a punctured disc; if $W$ is stabilized by an elliptic cyclic group, then $W$ projects on $S_0$ to a simple loop bounding a disc which contains exactly one branch point (i.e. the projection of an elliptic fixed point in $\Delta(H)$); if $W$ is stabilized by the identity, then $W$ projects to a simple loop on $S_0$ which bounds a disc containing no branch points.

**4.7** The factor subgroups are intrinsically defined. It was also shown in [**21**] that the stabilizers of the structure loops can be obtained as pairwise intersections of factor subgroups. However, the structure loops and structure regions are defined by the loops $w_1, \ldots, w_k$ on $S$. One would expect these loops to be unique in some sense, up to permutation, but there are no known conditions. For example, it is known that in general the loops are not unique up to permutation and up to homology with $Z_2$ coefficients.

**4.8** There are two special cases that deserve mention.

(i) If $G$ is a basic group, then $G$ is a factor subgroup of itself; in fact, $G$ is the only factor subgroup of itself. In this case, the set of loops $\{w_1, \ldots, w_s\}$ is empty, there are no structure loops, and there is only one structure region, namely $\Delta$.

(ii) At the other extreme, if $G$ is a Schottky group, then the identity is a factor subgroup of $G$; again, it is the only factor subgroup of $G$. The Schottky group is determined by $g$ homologously distinct simple non-dividing loops $w_1, \ldots, w_g$ on $S$, where $g$ is the genus of $S$. When we cut $S$ along these loops, we get one subsurface $Y$. If $A$ covers $Y$, then $A$ is a region bounded by $2g$ simple disjoint loops; i.e. the structure regions are the translates of the standard fundamental domain that defines the Schottky group. There are, of course, many essentially different standard fundamental domains that define the same Schottky group (see Chuckrow [7] and Hejhal [11]) and correspondingly there are many different sets of loops on $S$ which define the Schottky group.

In the two cases mentioned above the information about factor subgroups is characteristic of the groups in question. The basic groups are precisely those groups in $\mathscr{C}_1$ which are factor subgroups of themselves; the Schottky groups are precisely those groups in $\mathscr{C}_1$ for which the identity is a factor subgroup.

## 5. SIGNATURES

**5.1** Let $G$ be a group in $\mathscr{C}_1$ with invariant component $\Delta$. The signature of $G$ contains the following information. First of all $g$, the genus of $S = \Delta/G$. Then the signatures of the factor subgroups (i.e. there is one signature for each conjugacy class of factor subgroups). Then there is a pairing of certain of the $v_{ij}$ in these signatures. Each structure loop $W$ which is stabilized by a non-trivial subgroup of $G$ lies on the boundary of two structure regions with stabilizers $H$ and $H'$; $W$ picks out—hence pairs—a distinguished point on $\Delta(H)/H$ and a distinguished point on $\Delta(H')/H'$.

The precise definition of the signature appears in [18] (the definition appearing in [22] is unfortunately not quite correct).

**5.2** Theorem [18] *Two groups $G$ and $G^*$ in $\mathscr{C}_1$ are similar if and only if they have the same signature.*

**5.3** Theorem [18] *Two groups $G$ and $G^*$ in $\mathscr{C}_0$ are qc-conjugate if and only if they have the same signature.*

**5.4** As with Fuchsian groups, one can write down a set of necessary and sufficient conditions that the signature of a group in $\mathscr{C}_1$ must satisfy. These conditions are quite technical; the interested reader is referred to [**18**] for details.

## 6. OUTLINE OF PROOFS

**6.1** The principal ingredients in the proofs of these theorems are

(i) a theorem about planar covering surfaces,
(ii) the combination theorems,
(iii) existence and uniqueness of solutions to the Beltrami equation, and
(iv) generalizations of an estimate due to Koebe.

**6.2** The first problem is to characterize the covering $p:\Delta \to S$. It is easier to first delete the elliptic fixed points, and their projections. We then get a regular covering $p:\Delta' \to S'$; that is, the induced homomorphism from $\pi_1(\Delta')$ into $\pi_1(S')$ is a monomorphism and the image is a normal subgroup of $\pi_1(S')$. We also know that $S'$ is a finite Riemann surface (i.e. a closed orientable surface from which a finite number of points have been removed), and that $\Delta'$ is planar (i.e. $\Delta'$ is topologically equivalent to a plane domain).

Under the above circumstances, it was shown in [**24**] that there is a simple loop $w$ on $S'$ which when raised to some power $\alpha > 0$, lifts to a loop; that is, the homotopy class of $w^\alpha$ lies in the defining subgroup of the covering. In fact, more is true. There is a finite collection $w_1, \ldots, w_k$ of simple disjoint loops on $S'$, and there are positive integers $\alpha_1, \ldots, \alpha_k$ so that $w_1^{\alpha_1}, \ldots, w_k^{\alpha_k}$ all lift to loops, and $p:\Delta' \to S'$ is the highest regular covering of $S'$ for which these loops all lift to loops; i.e. the homotopy classes of the loops $w_1^{\alpha_1}, \ldots, w_k^{\alpha_k}$ normally generate the defining subgroup of the covering.

Some of the loops $w_1, \ldots, w_s$ bound punctured discs on $S'$. The puncture is artificial; the missing point $x$ is the image of an elliptic fixed point in $\Delta$, and the loop simply gives us information that we already knew; namely, that $x$ is a branch point of a certain order.

After removing the loops which only give us information about the branch numbers in the covering $p:\Delta \to S$, we are left with a set of loops $w_1, \ldots, w_j$. These loops are a subset of the loops mentioned in 4.5; they are precisely the loops whose covering subgroups are finite.

**6.3** The liftings of the loops divide $\Delta$ into regions. One easily sees that if $H$ is the stabilizer of one of these regions, then $\Delta(H)$ is simply connected.

Using the Riemann map, we see that the accidental parabolic elements of $G$ are defined by simple disjoint loops on $\Delta(H)/H$.

**6.4** The simplest combination theorem is Klein's original combination theorem [**12**], which roughly goes as follows. Let $G_1$ and $G_2$ be Kleinian groups. Let $W$ be a simple loop bounding two topological discs $B_1$ and $B_2$. Suppose that $B_i$ is precisely invariant under the identity in $G_i$, $i = 1, 2$. Then $G$, the group generated by $G_1$ and $G_2$ is Kleinian; the algebraic structure of $G$ is determined by the algebraic structure of $G_1$ and $G_2$ (in this case, $G$ is the free product of $G_1$ and $G_2$); a fundamental domain for $G$ can be constructed from appropriate fundamental domains for $G_1$ and $G_2$; every elliptic or parabolic element of $G$ is conjugate in $G$ to some element in $G_1$ or in $G_2$.

In the more general case [**25, 26, 27**] the hypotheses are somewhat more complicated, but the conclusions remain the same (there are also versions of these combination theorems which were developed by Fenchel and Nielsen [**8**] in their study of Fuchsian groups and are discussed in Chapter 9 Section 1.6).

Another conclusion of the combination theorems is approximately as follows (see [**18, 28**]). If $G$ has been constructed from $G_1$ and $G_2$ using a combination theorem, and $G_i$ is similar to $G_i^*$, $i = 1, 2$, and $G^*$ has been constructed from $G_1^*$ and $G_2^*$ using the same combination theorem, then $G$ and $G^*$ are similar.

**6.5** Using the construction mentioned above, we can show that if two groups in $\mathscr{C}_1$ have the same signature, then they are similar.

Using the combination theorems again one sees that if $G$ is a group in $\mathscr{C}_1$, then every component of $G$ other than the invariant component is stabilized by a quasi-Fuchsian factor subgroup of $G$. Using this, together with the Fenchel and Nielson realization theorem [**8**] (see Chapter 9 theorem 1.4.2 and [**29**]) we can extend a similarity between groups $G$ and $G^*$ in $\mathscr{C}_0$ to a homeomorphism, $\phi:\Omega(G) \to \Omega(G^*)$ of the entire set of discontinuity.

**5.5** Using some generalizations of Koebe's Theorem [**13**] (see also [**27**]) to obtain uniform estimates on structure loops [**28**], we obtain the following. If $G$ and $G^*$ are in $\mathbb{C}_0$ and if $\phi:\Omega(G) \to \Omega(G^*)$ is a conformal homeomorphism where $g \to \phi \circ g \circ \phi^{-1}$ is a type-preserving isomorphism, then $\phi$ is fractional linear (using deep results from the theory of 3-manifolds, Marden [**16**] has proven a more general uniqueness theorem described in Chapter 8 Section 6.6).

**6.7** Finally, we show that if $G$ and $G^*$ in $\mathscr{C}_0$ have the same signature, then $G$ and $G^*$ are $qc$-conjugate. The proof uses the existence and uniqueness of solutions to the Beltrami equation due to Ahlfors and Bers [**3**].

**6.8** In the other direction, it is almost immediate that similarities preserve signature. There is one difficulty involving elementary groups. For a Fuchsian group $G$ acting on $\mathscr{U}$, there is a one-to-one correspondence between conjugacy classes of maximal elliptic cyclic subgroups and branch points on $\mathscr{U}/G$. For finite groups there need not be such a one-to-one correspondence. For example, if $n$ is odd, the $(2, 2, n)$ triangle group (i.e. the elementary group with signature $(0, 3; 2, 2, n)$) has only one conjugacy class of involutions.

## REFERENCES

1. W. Abikoff and B. Maskit, Geometric decompositions of Kleinian groups. *Amer. J. Math.* to appear.
2. L. V. Ahlfors, Finitely generated Kleinian groups. *Am. J. Math.* **86** (1964), 413–429.
3. L. V. Ahlfors and L. Bers, Riemann's mapping theorem for variable metrics. *Ann. of Math.* **72** (1960), 385–404.
4. L. Bers, Spaces of Kleinian groups, Several complex variables, I (Proc. Conf. University of Maryland, College Park, Maryland 1970), pp. 9–34. Springer, Berlin, 1970.
5. L. Bers, Uniformization, moduli, and Kleinian groups. *Bull. London Math. Soc.* **4** (1972), 257–300.
6. L. Bers, On boundaries of Teichmüller spaces and on Kleinian groups, I. *Ann. Math.* (2) **91** (1970), 570–600.
7. V. Chuckrow, On Schottky groups with applications to Kleinian groups. *Ann. Math.* **88** (1968), 47–61.
8. W. Fenchel and J. Nielsen, Discrete groups, to appear.
9. L. R. Ford, "Automorphic Functions", 2nd ed. Chelsea Publishing Co. New York, 1951.
10. L. Greenberg, Fundamental polyhedra for Kleinian groups, *Ann. Math.* (2) **84** (1966), 433–441.
11. D. Hejhal, On Schottky and Teichmüller spaces. *Advances in Math.* **15** (1975), 133–156.
12. F. Klein, Neue Beitrage zur Riemannschen Functionentheorie. *Math. Ann.* **21** (1883), 141–218.
13. P. Koebe, Über die uniformisierung der algebraischen Kurven III. *Math. Ann.* **72** (1912), 437–516.
14. I. Kra, On spaces of Kleinian groups. *Comment. Math. Helv.* **47** (1972), 53–69.
15. I. Kra and B. Maskit, Involutions on Kleinian groups. *Bull. Amer. Math. Soc.* **78** (1972), 801–805.
16. A. Marden, The geometry of finitely generated Kleinian groups. *Ann. of Math.* **99** (1974), 383–462.

17. B. Maskit, Self-maps on Kleinian groups. *Amer. J. Math.* **93** (1971), 840–856.
18. B. Maskit, On the classification of Kleinian groups, II—signatures, *Acta Math.* **138** (1977) 17–42.
19. B. Maskit, Intersections of component subgroups of Kleinian groups, *in* "Discontinuous Groups and Riemann Surfaces". Annals of Math. Studies, No. 79, pp. 349–367, 1974.
20. B. Maskit, On boundaries of Teichmüller spaces and on Kleinian groups II. *Ann. of Math.* (2) **91** (1970), 607–639.
21. B. Maskit, Decomposition of certain Kleinian groups. *Acta Math.* **130** (1973), 243–263.
22. 22. B. Maskit, Classification of Kleinian groups, "Proc. International Congress, Vancouver", 1974, Vol. 2, pp. 213–216.
23. B. Maskit, Uniformization of Riemann surfaces. *in* "Contributions to Analysis", pp. 293–312. Academic Press, New York, 1974.
24. B. Maskit, A theorem on planar covering surfaces with applications to 3-manifolds, *Ann. of Math.* **81** (1965), 341–355.
25. B. Maskit, On Klein's combination theorem, *Trans. Amer. Math. Soc.* **120** (1965), 499–509.
26. B. Maskit, On Klein's combination theorem II, *Trans. Amer. Math. Soc.* **131** (1968), 32–39.
27. B. Maskit, On Klein's combination theorem III, *in* "Advances in the Theory of Riemann Surfaces (Proc. Conf. Stony Brook, N.Y., 1969), Ann. of Math. Studies No. 66. 1971, pp. 297–316.
28. B. Maskit, On the classification of Kleinian groups: I—Koebe groups, *Acta Math.* **135**, (1975), 249–270.
29. H. Zieschang. Uber Automorphismen ebener diskontinuerlicher Gruppen, *Math. Ann.* **169** (1966), 148–167.

# 11. Hecke Operators, Oldforms and Newforms

R. A. RANKIN

*University of Glasgow, Glasgow, Scotland*



## 1. NOTATION; MATRICES

We write throughout:

$$I = \begin{pmatrix} 1 & 0 \\ 0 & 1 \end{pmatrix}, \qquad J_n = \begin{pmatrix} 1 & 0 \\ 0 & n \end{pmatrix}, \qquad T = \begin{pmatrix} a & b \\ c & d \end{pmatrix}.$$

The entries are rational numbers. As usual,

$\Gamma(1)\colon = \mathrm{SL}(2, \mathbb{Z})$, the (homogeneous) modular group.

$\Gamma(N)\colon = \{T \in \Gamma(1)\colon T \equiv \pm I \pmod N\}$, the principal congruence group of level $N \in \mathbb{Z}^+$.

$$\Gamma_0(N) = \{T \in \Gamma(1) : c \equiv 0 \pmod{N}\},$$

$$\Gamma^0(N) = \{T \in \Gamma(1) : b \equiv 0 \pmod{N}\},$$

$$\Gamma_0(m, n) = \Gamma_0(m) \cap \Gamma^0(n) \qquad (m, n \in \mathbb{Z}^+)$$

Accordingly, $\Gamma_0(N) = \Gamma_0(N, 1)$.

All these groups are of finite index in $\Gamma(1)$ and

$$|\Gamma(1) : \Gamma_0(n)| = \psi(N), \qquad |\Gamma(1) : \Gamma_0(m, n)| = \psi(mn),$$

where

$$\psi(n) = n \prod_{p \mid n} \left(1 + \frac{1}{p}\right).$$

For $n \in \mathbb{Z}^+$, $\Omega_n$ is the set of all matrices of order $n$, i.e. the set of all $T$ with integral entries and determinant $n$. Thus $\Omega_n \subseteq GL(2, \mathbb{Q}) = \Omega$.

## 2. NOTATION; MAPPINGS

Each $T \in \Omega$ determines a bilinear map $T$ given by $z \mapsto T(z) = (az + b)/(cz + d)$. These map the upper half-plane $\mathscr{H} = \{z \in \mathbb{C} : \operatorname{Im} z > 0\}$ onto itself. For $T \in \Gamma(1)$ the associated maps form the (inhomogeneous) *modular group*, which is isomorphic to $\Gamma(1)/\Lambda$, where $\Lambda = \{I, -I\}$. For $T \in \Omega$,

$$T'(z) = (cz + d)^{-2} \det T.$$

For any function $f$ defined on $\mathscr{H}$, and any $k \in \mathbb{Z}$, define

$$f(z)|_k T = (\det T)^{\frac{1}{2}k} (cz + d)^{-k} f(T(z)).$$

When we use this notation $k$ will be a fixed integer (and omitted on left) and $\det T$ will be positive. We have, for $S, T \in \Omega$,

$$f|(T_1 T_2) = (f|T_1)|T_2.$$

## 3. DEFINITION OF MODULAR AND CUSP FORMS

Let $\Gamma$ be a group between $\Gamma(N)$ and $\Gamma(1)$, so that $-I \in \Gamma$. Such a group is called a *congruence group of level N* (if $N$ is the smallest such integer). A function $v: \Gamma \to \mathbb{C}$ is called a *multiplier system* (MS) *of weight k*, when $v(ST) = v(S)v(T)$ for all, $S, T \in \Gamma$ and when $v(T)$ is a root of unity satisfying

$$v(-I) = (-1)^k. \tag{3.1}$$

A meromorphic function $f: \mathscr{H} \to \mathbb{C}$ is called an *unrestricted modular form of weight k and* MS $v$ on $\Gamma$ if and only if

$$f \mid T = v(T)f \qquad (T \in \Gamma). \tag{3.2}$$

For each $L \in \Gamma(1)$, $f_L := f \mid L$ is also meromorphic on $\mathscr{H}$ and is a periodic function, in the sense that, for some divisor $m_L$ of $N$,

$$f_L(z + m_L) = f_L(z),$$

so that $f_L(z)$ can, for sufficiently large Im $z$, be expressed as a Laurent–Fourier series

$$f_L(z) = \sum_{r=-\infty}^{\infty} a(r) \exp(2\pi i r z/m_L). \tag{3.3}$$

We say that $f$ is a *modular form* of weight $k$ and MS $v$ on $\Gamma$ if 3.2 holds and if, for each $L \in \Gamma(1)$, there are only a finite number of non-zero coefficients $a(r)$ with $r < 0$ in 3.3. It is only necessary to check this for the finite number of $L$ in a right transversal of $\Gamma$ in $\Gamma(1)$.

$f$ is an *entire form* if, in addition, for each $L \in \Gamma(1)$, $a(r) = 0$ for all $r < 0$; $f$ is a *cusp form* if, for each $L \in \Gamma(1)$, $a(r) = 0$ for all $r \leqslant 0$. That is, if $f$ is a cusp form, then $f$ vanishes at each cusp $L(\infty)$.

$\{\Gamma, k, v\}$ is the set of all cusp forms on $\Gamma$ of weight $k$ with MS$v$. It is a finite-dimensional vector space over $\mathbb{C}$. The trivial MS which takes the value 1 on $\Gamma$ (so that $k$ must be even) is devoted by 1. We write $\{\Gamma, k\}$ in place of $\{\Gamma, k, 1\}$.

## 4. EXAMPLES

The set $\{\Gamma(1), k\}$ has positive dimension only when $k$ (which must be even) is 12 or $k \geqslant 16$. $\{\Gamma(1), k\}$ has dimension 1 for $k = 12, 16, 18, 20, 22, 26$ and $\{\Gamma(1), 12\}$ is generated by the modular discriminant function

$$\Delta(z) = e^{2\pi i z} \prod_{m=1}^{\infty} (1 - e^{2\pi i m z})^{24} = \sum_{n=1}^{\infty} \tau(n)\, e^{2\pi i n z}. \tag{4.1}$$

The Fourier coefficients $\tau(n)$ (Ramanujan's function) are multiplicative; i.e.

$$\begin{cases} \tau(mn) = \tau(m)\,\tau(n) \quad \text{when} \quad (m, n) = 1, \\ \tau(p^{r+1}) = \tau(p)\,\tau(p^r) - p^{11}\,\tau(p^{r-1}) \qquad (r \geqslant 1), \end{cases} \tag{4.2}$$

where $p$ is a prime. These facts were conjectured by Ramanujan [**11**] and proved by Mordell [**8**]. Accordingly the associated Dirichlet series has an Euler product (for sufficiently large $\sigma = Re\, s$):

$$\sum_{n=1}^{\infty} \frac{\tau(n)}{n^s} = \prod_p (1 - \tau(p)p^{-s} + p^{11-2s})^{-1}. \tag{4.3}$$

Clearly $\Delta^2 \in \{\Gamma(1), 24\}$ (which has dimension 2), while it can be shown that

$$\Delta^{\frac{1}{2}} \in \{\Gamma(2), 6\}, \Delta^{\frac{1}{3}} \in \{\Gamma(3), 4\},$$

$$\Delta^{\frac{1}{4}} \in \{\Gamma(4), 3, v\} \text{ for some MS } v,$$

etc.

## 5. ORTHOGONAL SUBSPACES

The space $\{\Gamma, k, v\}$ becomes a finite-dimensional Hilbert space under the *Petersson inner product*

$$(f, g) = \frac{1}{|\Gamma(1) : \Gamma|} \iint_F f(z)\, \overline{g(z)}\, y^{k-2}\, dx\, dy,$$

where $F$ is a fundamental region and $x = \text{Re}\, z$, $y = \text{Im}\, z$.

We study the space

$$M = \{\Gamma(N), k, v_0\}, \tag{5.1}$$

where $v_0(T) = 1$ for $T \equiv I \pmod N$, and $v_0(T) = (-1)^k$ for $T \equiv -I \pmod N$; for $k$ odd we must have $N > 2$.

Then, as shown by Hecke [**5**],

$$M = \bigoplus_{\chi \bmod N} M^{\chi} = \bigoplus_{t \mid N} M_t = \bigoplus_{\chi} \bigoplus_{t} M_t^{\chi}, \tag{5.2}$$

where

$$M^{\chi} = \{\Gamma_0(N, N), k, \chi\}, M_t^{\chi} = M^{\chi} \cap M_t \tag{5.3}$$

and $M_t$ is the subspace of all forms of divisor $t$ (to be defined). Here $\chi$ is the MS defined by

$$\chi(T) = \chi(d),$$

where, on the right, $\chi$ is a character modulo $N$ and $\chi(-1) = (-1)^k$. The different subspaces $M_t^{\chi}$ are all mutually orthogonal.

For any positive divisor $t$ of $N$ write

$$N = tt_1, \tag{5.4}$$

and call $t_1$ a *codivisor*. Then $M_t$ consists of all cusp forms $f$ in $M$ which have Fourier series of the form

$$f(z) = \sum_{r=1}^{\infty} a(r) \exp(2\pi irz/N) \qquad (y > 0), \tag{5.5}$$

where $a(r) = 0$ whenever $(r, N) \neq t$. Such a function can be expressed in the form

$$f(z) = \sum_{\substack{r=1 \\ (r,t_1)=1}}^{\infty} A(r) \exp(2\pi irz/t_1). \tag{5.6}$$

## 6. HECKE OPERATORS

For $n \in \mathbb{Z}^+$, put

$$[J_n] = \{T \in \Omega_n : T \equiv J_n \pmod{N}\}. \tag{6.1}$$

Clearly $[J_n]$ is a union of double cosets $\Gamma(N)T\Gamma(N)$ and so is a union of right cosets, namely

$$[J_n] = \Gamma(N) \cdot R_n. \tag{6.2}$$

Moreover $|R_n| = \psi(nN)/\psi(N)$, so that $|R_n| = \psi(n)$ when $(n, N) = 1$. (When $n$ is squarefree, $[J_n] = \Gamma(N)J_n\Gamma(N)$.)

For $f \in M_t^\chi$ define the *Hecke operator* $T_n$ by

$$f|T_n = n^{\frac{1}{2}k-1} \sum_{T \in R_n} f|J_t T J_t^{-1}. \tag{6.3}$$

It can be shown that this is independent of the transversal $R_n$ and that, moreover, $f|T_n \in M_t^\chi$ also. When $(n, N) = 1$, $J_t T J_t^{-1}$ can be replaced by $T$ in 6.3. When $p$ is a prime and $p \nmid N$, we may take

$$f(z)|T_p = p^{k-1}\chi(p) f(pz) + p^{-1} \sum_{\nu=0}^{p-1} f((z + \nu N)/p).$$

If $f(z)$ is given by 5.6, then

$$f(z)|T_n = \sum_{\substack{r=1 \\ (r, t_1)=1}}^{\infty} A_n(r) \exp(2\pi i r z/t_1)$$

where

$$A_n(r) = \sum_{d|(r,n)} \chi(d) d^{k-1} A(rn/d^2).$$

Note that, if $n$ is composed solely of primes dividing $N$ (i.e. if $n|N^\infty$), then $A_n(r) = A(nr)$.

For all $f$ and $g$ in $M^\chi$, Petersson [**10**] showed that

$$(f|T_n, g) = \chi(n)(f, g|T_n) \text{ when } (n, N) = 1. \tag{6.5}$$

Moreover, for all positive integers $m$ and $n$,

$$(6.6)\qquad (f\,|\,T_m)\,|\,T_n = (f\,|\,T_n)\,|\,T_m = \sum_{d\,|\,(m,n)} d^{k-1}\chi(d)f\,|\,T_{mn/d^2},$$

whenever $f \in M^\chi$. In particular, Hecke operators commute with each other.

## 7. EIGENFORMS AND EULER PRODUCTS

From 6.5, 6.6 and a general result [**3**] on commuting normal operators we deduce that each subspace $M_t^\chi$ possesses a basis of mutually orthogonal forms each of which is an eigenvector for all the operators $T_n$ with $(n, N) = 1$. Thus, if $f$ is such an *eigenform* then

$$(7.1)\qquad f\,|\,T_n = \lambda(n)f \quad \text{for} \quad (n, N) = 1$$

and so, from 5.6 and 6.4, we have

$$(7.2)\qquad \lambda(n)A(r) = \sum_{d\,|\,(r,n)} \chi(d)d^{k-1}A(rn/d^2) \quad \text{for} \quad (n, N) = 1$$

and all $r \in \mathbb{Z}^+$.

If we now write $F(s) = \sum_{n=1}^{\infty} A(n)n^{-s}$ for the corresponding Dirichlet series, 7.2 leads to the expression

$$F(s) = F_N(s) \prod_{p \nmid N} (1 - \lambda(p)p^{-s} + \lambda(p)p^{k-1-2s})^{-1}$$

for sufficiently great $\sigma$. Here

$$F_N(s) = \sum_{\substack{n=1 \\ n\,|\,N^\infty,\,(n,t_1)=1}}^{\infty} A(n)n^{-s}.$$

Thus, corresponding to each prime $p \nmid N$, we have an Euler factor, but no multiplicative properties are claimed for those coefficients $A(n)$ for which $n\,|\,N^\infty$. Moreover, $\lambda(n) = \chi(n)\,\overline{\lambda(n)}$ for $(n, N) = 1$.

Of course, if $N = 1$, or if $t_1$ is divisible by every prime dividing $N$ (e.g. if $t_1 = N$, $t = 1$) then $(n, t_1) = 1$ and $n\,|\,N^\infty$ imply that $n = 1$ so that $F_N(s)$

reduces to a constant and $F$ has an Euler factor (possibly equal to 1) for every prime $p$. This happens in particular for $f = \Delta$ (where $N = t = t_1 = 1$) and for $f = \Delta^{\frac{1}{2}}$ (where $N = t_1 = 2, t = 1$).

## 8. NEWFORMS

The basic eigenforms of Section 7 are not necessarily eigenvectors of the operators $T_n$ with $(n, N) > 1$. (If $(n, t_1) > 1$ then $f \mid T_n = 0$ so that $f$ has eigenvalue $\lambda(n) = 0$.) Eigenforms for all $T_n$ can be found by generalizing the theory of Atkin and Lehner **[1]**, which they applied to the subspace $M_N^{\chi_0}$, where $\chi_0$ is the principal character. The rough idea of this theory is to remove from the space of forms in which we are interested all forms that have lower level; for these, by an inductive argument, we can assume the problem of finding eigenforms for *all* the $T_n$ $(n \in \mathbb{Z})$ to have been solved.

We consider the space

$$M_t^\chi =: \{N, \chi, t\}. \tag{8.1}$$

Here, as indicated, $M_t^\chi$ is the space of cusp forms of level $N$, weight $k$, character (i.e. MS) $\chi$ and divisor $t$. We shall keep $k$ and $\chi$ fixed but vary the level and divisor in such a way that the codivisor $t_1$ remains fixed. Write

$$M(d) = \{N/d, k, \chi, t/d\}, \tag{8.2}$$

where $d$ is divisor of $t$. This makes sense if $\chi$, which is a character modulo $N$, is also a character modulo $N/d$; i.e. if $N_\chi$, the conductor of $\chi$, divides $N/d$. We also want $d$ to be prime to $t_1$ and therefore define $M(d)$ as in 8.2 only for $d \mid t_\chi$, where $t_\chi$ is the greatest factor of $N/N_\chi$ that is prime to $t_1$. In particular, we can take $d = 1$ and then $M(1) = M_t^\chi$. It can be shown that, for each $\delta$ dividing $d$ (where $d \mid t_\chi$), $M(d)\ J_\delta^{-1}$ is a subspace of $M(1)$.

We now define $M^-$ to be the vector sum of all the spaces $M(d) \mid J_\delta^{-1}$, where

$$d \mid t_\chi, d > 1, \delta \mid d; \tag{8.3}$$

it may, of course, happen that no integers $\delta$ and $d$ satisfy 8.3, so that $M^-$ consists solely of the zero form in this case. We define $M^+$ to be the orthogonal complement of $M^-$ in $M(1)$. It can be shown that both $M^-$ and $M^+$ are invariant under all the operators $T_n$ $(n \in \mathbb{Z}^+)$.

Now let $f$ be an eigenform in $M(1)$ for the operators $T_n$ with eigenvalues $\lambda(n)$, where $(n, N) = 1$. The set of all such forms constitutes an *eigenclass* which is a subspace of $M(1)$, and may have dimension exceeding 1. The trivial zero form belongs to every eigenclass. Because the operators commute, each eigenclass is invariant under all the operators $T_n$.

The key theorem asserts that, if $C$ is such a nontrivial eigenclass, then either (i) $C \subseteq M^+$, or (ii) $C \subseteq M^-$. Moreover, when (i) holds, then $\dim C = 1$ and each form in $C$ is a scalar multiple of a unique normalized eigenform $f$, which is called a *newform* in $M_t^\chi$. By saying that $f$ is normalized we mean that its first Fourier coefficient $A(1)$ is 1 and it then follows that

$$(8.4) \qquad f \mid T_n = A(n) f \qquad (n \in \mathbb{Z}^+);$$

thus the eigenvalues $\lambda(n)$ are just the Fourier coefficients of $f$ and they possess the multiplicative properties:

$$(8.5) \qquad A(mn) = A(m)\, A(n), \quad \text{when} \quad (m, n) = 1,$$

and

$$(8.6) \qquad A(p^{r+1}) = A(p)\, A(p^r) - \chi(p)\, p^{k-1}\, A(p^{r-1}),$$

for $r \geqslant 1$.

Accordingly, the associated Dirichlet series $F(s) = \sum_{n=1}^{\infty} A(n)\, n^{-s}$ has an Euler factor for each prime, namely

$$(8.7) \qquad F(s) = \prod_p (1 - A(p)\, p^{-s} + \chi(p) p^{k-1-2s})^{-1}$$

for sufficiently great $\sigma$.

The space $M^+$ is spanned by newforms, which are mutually orthogonal. Newforms $f_d$ in $M(d)^+$ can be defined similarly, for each $d \mid t_\chi$ $(d > 1)$ and then $f_d \mid J_\delta^{-1}$, for $\delta \mid d$, is called an *oldform*. The space $M^-$ is spanned by these oldforms.

Equation 8.6 holds also when $p \mid N$, although then $\chi(p) = 0$, of course. We can say more about $A(p)$ in this case. In fact $A(p) = 0$ if either (*i*) $p \mid t_1$, or (ii) $p \nmid t_1, p^2 \mid t$ and $p \mid (N/N_\chi)$. If (iii) $p \nmid t_1, p^2 \nmid t$ and $p \mid (N/N_\chi)$, then $|A(p)| = p^{\frac{1}{2}k-1}$. Finally if (iv) $p \nmid t_1$ and $p \mid N_\chi$, then $|A(p)| = p^{\frac{1}{2}(k-1)}$. When $(t, t_1) = 1$ these four cases exhaust the possibilities for $p \mid N$. Here (iii) generalizes a result of

Atkin and Lehner [1] and can be made slightly more precise, while (iv) has been given by Ogg [9] on the assumption that newforms exist. If $p \nmid N$, the work of Deligne and Serre indicates that

$$|A(p)| < 2p^{\frac{1}{2}(k-1)}.$$

The preceding results are based on the forthcoming book [13]. Professor J. Lehner informs me that Margaret Millington (née Ashworth) was working on an extension of the Atkin–Lehner theory of newforms at the time of her death in 1972. More recently there has appeared the work of Li [7] who has developed a theory of newforms on $\Gamma_0(M, N)$ for arbitrary character $\chi$, but without consideration of divisors.

It may be of interest to give a brief sketch of the most important ideas used in the proof of the key theorem. A member of an eigenclass $C$ is said to be *primitive* if its first Fourier coefficient $A(1)$ does not vanish. It is easily shown that, if $f$ is not primitive, then $A(n) = 0$ whenever $(n, N) = 1$. A fairly elaborate induction argument then shows that

$$\sum_{d \mid t_\chi} \mu(d) f \mid L_d = 0,$$

where $\mu(d)$ is the Möbius function. Here $L_d\colon M_t^\chi \to M_t^\chi$ is a projection operator with the property that the "trace operator" $L_d J_d$ maps $M(1)$ into $M(d)$. Accordingly $f = \sum_{d>1} f_d$, where

$$f_d = -\mu(d) f \mid L_d \in M(d)\, J_d^{-1} \subseteq M^- \qquad (d > 1).$$

Thus $f \in M^-$ whenever $f$ is not primitive.

If $M^+ \cap C \neq 0$ then clearly $\dim M^+ \cap C = 1$ and $M^+ \cap C$ is generated by a normalized eigenform $F$, say. It can be shown that if, in addition, $M^- \cap C \neq 0$, then this space contains a primitive (and therefore normalized) member $f$. Since $F - f$ is not primitive we deduce that $F \in M^-$, a contradiction. Thus $C \subseteq M^+$.

Other results can be obtained by generalizing the involution operators $W$ introduced by Atkin and Lehner and studying their interaction with Hecke operators.

## 9. EXAMPLES

When $N = 1$ all eigenforms are scalar multiples of newforms. Thus the discriminant function $\Delta$ is the sole newform of level 1 and weight 12. $\Delta$ may, however, be regarded as a cusp form of any level $N$, weight 12, divisor $N$ and character the trivial character $\chi_0$.

In particular, for $N = 2$ and $M = \{\Gamma(2), 12\}$, the subspace $M_2$ of forms of divisor 2 has dimension 2 and is spanned by the oldforms $\Delta(z)$ and $\Delta(2z)$; there are no newforms in this space. On the other hand, the subspace $M_1$ of forms of divisor 1 also has dimension 2 and is generated by the newforms $\Delta^{\frac{1}{2}}E_6$ and

$$\tfrac{1}{2}\{\Delta(\tfrac{1}{2}z) - \Delta(\tfrac{1}{2}z + \tfrac{1}{2})\} = \sum_{\substack{n=1 \\ (n\,\text{odd})}}^{\infty} \tau(n)\, e^{\pi i n z}.$$

Here $E_6$ is the "Eisenstein series" of dimension 6 belonging to $\Gamma(1)$. Note that the second newform has the same eigenvalues $\tau(n)$ ($n$ odd) as the oldform $\Delta$.

Atkin and Lehner showed that (for $t = N$ and $\chi = \chi_0$) two different newforms in the same space could not have the same eigenvalues, apart from a finite number of values of $n$. However, two newforms in the space can have an infinite number of common eigenvalues. Thus the two newforms that span the space $\{4, 4, \chi_0, 1\}$ have coefficients $P(n)$ and $(-1)^{\frac{1}{4}(n-1)}\, P(n)$ (in Glaisher's notation) and these agree for all $n \equiv 1 \pmod 4$.

## 10 FURTHER APPLICATIONS OF HECKE OPERATORS

Wohlfahrt [**16**] has defined Hecke operators for arbitrary real weight on Fuchsian groups of the first kind, but it is only in special cases that this has given rise to a fruitful theory. Thus, even for subgroups of the modular group, there are difficulties for non-integral weight $k$, although some results of great interest have been obtained by Shimura [**14**] when $2k$ is an odd integer; see also earlier work by van Lint [**15**] and Kløve [**6**].

When $\Gamma$ is an arbitrary congruence group of level $N$ and $(n, N) = 1$, the operator $T_n$ does not necessarily leave the space $\{\Gamma, k, v\}$ invariant, but maps it into a space $\{\Gamma_n, k, v_n\}$, where the group $\Gamma_n$ and MS $v_n$ need not be the same as $\Gamma$ and $v$, respectively. This can be used [**12**] to explain the existence of some curious partially multiplicative arithmetical functions used by Glaisher [**4**] in his work on the number of representations of a number as a sum of an even number of squares. In his formulae for 18 squares he used a cusp form belonging to $\{\Gamma, 9, v\}$ where $\Gamma$ has index 3 in $\Gamma(1)$ and $v$ is a certain multiplier system. The Fourier coefficients $G(n)$ of this cusp form are integers and possess the strange property that, for odd $mn$, with $(m, n) = 1$

$$G(mn) = \begin{cases} G(m)G(n) & \text{when} \quad m \equiv 1 \quad \text{or} \quad n \equiv 1 \pmod 4, \\ -\frac{39}{25}G(m)\,G(n) & \text{when} \quad m \equiv n \equiv -1 \pmod 4. \end{cases}$$

This was not proved by Glaisher, but can be explained by the fact that, although $\Gamma_n = \Gamma$ for all odd $n$, $v_n = v$ only when $n \equiv 1 \pmod 4$. There are, of course, no newforms in the space $\{\Gamma, 9, v\}$; however this space is a subspace of some space of level 2 and divisor 1 in which there are newforms with Fourier coefficients in the field $\mathbb{Q}([-39/25]^{\frac{1}{2}})$.

In conclusion it may be remarked that there are indications that results of great interest await discovery and proof for non-congruence groups and Hecke operators; see [**2**].

## REFERENCES


1. A. O. L. Atkin and J. Lehner, Hecke operators on $\Gamma_0(m)$. *Math. Ann.* **185** (1970), 134–160.
2. A. O. L. Atkin and H. P. F. Swinnerton-Dyer, Modular forms of non-congruence groups. *Combinatorics* (*Proc. Sympos. Pure Math., Vol XIX*), pp. 1–25. Amer. Math. Soc., Providence, R.I., 1971.
3. F. R. Gantmacher, "Matrix Theory" Vol. I, p. 291. Chelsea, 1960.
4. J. W. L. Glaisher, On the representation of a number as the sum of a number of squares. *Quart. J. Math.* **38** (1907), 1–62, 178–236, 289–351.
5. E. Hecke, Über Modulfunktionen und die Dirichletschen Reihen mit Eulerscher Produktentwicklung. *Math. Ann.* **114** (1937), 1–28, 316–351.
6. T. Kløve, Recurrence formulae for the coefficients of modular forms. *Math. Scand.* **26** (1970), 221–232.
7. W. C. W. Li, Newforms and functional equations. *Math. Ann.* **212** (1975), 285–315.
8. L. J. Mordell, On Ramanujan's empirical expansions of modular functions. *Proc. Cambridge Philos Soc.* **19** (1920), 117–124.
9. A. P. Ogg, On the eigenvalues of Hecke operators. *Math. Ann.* **179** (1969), 101–108.
10. H. Petersson, Konstruktion der sämtlichen Lösungen einer Riemannschen Funktionalgleichung durch Dirichlet-Reihen mit Eulerscher Produktentwicklung. *Math. Ann.* **116** (1939), 401–412; **117** (1939), 39–64; **117** (1940), 277–300.
11. S. Ramanujan, On certain arithmetical functions. *Trans. Cambridge Philos. Soc.* **22** (1916), 159–184,

12. R. A. Rankin, Hecke operators on congruence subgroups of the modular group. *Math. Ann.* **168** (1967), 40–58.
13. R. A. Rankin, "Modular Forms and Functions" Chap. 9 and 10. Cambridge University Press, 1977.
14. G. Shimura, On modular forms of half integral weight. *Ann. of Math.* **97** (1973), 440–481.
15. J. H. van Lint, "Hecke Operators and Euler Products", Dissertation, Leiden, 1957.
16. K. Wohlfahrt, Über Operatoren Heckescher Art bei Modulformen reeller Dimension. *Math. Nachr.* **16** (1957), 233–256.

# 12. Arithmetic Groups

H. P. F. SWINNERTON-DYER

*St. Catharine's College, Cambridge, England*



## 1. INTRODUCTION

Let $G$ denote a finite-dimensional real Lie group, and let $\Gamma$ be a discrete subgroup of $G$. This conference has been concerned with the study of the space $\Gamma\backslash G$, primarily in the case $G = \mathrm{SL}(2, \mathbb{R})$ so that $\Gamma$ is a Fuchsian group, but to some extent also in the case when $G = \mathrm{SL}(2, \mathbb{C})$ and $\Gamma$ is a Kleinian group. Have we confined ourselves to these two cases merely because they are the simplest non-trivial ones and were historically the first to be considered, or do they really have features which do not occur for more general $G$? The answer seems to depend on whether we assume that $\Gamma$ is a "good" subgroup of $G$ or not. No one has yet obtained interesting results about arbitrary discrete subgroups of a general $G$; and indeed the theory even for the case of Kleinian groups is so complicated that one is not tempted to try to generalize it. The extra condition which has led to an interesting theory is that $\Gamma$ is "large enough". The strong form of this is that $\Gamma\backslash G$ is compact; the weak form is that $\Gamma\backslash G$ has finite measure with respect to the invariant Haar measure on $G$. At present there are important theorems which have been proved only when $\Gamma\backslash G$ has finite measure but is not compact, and others which have only been proved when $\Gamma\backslash G$ is compact; however this is probably a consequence of the methods of proof, rather than a difference in what is actually true in the two situations.

Neither form of the extra condition translates into anything recognizable for Kleinian groups; and indeed if $\Gamma$ is a subgroup of $G = \mathrm{SL}(2, \mathbb{C})$ such that $\Gamma\backslash G$ has finite measure, then $\Gamma$ cannot be discontinuous on any domain in the complex plane. However, if $\Gamma$ is a Fuchsian group, that is a discrete subgroup of $G = \mathrm{SL}(2, \mathbb{R})$, then $\Gamma\backslash G$ is compact or has finite measure if and only if the same is true of a fundamental domain for the action of $\Gamma$ on the upper half plane. Moreover $\Gamma\backslash G$ has finite measure if and only if $\Gamma$ is of the first kind and has a fundamental domain with only finitely many sides, and $\Gamma\backslash G$ is compact if and only if in addition $\Gamma$ has no parabolic elements.

With this extra condition, the theory for $\mathrm{SL}(2, \mathbb{R})$ is certainly quite different from that for other Lie groups. This can already be seen by considering the closely related problems of embedding a given abstract group in a Lie group and of deforming a discrete subgroup of a Lie group. If $\Gamma \subset G = \mathrm{SL}(2, \mathbb{R})$ is a Fuchsian group such that $\Gamma\backslash G$ has finite measure, then $\Gamma$ will usually admit a large and interesting family of deformations; indeed the study of such deformations is virtually equivalent to the study of the relevant Teichmüller space. By contrast, in the general case we have the following global rigidity theorem, which combines the results of Margulis [**12**], Mostow [**14**] and Prasad [**17**].

1.1 THEOREM. *Let G be a semi-simple Lie group which has trivial centre and has no compact normal subgroup other than the identity; and let $\Gamma$ be a discrete subgroup of G such that $\Gamma\backslash G$ has finite measure. Suppose that* $\mathrm{PSL}(2, \mathbb{R})$ *is not a direct factor of G which is closed* mod $\Gamma$*—that is, suppose there is no factor H of G which is isomorphic to* $\mathrm{PSL}(2, \mathbb{R})$ *and is such that $\Gamma H$ is a closed subgroup of G. Then the pair $G, \Gamma$ is uniquely determined by $\Gamma$—that is, given two such pairs $G, \Gamma$ and $G', \Gamma'$ and an isomorphism $\theta: \Gamma \xrightarrow{\sim} \Gamma'$, there is a Lie group isomorphism $G \xrightarrow{\sim} G'$ which extends $\theta$.*

The analogous local rigidity theorem below, which is more relevant to this survey, combines the results of Weil [**26**] for case (i) and Garland and Raghunathan [**6**] for case (ii). The analogous problem for the case when $G$ has rank greater than 1 and $\phi\Gamma\backslash G$ is not compact is still unsolved. The reader may also usefully consult [**15**], [**19**] and [**20**].

1.2 THEOREM. *Let G be a connected semi-simple Lie group with no compact* *direct factor, and let $\Gamma$ be an abstract group and $\phi:\Gamma \to G$ an embedding such that $\phi\Gamma$ is discrete in G and $\phi\Gamma\backslash G$ has finite measure. Suppose that either*

(i) *$\phi\Gamma\backslash G$ is compact, and no group isogenous to G has a direct factor isomorphic to* PSL (2, $\mathbb{R}$) *or*

(ii) *G is of rank* 1 *and is not locally isomorphic to* SL (2, $\mathbb{R}$) *or to* SL (2, $\mathbb{C}$) *considered as a real Lie group.*

*Then any homomorphism $\psi:\Gamma \to G$ which is near enough to $\phi$ can be obtained from $\phi$ by composition with an inner automorphism of G. Moreover $\Gamma$ is finitely generated.*

Theorem 1.2 does not follow from Theorem 1.1, because in Theorem 1.2 it is not assumed that $\psi\Gamma$ is discrete; the importance of this will be seen below. This also explains the final clause in (ii). If for example we take $G = \mathrm{SL}(2, \mathbb{C})$ and $\Gamma = \mathrm{SL}(2, \mathbb{Z} + \mathbb{Z}i)$, with $\phi$ the natural embedding, then every $\psi:\Gamma \to G$ near $\phi$ for which $\psi\Gamma$ is discrete in $G$ is obtained from $\phi$ by composition with an inner automorphism of $G$; but there are $\psi$ arbitrarily near to $\phi$ for which $\psi\Gamma$ is not discrete in $G$.

A local rigidity theorem (or "isolation theorem" in older terminology) nearly always has number-theoretic implications. To express them in this case, we need a further definition. Let $K$ be a subfield of $\mathbb{R}$; the case of interest is when $K$ is $\mathbb{Q}$ or at worst an algebraic number field. A real Lie group $G$ is called a *linear algebraic group defined over K* if

(i) $G$ is a subgroup of GL $(n, \mathbb{R})$ for some $n$, and

(ii) if $g = (g_{ij})$ denotes an element of GL $(n, \mathbb{R})$, then there is a set $\mathscr{S}$ of polynomials in the $g_{ij}$ with coefficients in $K$ such that $g$ is in $G$ if and only if every polynomial in $\mathscr{S}$ vanishes at $g$.

In other words $G$ is both a subgroup of GL $(n, \mathbb{R})$ and a subvariety of it defined over $K$.

For many purposes we can confine ourselves to the study of linear algebraic groups; for it is known that any semi-simple Lie group is at least isogenous to a linear algebraic group defined over $\mathbb{Q}$. On the other hand PSL (2, $\mathbb{R}$), for example, is not isomorphic to a linear algebraic group; but it is isogenous to SL (2, $\mathbb{R}$) and so the study of subgroups of PSL (2, $\mathbb{R}$) is nearly the same as the study of subgroups of SL (2, $\mathbb{R}$). With this terminology Theorem 1.2 has the following immediate corollary.

1.3 THEOREM. *Let G, $\Gamma$, $\phi$ satisfy the conditions of Theorem* 1.2 *and assume* *that $G$ is a linear algebraic group defined over a field $K \subset \mathbb{R}$. Then there exist an element $g$ in $G$ and a finite extension $L$ of $K$ in $\mathbb{R}$ such that every element of $g(\phi\Gamma)g^{-1}$ is defined over $L$.*

For a group homomorphism $\psi:\Gamma \to G$ is determined by the images of some assigned set of generators of $\Gamma$, and we may take this set to be finite. These images are given by $n \times n$ matrices, whose elements we may regard as coordinates in some affine space $\mathbb{R}^N$. Necessary and sufficient conditions that a point of $\mathbb{R}^N$ should define a homomorphism $\psi$ are that the images of the generators should lie in $G$ and the relations between the images which correspond to the relations between the generators of $\Gamma$ should hold. These conditions are all polynomial equations between the coordinates, with coefficients in $K$; hence they determine a variety defined over $K$—or to be precise a Zariski open subset of such a variety. Denote by $V$ the component of this, irreducible over $\mathbb{R}$, which contains the point corresponding to $\phi$. By Theorem 1.2, the real points of $V$ correspond to the homomorphisms obtained by composing $\phi$ with an inner automorphism of $G$. On the other hand $V$ is defined over a real finite extension of $K$, and with a further finite extension we can find real points on $V$; if $P$ is such a point and is defined over a field $L$, the corresponding homomorphism $\psi$ has the property that every element of $\psi\Gamma$ is defined over $L$. This proves Theorem 1.3. The result is still true when $G$ is locally isomorphic to SL(2, $\mathbb{C}$), but the proof is no longer trivial; for details see [**6**].

This result suggests that in the general case we shall not get much further without the help of number theorists; but unfortunately the conclusion of Theorem 1.3 is not strong enough to be of much use to them—and would still not be, even if $L$ were replaced by $\mathbb{Q}$ in it. We need a similar theorem with a stronger conclusion, and to express this requires more definitions. If $G$ is a Lie group, then two subgroups of it are said to be *commensurable* if their intersection has finite index in each of them; clearly this is an equivalence relation. If $G$ is a Lie group and $\Gamma$ a subgroup of $G$, then Comm ($\Gamma$), the *commensurability group* of $\Gamma$ in $G$, consists of those $g$ in $G$ for which $g\Gamma g^{-1}$ is commensurable with $\Gamma$; thus it certainly contains $\Gamma$.

We are now in a position to define an *arithmetic subgroup* of a Lie group $G$. The original definition, due to Borel and Harish-Chandra, was as follows. Let $G$ be a linear algebraic group defined over $\mathbb{Q}$, and denote by $G_{\mathbb{Z}}$ the group of those $g$ in $G$ such that both $g$ and $g^{-1}$ have all their matrix elements in $\mathbb{Z}$. A subgroup $\Gamma$ of $G$ is called *arithmetic* if it is commensurable with $G_{\mathbb{Z}}$; an equivalent condition is that it is commensurable with the intersection of $G$

with SL $(n, \mathbb{Z})$. This condition appears to depend not merely on $G$ but on the choice of its representation as a linear algebraic group; but the following theorem of Borel [**1**] limits the extent of that dependence.

1.4 THEOREM. *Let $G$, $G'$ be linear algebraic groups defined over $\mathbb{Q}$, let $\theta: G \to G'$ be a surjective homomorphism which is defined over $\mathbb{Q}$ when considered as a morphism of varieties, and let $\Gamma$ be an arithmetic subgroup of $G$. Then $\theta\Gamma$ is an arithmetic subgroup of $G'$.*

This has led to various extensions of the original definition, of which the most sensible seems to be the following. Let $\mathscr{S}$ be the smallest set of pairs $(G, \Gamma)$, with $G$ a Lie group and $\Gamma$ a subgroup of $G$, which has the following two properties:

(i) If $G$ is a linear algebraic group defined over $\mathbb{Q}$ and $\Gamma$ is an arithmetic subgroup of $G$ in the sense of Borel and Harish-Chandra, then $(G, \Gamma)$ is in $\mathscr{S}$.
(ii) If $G$, $G'$ are Lie groups and $\theta: G \to G'$ is a surjective homomorphism with compact kernel, and if $\Gamma$ is a subgroup of $G$ whose intersection with the kernel of $\theta$ is finite, then if one of $(G, \Gamma)$ and $(G', \theta\Gamma)$ is in $\mathscr{S}$ so is the other.

Then $\Gamma$ is an *arithmetic subgroup* of $G$ if and only if $(G, \Gamma)$ is in $\mathscr{S}$. Note that with this definition two arithmetic subgroups of a given Lie group $G$ need not be commensurable; but they can be made so by applying to one of them a suitably chosen automorphism of $G$. It might appear that this definition could be further generalized by replacing $\mathbb{Q}$ and $\mathbb{Z}$ throughout by an arbitrary algebraic number field and its ring of integers; but Borel [**1**] has shown that in fact this produces no new arithmetic subgroups. Borel and Harish-Chandra [**2**] have shown that arithmetic subgroups are "good" subgroups in the sense of the first paragraph of this article:

1.5 THEOREM. *Let $G$ be a semi-simple Lie group and $\Gamma$ an arithmetic subgroup of $G$. Then $\Gamma$ is discrete and $\Gamma\backslash G$ has finite measure; moreover $\Gamma\backslash G$ is compact if and only if $\Gamma$ has no unipotent elements.*

If $G$ is a linear algebraic group defined over $\mathbb{Q}$ and $\Gamma$ is an arithmetic subgroup of $G$ in the sense of Borel and Harish-Chandra, then it is easy to see that Comm $(\Gamma)$ contains the intersection of $G$ and GL $(n, \mathbb{Q})$. Using the ideas of [**2**] it may be deduced that Comm $(\Gamma)$ is dense in $G$ whenever $\Gamma$ is an arithmetic subgroup of any Lie group $G$. It seems likely that this property characterizes arithmetic subgroups among those discrete subgroups $\Gamma$ for which $\Gamma\backslash G$ has finite measure. The most important case is when $G = \mathrm{SL}(2, \mathbb{R})$, for which Margulis has announced the following theorem but has not yet published a proof.

1.6 THEOREM. *Let* $G = \mathrm{SL}(2, \mathbb{R})$ *and let* $\Gamma$ *be a discrete subgroup of* $G$ *such that* $\Gamma\backslash G$ *has finite measure. If* $\Gamma$ *is arithmetic then* $\mathrm{Comm}(\Gamma)$ *is dense in* $G$*; if* $\Gamma$ *is not arithmetic then* $\mathrm{Comm}(\Gamma)$ *is commensurable with* $\Gamma$.

Suppose that $G$ is a semi-simple Lie group other than $\mathrm{SL}(2, \mathbb{R})$ or $\mathrm{PSL}(2, \mathbb{R})$, and let $\Gamma$ be an *irreducible* discrete subgroup of $G$ such that $\Gamma\backslash G$ has finite measure. ("Irreducible" means that the pair $G, \Gamma$ cannot be constructed from the product of two such pairs by means of isogeny and commensurability.) Until recently, the only such $\Gamma$ that were known were arithmetic subgroups; and this naturally led to the conjecture that all such $\Gamma$ were arithmetic. This is known to be true provided some additional hypotheses hold; case (i) of the following theorem is due to Margulis [**13**], and case (ii) to Selberg (see [**9**], p. 277). Note that case (i) fills the gap in Theorem 1.3.

1.7 THEOREM. *Let* $G$ *be a connected semi-simple Lie group which is either*

(i) *of rank greater than* 1, *with trivial centre and without compact factors, or*
(ii) *isogenous to* $(\mathrm{SL}(2, \mathbb{R}))^n$ *for some* $n > 1$.

*Let* $\Gamma$ *be a discrete irreducible subgroup of* $G$ *such that* $\Gamma\backslash G$ *has finite measure but is not compact. Then* $\Gamma$ *is arithmetic.*

On the other hand, if $G = \mathrm{SO}(f, \mathbb{R})$ where

$$f = X_1^2 - X_2^2 - \ldots - X_n^2 \quad \text{for} \quad n = 4, 5 \text{ or } 6,$$

then Vinberg [**25**] has shown how to construct non-arithmetic discrete subgroups $\Gamma$ such that $\Gamma\backslash G$ has finite measure; for $n = 4$ or 5 he has examples both with $\Gamma\backslash G$ compact and with $\Gamma\backslash G$ not compact, but for $n = 6$ all his examples have $\Gamma\backslash G$ not compact.

In view of all this, it is clear that arithmetic subgroups deserve special study even in the classical case $G = \mathrm{SL}(2, \mathbb{R})$; for they represent that part of the theory of Fuchsian groups which one may reasonably hope to extend to more general Lie groups. Nevertheless, since the classical work of Fricke and Klein (and with the obvious exception of the modular group and its subgroups) they have been neglected until the last fifteen years. There are essentially two problems to consider. First, to describe the field of automorphic functions for a given arithmetic group, and in particular the natural field of definition of a geometric model for it. (Note that the correct definition of "natural" here is not obvious; indeed even when the geometric model is a straight line, the natural field of definition may be larger than $\mathbb{Q}$. Also, to write down the Poincaré series associated with the group does not appear to be helpful.) Second, to find those points at which the values of the automorphic functions can be determined exactly, and to describe those values. These correspond to the two basic problems of the theory of elliptic modular functions. We now know three possible lines of attack on them: the direct method based on classical ideas, the method of Shimura and the method of Langlands. In [**11**], Langlands has tried to unify his approach and that of Shimura.

## 2. THE DIRECT METHOD

The direct method depends on combining two tools: Schwartz's differential equation for automorphic functions, and the modular equations. To simplify the exposition, I shall confine myself to the case when the compactification of $\Gamma\backslash\mathcal{H}$ has genus 0; thus there is an embedding of complex manifolds $\Gamma\backslash\mathcal{H} \to \hat{\mathbb{C}}$, the Riemann sphere, and we shall use $f$ to denote the induced univalent automorphic function on $\mathcal{H}$. (In the language of Klein and Fricke, $f$ is the *Hauptmodul*.) Since this only determines $f$ up to linear transformation, we may also assume that $f$ is finite at all the ramification points of the map $\mathcal{H} \to \Gamma\backslash\mathcal{H}$. The Schwarzian derivative is defined to be

$$\begin{aligned}\{z, w\} &= z'''/z' - \tfrac{3}{2}(z''/z')^2 \\ &= -\, \dddot{w}/\dot{w}^3 + \tfrac{3}{2}(\ddot{w}/\dot{w}^2)^2\end{aligned}$$

where the dashes and dots denote differentiation with respect to $w = f(z)$ and $z$ respectively. This has the property that

$$\{\gamma z, w\} = \{z, w\}$$

for any $\gamma$ in PSL (2, $\mathbb{C}$); it follows that $\{z, f\}$ is an automorphic function with respect to $\Gamma$ and hence in our case a rational function of $f(z)$:

$$\{z, f\} = \Phi(f). \tag{2.1}$$

Straightforward local arguments show that

(i) $\Phi$ is regular except possibly at ramification values of $f$,
(ii) $\Phi(f) = O(f^{-4})$ at infinity, and
(iii) if $f = c$ is a point of ramification of order $r > 1$ for $\mathscr{H} \to \Gamma\backslash\mathscr{H}$, then

$$\Phi(f) = \tfrac{1}{2}(1 - r^{-2})(f - c)^{-2}(1 + o(1)) \quad \text{near} \quad f = c.$$

Suppose that there are $n$ values of $f$ at which the map $\mathscr{H} \to \Gamma\backslash\mathscr{H}$ is ramified; and let them be $c_1, \ldots, c_n$ with orders of ramification $r_1, \ldots, r_n$ respectively, where necessarily $n \geqslant 3$. When $n = 3$ the conditions above are enough to determine $\Phi$ uniquely; but for $n > 3$ the $\Phi$ satisfying these conditions contain $(n - 3)$ undetermined constants and for general (non-arithmetic) $\Gamma$ there is no known way of obtaining further information about them. Moreover the values of the $c_v$ are usually unknown, though three of them can be assigned by making a suitable bilinear transformation on $f$. But for given $\Gamma$ we do know $n$ and the $r_v$.

Now choose an element $g$ in Comm $(\Gamma)$, and write

$$m = [\Gamma : \Gamma \cap g^{-1}\Gamma g] = [g^{-1}\Gamma g : \Gamma \cap g^{-1}\Gamma g];$$

that these last two expressions are equal is shown by considering measures of fundamental domains of the three groups involved. The functions $f(z)$ and $f(gz)$ are both automorphic functions for $\Gamma \cap g^{-1}\Gamma g$, and there is therefore a polynomial $\Psi(X, Y)$, of degree $m$ in each of $X$ and $Y$, such that

$$\Psi(f(z), f(gz)) = 0. \tag{2.2}$$

This is called the *modular equation* for $g$. For an example of how to determine it in practice, see Fricke and Klein **[5]**, pp. 553 *et seq.*

Suppose that $X$ and $Y$ are any well-behaved functions of $z$ satisfying $\Psi(X, Y) = 0$. Direct calculation shows that

$$\{gz, Y\} = \{z, Y\} = (\Psi_2/\Psi_1)^2\{z, X\} + \Theta(X, Y)$$

where the subscripts on $\Psi$ denote partial derivatives and $\Theta$ is a rational function of the partial derivatives of $\Psi$ whose exact shape is not important. Compatibility between (2.1) and (2.2) therefore requires that $\Psi(X, Y) = 0$ implies

$$\Phi(Y) = (\Psi_2/\Psi_1)^2\Phi(X) + \Theta(X, Y). \tag{2.3}$$

Note that this is a purely algebraic condition on the $c_v$ and the unknown coefficients in $\Phi$ and $\Psi$. We shall show that it imposes enough constraints to determine them up to finitely many possibilities (subject to the allowable arbitrary bilinear transformation on $f$), at least provided $g$ is suitably chosen.

To show this, we shall assume that $g$ is elliptic and not of finite order; this is possible because Comm $(\Gamma)$ is dense in $SL(2, \mathbb{R})$. Denote by $z_0$ the one fixed point of $g$ in $\mathscr{H}$. Putting $z = z_0$ in (2.2) we obtain $\Psi(f(z_0), f(z_0)) = 0$, which gives the value of $f(z_0)$; similarly, differentiating (2.2) two or three times and putting $z = z_0$ gives equations for $f'(z_0)$ and $f''(z_0)$. (The equation obtained by differentiating once and putting $z = z_0$ turns out to be a platitude.)

We now turn this process back to front. Given $g$ and $\Gamma$, we know $n$, $m$ and the $r_\nu$. Let $\Phi^*$, which depends on the unknown parameters $c_\nu^*$, be the most general polynomial satisfying the conditions stated below (2.1) and let $\Psi^*(X, Y)$ be the most general polynomial of degree $m$ in each of $X$ and $Y$. The condition that $\Psi^*(X, Y) = 0$ should imply the analogue of (2.3) is an algebraic condition on the $c_\nu^*$ and the undetermined coefficients of $\Phi^*$ and $\Psi^*$; viewing these as coordinates in some affine space, we see that the set of $(c_1^*, \ldots, c_n^*, \Phi^*, \Psi^*)$ satisfying the above conditions is parametrized by a variety $V$, not necessarily irreducible. $V$ is not empty because it contains the point corresponding to $(c_1, \ldots, c_n, \Phi, \Psi)$. Choose any point of $V$ defined over $\mathbb{C}$ and any solution $b$ of $\Psi(b, b) = 0$. Let $f_1(z)$ be the solution of

$$\{z, f_1\} = \Phi^*(f_1) \tag{2.4}$$

with $f_1(z_0) = b$ and with values of $f_1'(z_0)$ and $f_1''(z_0)$ which will be defined in a moment; and let $f_2(z)$ be defined by $f_2(z_0) = b$ and

$$\Psi^*(f_1(z), f_2(gz)) = 0. \tag{2.5}$$

We now choose the values of $f_1'(z_0)$ and $f_1''(z_0)$ so that (2.5) implies

$$f_1'(z_0) = f_2'(z_0) \quad \text{and} \quad f_1''(z_0) = f_2''(z_0)$$

by calculations like those described in the previous paragraph. We have arranged already that (2.4) and (2.5) imply

$$\{z, f_2\} = \Phi^*(f_2);$$

the agreement of the initial conditions now shows that $f_1(z) = f_2(z)$.

The differential equation (2.4) with the initial conditions chosen at $z = z_0$ determines a Riemann surface $\mathscr{R}$, and the conditions below (2.1) ensure that

$\mathscr{R}$ is unramified over the $z$-plane. Equation (2.5) gives an action of $g$ on $\mathscr{R}$ which extends the action of $g$ on the $z$-plane. Because $g$ is elliptic and not of finite order this is only possible if $\mathscr{R}$ is a subdomain of the $z$-plane which is mapped onto itself by $g$, and the boundary of such a domain must be either

(i) a circle which is mapped to itself by $g$, or
(ii) the fixed point of $g$ other than $z_0$, or
(iii) empty.

Moreover $f_1$ is an automorphic function for a group $\Gamma^*$ acting on this domain. We can now reject (ii) and (iii) as possibilities, because we know the possible $\Gamma^*$ in those cases and they cannot have the same values for the $r_v$ as those we started with. Hence $\Gamma^*$ is a principal-circle group. But Comm($\Gamma^*$) contains $g$, by (2.5), and therefore it cannot be discrete; so $\Gamma^*$ is an arithmetic group by Theorem 1.6 and it is possible to replace the appeal to Theorem 1.6 by an elementary argument. Clearly $\Gamma^*$ depends continuously on the point $P$ on $V$ which we started from. But any maximal continuous family of arithmetic groups is obtained from any one member of the family by operating with an arbitrary element of SL(2, $\mathbb{C}$); and in this case the element must also commute with $g$ and must therefore have the same fixed points as $g$. Hence each component of $V$ essentially corresponds to a single arithmetic group, and as $V$ has only finitely many components this completes the proof of our assertion. Since $V$ is clearly defined over $\mathbb{Q}$, we have given a proof of the following theorem.

2.6 Theorem. *Let $\Gamma$ be an arithmetic subgroup of* SL(2, $\mathbb{R}$) *and suppose that $\Gamma\backslash\mathscr{H}$ has genus* 0. *Let $f$ be a Hauptmodul for $\Gamma$, suitably normalized. Then the differential equation for $f$, the modular equations, and the values of $f$ at fixed points of elliptic elements of* Comm($\Gamma$) *can be constructively determined; and the coefficients and values involved are algebraic over $\mathbb{Q}$.*

There is a similar result without any condition on the genus of $\Gamma\backslash\mathscr{H}$. However the disadvantage of this method is that it gives no general information about the algebraic number fields involved. In the classical case of elliptic modular functions, the values of $j$ at complex quadratic points are known to generate class fields. But the proofs of this by modular function theory are very intricate; and it was only when the argument was rewritten in geometric terms, treating SL(2, $\mathbb{Z}$)$\backslash\mathscr{H}$ as the moduli space of isomorphism classes of elliptic curves, that the underlying ideas became clear. (See for example [**23**], Section 13). This was the seed of the profound investigations of Shimura, published in the Annals of Mathematics between 1960 and 1970, to which we now turn.

## 3. THE METHOD OF SHIMURA

At the heart of Shimura's work is the question of classifying all Abelian varieties of dimension $n$ whose rings of endomorphisms contain (up to isomorphism) a given ring $R$. We define these terms below; for more details see [**24**]. Let $\Lambda$ be a *lattice* in $\mathbb{C}^n$—that is, a subgroup of $\mathbb{C}^n$ which is free of rank $2n$ and is discrete in $\mathbb{C}^n$. Then $T = \mathbb{C}^n/\Lambda$ is a compact complex manifold of dimension $n$, which we call a *complex torus*. If $z_1, \ldots, z_n$ are coordinates in $\mathbb{C}^n$ then $\mathrm{d}z_1, \ldots, \mathrm{d}z_n$ is a base for the $\mathbb{C}$-vector space of holomorphic 1-forms on $T$. Any endomorphism of $T$ must map this space into itself; and it follows easily that the endomorphisms of $T$ correspond to those elements of $\mathrm{End}(\mathbb{C}^n)$ which map $\Lambda$ into itself. In particular, $\mathrm{End}(T)$ has two natural faithful representations: the *complex representation* coming from

$$\mathrm{End}(T) \subset \mathrm{End}(\mathbb{C}^n) \cong M_n(\mathbb{C})$$

and the *rational representation* coming from

$$\mathrm{End}(T) \subset \mathrm{End}(\Lambda) \cong \mathrm{End}(\mathbb{Z}^{2n}) \cong M_{2n}(\mathbb{Z}).$$

It can be shown that the rational representation is equivalent to the sum of the complex representation and its complex conjugate.

A complex torus is only really interesting if there are "enough" meromorphic functions on it. The well-known case $n = 1$ is now misleading; if $n > 1$ and $\Lambda$ is in general position, the only meromorphic functions on $T$ are constants. In more detail, there is an $r = r(n, \Lambda)$ with $0 \leqslant r \leqslant n$ and a subspace $\mathbb{C}^r \subset \mathbb{C}^n$ with the following properties. The meromorphic functions on $T$ form a finitely generated field of transcendence degree $(n - r)$ over $\mathbb{C}$. The intersection $\mathbb{C}^r \cap \Lambda$ is a lattice in $\mathbb{C}^r$, so that $T' = \mathbb{C}^r/\mathbb{C}^r \cap \Lambda$ is a sub-torus of $T$ and the quotient $T/T'$ can be regarded as a complex torus of dimension $(n - r)$; and every meromorphic function on $T$ is constant on each coset of $T'$, so that it is induced by a meromorphic function on $T/T'$. If $r = 0$ the meromorphic functions separate points on $T$— that is, for any two distinct points $P_1$ and $P_2$ on $T$ there is a meromorphic function $f$ on $T$ such that $f(P_1) \neq f(P_2)$.

Clearly the case of interest is when $r = 0$; when this holds we call $T$ an *abelian manifold* or an *abelian variety*, according as our point of view is primarily analytic or primarily geometric. A necessary and sufficient condition for $r = 0$ is that $T$ should admit an alternating Riemann form $E$—that is, an $\mathbb{R}$-bilinear skew-symmetric function $E\colon \mathbb{C}^n \times \mathbb{C}^n \to \mathbb{R}$ such that

(i) $E(iz, iw) = E(z, w)$,
(ii) $E(iz, z) > 0$ for all $z \neq 0$,
(iii) $E$ is $\mathbb{Z}$-valued on $\Lambda \times \Lambda$.

An abelian manifold may admit several essentially different Riemann forms; so we define a *polarized abelian manifold* to be a pair $(T, E)$ with the proviso that for this purpose we do not distinguish between two Riemann forms one of which is a constant multiple of the other.

Any polarized abelian manifold can be viewed as a projective abelian variety, More precisely, given $(T, E)$ there are a projective variety $A$, defined over $\mathbb{C}$, and a homeomorphism $T \to A$ given by meromorphic functions such that there is a one-to-one correspondence between the field of meromorphic functions on $T$ and the function field of $A$. Moreover the group law on $A$ (induced by that on $T$) is everywhere defined and is given by regular maps in the sense of algebraic geometry. Given $(T, E)$ there are countably many essentially different $A$, corresponding to the possibility of replacing $E$ by a multiple of itself; but apart from this $A$ is determined up to a linear transformation of the ambient space. Conversely, let $A$ be a projective variety whose points form a group under a law defined by regular maps; we do not need to require that the group is commutative because that follows from the other conditions. If $A$ is defined over $\mathbb{C}$ then $A$ can be identified with a polarized abelian manifold.

Thus a polarized abelian manifold can be described in analytic and in geometric terms, the former description being a good deal more explicit than the latter. One therefore expects to be able to classify any family of such manifolds both by the points of an analytic set and by the points of a geometric variety; and so each such family will give geometric information about a certain analytic set. Much the simplest case is that of dimension 1, which we now outline.

When $n = 1$, the lattice $\Lambda$ is determined by a base $\tau_1, \tau_2$; so it is determined up to isomorphism by the complex number $\tau = \tau_1/\tau_2$. Since $\Lambda$ is discrete, $\tau$ is not real; by interchanging $\tau_1$ and $\tau_2$ if necessary we impose the convention

$\operatorname{Im} \tau > 0$. There is just one polarization on $T = \mathbb{C}/\Lambda$, given by $E(z, w) = cz.w$ for a suitable constant $c$; in this formula $z$ and $w$ are regarded as real 2-vectors and the dot denotes the scalar product. (This is why the Riemann form is not mentioned in standard expositions of this case.) So each point of the upper half-plane $\mathscr{H}$ determines a polarized abelian manifold. Two distinct points $\tau$ and $\tau'$ correspond to the same $T$ if and only if they correspond to different bases of the same lattice $\Lambda$—that is, if and only if $\tau'$ is equivalent to $\tau$ under the action of $\mathrm{SL}(2, \mathbb{Z})$ on $\mathscr{H}$. So the family of (isomorphism classes of) complex tori of dimension 1 is classified by the points of the analytic set $\mathrm{SL}(2, \mathbb{Z})\backslash\mathscr{H}$. On the other hand, a geometric model for $T$ is given in traditional notation by the curve

$$Y^2 = 4X^3 - g_2 X - g_3 \tag{3.1}$$

and the isomorphism classes of such curves correspond to the values of

$$j = g_2^3/(g_2^3 - 27g_3^2).$$

By means of $j$ we identify $\mathrm{SL}(2, \mathbb{Z})\backslash\mathscr{H}$ with the affine line.

So far we have made no assumptions about $\operatorname{End}(T)$. It is well-known that for $n = 1$ $\operatorname{End}(T)$ must be either $\mathbb{Z}$ or an order in a complex quadratic field. We ask when $\operatorname{End}(T) = R$, where for simplicity we take $R$ to be the ring of integers of the complex quadratic field $K$. The embedding $R \subset \mathbb{C}$ is just the complex representation of $\operatorname{End}(T)$, and this gives $\Lambda$ a structure as an $R$-module. Thus $\Lambda$ is isomorphic to an ideal in $R$, and the possible values of $\tau$ come from choosing $\tau_1$, $\tau_2$ to be any base for an ideal of $R$ considered as a $\mathbb{Z}$-module. In particular, there are just $h$ relevant values of $j(\tau)$, where $h$ is the class-number of $K$; and they correspond to the ideal classes of $K$. But if the curve (3.1) has endomorphism ring $R$, the same is true of any conjugate of (3.1) over $\mathbb{Q}$; for the endomorphisms and their laws of composition are described by maps given by algebraic formulae, and we can extend any isomorphism of $\mathbb{Q}(g_2, g_3)$ over $\mathbb{Q}$ to the coefficients of these formulae. Hence the relevant values of $j(\tau)$ are algebraic over $\mathbb{Q}$ and form complete sets of conjugates; in fact they are all conjugate over $\mathbb{Q}$, but this lies deeper.

To identify the fields generated by the relevant values of $j$ we must appeal to class field theory. Note that if $z$ is the coordinate on $\mathbb{C}$ and $T = \mathbb{C}/\Lambda$ corresponds to the curve (3.1) then $\mathrm{d}z = Y^{-1}\,\mathrm{d}X$ is the differential of the first kind; this is the vital link between analysis and geometry. In particular suppose (3.1) is defined over $K(j)$ and $\alpha$ is in $R$; then the endomorphism of (3.1) corresponding to $\alpha$, which is induced by $z \mapsto \alpha z$ on $\mathbb{C}$, is the only endomorphism which multiplies the differential of the first kind by $\alpha$, and hence by Galois theory it is defined over $K(j)$. Thus so is its kernel. Now if $\mathfrak{a} = (\alpha, \beta)$ is any ideal of $R$ the kernel of the map $\mathbb{C}/\Lambda \to \mathbb{C}/\mathfrak{a}^{-1}\Lambda$ induced by the identity map on $\mathbb{C}$ is the intersection of the kernels of $\alpha$ and $\beta$; thus the corresponding geometric map is defined over $K(j)$ and hence so is its image. It follows that the field $K(j)$ does not depend on the ideal class associated with $\Lambda$, and in particular $K(j)$ is a normal extension of $K$. Now with $\Lambda$ as before, let the $\mathfrak{a}_\nu$ for $\nu = 1, 2, \ldots, h$ be ideals in $R$, one in each ideal class; and let $C_\nu$ and $j_\nu$ be the curve (3.1) and the value of $j$ corresponding to the torus $\mathbb{C}/\mathfrak{a}_\nu\Lambda$. Let $p$ be a rational prime which splits in $K$, as $(p) = \mathfrak{p}\mathfrak{p}'$, and let $\mathfrak{P}$ be a prime in $K(j)$ which divides $\mathfrak{p}$. Assume that each $C_\nu$ has good reduction mod $\mathfrak{P}$ (so that the $j_\nu$ are all integers for $\mathfrak{P}$); this means that $C_\nu$ can be written as a cubic equation with integer coefficients in such a way that if the coefficients are replaced by their residues mod $\mathfrak{P}$ the resulting equation is that of a non-singular cubic (over the finite field of $N\mathfrak{P}$ elements). In particular, if $C_\nu$ is given in the classical form (3.1) a sufficient condition for $C_\nu$ to have good reduction mod $\mathfrak{P}$ is that $\mathfrak{P}$ does not divide 2 nor the denominators of $g_2$ and $g_3$ nor the numerator of $g_2^3 - 27g_3^2$. Assume also that no two of the $j_\nu$ are congruent mod $\mathfrak{P}$ and that $\mathfrak{p}$ is prime to each $\mathfrak{a}_\nu$; all these conditions only exclude finitely many $\mathfrak{p}$. Let $i$ be such that $\mathfrak{a}_i$ is in the class of $\mathfrak{p}^{-1}\mathfrak{a}_1$, and write $\mathfrak{a}_i\mathfrak{p}\mathfrak{a}_1^{-1} = (\alpha)$ for some $\alpha$ in $K$. Then $\mathfrak{p}$ defines a map $C_1 \to C_i$ by means of the map of degree $p$ on complex tori

$$\mathbb{C}/\mathfrak{a}_1\Lambda \to \mathbb{C}/\mathfrak{p}^{-1}\mathfrak{a}_1\Lambda \to \mathbb{C}/\mathfrak{a}_i\Lambda$$

induced by $z \mapsto \alpha z$; and this map multiplies the canonical differential of the first kind by $\alpha$. Now reduce all geometric objects mod $\mathfrak{P}$ and denote the results by a tilde. The induced map $\tilde{C}_1 \to \tilde{C}_i$ is of degree $p$, and it annihilates the differential of the first kind because $\tilde{\alpha} = 0$; so it is purely inseparable. Thus

$$\tilde{j}_i = \tilde{j}_1^p, \qquad \text{whence} \qquad j_i \equiv j_1^p \bmod \mathfrak{P}.$$

Recall that in algebraic number theory the *Frobenius element* of $\mathfrak{p}$ for the normal extension $K(j)/K$ is the unique $\sigma$ in the Galois group of the extension such that $\sigma\beta \equiv \beta^p \bmod \mathfrak{P}$ for all $\beta$ in $K(j)$ which are integral at $\mathfrak{p}$. Since the $j_\nu$ are all integral at $\mathfrak{P}$ and all incongruent mod $\mathfrak{P}$, the Frobenius element of $\mathfrak{P}$ must satisfy $\sigma j_1 = j_i$. In particular, it is the identity if and only if $i = 1$—that is, if and only if $\mathfrak{p}$ is a principal ideal. But by general theory $\mathfrak{p}$ splits completely in $K(j)$ if and only if its Frobenius element is the identity, so (with at most finitely many exceptions) $\mathfrak{p}$ splits completely if and only if it is principal. This property is enough to determine $K(j)$ as the *absolute class field* of $K$—that is, the maximal unramified abelian extension of $K$. A similar but more complicated argument identifies the field $K(j(\tau))$ if $\tau$ is any element of $K$.

What is special about these values of $\tau$ is that the corresponding $T$ have unusually large rings of endomorphisms. Indeed, let $\phi$ be in $\mathrm{GL}(2, \mathbb{Z})$ with $\det \phi > 0$; then the map of $\mathscr{H}$ to itself given by $\tau \to \tau' = \phi\tau$ corresponds to a map $T \to T'$ of degree $\det \phi$, and every non-constant map can be described in this way. This map is an endomorphism if and only if $\tau'$ is equivalent to $\tau$ under $\mathrm{SL}(2, \mathbb{Z})$; and the $\tau$ for which this happens non-trivially are just those which lie in some complex quadratic field. Note that $\mathrm{GL}(2, \mathbb{Z})$ appears here as the commensurability group of $\mathrm{SL}(2, \mathbb{Z})$, in a rather loose sense; in fact the groups which really matter are projective ones, and $\mathrm{PGL}^+(2, \mathbb{Z})$ is the commensurability group of $\mathrm{PSL}(2, \mathbb{Z})$ in $\mathrm{PGL}^+(2, \mathbb{R})$. Moreover $\mathscr{H}$ can be viewed as the quotient of $\mathrm{PGL}^+(2, \mathbb{R})$ by a maximal compact subgroup, and this is consistent with the usual action of $\mathrm{PGL}^+(2, \mathbb{R})$ on $\mathscr{H}$.

This concludes our discussion of the special case $n = 1$, and we now return to the general case. For given $n$, there are only countably many rings $R$ which can occur; they are listed for example in [**24**], Section 10. Without much loss of generality we can assume that $R$ has no divisors of zero, for otherwise $T$ would be isogenous to the product of two abelian manifolds of lower dimension and we could classify the two factors separately.

A complete account of the analytic part of the classification problem is given in [**21**], and only the briefest outline can be given here. First, one makes some further discrete choices. The rational representation of $R$ is unique up to equivalence, but there may be a finite choice for the complex representation. As explained above, we can assume that $D = R \otimes \mathbb{Q}$ is a division algebra; thus $V = \Lambda \otimes \mathbb{Q}$ is a vector space over $D$ and is uniquely determined thereby. Up to isomorphism, there are finitely many ways of embedding $\Lambda$ (with its structure as an $R$-module) in $V$; and there are a discrete infinity of possible choices for the restriction of $E$ to $\Lambda \times \Lambda$. It is convenient to define a "marked lattice $\Lambda$" to consist of a base for $V$ as a vector space over $D$, together with a description of $\Lambda$ in terms of this base.

To determine $T$, it only remains to impose a complex structure on $V$; and this must be done in a way compatible with the chosen complex representation of $R$ (and hence of $D$) and with properties (i) and (ii) of the Riemann form $E$. So far we have only chosen the equivalence class of the complex representation, but we can vary it within this class by a complex linear transformation on $\mathbb{C}^n$, so we may assume it completely known. Initially we consider the possible complex structures on the marked lattice $\Lambda$—that is, the assignment of complex coordinates to the elements of the given base for $V$. Because we know the complex representation of $D$, this determines the complex coordinates of every point of $V$. The remaining conditions are properties (i) and (ii) of $E$; and these determine a complex manifold $\mathscr{M}$ whose points classify the admissible complex structures on the marked lattice $\Lambda$. This manifold $\mathscr{M}$ belongs to one of three families, each of which has a very simple description; one consists of the Siegel half-spaces, and the other two also generalize $\mathscr{H}$. However, we can have more than one base for $V$ with respect to which $\Lambda$ and $E$ have the same description. So let $\Gamma$ be the group of those transformations of the base of $V$ which leave $\Lambda$ and $E$ invariant; then there is a natural action of $\Gamma$ on $\mathscr{M}$, and the admissible complex structures on $V$ are classified by the points of the analytic set $\Gamma\backslash\mathscr{M}$. (This may not be a manifold, because some elements of $\Gamma$ may have fixed points in $\mathscr{M}$.) Clearly $\Gamma$ is arithmetically defined in a rather loose sense, and detailed calculation shows that it acts discontinuously on $\mathscr{M}$. But much more than this is true, though the calculations give no clue why it should happen. In fact, if one forgets its complex structure, $\mathscr{M}$ can always be identified with the quotient of a real Lie group $G$ by a maximal compact subgroup; and $\Gamma$ then becomes an arithmetic subgroup of $G$, in the sense of the definition following Theorem 1.4 in Section 1.

We now turn to the geometric side. An essential technical tool here is the theory of the *Chow point* of a variety. This associates to each variety (or more generally each positive cycle) of degree $d$ and dimension $r$ in $\mathbb{P}^m$ a point of $\mathbb{P}^N$, where $N$ depends only on $d$, $r$ and $m$. (Here $\mathbb{P}^m$ denotes projective $m$-dimensional space.) From the Chow point one can recover the original variety, and the Chow point and the variety have the same least field of definition. The Chow points of all positive cycles with given $d$, $r$ and $m$ form a variety in $\mathbb{P}^N$, called the *Chow variety*. Any geometric relation between two varieties, for example inclusion, can be expressed as an algebraic relation between the coordinates of their Chow points. For a detailed account, see for example [**18**], pp. 44–54.

An abelian manifold $T$ of dimension $n$ can be realized as an abelian variety $A$ of degree $d$ and dimension $n$ in $\mathbb{P}^m$; here $d$ and $m$ depend only on $n$ and the isomorphism class of the restriction of the Riemann form $E$ to $\Lambda \times \Lambda$. This $A$ is determined up to an arbitrary linear transformation of $\mathbb{P}^m$. The law of composition on $A$ is geometrically described by its graph, which is a subvariety of $A \times A \times A$ of dimension $2n$ and known degree; and the fact that $A$ is an abelian variety is equivalent to certain geometric properties of this graph, which translate the associative law and the fact that composition is everywhere defined. Similar remarks apply to the graphs of the finitely many endomorphisms which generate $R$ as a subring of $\mathrm{End}(A)$.

For each point of the analytic manifold $\mathcal{M}$ we form the associated $T$ and thence all the abelian varieties $A$ and their associated graphs which correspond to this $T$; we recall that they all lie in $\mathbb{P}^m$ and can be derived from any one of them by the action of $\mathrm{PGL}(m + 1, \mathbb{C})$. By abuse of language, we may assume that each $A$ and its associated graphs is described by a single Chow point P, lying in a fixed $\mathbb{P}^N$. If $P'$ is a general point of $\mathbb{P}^N$, then necessary and sufficient conditions for $P'$ to be the Chow point of some abelian variety $A'$ and its associated graphs are that certain polynomials in the coordinates of $P'$ should vanish and that certain others should not—the latter conditions corresponding to $A'$ being irreducible, the laws of composition being everywhere defined, and so on. Since we have included the graphs of some endomorphisms, these conditions ensure that $\mathrm{End}(A')$ contains $R$ up to isomorphism. These conditions are all defined over $\mathbb{Q}$ so they determine a Zariski-open subset $U$ of a variety in $\mathbb{P}^N$; and $U$ is defined over $\mathbb{Q}$. (Of course $U$ need not be irreducible, even over $\mathbb{Q}$.) There is a natural action of $\mathrm{PGL}(m + 1)$ on $U$, obtained by pulling back any point $P'$ of $U$ to the associated $A'$ and letting $\mathrm{PGL}(m + 1)$ act on the ambient space of $A'$. Write $U^* = \mathrm{PGL}(m + 1)\backslash U$, so that $U^*$ is also a Zariski-open subset of a projective variety and $U^*$ is defined over $\mathbb{Q}$. To each point of $U^*$ defined over $\mathbb{C}$ corresponds just one abelian manifold $T$, and $T$ has the correct value of $n$ and satisfies $\mathrm{End}(T) \supset R$. Conversely, each point of $\mathcal{M}$ gives rise to a point of $U^*$, and the embedding $\Gamma\backslash\mathcal{M} \to U^*$ thus defined can be shown to be holomorphic. Moreover $U^*$, or more precisely the set of points of $U^*$ defined over $\mathbb{C}$, is the union of the images of at most countably many manifolds like $\mathcal{M}$; and since the general point of $\mathcal{M}$ corresponds in all interesting cases to a manifold $T$ with $\mathrm{End}(T) = R$, its image in $U^*$ is not contained in the union of the images of the other manifolds like $\mathcal{M}$. If $W$ is the image of $\mathcal{M}$ in $U^*$, we see from the last few remarks that $W$ must be an irreducible component of $U^*$; hence it is defined over some algebraic number field $K$. The map $\Gamma\backslash\mathcal{M} \to W$ is one-to-one and holomorphic; it is not known whether it is biholomorphic. Moreover the field $K$ can be identified as a class-field, by arguments similar to those sketched above in the case $n = 1$.

In the form outlined above, the geometric classification is not constructive. But a great deal is known about explicit equations of abelian varieties (see in particular Mumford [**16**]) and about theta series and theta constants; so in particular cases it should be possible to construct $W$ and the map $\Gamma\backslash\mathcal{M} \to W$ explicitly. In particular this would give the automorphic functions on $\mathcal{M}$ with respect to $\Gamma$, for they correspond to the elements of the function field of $W$.

Now let $R_1 \supset R$ be another ring, and repeat the constructions above with $R_1$ for $R$ but with the same value of $n$; we use a subscript one to distinguish the resulting objects. To each point of $\Gamma_1\backslash\mathscr{M}_1$ corresponds an abelian manifold $T_1$ with $\mathrm{End}(T_1) \supset R_1 \supset R$; so $T_1$ in its turn determines a point of $\Gamma\backslash\mathscr{M}$. Thus we have defined an embedding

$$\Gamma_1\backslash\mathscr{M}_1 \hookrightarrow \Gamma\backslash\mathscr{M}$$

which is holomorphic except perhaps at fixed points of elements of $\Gamma_1$. The corresponding map $W_1 \hookrightarrow W$ is at least rational, by the geometric version of the argument above. In particular, if $\mathscr{M}_1$ is just a point then the corresponding point on $\Gamma\backslash\mathscr{M}$ is one at which the automorphic functions on $\mathscr{M}$ can be evaluated; and their values are algebraic numbers which lie in readily identifiable class fields. However, these may not correspond to the points described in Section 2—that is, the fixed points of elements of $\mathrm{Comm}(\Gamma)$ acting on $\mathscr{M}$. (To define $\mathrm{Comm}(\Gamma)$ we regard $\Gamma$ as embedded in the group of automorphisms of $E$ as a bilinear form on $\Lambda \otimes \mathbb{R}$ or even on $V$; this group has a natural action on $\mathscr{M}$.) In fact if $\rho_1$ is an element of $R_1$ then $\rho_1$ induces a map $\Lambda_1 \to \Lambda_1$ which takes $E_1$ to a multiple of itself; since we can identify $\Lambda_1$ with $\Lambda$ this means that $\rho_1$ induces an isogeny on every $T$ corresponding to a point of $\mathscr{M}$. Thus $\rho_1$ determines a map from $\mathscr{M}$ to some manifold like $\mathscr{M}$; but it is only when the image is $\mathscr{M}$ itself that $\rho_1$ can be compared with $\Gamma$.

Another survey of the material in this section, from a rather different point of view, can be found in Shimura [**22**].

## 4. THE METHOD OF LANGLANDS

In the last fifteen years a new approach to the theory of automorphic functions has emerged, which is based on representation theory. The major texts for this are Gelfand, Graev and Pyatetskii-Shapiro [**8**] and Jacquet and Langlands [**10**], both of which are rather daunting. The only readable introductions to the subject are the talk by Borel [**3**] and the book of Gelbart [**7**]. Also important is Weil's paper [**27**].

To reduce the complication, we consider modular forms of weight $k$ for the full group $\mathrm{SL}(2, \mathbb{Z})$ and belonging to the principal character; the generalizations to a congruence subgroup and a non-principal character involve no new ideas. However it must be said that one major advantage of introducing the adelic group, as is done below, is that it allows the simultaneous treatment of cusp forms for all congruence subgroups.

The stabilizer of the point $i$ in $\mathrm{SL}(2, \mathbb{R})$ is $\mathrm{SO}(2, \mathbb{R})$, which gives the identification

$$\mathscr{H} \cong \mathrm{SL}(2, \mathbb{R})/\mathrm{SO}(2, \mathbb{R}).$$

If $f$ is a modular form of weight $k$ for $\mathrm{SL}(2, \mathbb{Z})$, it induces a function $\phi$ on $\mathrm{SL}(2, \mathbb{R})$ by the formula

$$\phi(g) = (ci + d)^{-k} f\left(\frac{ai + b}{ci + d}\right) \quad \text{where} \quad g = \begin{pmatrix} a & b \\ c & d \end{pmatrix}. \tag{4.1}$$

It is easy to verify that

$$\phi(\gamma g) = \phi(g) \text{ for all } \gamma \text{ in } \mathrm{SL}(2, \mathbb{Z}), \tag{4.2}$$

which corresponds to the functional equation for $f$; and

$$\phi(gh) = \mathrm{e}^{-ik\theta}\, \phi(g) \text{ if } h = \begin{bmatrix} \cos\theta & -\sin\theta \\ \sin\theta & \cos\theta \end{bmatrix}. \tag{4.3}$$

There is a second-order differential operator $\Delta$, called the Laplace or the Casimir operator, such that

$$\Delta\phi = -\tfrac{1}{4}k(k-2)\,\phi; \tag{4.4}$$

the verification of this depends only on (4.1) and the fact that $f$ is holomorphic on $\mathscr{H}$. Finally, the usual condition on $f$ at cusps corresponds to a boundedness condition on $\phi$; and if $f$ is a cusp form this too can be translated into a property of $\phi$. Conversely, let $\phi$ be any well-behaved function on $\mathrm{SL}(2, \mathbb{R})$ satisfying these four conditions; then (4.1) defines a function $f$ on $\mathscr{H}$ which is well-determined because of (4.3) and it can be shown that $f$ must be a modular form of weight $k$. It is now natural to drop the condition (4.3) and to weaken (4.4) to the condition that $\phi$ should be an eigenform of $\Delta$; in this way we obtain a more general definition of an automorphic form on $\mathrm{SL}(2, \mathbb{R})$ with respect to $\mathrm{SL}(2, \mathbb{Z})$. The forms having a given eigenvalue constitute a linear subspace $\mathscr{L}$ of $L^2(\mathrm{SL}(2, \mathbb{Z})\backslash\mathrm{SL}(2, \mathbb{R}))$ where $L^2$ denotes square-integrable functions with respect to the natural Haar measure.

The *right regular representation* $R$ of $\mathrm{SL}(2, \mathbb{R})$ on the $L^2$ above is defined by

$$R(g)\,\phi(h) = \phi(hg).$$

Since the actions of $\Delta$ and $R$ commute, this representation is the sum of representations on the various spaces $\mathscr{L}$; and since all the irreducible unitary representations of $\mathrm{SL}(2, \mathbb{R})$ are known, it is natural to ask which of them occur, and with what multiplicities, in the representation on $\mathscr{L}$. In particular, we can consider a classical cusp form $f$ and use instead of $\mathscr{L}$ the subspace generated by the $\phi$ corresponding to $f$ and its right translates by elements of $\mathrm{SL}(2, \mathbb{R})$.

It turns out that this is not quite the correct question—essentially because in working with $\mathrm{SL}(2, \mathbb{R})$ we have laid special emphasis on $\mathbb{R}$ as a completion of $\mathbb{Q}$, whereas we should pay equal attention to the $p$-adic fields $\mathbb{Q}_p$. For this purpose write

$$G_p = \mathrm{GL}(2, \mathbb{Q}_p) \quad \text{and} \quad K_p = \mathrm{GL}(2, \mathbb{Z}_p)$$

for finite primes $p$, and analogously $G_\infty = \mathrm{GL}(2, \mathbb{R})$. Now define the corresponding *adelic* group to be the restricted direct product

$$\mathrm{GL}(2, \mathbb{A}) = G_\infty \Pi\, G_p;$$

here "restricted" means that we include only those elements $g_\infty \Pi\, g_p$ for which $g_p$ belongs to $K_p$ for all but finitely many $p$. Let $f$ as before be a modular form of weight $k$. It induces a function $\phi$ on $\mathrm{GL}^+(2, \mathbb{R})$ by means of

$$\phi(g) = (ci + d)^{-k}\,(ad - bc)^{k/2}\, f(g(i)),$$

which is the natural modification of (4.1); and this can be extended to $G_\infty$ by requiring (4.2) to hold for all $\gamma$ in $\mathrm{GL}(2, \mathbb{Z})$. We can regard $\phi$ as defined on $G_\infty \Pi K_p$ by ignoring all but the first factor. But the strong approximation theorem implies that

$$\mathrm{GL}(2, \mathbb{A}) = \mathrm{GL}(2, \mathbb{Q}) . \{G_\infty \Pi K_p\} \tag{4.5}$$

with the obvious diagonal map $\mathrm{GL}(2, \mathbb{Q}) \hookrightarrow \mathrm{GL}(2, \mathbb{A})$; and the intersection of the two factors on the right in (4.5) is $\mathrm{GL}(2, \mathbb{Z})$. Hence there is just one way of extending $\phi$ to a function on $\mathrm{GL}(2, \mathbb{A})$ which satisfies

$$\phi(\gamma g) = \phi(g) \quad \text{for all} \quad \gamma \text{ in } \mathrm{GL}(2, \mathbb{Q}). \tag{4.6}$$

To sum up, we have shown how to derive from $f$ a function $\phi$ satisfying (4.3) and (4.6); viewed as a function of $G_\infty$ alone, it still satisfies (4.4), and it is easy to check that

$$\phi(hg) = \phi(gh) = \phi(g) \text{ for } h \text{ in the centre of } \mathrm{GL}(2, \mathbb{A}).$$

Moreover, as before $\phi$ satisfies some well-behavedness conditions.

We again have a right regular representation of $\mathrm{GL}(2, \mathbb{A})$, this time on

$$L^2((Z(2, \mathbb{A})\,\mathrm{GL}(2, \mathbb{Q})\backslash \mathrm{GL}(2, \mathbb{A}))$$

where $Z(2, \mathbb{A}) \cong \mathbb{A}^*$ is the centre of $\mathrm{GL}(2, \mathbb{A})$; and the functions $\phi$ just obtained lie in this space. All the irreducible unitary representations of $\mathrm{GL}(2, \mathbb{A})$ are known; and indeed all the relevant ones arise as products of irreducible representations of the factors $G_\infty$ and $G_p$. So we may again ask, for a given cusp form $f$, which irreducible representations occur in the right representation on the space generated by $\phi$ and its right translates. The first major achievement of the theory is the complete answer to this question—of course in a much more general form than is given here. It can be found in Theorems 5.19 and 5.21 of [7], and the parts of it which we are able to state in terms of our previous description are as follows.

4.7 THEOREM. *Let $f$ be a cusp form of weight $k$ which is an eigenfunction of every Hecke operator $T_p$; then the right representation of* $\mathrm{GL}(2, \mathbb{A})$ *on the space generated by the corresponding $\phi$ and its right translates is irreducible. The map from cusp forms to irreducible representations thus defined is a one-to-one correspondence between cusp forms which are eigenfunctions for every $T_p$ and a certain family of irreducible representations which can be explicitly described. Moreover if $\pi$ denotes the representation thus corresponding to $f$, and $\pi_p$ is its component at $p$ (that is, the corresponding representation of $G_p$), then each of $\pi_p$ and the eigenvalue of $T_p$ acting on $f$ determines the other.*

To explain the relevance of this, we must give an explicit description of the arithmetic subgroups of $\mathrm{SL}(2, \mathbb{R})$, at least up to commensurability; this is due to Weil—see [**8**], Appendix to Chapter I.

Let $K$ be a totally real algebraic number field with $[K : \mathbb{Q}] = n$, and let $D$ be a quaternion algebra over $K$. Denote by $\mathbb{H}$ the classical Hamiltonian quaternions and by $M_2(\mathbb{R})$ the ring of real $2 \times 2$ matrices; these are the only two quaternion algebras over $\mathbb{R}$. The ring $D \otimes \mathbb{R}$ is a product of $n$ factors, corresponding to the $n$ embeddings of $K$ in $\mathbb{R}$; it follows that there is an $r$ with $0 \leqslant r \leqslant n$ such that

$$D \otimes \mathbb{R} \cong \{M_2(\mathbb{R})\}^r \times \mathbb{H}^{n-r}.$$

Let $\delta$ be in $D$; then the images of $\delta$ in the factors $M_2(\mathbb{R})$ all have positive determinant if and only if the reduced norm of $\delta$ is totally positive. If this is so, then $\delta$ acts in an obvious way on $\mathscr{H}^r$.

An *order* of $D$ is a subring of $D$ which contains 1 and is a free $\mathbb{Z}$-module of rank $4n$; thus if $S$ is an order, $S \otimes \mathbb{Q} = D$. Let $\Gamma = \Gamma(S)$ be the group of those units in $S$ whose reduced norm is totally positive; this is also the group of those elements of $S$ whose reduced norm is a totally positive unit of $K$, at least provided $S$ is a maximal order. Then $\Gamma$ acts on $\mathscr{H}^r$ discontinuously, and with some abuse of language $\Gamma$ can be regarded as an arithmetic subgroup of $\{\mathrm{SL}(2, \mathbb{R})\}^r$. Moreover $\Gamma \backslash \mathscr{H}^r$ is compact if and only if $D$ is a division algebra. If we replace $S$ by another order in $D$, then $\Gamma$ is replaced by another group commensurable with it. Obviously, if $S$ is increased then so is $\Gamma$; but the reader should be warned that even if $S$ is a maximal order the corresponding $\Gamma$ need not be maximal among arithmetic subgroups of $\{\mathrm{SL}(2, \mathbb{R})\}^r$. Also, different $D$ can give rise to commensurable $\Gamma$.

Now suppose that $r = 1$; then Weil proved that every arithmetic subgroup of $\mathrm{SL}(2, \mathbb{R})$—that is, every arithmetic Fuchsian group—can be obtained in this way up to commensurability. For many purposes, therefore, it is enough to study such $\Gamma$. Groups obtained in this way had already been studied by Poincaré and by Klein (see for example [**5**]) though they described them by means of ternary quadratic forms rather than quaternion algebras.

Theorem 4.7 suggests the following programme. Let $\Gamma$ be an arithmetically defined linear group, defined in terms of some number field $K$ which is not necessarily the rationals. By considering all the completions of $K$, associate with $\Gamma$ an adelic group $G$ such that $\Gamma$ is discrete in $G$. Pick out in some suitable way elements of $L^2(\Gamma \backslash G)$ which deserve to be called automorphic functions on $G$ with respect to $\Gamma$; this will certainly involve both a "well-behavedness" condition and being an eigenfunction with respect to a suitably chosen invariant differential operator. There are now two tasks:

(i) If $\mathscr{L}$ denotes the subspace of $L^2(\Gamma \backslash G)$ spanned by an automorphic function $\phi$ and its right translates, describe in terms of the irreducible unitary representations of $G$ its right representation on $\mathscr{L}$.

(ii) When $\Gamma$ is also an arithmetic group in the sense of Section 1, relate the automorphic functions just defined to the classical automorphic forms with respect to $\Gamma$.

In particular, a satisfactory solution of the first task should establish a one-to-one correspondence between a large (and explicitly described) class of representations $\pi$ and a set of automorphic functions $\phi$ (which should span the space of all automorphic functions) such that the representation of $G$ on the space $\mathscr{L}$ associated with $\phi$ is equivalent to $\pi$. Once this is done, we can identify $\pi$ with $\phi$ and can say that an automorphic function in the new sense is just an irreducible representation of $G$ belonging to a certain class.

The methods of Jacquet and Langlands depend absolutely on being able to describe $\Gamma$ by means of $2 \times 2$ matrices. With this restriction, they have largely carried out the programme described above; and where there are deficiencies they seem to be caused mainly by gaps in the classical theory. For example, the correspondence between irreducible representations and classical automorphic forms is in general not at all explicit; but this is because, when $\Gamma$ is an arithmetic group without cusps, there is no satisfactory description of an automorphic form for $\Gamma$ in terms of its behaviour under the action of the Hecke operators.

The general philosophy of Langlands, which has its roots in the study of $L$-series, also suggests that it would be profitable to study the following question. Let $\Gamma$, $\Gamma'$ be two arithmetic groups and $G$, $G'$ the corresponding adelic groups; what natural relations are there between automorphic forms on $G$ with respect to $\Gamma$ and automorphic forms on $G'$ with respect to $\Gamma'$? In view of the remarks above, this can be treated as a question about irreducible representations; and a number of theorems of this type are known. For our present purpose the relevant one is the following; for more details see [7], Section 10 and [**10**], Chapter III.

4.8 THEOREM. *Let $K$ be an algebraic number field and $D$ a quaternion algebra over $K$ which has no divisors of zero. Let* $\Gamma = \mathrm{GL}(2, K)$ *and let $\Gamma'$ be the multiplicative group of $D$; and denote by $G$, $G'$ the adelizations of* $\Gamma$, $\Gamma'$ *respectively. To every irreducible unitary representation of $G'$ there corresponds an irreducible unitary representation of $G$; and under this correspondence automorphic forms on $G'$ with respect to $\Gamma'$ give rise to automorphic forms on $G$ with respect to* $\Gamma$.

The importance of [**10**], Chapter III is that as well as containing a proof of this theorem it completely answers the converse problem: given $\Gamma$ and $\Gamma'$ as above, which irreducible representations of $G$ correspond to irreducible representations of $G'$? In view of Weil's representation of arithmetic Fuchsian groups, any such group is at least commensurable with the group of units of a maximal order of a quaternion algebra $D$ over a totally real field $K$; so Theorem 4.8 provides a connection between the automorphic forms (in the classical sense) with respect to an arithmetic Fuchsian group and the Hilbert modular forms associated with $K$. In particular, if $K = \mathbb{Q}$ then Theorem 4.8 states that the theta-function associated with a quaternary quadratic form whose determinant is a square is an elliptic modular form; and the work of Jacquet and Langlands enables us to say precisely which modular forms turn up in this way. This particular problem has also been solved by Eichler [**4**]; his paper is written in classical language but his methods are closely related to those of Jacquet and Langlands.

Unfortunately this correspondence is not explicit enough to answer all the questions which the classical analyst would like to ask. Its most hopeful classical application seems to be to the modular equations associated with $\Gamma'$. An irreducible representation of $G'$, because it is made up of local components, gives rise to an $L$-series. If the representation corresponds to an automorphic form $f$ in the classical sense, then $f$ is an eigenfunction for every Hecke operator and the associated eigenvalues give rise to an $L$-series. These two $L$-series are of course the same. But by the theorem of Eichler–Shimura, the product of the $L$-series corresponding to the forms $f$ of weight 2 is essentially the same as the zeta-function of the field of classical automorphic functions with respect to $\Gamma'$. Hence one expects, in the situation of Theorem 4.8, that the zeta-function of the field of classical automorphic functions for $\Gamma'$ should be a factor of that for $\Gamma$. Such a situation should only naturally arise from a monomorphism of the corresponding function fields. The same argument can be applied to congruence subgroups of $\Gamma'$ and $\Gamma$; so it should suggest the existence of a natural map from the variety defined by a modular equation for $\Gamma$ to that defined by a modular equation for $\Gamma'$. But as far as I know, no detailed work has been done in this direction.

## REFERENCES

1. A. Borel, Density and maximality of arithmetic subgroups. *J. Reine Angew Math.* **224** (1966), 78–89.
2. A. Borel and Harish-Chandra. Arithmetic subgroups of algebraic groups. *Ann. of Math.* **75** (1962), 485–535.
3. A. Borel, Formes automorphes et séries de Dirichlet, Séminaire Bourbaki 466 (June 1975).

4. M. Eichler, The basis problem for modular forms and the traces of the Hecke operators, *in* "Modular Functions of One Variable, I", pp. 75–151. Springer Lecture Notes No. 320. Berlin, 1973.
5. R. Fricke and F. Klein, "Vorlesungen über die Theorie der Automorphen Functionen", Vol. II. B. Teubner, Leipzig, 1926.
6. H. Garland and M. S. Raghunathan, Fundamental domains for lattices in (**R**$^-$) rank 1 semi-simple Lie groups. *Ann. of Math.* **92** (1970), 279–326.
7. S. Gelbart, "Automorphic Forms on Adele Groups". Princeton, 1975.
8. I. M. Gelfand, M. I. Graev and I. I. Pyatetskii-Shapiro, "Representation Theory and Automorphic Functions" (English translation). W. B. Saunders, Philadelphia, 1969.
9. F. E. P. Hirzebruch, Hilbert modular surfaces. *L'Enseignement Math.* **19** (1973), 183–281.
10. H. Jacquet and R. P. Langlands, "Automorphic Forms on GL(2)". Springer Lecture Notes No. 114, Berlin, 1970.
11. R. P. Langlands, Modular forms and *l*-adic representations, *in* "Modular Functions of One Variable II", pp. 361–500. Springer Lecture Notes No. 349, Berlin, 1973.
12. G. A. Margulis, Non-uniform lattices in semisimple algebraic groups, *in* "Lie groups and their representations" (I. M. Gelfand, Ed.), pp. 371–553. Adam Hilger Ltd. London, 1975.
13. G. A. Margulis, Arithmetic properties of discrete subgroups. *Uspehi Mat. Nauk.* **29** (1974), 49–98.
14. G. D. Mostow, "Strong Rigidity of Locally Symmetric Spaces". Princeton, 1973.
15. G. D. Mostow, Discrete subgroups of Lie groups, *Advances in Math.* **16** (1975), 112–123.
16. D. Mumford, On the equations defining abelian varieties. *Invent. Math.* **1** (1966), 287–354; **3** (1967), 75–135, 215–244.
17. G. Prasad, Strong rigidity of **Q**-rank 1 lattices. *Invent. Math.* **21** (1973), 255–286.
18. P. Samuel, "Méthodes d'Algèbre abstraite en Géométrie algébrique", Springer, Berlin, 1955.
19. A. Selberg, On discontinuous groups in higher dimensional symmetric spaces, *in* "Contributions to Function Theory", pp. 147–164. Bombay, 1960.
20. A. Selberg, "Recent Advances in the Theory of Discontinuous Groups of Motions of Symmetric Spaces", pp. 99–120. Springer Lecture Notes No. 118, Berlin, 1970.
21. G. Shimura, On analytic families of polarized abelian varieties and automorphic functions. *Ann. of Math.* **78** (1963), 149–192.
22. G. Shimura, "Automorphic Functions and Number Theory", Springer Lecture Notes, No. 58, Berlin 1968.
23. H. P. F. Swinnerton-Dyer, Applications of algebraic geometry to number theory, *in* "Amer. Math. Soc. Symposia", Vol. XX. Providence, 1971.
24. H. P. F. Swinnerton-Dyer, "Analytic Theory of Abelian Varieties". Cambridge University Press, 1974.
25. E. B. Vinberg, Discrete groups generated by reflections in Lobachevski space. *Mat. USSR-Sb.* **72** (1967), 471–488.
26. A Weil, On discrete subgroups of Lie groups II, *Ann. of Math.* **75** (1962), 578–602.
27. A. Weil, Sur certains groupes d'operateurs unitaires. *Acta Math.* **111** (1964), 143–211.
28. A. Weil, Adeles and Algebraic Groups (Mimeographed lecture notes), Princeton, 1961.

# Index